T0142451

Ecology and Conservation of Tropical Ungulates in Latin America

Sonia Gallina-Tessaro
Editor

Ecology and Conservation of Tropical Ungulates in Latin America

Editor
Sonia Gallina-Tessaro
Red de Biología y Conservación de Vertebrados
Instituto de Ecología, A.C.
Xalapa, Veracruz, Mexico

ISBN 978-3-030-28870-9 ISBN 978-3-030-28868-6 (eBook)
https://doi.org/10.1007/978-3-030-28868-6

This Springer imprint is published by the registered company Springer Nature Switzerland AG
The registered company address is: Gewerbestrasse 11, 6330 Cham, Switzerland

Acknowledgments

I would like to thank my family first of all, my husband, Alberto; my sons, Alberto and Alejandro; and my granddaughters, Minerva and Leonora, as well as my sisters, María Pia and Patricia, to whom I dedicate this book, because they have improved my academic performance; the Institute of Ecology, A.C., to which I have dedicated more than 40 years of my professional development with great pride; and all the students and colleagues who have helped me in the field during all these years for their invaluable friendship. I also thank the Springer team for giving me this wonderful opportunity to spread the knowledge of ecology and conservation of tropical ungulates, which has involved a great effort by all the authors involved and represents a great satisfaction.

Contents

About the Editor

Sonia Gallina-Tessaro was awarded her Bachelor, Master, and Doctorate of Science by the Faculty of Sciences of the Universidad Nacional Autónoma de México (UNAM). She is a tenured researcher listed on the National System of Researchers (SNI II). She is a pioneer in the long-term eco-ethological studies of deer in areas such as population dynamics, habitat use, feeding behavior, activity patterns, and conservation strategies in a variety of protected areas in Mexico. She set up and coordinated the Master of Science degree program in Wildlife Management at the Instituto de Ecología, A.C. She has served as president of the Mexican Association of Mammalogy (AMMAC) and is currently the coordinator of the North American Region for the IUCN-SSC Deer Specialist Group and a member of the Mexican Academy of Sciences. She is on the editorial boards of *Acta Zoológica Mexicana (n.s.)* and *Acta Biológica Colombiana.* She has published 73 scientific articles, 69 book chapters, and 44 articles for magazines and newspapers and has coedited 9 books and coauthored 1.

Part I
General Topics of Tropical Ungulate Communities

Chapter 1
Introduction

Sonia Gallina-Tessaro

1.1 Diversity of Ungulates in Latin America

Of the 34 species of American ungulates, 10 species are restricted to the Nearctic zoogeographic region, while 22 species inhabit the Neotropical region. The remaining two species, with the largest geographical distribution in America are the white-tailed deer (*Odocoileus virginianus*) and collared peccary (*Pecari tajacu*), which inhabit very different vegetation types in the Nearctic and Neotropics.

The book of Groves and Grubb (2011) explains in detail the ungulate taxonomy, but new genetic knowledge has made changes to it; thus, only some details will be presented on this topic. All perissodactyls are considered hindgut fermenters because none has a complex stomach, all have a large caecum and colon. This group was the most diverse ungulates in the Oligocene, but they began to decline in the Miocene, presumably in the face of competition from the rising artiodactyls, more efficient because their complex stomach with a rumen that allow them to obtain and exploit a more diverse plant resources. Today, perissodactyls are greatly reduced in both diversity and abundance, and in Latin America there is only a single family: Tapiridae. Artiodactyls seem to have shown a rapid radiation in the Late Oligocene or Early Miocene.

Gilbert et al. (2006) for the family Cervidae include 40 species of deer distributed throughout the northern hemisphere, as well as in South America and Southeast Asia. In this publication the authors examine the phylogeny of the family by analyzing two mitochondrial protein-coding genes and two nuclear introns for 25 species of deer representing most of the taxonomic diversity of the family. They propose a new classification where the family Cervidae is divided in two subfamilies and five tribes. The subfamily Cervinae is composed of two tribes: the tribe Cervini groups

S. Gallina-Tessaro (✉)
Red de Biología y Conservación de Vertebrados, Instituto de Ecología, A.C.,
Xalapa, Veracruz, Mexico
e-mail: sonia.gallina@inecol.mx

S. Gallina-Tessaro (ed.), *Ecology and Conservation of Tropical Ungulates in Latin America*, https://doi.org/10.1007/978-3-030-28868-6_1

the genera *Cervus*, *Axis*, *Dania*, and *Rucervus*, with the Pere David's deer (*Elaphurus davidianus*) included in the genus *Cervus*, and the swamp deer (*Cervus duvauceli*) placed in the genus *Rucervus*; the tribe Muntiacini contains Muntiacus and Elaphodus. The subfamily Capreolinae consists of the tribes Capreolini (*Capreolus* and *Hydropotes*), Alceini (*Alces*), and Odocoileini (Rangifer + American genera). Deer endemic to the New World fall in two biogeographic lineages: the first one groups *Odocoileus* and *Mazama americana* and is distributed in North, Central, and South America, whereas the second one is composed of South American species only and includes *Mazama gouazoubira*. This implies that the genus *Mazama* is not a valid taxon. Molecular dating suggests that the family originated and radiated in central Asia during the Late Miocene and that Odocoileini dispersed to North America during the Miocene–Pliocene boundary and underwent an adaptive radiation in South America after their Pliocene dispersal across the Isthmus of Panama. The phylogenetic inferences show that the evolution of secondary sexual characters (antlers, tusk-like upper canines, and body size) has been strongly influenced by changes in habitat and behavior.

Ruiz-García et al. (2009) studied the genetic variability in Neotropical deer genera (Mammalia: Cervidae) according to DNA microsatellite loci. Species conservation programs are highly based on analyses of population genetics. They compared eight Neotropical Cervidae species (*Mazama americana*, *M. gouzaoubira*, *M. rufina*, *Odocoileus virginianus*, *Hippocamelus antisensis*, *Pudu mephistopheles*, *Ozotoceros bezoarticus*, and *Blastoceros dichotomus)* and some European and Asian Cervidae species (*Cervus elaphus*, *C. nippon*, *Capreolus capreolus*, *C. pygargus*, and *Dama dama*). The *M. americana*, *M. gouzaoubira*, and *O. virginianus* samples had high diversity values close to our *C. elaphus* population ($H = 0.64$, 0.70, and 0.61, respectively), while *M. rufina* was very low, close to *C. nippon.* Several sample sets of *Mazama* and *Odocoileus* yielded a homozygote excess, probably due to the Wahlund (subdivision) effect.

Groves and Grubb (2011) prefer to retain the Artiodactyla as an order with Suborder Tylopoda (Family Camelidae*)*, Suborder Suina (Family Tayassuidae), and Suborder Ruminantia (Family Cervidae). The family Cervidae has three subfamilies—the Hydropotinae, the Odocoileinae, and the Cervinae—based on the lack of antlers in the first family and the way the lateral metacarpals have been retained in the other two: the Odocoileinae maintain the distal ends (telemetacarpals) and the Cervinae the proximal ends (plesiometacarpals). More detailed information on the Neotropical representatives of the Cervidae will be found in Duarte and González (2010); their evolution, both in the fossil record and in their adaptive biology, is treated by Merino and Rossi (2010), so we do not repeat the complete data of all these species and we exhort interested people to check the different chapters on this excellent book.

Araujo Absolon et al. (2016) studied the distribution of 24 species of living ungulates (families Camelidae, Cervidae, Tapiridae, and Tayussuidae) by the panbiogeographical method of track analysis. The results have shown five generalized tracks (GTs) explainded mainly by geological events such as tectonism and volcanism: GT1: Mesoamerican/Chocó (composed of *Mazama pandora* merriam, 1901,

Mazama temama (Kerr, 1792), *Odocoileus virginianus* zimmerman, 1780, and *Tapirus bairdii* (Gill, 1785) biota distributed from the southern Mexico to Northwestern south American; GT2, Northern Andes (*Mazama rufina* (Pucheran, 1951), *Pudu mephistophiles* (de Winton, 1896), and *Tapirus pinchaque* (Roullin, 1829), a biota distributed from the cordillera Occidental (Colombia) to Northernmost Peru; GT3, Central Andes (*Hippocamelus antisensis* (d'Orbigny, 1834), *Lama guanicoe* (Muller, 1776); *Mazama chunyi* (Hershkovitz, 1959); and *Vicugna vicugna* (Molina, 1782) represents a biota distributed in the Central Andes which includes Central Western Peru, Northern Chile, Western Bolivia and Western Argentina; GT4, Chilean Patagonia (*Hippocamelus bisulcus* (Molina, 1782) and *Pudu puda* (Molina, 1782) shows a strictly distributional pattern in the Chilean Patagonia; and GT5, Chaco/Central western Brazil (*Blastocerus dichotomus* (Illiger, 1815); *Catagonus wagneri* (Rusconi, 1930), and *Ozotocerus bezoarticus* (Linnaeus, 1758), shows a distributional pattern of biota mainly in lowlands from South America (Cerrado, Pantanal and Chaco).). The biogeographic node was recognized in the intersection of GT1 and GT2 in Northwestern Colombia, possesses great importance in the evolutionary context of the ungulates. It is particularly important as an indicator of a priority conservation area due to the presence of biotic elements of distinct origins. The geological events that occurred through the Neogene and mainly in the Pliocene–Pleistocene caused fragmentation, diversification, and endemism among biota.

Navas-Suárez et al. (2018) analyzed retrospective data on epidemiological (season, circumstances of death, history of trauma, ectoparasitosis) and biological variables (sex, age, body condition) of marsh deer (MD, *Blastocerus dichotomus*) and brown brocket deer (BBD, *Mazama gouazoubira*). Pneumonia is one of the most important ailments in cervid medicine, and causes may include viruses, bacteria, fungi, metazoan parasites, and physicochemical agents. It is the most frequent inflammatory process in both species. In deer, the most frequently isolated bacteria from pneumonic lung tissues are *Arcanobacterium pyogenes*, *Escherichia coli*, *Fusobacterium necrophorum*, *Klebsiella pneumoniae*, *Manheimia haemolytica*, *Mycobacterium* spp., and *Streptococcus gallolyticus*. Several species of ticks have been identified in MD and BBD, the most common being *Amblyomma* spp. and *Rhipicephalus microplus*.

1.2 Aims and Proposal

This book aims to gather recently generated knowledge about tropical ungulates in several countries of Latin America, which not only are important from the point of view of its role in different ecosystems, for their feeding habits and as important prey for carnivores, but also have a cultural value for human populations since ancient times. Tropical ungulates suffer different pressures mainly from accelerated habitat transformation and hunting. In addition to the rapid destruction of their habitat, in many places they are competing with domestic species for food and space, and they play an important role in disease transmission in both directions. Moreover, many tropical ungulates are important to indigenous and local populations in Latin America, since they are considered as an important food resource, in most cases as the only source of protein.

The approach of this book encompasses the knowledge of ungulates in the ecological community of diverse ecosystems from some countries as well as information of the most representative tropical ungulates species. The book will help you to better understand the biology, ecology, and other issues and experiences to develop management and conservation strategies, that may be applied in different countries for long-term studies, as occurring in some Protected Natural Areas as the Biosphere Reserves of Mexico (Fig. 1.1).

Other proposals of this book include recent information about tropical ungulates of Central and South America (Fig. 1.2). The information including their distribution, population (dynamics, structure by age and sex, sexual segregation) and their role in ecosystems (aspects of herbivory, frugivory, and dispersal of seeds), behavioral ecology, habitat requirements, and fragmentation, among others, which are indicators to assess the changes that have affected these species in the countries

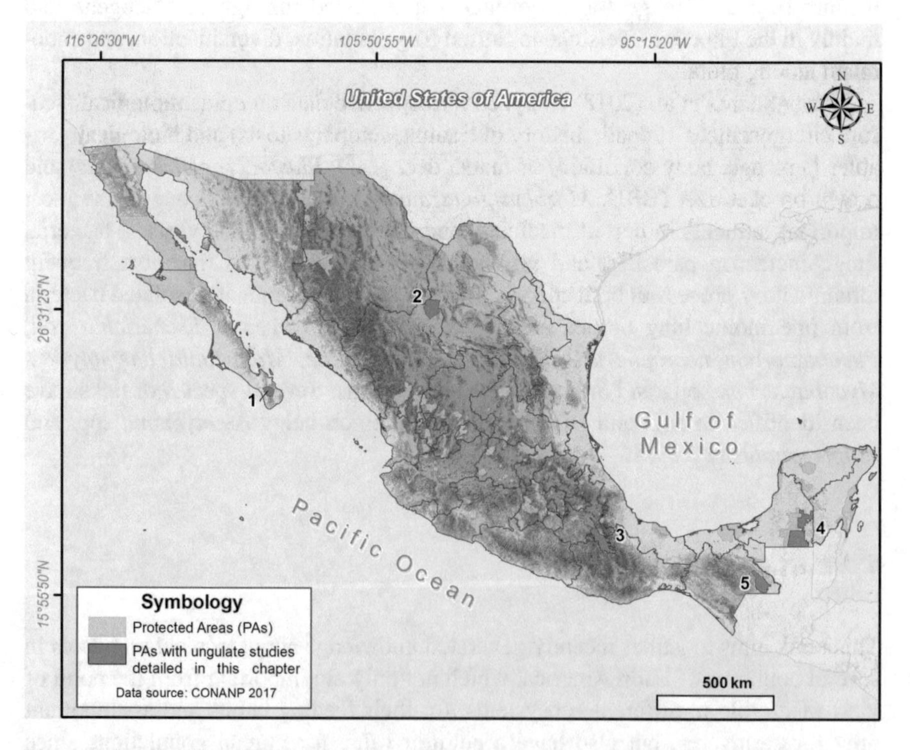

Fig. 1.1 Mexican Natural Protected Areas where long-term research studies on ungulate communities are underway: (1) Sierra de La Laguna Biosphere Reserve, Baja California Sur, (2) Mapimi Biosphere Reserve, Durango, (3) Tehuacán-Cuicatlán Biosphere Reserve, Puebla-Oaxaca, (4) Calakmul Biosphere Reserve, Campeche, (5) Montes Azules Biosphere, Chiapas

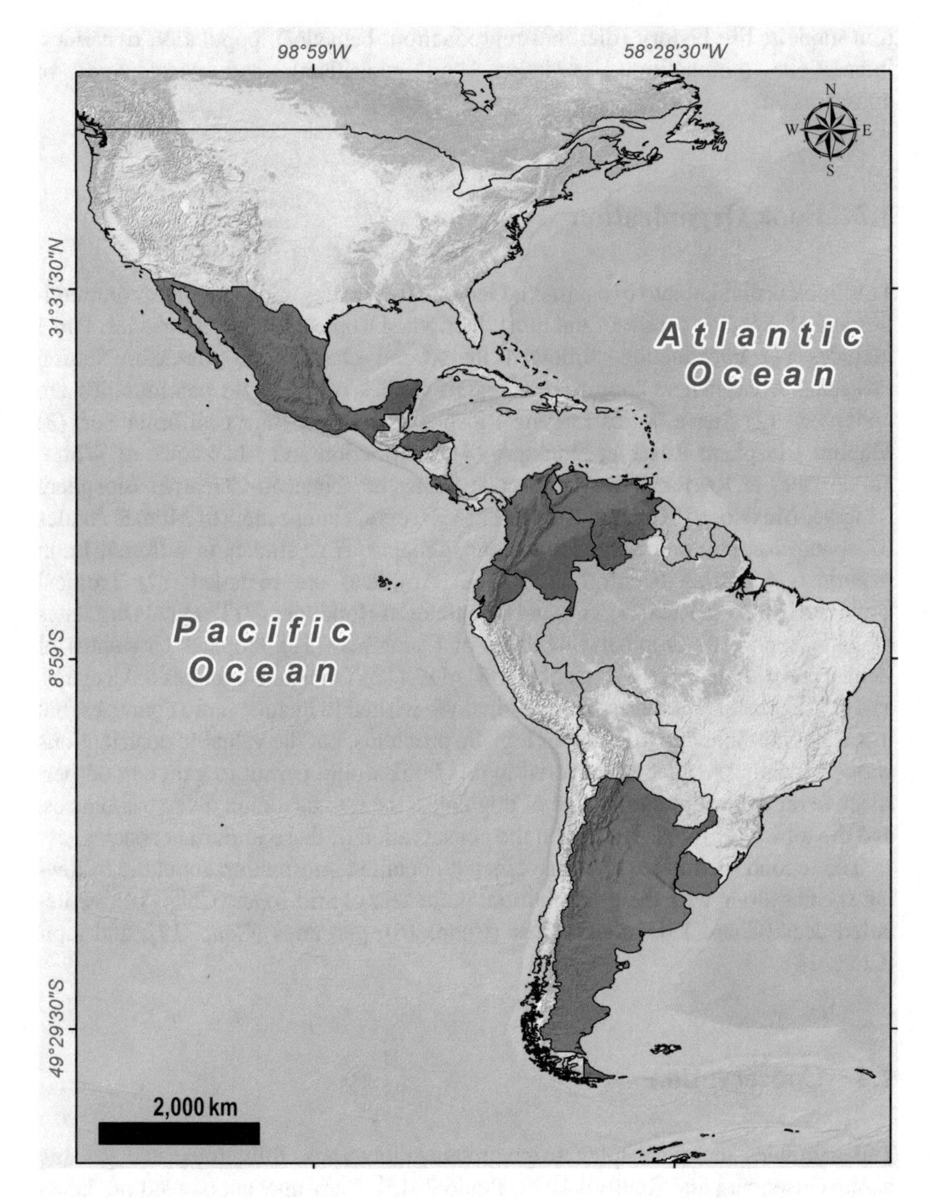

Fig. 1.2 Latin American countries that contributed in this book with the knowledge generated about the tropical ungulates

where they are distributed. We would have included all Latin American countries in this book, which we considered important at the beginning, but time and various commitments did not allow the researchers to collaborate on this occasion.

In addition, chapters on species with large distribution would also be included, with relevant aspects such as their biological description, distribution and occupation models, life history (diet and reproduction, behavior), population dynamics, interactions (competition, predation, and parasitism), and perspectives of conservation.

1.3 Book Organization

This book is divided into two parts: I: General Topics of tropical ungulate communities and II. Specific topics of the most distributed tropical ungulated species. Part I includes (1) Introduction, followed by several chapters on Mexican Natural Protected Areas where long-term research studies on ungulate communities are underway: (2) Sierra de La Laguna Biosphere Reserve, Baja California Sur, (3) Mapimi Biosphere Reserve, Durango, (4) Distribution and Abundance of White-Tailed Deer at Regional and Landscape Scales at Tehuacán-Cuicatlán Biosphere Reserve, Mexico, (5) Calakmul Biosphere Reserve, Campeche, (6) Montes Azules Biosphere Reserve and Selva Lacandona, Chiapas. Also studies in different Latin American countries (Central and South America) are included: (7) Tropical Ungulates of Costa Rica, (8) Tropical Ungulates of Honduras, (9) Tropical Ungulates of Venezuela, (10) Tropical Ungulates of Colombia, (11) Tropical Ungulates of Ecuador: An Update of the State of Knowledge, (12) Tropical Ungulates of Uruguay, and (13) Tropical Ungulates of Argentina. We wished to include more countries, but it was not possible due to different logistic problems, but the valuable contributions about the countries that are covered in this book would permit to gain a good perspective on the recent knowledge of ungulates, the risk factors in diverse countries, and the topics for future studies on the conservation of these important species.

The second section has five chapters with detailed information about the following species along with their distribution: mule deer of arid zones (Chap. 14), white-tailed deer (Chap. 15), brocket deer (Chap. 16), peccaries (Chap. 17), and tapir (Chap. 18).

1.4 Conservation

The ungulates mainly display frugivorous, granivorous, folivorous, and grazing habits (Eisenberg and Redford 1999; Prado 2013). They may act as seed predators or dispersers or as seedling predators, and they often help control the growth and primary productivity of herbaceous and woody elements in different ecosystems (McNaughton et al. 1988; Augustine and Frelich 1998).

In addition to hunting, the main threats to the Neotropical ungulates are habitat degradation, loss, and fragmentation (Cullen Jr et al. 2000; Robinson and Bennett 2000; Peres 2001). Habitat fragmentation not only isolates local populations but also directly affects the amount of plant food resources available for ungulate populations by reducing plant species richness (Collinge 1996). The predominance of secondary forests in tropical regions and their effects on regional fauna have also been widely discussed in the conservation literature (Wright and Muller-Landau 2006; Gardner et al. 2007; Prado et al. 2014). Further investigation of the foraging patterns of animals in these novel environmental contexts is needed.

Examining edge effects is imperative to developing effective conservation and management strategies in fragmented landscapes, as edge effects are a key component of how landscape change influences habitat quality. Although medium- to large-bodied mammals are recognized as key components of tropical forests, their responses to forest edges remain poorly documented.

In general, ungulates have been highly appreciated by man as stated earlier, from the cultural point of view as species that have been used both for self-consumption, as an important source of proteins in the tropics, and for sport hunting use. Some of the species can be faced with conservation problems, especially those that are habitat specialists, that need less disturbed tropical forests, which due to anthropogenic activities are getting transformed rapidly. In addition, across all the chapters of the book species that have problems in different countries are covered; in addition, there are several species that lack biological and ecological knowledge. Therefore, there is urgency to carry out studies that provide needed information and that may allow to propose actions for its conservation.

We hope that this effort to compile recently generated knowledge about the tropical ungulates, their ecology and problems faced by conservation, will be an incentive to continue studying this important group of tropical fauna to avoid its disappearance for the sake of humanity. It is my wish that this valuable opportunity to spread this information would be used by local populations without any distinction as to countries.

Acknowledgements I want to thank all the authors for their efforts to contribute their important knowledge on ungulates in different Latin American countries, Beatríz Bolívar for revision of some of the chapters, Rolando González Trápaga for his help with formatting to unify the chapters, Adriana Sandoval Comte for the maps, and specially my son Alejandro González Gallina for the draw of the main ungulate species for the Part II.

References

Araujo Absolon B, Gallo V, Avilla LS (2016) Distributional patterns of living ungulates (mammalia: Cetartiodactyla and Perissodactyla) of the Neotropical region, the South American transitional zone and Andean region. J South Am Earth Sci 71:63–70

Augustine DJ, Frelich LE (1998) Effects of White-tailed deer on populations of an understory forb in fragmented deciduous forests. Conserv Biol 12:995–1004

Cullen L Jr, Bodmer RE, Valladares-Padua C (2000) Effects of hunting in habitat fragments of the Atlantic forests, Brazil. Biol Conserv 95:49–56

Duarte JMB, González S (eds.) (2010) Neotropical Cervidology: Biology and Medicine of Latin American Deer. Jaboticabal:Funep/IUCN

Eisenberg JF, Redford KH (1999) Mammals of the neotropics: Ecuador, Peru, Bolivia, and Brazil, vol 3. University of Chicago Press, Chicago, IL

Gardner TA, Barlow J, Parry LW, Peres CA (2007) Predicting the uncertain future of tropical forest species in a data vacuum. Biotropica 39:25–30
Gilbert C, Ropiquet A, Hassanin A (2006) Mitochondrial and nuclear phylogenies of Cervidae (Mammalia, Ruminantia): systematics, morphology, and biogeography. Mol Phylogenet Evol 40:101–117
Groves C, Grubb P (2011) Ungulate taxonomy. The Johns Hopkins University Press, Baltimore
McNaughton SJ, Ruess RW, Seagle SW (1988) Large mammals and process dynamics in African ecosystems. BioScience 38:794–801
Merino ML, Rossi RV (2010) Origin, Systematics, and Morphological Radiation. Pp. 2-11. In:Duarte, J.M.B. and S. González (eds.). Neotropical Cervidology:Biology and Medicine of Latin American Deer. Jaboticabal:Funep/IUCN
Navas-Suárez PE, Doaz-Delgado JE et al (2018) A retrospective pathology study of two Neotropical deer species (1995–2015), Brazil: Marsh deer (*Blastocerus dichotomus*) and brown brocket deer (*Mazama gouazoubira*). PLoS One 13(6):e0198670. https://doi.org/10.1371/journal.pone.0198670
Peres CA (2001) Synergistic effects of subsistence hunting and habitat fragmentation on Amazonian forest vertebrates. Conserv Biol 15:1490–1505
Prado HM (2013) Feeding ecology of five Neotropical ungulates: a critical review. Oecol Aust 17:459–473
Prado HM, Murrieta RSS, Adams C, Brondizio ES (2014) Local and scientific knowledge for assessing the use of fallows and mature forest by large mammals in SE Brazil: identifying singularities in folk ecology. J Ethnobiol Ethnomed 10:7. https://doi.org/10.1186/1746-4269-10-7
Robinson J, Bennett E (2000) Hunting for sustainability in Tropical Forest. Columbia University Press, New York
Ruiz-García M, Martinez-Aguero M, Alvarez D (2009) Genetic variability in Neotropical deer genera (Mammalia: Cervidae) according to DNA microsatellite loci. Rev Biol Trop 57:879–904
Wright SJ, Muller-Landau HC (2006) The uncertain future of tropical forest species. Biotropica 38:443–445

Chapter 2
Ungulates in Sierra La Laguna Biosphere Reserve, Baja California Sur, México

Patricia Galina-Tessaro, Gustavo Arnaud, Aurora Breceda Solis-Cámara, and Sergio Álvarez-Cárdenas

Abstract Sierra La Laguna Biosphere Reserve (SLLBR), located in the southern tip of the Baja California Peninsula in the Cape Region of the State of Baja California Sur (BCS), México, harbors both the unique pine-oak forest in BCS and tropical dry forest of the peninsula with palms and oases throughout riparian vegetation. It concentrates numerous endemic species and a great biological diversity. Seven ungulate species inhabit the reserve: feral pigs, cattle, sheep, goats, horses, mules as introduced species and only the mule deer (*Odocoileus hemionus peninsulae*) as native species. This chapter provides knowledge of wild and feral ungulates including data obtained by trap-cameras. The mule deer in the Reserve is a common species although feral pigs and cattle are distributed in all the vegetation types, including the pine-oak forest. Cattle is the introduced species with more records in trap-cameras and sometimes more frequent than mule deer. Despite the presence of introduced species and poaching that occur in some localities, reported mainly in the lowlands surrounding the mountains, the reserve in the oak-pine forest has been a refuge for mule deer and other species. Livestock has been identified as a problem; nonetheless, more research is needed in this field, as well as in determining the effect that these species have, directly or indirectly, on wildlife using these resources, which is still unknown. Environmental education programs at a general level (including locals) are a priority to increase efficiency of conservation strategies in the reserve and its surroundings.

2.1 Introduction

Sierra La Laguna Biosphere Reserve (SLLBR) is a mountain chain decreed as a protected natural area (Biosphere Reserve) in 1994 and incorporated into Man and Biosphere Programme (MAB-UNESCO) in 2003. It is located in the southern tip of the Baja California Peninsula in the Cape Region of the State of Baja California

P. Galina-Tessaro · G. Arnaud · A. B. Solis-Cámara · S. Álvarez-Cárdenas (✉)
Centro de Investigaciones Biológicas del Noroeste S.C., Baja California Sur, Mexico
e-mail: salvarez04@cibnor.mx

S. Gallina-Tessaro (ed.), *Ecology and Conservation of Tropical Ungulates in Latin America*, https://doi.org/10.1007/978-3-030-28868-6_2

Sur (BCS), Mexico, one of the regions in the peninsula with the greatest biological diversity that shelters a great variety of ecosystems, and it concentrates numerous endemic species of flora and fauna, some of tropical affinity product of long periods of geographical and ecological isolation. The SLLBR harbors the unique pine-oak forest in BCS and the unique tropical dry forest of the peninsula with palms and oases throughout riparian vegetation.

Seven species of ungulates inhabit the reserve, but only is the mule deer (*Odocoileus hemionus*) native to the region; the other species have been introduced as domestic animals: pigs (*Sus scropha*), sheep (*Ovis aries*), goats (*Capra aegagrus hircus*), cattle (*Bos taurus*), horses (*Equus caballus*) and mules (*Equus assinus* × *Equus caballus*) donkeys; these last ones are utilized by ranchers for personal and tourism transportation, as well as pack animals (CONANP 2003).

The presence of exotic ungulates in BCS dates before missionary work started in the southern portion (1697–1768); although Hernán Cortés introduced the first domestic animals in 1535, it was not until the establishment of the Jesuit missionaries that livestock was extended to this territory (INEGI 1997; Martínez-Balboa 1980). Particularly, the Sierra La Laguna started to be populated by cattle ranchers towards the end of the eighteenth century (Trejo-Barajas and González-Cruz 2002) since cattle raising is one of the main economic activities in the region.

To date, many ranches persist in the mid-areas of the Reserve, which are mainly dedicated to extensive cattle ranching, but some ranches also raise pigs and sheep in smaller numbers. Domestic animals that have escaped from the care of their owners, mainly cattle and pigs, have reproduced wildly, which constitutes a challenge for conservation and management of the reserve ecosystems.

Native and introduced ungulates are important elements in the different ecosystems of the Reserve and share resources; competition for space and food is one of the most common interactions (Chirichela et al. 2014) although it has not been assessed in the Reserve yet.

Several mammal studies have been published where the presence of the mule deer is reported in the Peninsula (Hall 1981; Huey 1964; Wallmo 1981) and one in particular about the mammals of Sierra La Laguna (Woloszyn and Woloszyn 1982). The authors considered mule deer abundant in relation to pine-oak and oak and already mentioned: "*The lack of food in the dry season is a factor that greatly limits the populations of mule deer; the situation is aggravated by competition with domestic livestock, especially in the vicinity of water tanks, where cattle eat all the usable forage. Despite the protective laws, the mule deer is hunted practically all year round. In the vicinity of most ranches in the mountains, you can find an abundance of deer bones as a result of that hunting, because this persecution without limit is the cause of its shortage*" (Woloszyn and Woloszyn 1982: 151), highlighting the presence of cattle and their possible competition with mule deer.

The first study about the population and distribution pattern of the mule deer in the relict oak-pine forest of Sierra La Laguna was performed by Galina-Tessaro et al. (1988). Monitoring had continuity for 7 years in which density and population structure were analyzed and related with habitat characteristics (Alvarez-Cárdenas 1995; Alvarez-Cárdenas et al. 1999a; Gallina et al. 1991).

The presence of feral pigs within the reserve, particularly in oak-pine forest, attracted attention and caused concern to the Reserve staff and several researchers on forest regeneration due to knowledge of the problems that this species has caused in other ecosystems of Mexico and worldwide (March 2007). Therefore, the Comisión Nacional de Áreas Naturales Protegidas (CONANP) in collaboration with The Nature Conservancy (TNC), Comisión Nacional para el Conocimiento y Uso de Biodiversidad (CONABIO), and Centro de Investigaciones Biológicas del Noroeste (CIBNOR) performed specific studies to assess the situation of the species in the Reserve, particularly in the pine-oak forest and its effect on forest regeneration (Arnaud-Franco et al. 2012, 2014; Breceda et al. 2009). Research on reproduction and diet of feral pigs, as well as health aspects between feral and domestic animals in the ranches were performed by Montes-Sánchez et al. (2012) and Pérez (2014).

This chapter describes knowledge about wild and feral ungulates in Sierra La Laguna Biosphere Reserve; research results on the presence of these ungulates are summarized, and the data obtained by trap camera are analyzed to finally discuss their impact on the reserve to propose the main challenges for management and conservation priorities.

2.2 Description of the Study Area

The Sierra La Laguna Biosphere Reserve is located in the highest zone of this mountain range between 23° 42′–23° 20′ N and 109° 46′–110° 11′ W, occupying an area of 112,437 ha and an altitudinal gradient from 300 to 2080 m. The highest areas are considered as the core zone surrounded by a buffer zone (Fig. 2.1).

In lower parts, on plateaus and hills with arid and warm climate, the sarcocaule scrub develops. It is characterized by abundant columnar cacti, shrubs and few trees of short stature, among which the most conspicuous species are *Pachycereus pringlei*, *Stenocereus thurberi*, *Jatropha cinerea*, and *Fouquieria diguetii*. In the mid-mountain area from 400 to 1000 masl, the unique tropical dry forest of the peninsula develops. Its physiognomy is characterized by being exuberant and impenetrable during the rainy season, which contrasts with a landscape devoid of leaves during the long dry season that could last up to more than 6 months. Among the most abundant arboreal and shrub species are *Lysiloma microphyllum*, *Tecoma stans*, and *Jatropha vernicosa*. In the highest areas with temperate climate, *Quercus tuberculata*, *Q. devia* and *Pinus lagunae* are abundant in oak and pine-oak forests. Another type of very conspicuous vegetation growing along the arroyos is riparian vegetation, which can reach true oasis. Tall and slender palm trees can be found in these places, such as *Washingtonia robusta* and *Erythea brandegeei*.

According to the 2000 census developed by Reserve authorities and CIBNOR researchers, there were 146 ranches with 641 inhabitants. This number of inhabitants has decreased in the last decade since a marked abandonment process of the ranches has been observed and with it a loss of valuable traditional knowledges of natural management. The ranches have been small horticultural and livestock productive units; they are isolated and dispersed in an extensive territory, located in places near

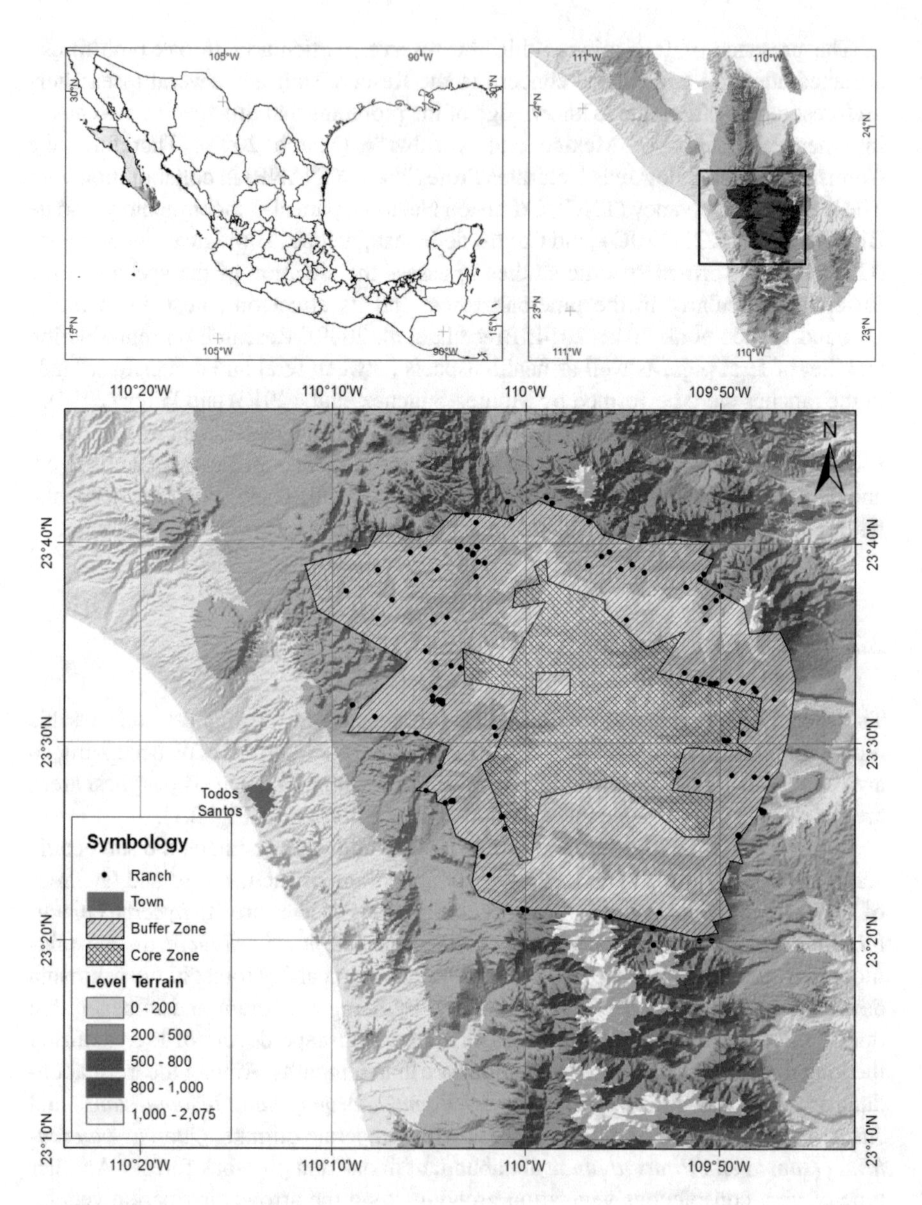

Fig. 2.1 Sierra La Laguna Biosphere Reserve in Baja California Sur, Mexico and ranches within its limits (INEGI, CONANP, SRTM (NASA))

permanent water sources that have allowed them to develop a modest but varied horticulture and an extensive creole regional livestock (Castorena and Breceda 2008; Breceda et al. 2012). The productive activities of the ranchers have transformed the nature of this mountainous area and built a particular landscape of great identity and cultural interest in the region. To our knowledge no recent data are available regard-

Fig. 2.2 Mule deer in the oak-pine forest in Sierra La Laguna Biosphere Reserve in Baja California Sur, Mexico

ing the number of heads of cattle that exist in the reserve. In the year 2000, the census counted 7243 heads of cattle, 1254 goats, 538 pigs and 380 sheep (Breceda 2005; Breceda et al. 2005).

2.3 Results

Information on native and introduced ungulates in the Sierra La Laguna Biosphere Reserve is the result of the bibliographic review and information not yet published on research performed by the authors, showing the most relevant results, as well as the analysis obtained from the use of trap cameras from 2010 to 2011.

2.3.1 Wild Ungulate

2.3.1.1 Peninsular Deer *Odocoileus hemionus peninsulae* (Lydekker, 1898)

The subspecies of the mule deer *O. hemionus peninsulae* (Fig. 2.2) is distributed in Baja California Sur and native to the Baja California Peninsula (Latch et al. 2009). It is found in a great variety of peninsular ecosystems: desert scrub, low deciduous and oak-pine forests from the plains of Vizcaino in the north to the southern mountains of BCS as the Sierra La Laguna and Sierra La Trinidad from sea level to about 2000 masl.

In the Sierra La Laguna, it is common to observe males with antlers from 6 to 10 branches and growing up to 14. Males lose the antler velvet in the months of October and November and pull the antler in May (Gallina et al. 1992); other subspecies (*O. h. hemionus* do in September–October and February–April, respectively; *O. h. cerroensis* in July–August and February–April, respectively, and *O. h. columbianus* in July–October and December–March, respectively) (Anderson 1981; Perez-Gil 1981).

This phase shift also happens with the heat season that runs from December to February while for other subspecies it starts in October (*O. h. fuliginatus* and *O. h. cerrosensis*) or in November (*O. h. hemionus*); the highest percentage of births in Sierra La Laguna occurs in August after a period of approximately 7 months while for *O h. hemionus* it occurs between May and August (Gallina et al. 1992), which could be due to synchronization with the period of greater food quality and availability.

The mule deer has great importance for the ecological role it plays in the ecosystem, as one of the great wild herbivores since it is the main prey of larger carnivores as the cougar and coyote, an important food resource for rural communities and one of the most important game species in BCS, Mexico; only in the last 9 years has the Ministry of the Environment and Natural Resources (SEMARNAT for its abbreviation in Spanish) granted 790 headbands (legal permits) for deer hunting in Baja California Sur to 31 Management Units for Wildlife Conservation (UMA in Spanish: Unidades de Manejo para la Conservación de la Vida Silvestre), whose number varies each year (Table 2.1). Unfortunately, the percentage data of headbands charged was not provided to confirm the number of animals hunted. In the last 3 years, up to 25 headbands/year have been granted to some hunters, and the maximum total number granted yearly were 143 headbands in the State during that period. It is worth to mention that due to a long drought period that caused an important mortality of deer in BCS, a consensual closure was established for the 2012–2013 season among the State government, UMA and deer hunters to allow the recovery of the species.

According to the management plan of the SLLBR (CONANP 2003) the activities for better use of fauna (sport hunting and self-consumption) through the wildlife management unit (UMA) are permitted in the buffer zone and only in the ranches of Cañón de la Zorra within the core zone, complying with SEMARNAT requirements and permits.

The first analysis on deer population in Sierra de La Laguna was published in the book "*La Sierra de La Laguna of Baja California Sur*" (Arriaga and Ortega 1988) where biotic and abiotic aspects of the Sierra were evaluated. According to an indirect census, the population estimated was 1.25 individuals/ha + 0.45 by counting pellets (considering daily defecation rate of 12.7 groups per animal per day) in six transects in the oak-pine forest inhabited by deer in the areas of El Madroñito, San Antonio, La Torre, Chuparrosa, Cieneguita, and Cerro Verde (located from 1600 to 1950 masl) (Galina-Tessaro et al. 1988). In the study area the mule deer was found in three plant associations: (1) Black oak (*Quercus devia*) –pinyon pine (*Pinus lagunae*)–madrone (*Arbutus peninsularis*); (2) Pinyon pine–black oak–madrone and, (3) Black oak–oak (*Quercus tuberculata*)–Pinyon pine.

Table 2.1 Number of headbands (hunting permits) for mule deer granted by UMA during the 2009–2010 to 2017–2018 seasons in Sierra La Laguna, Baja California Sur, Mexico

UMA	2010	2011	2012		2014	2015	2016	2017	2018	Total/UMA
1	7	12	12	–	8	13	15	15	15	97
2	2	2	2	–	2	4	2	1	2	17
3	9	10	10	–	8	10	10	5	8	70
4	1	1	2	–	1	1	1	0	2	9
5	20	3	5	–	3	3	4	3	4	45
6	1	1	2	–	0	1	1	1	2	9
7	4	4	4	–	4	4	4	1	4	29
8	2	2	4	–	2	2	3	2	0	17
9	14	8	0	–	10	20	20	0	0	72
10	10	10	10	–	10	13	13	0	0	66
11	0	1	2	–	1	1	1	0	1	7
12	15	9	8	–	0	4	0	0	0	36
13	15	9	6	–	0	0	0	0	0	30
14	14	0	0	–	0	0	0	0	0	14
15	2	2	0	–	0	0	0	2	0	6
16	3	0	0	–	0	0	0	0	0	3
17	14	0	10	–	0	0	0	0	0	24
18	10	14	12	–	0	0	0	8	0	44
19	0	0	2	–	0	0	8	0	0	10
20	0	0	6	–	0	0	0	0	0	6
21	0	0	0	–	8	15	15	0	0	38
22	0	0	0	–	2	2	2	2	0	8
23	0	0	0	–	0	10	8	20	0	38
24	0	0	0	–	0	2	2	2	2	8
25	0	0	0	–	0	10	25	25	0	60
26	0	0	0	–	0	1	2	1	0	4
27	0	0	0	–	0	2	2	2	0	6
28	0	0	0	–	0	1	2	2	2	7
29	0	0	0	–	0	1	1	0	0	2
30	0	0	0	–	0	1	0	1	0	2
31	0	0	0	–	0	0	2	2	2	6
Total	143	88	97	0	59	121	143	95	44	790

More detailed analyses evaluating different parameters of the population allowed estimating for the years 1987–1990 a density of $20 + 8$ deer/km^2, calculating an abundance of $1007 + 273$ individuals for the 5050 ha of forest with similar vegetation characteristics (Gallina et al. 1991, 1992). Subsequently, as a result of 14 seasonal censuses (1987–1993) and pellet group count analyses, an average density of $19 + 5$ deer/km^2 was obtained, as well as growth and survival rates that indicated a stable population (Alvarez-Cárdenas 1995). Through another sampling done in 1997, an average of $20 + 5$ deer/km^2 was obtained (completing 8 years of census), confirming population stability (Alvarez-Cárdenas et al. 1999b).

For the analysis of the mule deer population structure in the Sierra La Laguna, only were the censuses carried out in the fall taken into account since they covered the 7 years sampled (1987–1993); no significant differences were observed in the groups of excreta found (Fig. 2.3a). In general, a greater number of yearlings was observed, followed by fawns and adults except in 1988 when the number of adults was double than that of the other categories. The annual sex proportions for adults are shown in Fig. 2.3b where except for 1987 and 1993, sex ratio was almost 1:1, varying annually (Alvarez-Cárdenas 1995).

As results of the habitat use research, Alvarez-Cárdenas et al. (1999a) found that mule deer at the Sierra La Laguna seemed to prefer habitats with steep slopes and rough, rocky terrain. They had a preferred index of 1.28 for rocky terrain and 1.26 for slopes steeper than 30%.

2.3.2 *Livestock and Feral Species*

Most ranches in the tropical dry forest prevail breeding livestock composed of cattle, goats, pigs and poultry, and those that only have one type of livestock, commonly raise bovine (Arriaga and Cancino 1992).

2.3.2.1 Feral Pig *Sus scrofa* Linnaeus, 1758

The feral pig (*Sus scrofa*) (Fig. 2.4) is distributed in different environments of the Sierra La Laguna, both in the deciduous forest and in the oak-pine forest without concentrating on a specific habitat. It also uses gullies, canyons, hillsides and plains, open areas or with a wide vegetation cover of shrubs and trees. Apparently their seasonal movements respond to availability of essential resources (water, food, plant cover) (Arnaud-Franco et al. 2012).

The pig represents a potential danger for the Sierra biodiversity due to its habits and feeding behavior since it removes soil in search for food, causing eventual changes in vegetation. They are also transmitters of viral and bacterial diseases, as well as parasites (Choquenot et al. 1996; Everitt and Alanis 1980; Hutton et al. 2006; Kotanen 1995; Lacki and Lancia 1986; March 2007; Massei and Genov 2004; Meng et al. 2009; Villarreal and Alanís 2015; Zanin and Marini 2013) which is why it is considered a high-impact invasive species in Mexico (Álvarez-Romero et al. 2008; CONABIO, Aridoamérica, GECI, TNC 2006).

Feral pig abundance was estimated by using trap cameras arranged in five transects (Fig. 2.5) and separated from each other approximately 250–300 m in the high part of the Sierra. Samplings were made from February 2010 to August 2011. The highest values were in autumn–winter with a total of 107 feral pigs recorded in the cameras while only 21 were recorded in spring-summer.

The differences in the feral pig records apparently reflected the mobility of these animals in response to climatic conditions that determined the availability of

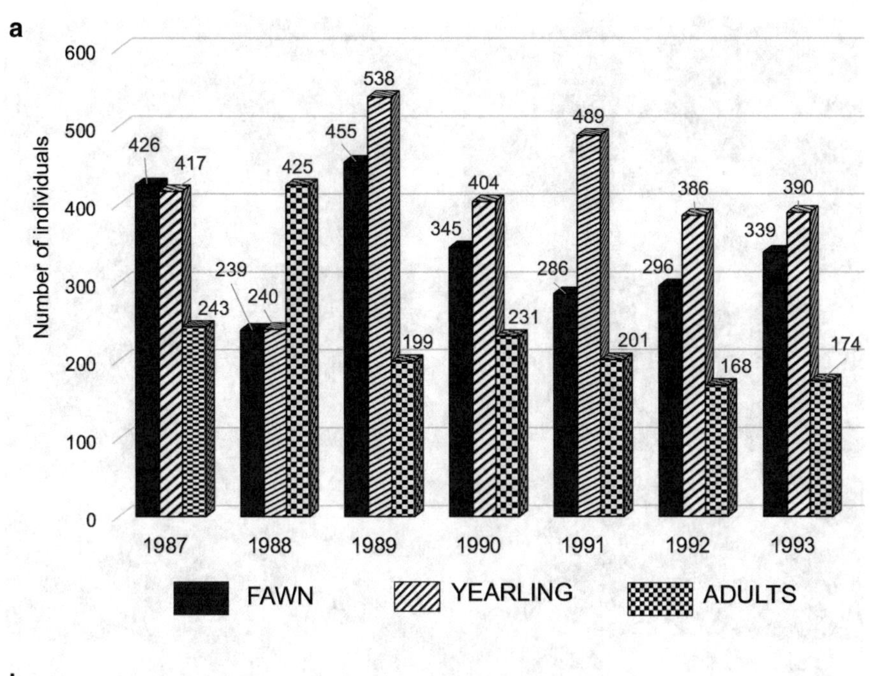

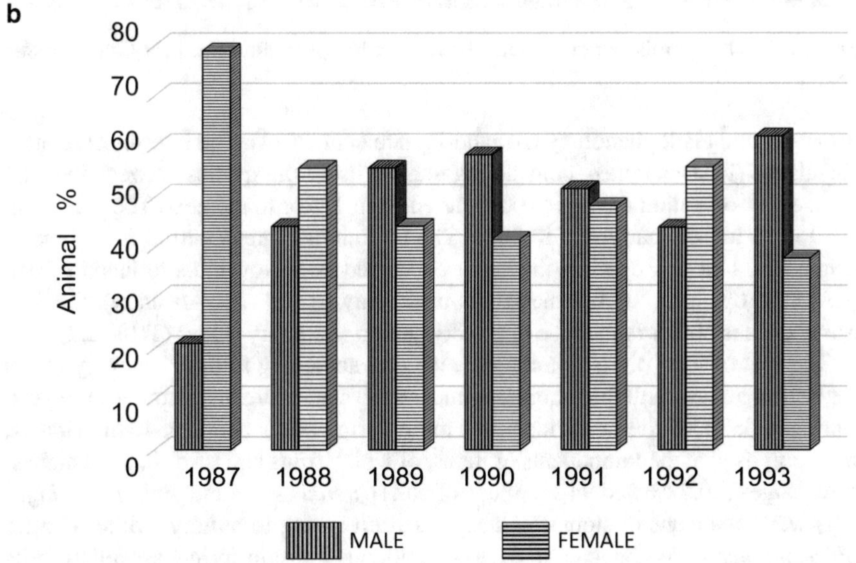

Fig. 2.3 Population structure of mule deer of the oak-pine and pine-oak forest of Sierra La Laguna Biosphere Reserve, Baja California Sur Mexico. (**a**) Annual changes (in autumn) in the age-groups in the population of mule deer, (**b**) Sex ratio of mule deer adults of the forest from 1987 to 1993 (Álvarez-Cárdenas 1995)

Fig. 2.4 Feral pig in the forest in Sierra La Laguna Biosphere Reserve, Baja California Sur, Mexico

resources and plant phenology (Arnaud-Franco et al. 2014). The indexes of relative abundance (RAI) obtained, considering total visits by the species between the total numbers of operation days per 1000, varied from 1 year to the next. The records in 2010 were higher than those in 2011. The minimum value occurred from June to August 2011 (RAI = 6) while the highest occurred from November to January 2011 (RAI = 89). During the summer and winter rainy season (July–January), the RAI was six times higher than those of the dry season (February–June) (Table 2.2).

The diet of feral pig is omnivorous, varying according to the phenology of the vegetation related with the rainfall regime, which can in turn vary from year to year. During 2008–2009 their diet, identified through stomach contents of 40 individuals, was composed of 59 components or items, of which fruits and seeds had the highest percentages (20.52% and 94%). Seeds of oak (*Quercus* spp.) and pine (*Pinus lagunae*) were abundant in stomachs analyzed from spring to autumn while wild fig (*Ficus palmeri*) was abundant in the low deciduous forest in spring, as well as fruits and seeds of palms (*Erythea brandegeei* and *Washingtonia robusta*), western bumelia known in Spanish as "bebelama" (*Sideroxylum occidentalis*) and "mauto" (*Lysiloma microphyllum*). During summer and winter the prickly pears or "nopal" (*Opuntia* spp.) and the tubers of "chuchupate" (*Arracacia brandegeei*) were the most important (Montes-Sánchez et al. 2012).

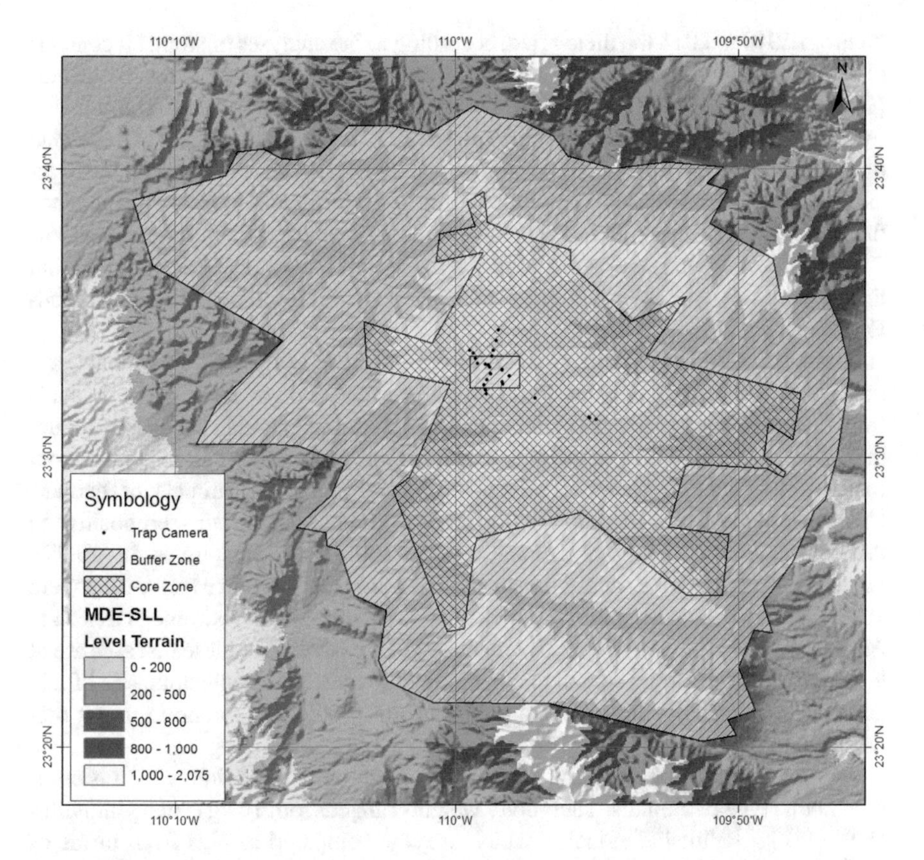

Fig. 2.5 Geographic location of the trap cameras used to register pigs and other species in Sierra La Laguna Biosphere Reserve, Baja California Sur, Mexico (CONANP, SRTM (NASA))

Table 2.2 Relative abundance index (RAI) of feral pigs estimated with trap cameras in the pine-oak forest in Sierra La Laguna Biosphere Reserve, Baja California Sur, Mexico, during the period February–March 2010 to June–August 2011

	2010–2011					
	Feb–Mar	May–Jun	Sep–Oct	Nov–Jan	Mar–Apr	Jun–Aug
Cameras used	24	25	25	18	17	20
Cameras with feral pig records (%)	6 (25%)	7 (28%)	15 (60%)	11 (61%)	2 (12%)	4 (20%)
Total visits	8	13	30	68	4	5
Days of operation	531	690	489	761	530	876
RAI	15	19	61	89	8	6

From 2010 to 2011 the diet varied, according to the analyses of stomach contents of 25 adult individuals from the highlands of the Sierra. In these years, the cactus (*Opuntia lagunae*) also had high frequency of occurrence (FO) (12%), and it was the second most represented item in volume (22%); the "chuchupate" (*Arracacia brandegeei*) followed the prickly pear in FO (11%), but it was the one with the highest volume (28%). Insect larvae of the families Tenebrionidae and Scarabeidae had the same FO percentage as the chuchupate (11%); however, their volume was low (2% and 3%, respectively). Oak acorns (*Quercus* sp.) showed a low FO (5%), but they represented the fourth highest percentage in diet volume (10%) (Arnaud-Franco et al. 2014).

In relation to the zoo-sanitary status of feral pigs, an analysis of 25 animals captured in the upper parts of the Sierra was performed, as well as of 70 individuals from the lower parts that were maintained in semi-captivity and in backyard production. The results showed that the pigs present in the area of pine-oak forest were not focus transmitters of bacterial infectious agents, either for animals or humans (Arnaud-Franco et al. 2012). In contrast, pigs from lower parts were seropositive to different antigens: Swine influenza virus in 30.7%, Leptospirosis in 25.9%, Salmonellosis in 87.1%, and Brucellosis in 14.3%. On the other hand, they were 100% negative in the porcine reproductive respiratory syndrome virus and Aujeszky's disease virus (Pérez-Rivera et al. 2017). Given that these diseases are at high risk of infection and spread to other animal species in free and domestic life, in addition to humans, it represents a public health problem of zoonotic origin that must be addressed.

All the wild pigs observed had a lean or slender body condition, which is common in this type of animals. Their body weights ranged from 16 kg (young individuals) to 90 kg (adult individuals), which are low compared to pigs from farms of similar ages (Pinelli Saavedra et al. 2004). The scarce body fat is due to the fact that it is used as an energy source in its movements to find food and water, as well as to prolonged fasts during drought periods, in which food availability is scarce. On the other hand, the animals kept in semi-captivity and in backyard production showed from regular body condition to bad.

Given the unusual good health condition of the pigs in the upper parts of the Sierra, coupled with their lean meat and appreciable taste, they are preferred by ranchers in relation to farm pigs; thus, they have been subjected to traditional exploitation as a source of protein or as a breeding ground for ranchers in the lower parts of the Sierra. In this context, the feral pig is a resource that has the potential to provide opportunities for development to the communities of the Sierra despite being an exotic species.

The feral pig is also part of the diet of the carnivores of the Sierra. Young pigs are prey to the coyote (*Canis latrans*), one of the predators with the highest records in trap cameras (Arnaud-Franco et al. 2014).

2.3.2.2 Cattle (*Bos taurus*) (Linnaeus, 1758); Synonymous *B. indicus*

The cattle found in the Reserve are the so-called "Criollo de Baja California" or "chinampo" in Spanish (Fig. 2.6), which were introduced in Mexico and the peninsula during the sixteenth century around 1697. This situation has been managed in a feral way in the particular environment of the peninsula for generations, adapting cattle to the environmental conditions of the region and giving rise to an extremely resistant animal capable of subsisting on cacti. Some cattle have been mainly crossed with zebu, Swiss brown-zebu, "criollo"-Swiss brown-Jersey and in smaller numbers pure-bred are also found (Arriaga and Cancino 1992; Martínez-Balboa 1980; Quiróz 2007).

Arriaga and Cancino (1992) reported 102 plant species used in livestock activity in the Sierra La Laguna; however, livestock preferably consumes 40 species, among which the most important are *Antigonon leptopus* (coral vine or "San Miguelito"), *Merremia aurea* (yellow morning-glory or "Yuca"), *Amaranthus* spp. (amaranth or "quelite"), *Lysiloma divaricata* ("Mauto"), *Viguiera deltoidea* (Parish's golden-eye "tacote"), *Phoradendron digeutianum* ("Toji"). Cacti, basically *Ferocactus* spp. (barrel cactus or "biznagas"), *Machaerocereus gummosus* (galloping cactus "pitaya agria") and *Stenocereus thurberi* (organ pipe cactus or "pitayas") and *Opuntia* spp. (prickly pear or "nopales") and an important part of

Fig. 2.6 Cattle near a stream in the oak-pine forest in Sierra La Laguna Biosphere Reserve, Baja California Sur, Mexico

the cattle diet during the dry season prepared by the farmers by burning the spines or prickles and fractioning them.

Some other plants used as cattle feed in the San Antonio area are herbaceous as amaranth or "quelite," chamomile, Caribbean (*Cnidoscolus maculatus*); shrubs, such as Mexican oregano (*Lippia palmeri*), damiana (*Turnera diffusa*), rathany or "mesquitillo" (*Krameria* sp.); grasses, such as white grass, three-edged grass, and crow feet (*Cynodon dactylon*); cacti, such as cholla (*Cylindropuntia* sp.), barrel cactus or "biznaga" (*Ferocactus* sp.), and cardon cactus and trees with pod like Brandegee cottonwood or "güéribo" (*Populus brandegeei*), mesquite (*Prosopis* sp.), palo verde (*Parkinsonia* sp.) (Rochín-Sánchez 2003).

According to Lagunas-Vázques et al. (2013) from the surveys applied to 95 ranches within the reserve, most of them had up to 100 heads of cattle and only a few had 400–500, which were the highest figures per ranch in terms of cattle. Although the majority is "criollo" in some ranches, crosses have been carried out with different cattle breeds as experiments of adaptation and production of the different breeds.

2.3.2.3 Goats (*Capra aegagrus hircus*) (Linnaeus, 1758); Synonymous *Capra hircus*

The domestic goat is believed to have originated directly from *Capra aegagrus* or they are conspecific. The first breeds introduced in Mexico were Spanish (Celtibérica, Castellana, Extremeña, and by the twentieth century Murciana had been introduced); Criolla goat of America and Peninsular South Californian goat were originated from uncontrolled breed-crossing. These goats have been managed through the traditional extensive system, mainly in the mountainous areas where grazing is mainly diurnal with night enclosure (Fig. 2.7). They are used for direct consumption and for cheese-making. A great variety of breed-crossing exists in the ranches of the reserve: Creole, Nubian, Saanen (French and English), Alpine, Toggemburg, Grenadine (which arrived as Murciana) and Ibias for milk production; some for meat production, such as Boer (pers. Victor Anguiano CONANP).

Although feral populations in the Reserve have not been reported, some were found in islands in the Gulf of California, such as Espiritu Santo and Cerralvo (Álvarez-Romero et al. 2008) where eradication work has been performed. In the reserve the goats are managed in a semi-hinged way with the support of dogs called "chiveros." Goats usually graze near ranches in the jungle and in the bush, especially close to intermittent surface runoff; according to the 2000 Census, goat livestock ranked second in number of heads (12.6% of total cattle) accounted for in the Reserve, and its number has increased due to programs implemented by the government through the Secretaria of Agricultura, Ganaderia y Desarrollo (Ministry of Agriculture, Livestock and Development) (CONANP 2003). For the Reserve area according to the census conducted in 2000, about 43 ranches were registered with goats, of which most of them reported less than 50 animals with one up to 100 animals (Castorena and Breceda 2008).

Fig. 2.7 Goat enclosure in a ranch in the tropical dry forest in Sierra La Laguna Biosphere Reserve, Baja California Sur, Mexico

Despite the relative control of goats, their effect should be evaluated by how both goats and sheep have become destructive in the ecosystems in which they are found, particularly in islands and the Reserve that could be considered an "island" by its peculiar isolation characteristics (Hess et al. 2017). Disappearance of plant species where goats and sheep had been introduced, as well as modification in vegetation structure and composition have in turn affected the fauna that lives there.

2.3.2.4 Sheep (*Ovis aries*) Linnaeus, 1758

Most of the sheep found in the ranches in the reserve are Pelibuey breed, which is a purely hair breed (without wool) and of great adaptation to tropical environments (Fig. 2.8). They are semi-hinged, released in the morning, and the return to the ranch in the afternoon with the help of chiveros. To our knowledge, no studies of their effect on the ecosystem are available.

2.3.2.5 Equine: Horse (*Equus caballus*), Mules (*Equus asinus* × *Equus caballus*) Donkey (*Equus asinus*) Linnaeus, 1758

The horses are chinampos (crosses of different breeds) except for some ranches that have fine horses (1/4 mile) (Fig. 2.9). The mules and donkeys are also used for loading and transport and usually semi-stacked, especially near the ranches. These

Fig. 2.8 Sheep in the tropical dry forest in Sierra La Laguna Biosphere Reserve, Baja California Sur, Mexico

ungulates are generally retained near ranches, so their records in the trap cameras were rare. Their numbers are small compared to other domestic ungulates.

2.3.3 Evaluation of Ungulates by Trap Camera

As a result of six photo-trapping surveys performed in the pine-oak forest for recording ungulate presence from 2010 to 2011 (Table 2.3, Fig. 2.10), the highest number of records usually corresponded to cattle and deer. The percentage of cameras with cattle records varied seasonally from 48% to 100%; deer records were also high from 58% to 100% while pigs were recorded from 12% and 61%. The relative abundance (estimated as the total number of species visits between the total number of camera days per 1000) (Fig. 2.10) varied per sampling with deer notoriously more abundant from September to October 2010 while cattle was from June to August 2011 and pigs from November to January 2010.

These results do not necessarily indicate that cattle population is very high; however, in addition to being large groups, they were more associated with the habitats where the traps were placed, which were gaps and proximity to water pools and springs.

Grouping information by rainy season and dry season (Table 2.4) shows a greater presence of cattle during the dry season, possibly looking for water that is always

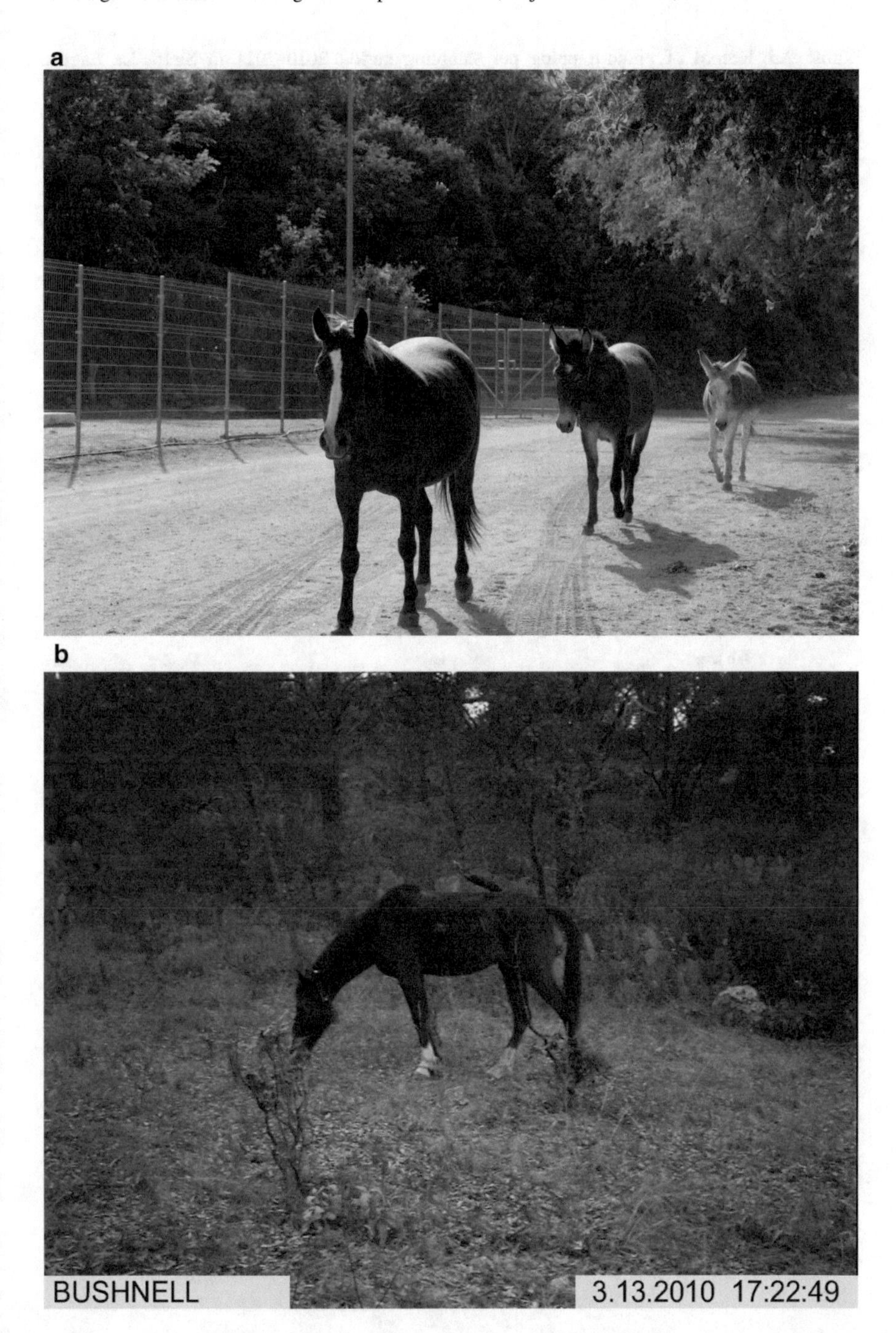

Fig. 2.9 Horse, mule and donkey walking freely in a ranch in Sierra La Laguna Biosphere Reserve, Baja California Sur, Mexico

Table 2.3 Record of photo-trapping per sampling period 2010–2011 in Sierra La Laguna Biosphere Reserve, Baja California Sur, Mexico

Sampling period		Feb–Mar	May–June	Sept–Oct	Nov–Jan	Mar–Apr	June–Aug
Total camera operative days		531	690	489	761	530	876
Total cameras		25	25	25	18	17	20
# of cameras where the species appears	Feral pigs	6 (25%)	7(28%)	15 (60%)	11 (61%)	2 (12%)	4 (20%)
	Cattle	12 (48%)	21 (84%)	14 (56%)	18 (100%)	13 (76%)	18 (90%)
	Mule deer	14 (58%)	19 (76%)	22 (88%)	18 (100%)	17 (100%)	17 (85%)
Total visits							
Feral pigs		8	13	30	68	4	5
Cows		78	229	106	324	118	397
Mule deer		49	131	478	297	163	183

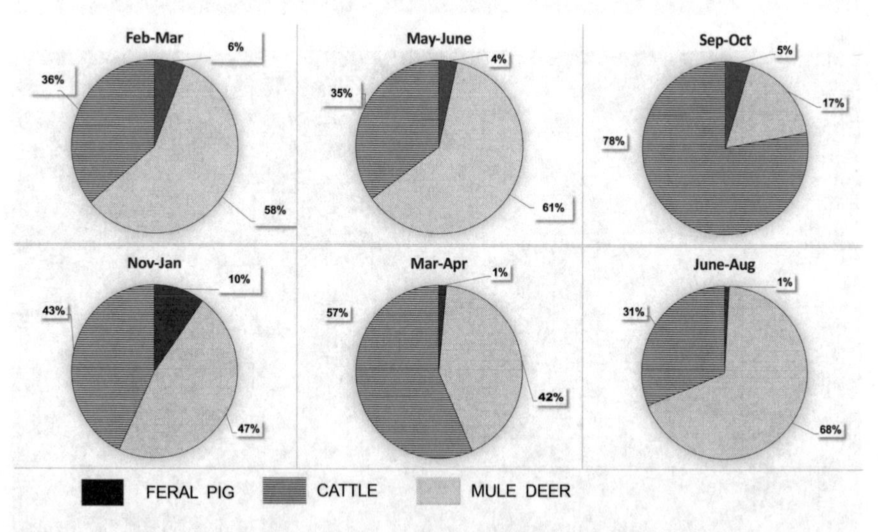

Fig. 2.10 Relative abundance of deer, cattle and feral pigs in Sierra La Laguna Biosphere Reserve, Baja California Sur, Mexico, per sampling period (Feb–Mar 2010 to June–Aug 2011)

present in the forest, unlike deer and pigs that are more present in the rainy season perhaps in response to the availability of a particular food resource. Deer was detected by most of the cameras same as cattle in the dry season while pigs were more restricted.

2.3.4 Mule Deer Records with Trap Camera

In a more detailed analysis of the photographic records of the mule deer, it was possible to differentiate between sexes in adult individuals when males showed antlers or the scar of lost antlers, and adult females were differentiated by size; those individuals of smaller size or corpulence were considered within the fawn–yearling group, including fawns whose speckled coloration pattern evidenced age. Photographs, in which the photographic record only included part of the body of the animal, were considered as unidentified individuals. The difference in time between each photograph determined its consideration as an independent record. The records grouped by age classes showed differences between sampling periods (Fig. 2.11). Superiority in the number of females was clear in all the samplings. In general, all the photographs showed animals with good body condition.

These records have allowed to verify that reported by Gallina et al. (1992) about horn temporality of adult males and the presence of offspring (Fig. 2.12).

The largest number of animals recorded by photography in this case was four individuals with one or two that were the most common; cattle, whose breeds and color patterns facilitated differentiation of individuals (even in partial images), were detected in larger groups (the maximum was eight animals per photograph, but following the photographic sequence the number of individuals was greater in some

Table 2.4 Records of trap camera by seasons (rainy and dry) in Sierra La Laguna Biosphere Reserve, Baja California Sur, Mexico

	Rainy season	Dry season
Total cameras used	67	62
Total camera operational days	1781	2096
Total cameras		
Feral pigs	32 (48%)	13 (21%)
Cattle	44 (66%)	52 (84%)
Mule deer	54 (81%)	53 (85%)
Total visits		
Feral pigs	106	22
Cattle	508	744
Mule deer	824	477
RAI		
Feral pigs	60	10
Cattle	285	355
Mule deer	463	228

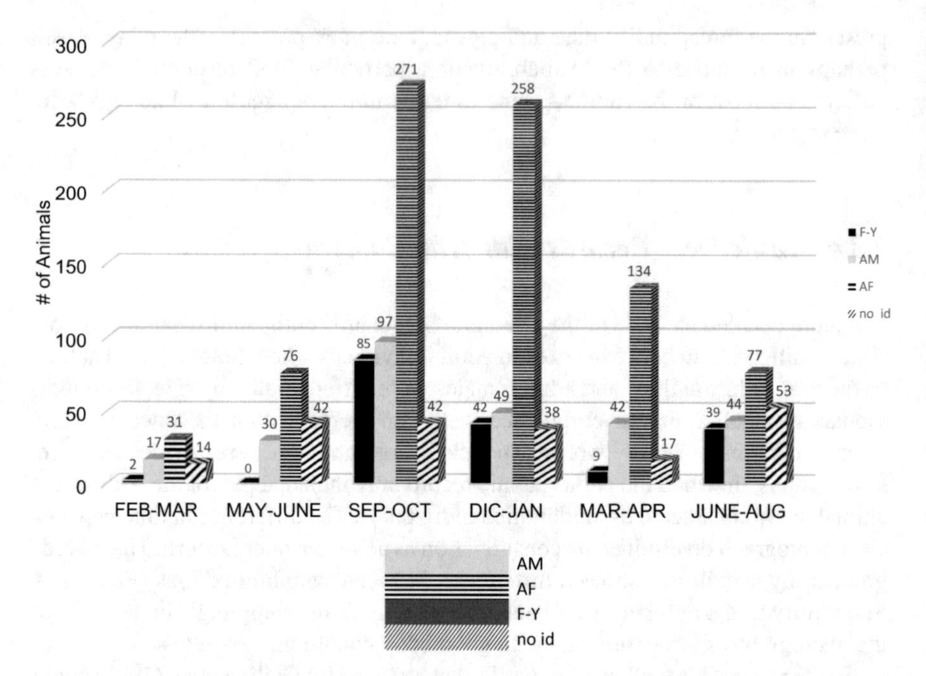

Fig. 2.11 Age classes of mule deer in Sierra La Laguna Biosphere Reserve, Baja California Sur, Mexico by photo-trapping from February–March 2010 to June–August of 2011. *AM* adult males, *AF* adult females, *J-F* juveniles-fawns, *not id* not identified

groups). In the case of pigs, it was also feasible to differentiate them by size, color pattern and bristle type; although these animals moved in groups or herds of different sizes, the records were mostly solitary individuals or groups of no more than four individuals.

The results reflected in some way species mobility in the ecosystem, perhaps in response to resource availability, such as water and food and seasonal variations in them, as well as behavioral pattern characteristic of the species. There were several photographs showing the use of the same water pool by the three species (Fig. 2.13).

2.4 Discussion and Final Considerations

The studies performed on mule deer and introduced ungulate populations in Sierra La Laguna should be taken as reference point to be able to determine whether favorable changes have occurred or not in the ecosystem with the generation of new information. Although deer density in the reserve has not been recorded recently, according to the reports from local ranchers and reserve authorities besides the records obtained by photo trapping, it is possible to consider that the mule deer is a common species. Despite the presence of introduced species and poaching that

Fig. 2.12 Records of Mule deer showing different characteristics and behavior of the species in the forest of Sierra La Laguna. (**a**) A group of deer drinking water, (**b**) Female with her fawn, (**c**) Yearling, (**d**) Male with velvenet antlers, (**e**) Male with antlers without velvet, (**f**) Male without antlers

Fig. 12 (continued)

Fig. 12 (continued)

Fig. 2.13 The three ungulates using the same water pool at Sierra La Laguna. (**a**) Cow, (**b**) Feral pig, (**c**) Mule deer

occurs in some localities reported mainly in the lowlands around the mountains, the reserve in general and core zone (oak-pine forest) have been a refuge for mule deer and other species.

Even though the deer density recorded in the Sierra La Laguna is not as high as it is in other productive areas in California, U.S.A. (about 84 deer/km^2), it can be compared to those populations found in some mountainous areas of Colorado and California where they vary from 20 deer/km^2 and 29.6 + 11.3 deer/km^2, respectively, and even surpass those found in other sites of the same states (4–12 deer/km^2) or in the Chihuahua desert where population densities are lower, 0.70–4.21 deer/km^2 (Bergman et al. 2015; Kufeld et al. 1987; Sánchez-Rojas and Gallina 2000; Stapp and Guttilla 2006).

Studies have shown the negative impact of cattle and feral pigs on wild ungulate population when they compete for seasonal resources, such as acorns or other plants, causing them to move from the prime feeding areas (Austin and Urness 1986; Bowyer and Bleich 1984; Chaikina and Ruckstuhl 2006), and the pressure that overgrazing exerts on native vegetation, which cause decrease of available resources for native species and changes in plant structure that affect the available refuges. On the other hand, some studies have not addressed a relevant competition between cattle and deer with respect to diet since cattle searches for more grasses and herbaceous plants while deer consumes mainly shrubs and some herbaceous plants (Torstenson et al. 2006). Therefore, further studies on long-term assessment of diet and feeding behavior of these three species within the Reserve should be performed to know their preferences, variations according to time of the year and their effect on vegetation, habitat in general, and consequently, on mule deer population.

In estimations performed on the degree of grazing in the reserve, the results showed that 44% of the surface was overgrazed, 37% undergrazed and 19% was not recorded; the most affected areas were in the north and west-southeast (Breceda Solis-Cámara and Vázquez-Miranda 2013) although this overgrazing should not be considered exclusively by cattle, given the presence of herds of goats and sheep that are in enclosures just part of the time. These results show, in some manner, the effect of livestock presence in general and grazing intensity, which allow us to glimpse the possible effects on wild flora and fauna.

Although studies on the effect of grazing on forest regeneration are not conclusive regarding their deterioration due to the presence of domestic ungulates, further research is essential to determine the effect of such species on deer and other wild animal and plants in the Reserve, especially in the core area. Monitoring with trap cameras was necessary because the results obtained reflected a greater presence, both in frequency in photographs and in group size of the cattle observed. Now that the degree of grazing has been locally assessed in the area where important data was related to considerable overgrazing (Breceda Solis-Cámara and Vázquez-Miranda 2013; Ortiz-Avila 1999), analyses should be performed to determine the most affected plant species, as well as the impact produced by soil compaction and alteration. Consistent information should also be obtained from native species to use it as a reference in detecting any changes or disturbances, given the known cases of the problem that exotic species have caused in other parts of the country and worldwide. Loss of biodiversity has been attributed to this problem with severe impacts on ecological functions and modification of community composition of plants and animals (Álvarez-Romero et al. 2008).

Indirect estimates of the population make it easier to determine age, structure and distribution of a given species, but the use of camera traps allows assessing physical condition of the specimens, seasonality and periodicity of activity, development (time of birth, antler molt, and so on), as well as interactions between species and individuals. Photo-trapping is a non-invasive and relatively inexpensive method that requires a first investment for equipment but with very important results in monitoring and knowledge of biodiversity.

The results of photo trapping reflected in some way species mobility in the ecosystem, probably in response to resource availability, such as water, food quality and abundance and seasonal variations in them (phenology of the species) besides responding to characteristic behavioral patterns of the species. In this project, the three species used the same water and sites with variations in the presence of each one over time (Fig. 2.13).

Although no studies in the Reserve have been available on overlap and competition for the use of different resources among these species, Barret (1982), who studied their preferences in their habitat in Sacramento, California, found that the maximum point of overlap and possible competition between them occurred between mule deer and cattle in some vegetation types, such as the oak forest and shrub lands; a direct interaction was observed where in all the cases, the deer was dominated or excluded by cattle. However, if competition exists between deer and feral pig, it is minimal in certain types of vegetation. These habitat preferences also varied with the season. Hosten et al. (2007) mentioned that native ungulates decreased their visits to water sources and springs that were frequented by livestock or changed their behavior by reducing the number of stalls and excreta in the places where the presence of cattle was observed. In addition, when species of native populations face adverse conditions in the ecosystem due to climate change or inadequate management of resources, they will be more affected if competition for scarce resources (water, food, refuge) appearance and transmission of diseases (of the same introduced species, which even be transmitted to humans). Therefore, it is urgent to initiate monitoring these factors with the support of citizens and relevant authorities.

Before the decree of Sierra La Laguna as Biosphere Reserve, wildlife was threatened by illegal hunting, and pressure was exerted on its habitat in the use of resources (Arriaga and Ortega 1988). Nonetheless, the current status of Protected Natural Area has meant that inhabitants of the surrounding rancherias, as well as visitors, should be more aware of the importance of the conservation of this natural area. Management programs related in particular to greater control of cattle and feral pigs, and their eradication if necessary, are essential to diminish the negative effects of their presence, especially in relation to forest regeneration, competition for resources with wild species and presence of diseases (Arnaud-Franco et al. 2012; Pérez 2014; Pérez-Rivera et al. 2017). Porcine activity is not as widespread as that of cattle; however, it has given rise to a feral population that is difficult to recapture and displaced throughout the area, including the core area of the reserve (Breceda et al. 2009). Feral pigs (*Sus scrofa*) are hunted for self-consumption because they are considered a tastier and healthier meat than that of individuals in captivity, mainly because of their diet with acorns, turning them into a good exploitable resource.

The management plan of the reserve has already identified livestock (bovine, caprine, ovine and porcine) as a problem because it is considered one of the main factors that prevents natural regeneration of plant communities, particularly those distributed in the low deciduous forest and pine-oak and oak-pine forests (CONANP 2003). More research is needed in this area, as well as determining the effect that this problem has, directly or indirectly, on wildlife that uses these resources, which is still unknown.

Although the Reserve staff conducted a ranching census in 2000 (CONANP 2003), no studies are available on some of the ungulates present and the effect of cattle and their distribution in the Reserve; few actions have been taken to reduce or control their presence in the core area and their effect on the ecosystem. In the same manner, it is necessary to work with the ranchers to improve livestock management, and at the same time allow them to continue with their economic activity without affecting the resources of the area.

The costs to carry out a management program are high, and the required staff seems to be insufficient; however, the local population and visitors should be involved, making them aware of the importance of acting as monitors of natural resources and promoting mechanisms for their conservation. Environmental education programs at a general level are of utmost importance to increase efficiency of the area and its surroundings.

2.5 Perspectives for the Management and Conservation of Mule Deer

The mixed forest of pine-oak and oak-pine of the Sierra de La Laguna with its diverse plant associations constitutes an adequate habitat for mule deer. Despite being a species adapted to the xeric conditions prevalent in the Baja California Peninsula, this great herbivore has found the habitat elements in the Sierra indicated as basic to its presence, essentially: land area, food availability, vegetation and water coverage, together with favorable climate and topographic conditions. These habitat attributes have favored good population density, together with the amount of food especially provided by the shrub layer, which is also relevant for the protection coverage it provides against climate and predators.

As far as the mule deer is concerned throughout its distribution in Baja California Sur, population studies are needed inside and outside of the protected areas and in the different types of vegetation in coordination with the UMA. Population and habitat monitoring should be maintained to issue headbands or permits to hunt them; all this information should be collected, as well as that of the specimens hunted (measurements, photographs, tissue samples, parasites, etc.) in a systematic way to generate information banks and establish appropriate management plans. Unfortunately, much of this information is still missing, particularly that related to health status to avoid epidemics or overexploitation. The information gathered needs to be analyzed by the authorities in charge of granting permits (of which much is not accessible) to detect trends and assess whether the actions taken by the different UMA for use and management of resources have been effective or not. In the same manner, long-term censuses should be performed following standardized methodologies.

Moreover, genetic research of the mule deer in the State is necessary since the deer in the peninsula has shown high levels of genetic variation in comparison to that reported (Bazaldúa 2016). These measures should be reinforced for an efficient management of the species taking advantage of the existence of protected natural areas as shelters.

Acknowledgments These studies were performed with the support of Centro de Investigaciones Biológicas del Noroeste (CIBNOR), Consejo Nacional de Ciencia y Tecnología (CONACYT), Comisión Nacional de Área Natural Protegidas (CONANP), Comisión Nacional para el Uso y Conocimiento de la Biodiversidad (CONABIO), The Nature Conservancy (TNC) jointly with the Institute of Ecology. The authors thank the technicians Amado Cota, Franco Cota, Raymundo Dominguez, Abelino Cota, Israel Guerrero; all students involved, Maria del Rosario Vázquez Miranda, Juan José Móntes-Sánchez, Claudia M. Pérez-Rivera; Joaquín Rivera for support with map preparation; and Diana Fischer for English editing.

References

Alvarez-Cárdenas S, Gallina S, Galina-Tessaro P, Domínguez-Cadena R (1999a) Habitat availability for the mule deer (Cervidae) population in a relictual oak-pine forest in Baja California Sur, Mexico. Trop Zool 12:67–78. https://doi.org/10.1080/03946975.1999.10539378

Alvarez-Cárdenas S, Gallina S, Galina-Tessaro P, Díaz-Castro S (1999b) Mule deer population dynamics in a relictual oak – pine forest in Baja California Sur, Mexico. In: Folliott PF, Ortega-Rubio A (eds) Ecology and management of forests, woodlands, and shrublands in the dryland region of the United States and México: perspectives for the 21st century. University of Arizona-Centro de Investigaciones Biológicas del Noroeste, Tempe, pp 197–210

Alvarez-Cárdenas S (1995) Estudio poblacional y hábitat del venado bura *Odocoileus hemionus peninsulae* en la Sierra de La Laguna, BCS. Tesis maestría, Universidad Nacional Autónoma de México

Álvarez-Romero JG, Medellín RA, Oliveras de Ita A, Gómez de Silva H, Sánchez O (2008) Animales exóticos en México: una amenaza para la biodiversidad. Comisión Nacional para el Conocimiento y Uso de la Biodiversidad. Instituto de Ecología, UNAM, Secretaría de Medio Ambiente y Recursos Naturales, México

Anderson AE (1981) Morphological and physiological characteristics. In: Wallmo O (ed) Mule and black tailed deer of North America. A Wildlife Management Institute Book. University of Nebraska Press, Lincoln and London, pp 27–97

Arnaud-Franco G, Breceda Solis-Cámara A, Alvarez-Cárdenas S, Cordero-Tapia A (2012) Implicaciones de la presencia del cerdo asilvestrado (*Sus scrofa*) en la Sierra la Laguna.

In: Ortega-Rubio A, Laguna-Vazques M, Beltrán-Morales LF (eds) Evaluación biológica y ecológica de la Reserva de la Biosfera Sierra La Laguna, Baja California Sur: avances y retos. Centro de Investigaciones Biológicas del Noroeste SC, La Paz, BCS, México, pp 205–219

Arnaud-Franco G, Breceda A, Alvarez-Cárdenas S, Cordero A, Bonfil C, Galina P (2014) Cerdos asilvestrados (*Sus scrofa*) en la reserva de la Biosfera Sierra La Laguna: evaluación e impacto sobre la biodiversidad. Final report SNIB-CONABIO, proyect No. GN016. Centro de Investigaciones Biológicas del Noroeste, México

Arriaga L, Cancino C (1992) Prácticas pecuarias y caracterización de especies forrajeras en la selva baja caducifolia. In: Ortega A (ed) Uso y manejo de los recursos naturales en la Sierra de la Laguna, Baja California Sur. Centro de Investigaciones Biológicas de Baja California Sur, World Wildlife Found, La Paz, BCS, México, pp 155–184

Arriaga L, Ortega A (eds) (1988) La Sierra de La Laguna Baja de California Sur. Centro de Investigaciones Biológicas de Baja California Sur, México

Austin DD, Urness PJ (1986) Effects of cattle grazing on mule deer diet and área selection. J Range Manag 39(1):18–21

Barret RH (1982) Habitat preference of feral hogs, deer and cattle on a sierra foothill range. J Range Manag 35(3):342–346

Bazaldúa LA (2016) La importancia de la variación genética para la conservación del venado bura (*Odocoileus hemionus*) en México. Tesis licenciatura, Universidad Nacional Autónoma de México

Bergman EJ, Doherty PF Jr, White GC, Freddy DJ (2015) Habitat and herbivore density: response of mule deer to habitat management. J Wildl Manag 79(1):60–68

Bowyer RT, Bleich VC (1984) Effects of cattle grazing on selected habitats of southern mule deer *Odocoileus hemionus fuliginatus*. Calif Fish Game 70:240–247

Breceda A (2005) El mosaico de vegetación de una selva baja caducifolia. Tesis doctoral, Universidad Nacional Autónoma de México

Breceda Solis-Cámara A, Vázquez-Miranda MR (2013) La ganadería: retos para la conservación. In: Lagunas-Vázques M, Beltrán-Morales LF, Ortega-Rubio A (eds) Diagnóstico y análisis de los aspectos sociales y económicos en la reserva de la biósfera Sierra La Laguna, Baja California Sur, México. Centro de Investigaciones Biológicas del Noroeste, SC, La Paz, BCS, México, pp 49–68

Breceda A, Arriaga L, Bojórquez L, Rodríguez M (2005) Defining critical areas for conservation and restoration in a Mexican biosphere reserve: a case study. Nat Areas J 25(2):123–129

Breceda A, Arnaud-Franco A, Alvarez-Cárdenas S, Galina-Tessaro P, Montes-Sánchez JJ (2009) Evaluación de la población de cerdos asilvestrados (*Sus scrofa*) y su impacto en la Reserva de la Biosfera Sierra La Laguna, Baja California Sur, México. Trop Conserv Sci 2:173–188

Breceda A, Castorena L, Maya Y (2012) Transformaciones de una selva seca por actividades humanas en el paisaje rural de Baja California Sur, México. Investigación Ambiental 4(1):141–150

Castorena L, Breceda A (2008) Remontando el Cañón de la Zorra. Ranchos y rancheros de la Sierra La Laguna. Instituto Sudcaliforniano de Cultura del Gobierno del Estado de Baja California Sur, BCS, México

Chaikina NA, Ruckstuhl KE (2006) The effect of cattle grazing on native ungulates: the good, the bad and the ugly. Rangelands 28:8–14

Chirichela R, Apollonio M, Putman R (2014) Competition between domestic and wild ungulates. In: Putman R, Apollonio M (eds) Behaviour and management of European ungulates. Whittles Publishing, Dunbeath Caithness, Scotland, pp 110–123

Choquenot D, McIlroy J, Korn T (1996) Managing vertebrate pest: feral pigs. Bureau of Resource Sciences, Australian Government Publishing Service, Canberra

CONABIO, Aridoamérica, GECI, TNC (2006) Especies invasoras de alto impacto a la biodiversidad: prioridades en México. Ciudad de México, Mexico

CONANP (2003) Programa de Manejo Reserva de la Biosfera Sierra La Laguna. Comisión Nacional de Áreas Naturales Protegidas, Mexico

Everitt JH, Alanis MA (1980) Fall and winter diets of feral pigs in south Texas. J Range Manag 33(2):126–129

Galina-Tessaro P, González A, Arnaud G, Gallina S, Alvarez S (1988) Mastofauna. In: Arriaga L, Ortega A (eds) La Sierra de la Laguna de Baja California Sur. Centro de Investigaciones Biológicas de Baja California Sur, Mexico

Gallina S, Galina-Tessaro P, Alvarez-Cárdenas S (1991) Mule deer density and pattern distribution in the pine-oak forest al the Sierra de la Laguna in Baja California Sur, Mexico. Ethol Ecol Evol 3:27–33

Gallina S, Galina-Tessaro P, Alvarez-Cárdenas S (1992) Hábitat y dinámica poblacional del venado bura. In: Ortega A (ed) Uso y manejo de los recursos naturales en la Sierra de La Laguna, Baja California Sur. Centro de Investigaciones Biológicas de Baja California Sur, AC y WWF, Mexico

Hall ER (1981) The mammals of North America, vol 2, 2nd edn. Wiley, New York

Hess SC, Van Vuren DH, Witmer G (2017) Feral goats and sheep. In: Pitt WC, Beasley J, Witmer GW (eds) Ecology and management of terrestrial vertebrate invasive species in the United States. USDA, APHIS WS National Wildlife Research Center, Ft Collins, CO, pp 289–309

Hosten PE, Whitridge H, Broyles M (2007) Diet overlap and social interactions among cattle, horses, deer and elk in the Cascade-Siskiyou National Monument, southwest Oregon USDI. Bureau of Land Manage, Medford District

Huey LM (1964) The mammals of Baja California, Mexico. Trans San Diego Soc Nat Hist 13:85–168

Hutton T, DeLiberto T, Owen S, Morrison B (2006) Disease risks associated with increasing feral swine numbers and distribution in the United States. Midwest Association of Fish and Wildlife Agencies, Wildlife and Fish Health Committee, Washington, DC

INEGI (1997) Ganado bovino en el estado de Baja California Sur. Instituto Nacional de Estadística, Geografía e Informática, Mexico

Kotanen PM (1995) Responses of vegetation to a changing regime of disturbance: effects of feral pigs in a Californian coastal prairie. Ecography 18:190–199

Kufeld RC, Bowden DC, Schrupp DL (1987) Estimating mule deer density by combining mark recapture and telemetry data. J Mammal 68(4):818–825

Lacki MJ, Lancia RA (1986) Effects of wild pigs on beech growth in Great Smoky Mountains National Park. J Wildl Manag 50:655–659

Lagunas-Vázques M, Acevedo-Beltrán M, Cervantes-Martínez EF, Beltrán-Morales LF, Ortega-Rubio A (2013) Sociohistoria de la ganadería y su importancia en la seguridad alimentaria para las familias rancheras de la REBIOSLA. In: Lagunas-Vázques M, Beltrán-Morales LF, Ortega-Rubio A (eds) Diagnóstico y análisis de los aspectos sociales y económicos en la reserva de la biósfera Sierra La Laguna, Baja California Sur, México. Centro de Investigaciones Biológicas del Noroeste SC, La Paz, BCS, pp 31–48

Latch EK, Heffelfinger JR, Fike JA, Rhodes OE Jr (2009) Species-wide phylogeography of North American mule deer (*Odocoileus hemionus*): cryptic glacial refugia and postglacial recolonization. Mol Ecol 18:1730–1745

Linnaeus C (1758) Systema Naturae per regna tria naturae, secundum classis, ordines, genera, species cum characteribus, differentiis, synonymis, locis. Tenth ed. Vol. 1. Laurentii Salvii, Stockholm, 824pp

Lydekker R (1898) On a new mule deer. Proceedings of the Zoological Society of London 1897, Pt 4:899–900

March IJ (2007) Evaluación rápida de especies invasoras en la Reserva de la Biosfera Sierra La Laguna. Reporte Ejecutivo. The Nature Conservancy, 11 p

Martínez-Balboa A (1980) La Ganadería en Baja California Sur, vol 1. J.B., La Paz, BCS

Massei G, Genov PV (2004) The environmental impact of wild boar. Galemys 16:135–145

Meng XJ, Lindsay DS, Sriranganathan N (2009) Wild boars as sources for infectionus diseases in livestock and humans. Philos Trans R Soc Lond Biol 364(1539):2697–2707

Montes-Sánchez JJ, León-de la Luz JL, Buntinx-Dios SE, Huato-Soberanis L, Blázquez-Moreno MC (2012) Dieta, Crecimiento y reproducción del cerdo asilvestrado *Sus scrofa* en la Reserva de la Biosfera Sierra La Laguna. In: Ortega-Rubio A, Laguna-Vazques M, Beltrán-Morales LF (eds) Evaluación biológica y ecológica de la Reserva de la Biosfera Sierra La Laguna, Baja California Sur: Avances y retos. Centro de Investigaciones Biológicas del Noroeste, SC, La Paz, BCS

Ortiz-Avila V (1999) Efecto del pastoreo sobre el establecimiento de juveniles en la selva baja caducifolia de la Reserva de la Biósfera: Sierra de La Laguna, Baja California Sur, México. Tesis licenciatura, Benemérita Universidad Autónoma de Puebla

Pérez RCM (2014) Diagnóstico y prevalencia de enfermedades de importancia epidemiológica en cerdos (*Sus scrofa*) asilvestrados y domésticos de la Reserva de la Biósfera Sierra La Laguna, BCS. Tesis maestría, Centro de Investigaciones Biológicas del Noroeste, SC La Paz, BCS

Perez-Gil R (1981) A preliminary study of the deer from Cedros Island, Baja California, Mexico. MS thesis, University of Michigan, School of Natural Resources

Pérez-Rivera CM, Sanvicente-López M, Arnaud-Franco G, Carreón-Nápoles R (2017) Detection of antibodies against pathogens in feral and domestic pigs (*Sus scrofa*) at the Sierra La Laguna Biosphere Reserve, Mexico. Vet Mex 4(1):1–11

Pinelli Saavedra A, Acedo Félix E, Hernández López J, Belmar R, Beltrán A (2004) Manual de buenas prácticas de producción en granjas porcícolas. SAGARPA, SENASICA, Centro de Investigación en Alimentación y Desarrollo AC, CIAD, AC, Unidad Hermosillo, Depto Nutrición Animal, Mexico

Quiróz J (2007) Caracterización genética de los bovinos criollos mexicanos y su relación con otras poblaciones bovinas. Tesis doctoral, Universidad de Córdoba

Rochín-Sánchez S (2003) El desarrollo de la ganadería en la delegación de San Antonio, Baja California Sur 1926–1961. Tesis licenciatura, Universidad Autónoma de Baja California Sur

Sánchez-Rojas G, Gallina S (2000) Mule deer (*Odocoileus hemionus*) density in a landscape element of the Chihuahuan Desert, México. J Arid Environ 44:357–368

Stapp P, Guttilla DA (2006) Population density and habitat use of mule deer (*Odocoileus hemionus*) on Santa Catalina Island, California. Southwest Nat 51(4):572–578

Torstenson WLF, Mosley JC, Brewer TK, Tess MW, Knight JE (2006) Elk, mule deer and cattle foraging relationships on foothill and mountain rangeland. Rangeland Ecol Manage 59:80–87

Trejo-Barajas D, González-Cruz E (eds) (2002) Historia general de Baja California Sur. I. La Economía regional. Universidad Autónoma de Baja California Sur, México

Villarreal J, Alanís GJ (2015) Impacto del marrano alzado y el jabalí europeo en hábitats del matorral espinoso tamaulipeco en el noreste de México. Ciencia UANL 18(72):23–29

Wallmo O (ed) (1981) Mule and black-tailed deer of North America. A Wildlife Management Institute Book. University of Nebraska Press, Lincoln and London

Woloszyn D, Woloszyn BW (1982) Los mamíferos de la Sierra de La Laguna, Baja California Sur. Consejo Nacional de Ciencia y Tecnología, México

Zanin CG, Marini MA (2013) Impact of the wild boar, *Sus scrofa* on a fragment of Brazilian Atlantic forest. Neotrop Biol Conserv 8(1):17–24

Chapter 3
The Mule Deer of the Mapimí Biosphere Reserve

Sonia Gallina-Tessaro, Gerardo Sánchez-Rojas, Dante Hernández-Silva, Luz A. Pérez-Solano, Luis García-Feria, and Juan Pablo Esparza-Carlos

Abstract In this chapter we present information collected from the population of mule deer in Mapimí, the first Biosphere Reserve in Mexico and Latin America. The reserve was decreed in 1979, covers 342,388 ha and is located at the junction of the states of Durango, Coahuila, and Chihuahua. Mule deer population dynamics data from 14 nonconsecutive years were used to analyze population density, using indirect methods such as counting fecal pellet groups, population structure by age and sex, population growth rate, and their relationship to precipitation in this arid environment. Feeding habits are presented, with mention of the main species in their diet as well as the nutrient content of the latter, diet quality was estimated by sex and age as a function of fecal nitrogen to determine if there were any differences and the relationship of habitat use and diet to sexual segregation. Data on home range and core area size are presented, as are the way in which habitat in the Chihuahuan Desert is being used by the mule deer taking the dry and rainy seasons into account, and how the risk of predation affects behavior using radiotelemetry techniques. Extensive cattle pasturing is the main economic activity in this arid zone and our data indicate that there is little overlap of habitat and little zoonosis, which is a rare occurrence between deer and cattle. The information compiled demonstrates that the population of mule deer in Mapimí is the best studied population in Mexico and

S. Gallina-Tessaro (✉) · L. A. Pérez-Solano · L. García-Feria
Red de Biología y Conservación de Vertebrados, Instituto de Ecología A.C., Veracruz, Mexico
e-mail: sonia.gallina@inecol.mx; luis.garcia@inecol.mx

G. Sánchez-Rojas
Centro de Investigaciones Biológicas, Universidad Autónoma del Estado de Hidalgo, Pachuca, Hidalgo, Mexico
e-mail: ganchez@uaeh.edu.mx

D. Hernández-Silva
Research and Management, Wild Forest Consulting S.C., Morelos, Mexico
e-mail: dal_silva@hotmail.com

J. P. Esparza-Carlos
Departamento de Ecología y Recursos Naturales, Universidad de Guadalajara, Guadalajara, Mexico
e-mail: juan.esparza@cucsur.udg.mx

S. Gallina-Tessaro (ed.), *Ecology and Conservation of Tropical Ungulates in Latin America*, https://doi.org/10.1007/978-3-030-28868-6_3

should be used as a model to develop suitable strategies to improve the sustainable use of this species. The results also highlight the importance of protected areas as sources of knowledge creation and transfer.

3.1 Introduction

The historic distribution of populations of mule deer in the Chihuahuan Desert of Mexico includes the state of Coahuila, the extreme southwest of the state of Nuevo León, the northern region of San Luis Potosí and Zacatecas, a small region in western Tamaulipas, the desert region of the states of Chihuahua and Durango and the northeast region of Sonora. Unfortunately, to date the situation has not been systematically evaluated, but there is evidence that in the state of Nuevo León, the center of Coahuila, the north of San Luis Potosí and Zacatecas, mule deer populations have been extirpated (Sánchez-Rojas and Gallina 2007; Sánchez-Rojas and Gallina-Tessaro 2016), mainly by poaching, habitat destruction (Martínez-Muñoz et al. 2003), and the expansion of cattle ranching (Galindo-Leal 1993; Sánchez-Rojas and Gallina 2006, 2007). The apparent reduction in the distribution area of this species has mainly been recorded at the southern and eastern limits of its distribution in Mexico. Therefore, the only mule deer population confirmed in the southern part of its distribution is in the Mapimí Biosphere Reserve (MBR) (Fig. 3.1) and adjacent areas. This reserve is the only protected area in the most arid region of the Chihuahuan Desert, and the first Biosphere Reserve to be decreed in Mexico and Latin America, in 1979 (Barral and Hernández 2001; Sánchez-Rojas and Gallina 2006).

The mule deer population of the Mapimí Biosphere Reserve is the most systematically studied population (i.e., not anecdotal). All ecological studies of the mule deer of Mapimí have been carried out in the approximately 64 km^2 around San Ignacio Hill (Figs. 3.2 and 3.3). At this site, the climate is characteristic of arid zones that include cool winters and warm summers with a mean annual temperature of 20.8 °C (range: 3.9–37.41 °C). Average annual rainfall is 264 mm, most of which occurs from June through September (CONANP 2006). There are three main types of vegetation: (1) desert rosette shrubs with combinations of agaves (*Agave asperrima* and *A. lechuguilla*), candelilla (the wax plant; *Euphorbia antisyphilitica*) and nopales or prickly pears (*Opuntia rastrera*, *O. microdasys*); (2) natural grassland that includes toboso grass (*Pleuraphis mutica*) and alkali zacaton (*Sporobolus aireoides*); and (3) desert microphyll shrubs such as mesquite (*Prosopis glandulosa*), creosote bush (*Larrea tridentata*), bitterbush (*Castela texana*), goatbush (*C. erecta*) and tarbush (*Flourensia cernua*); there is also grassland mixed with mesquite, locally known as *mogotes* (Fig. 3.4, Montaña 1988; Barral and Hernández 2001; CONANP 2006). These environmental features offer conditions for evaluating different aspects of mule deer ecology, including population dynamics, feeding ecology, behavioral ecology, habitat selection and use, and spatial ecology.

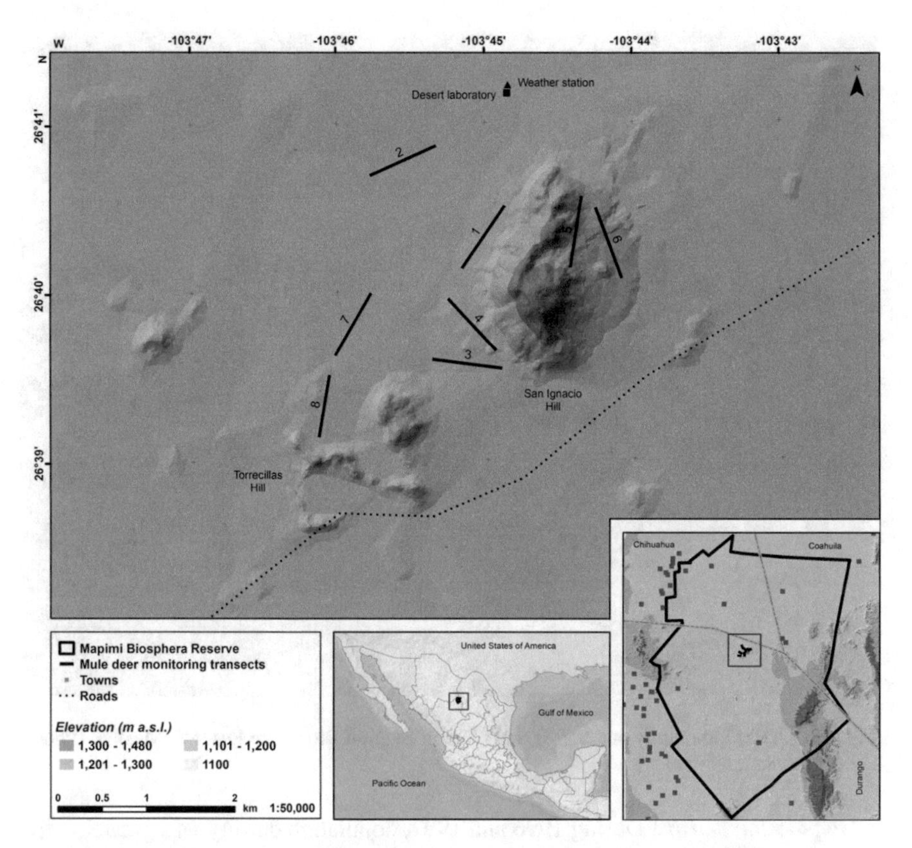

Fig. 3.1 Location of the Mapimí Biosphere Reserve (area: 342,388 ha) in the states of Durango, Chihuahua, and Coahuila, Mexico. The area where mule deer were monitored at the San Ignacio Hill site is shown

3.2 Population Dynamics

The members of a population share a geographic area characterized by specific physical and biological factors that interact with the intrinsic characteristics of the population. Population dynamics offers a means looking at the changes that occur in a population of a given species over certain temporal and spatial scales in their density, effective number, growth rate, age and sex structure, both within and between populations. The first study of mule deer in the Mapimí Biosphere Reserve began in 1996, and over 2 years the method of counting fecal groups along linear transects (Fig. 3.1) was used, adapted to the desert matorral conditions of the study sites (Eberhardt and Van Etten 1956; Sánchez-Rojas 2000). The population dynamics data obtained over the course of that study were estimated population density (deer per km^2), age and sex structure, proportion of males, females, and juveniles, and population growth rate (lambda: λ).

Fig. 3.2 Desert Laboratory and San Ignacio Hill in Mapimí Biosphere Reserve, Mexico. (Photo: Dante Hernández-Silva)

Population density. During 1996 and 1997, population density was greater at the San Ignacio Hill site (2.64 ± 1.12 deer per km^2), and lower at the Coronas Hill site (1.7 ± 0.65); the two sites are separated by a plain that spans about 17 km. These results show that the distribution of mule deer across the landscape studied was not uniform and could vary as a function of site quality, particularly with respect to water availability and terrain heterogeneity (Sánchez-Rojas and Gallina 2000a).

The members of the subpopulations of mule deer select, within available habitat, the fragments with the most heterogeneous terrain and, specifically during the dry season, the fragments closest to water (Sánchez-Rojas and Gallina 2000b). These data support the idea that the population of mule deer in Mapimí has a spatial structure, and also suggest, that it is a metapopulation (Sánchez-Rojas and Gallina 2007; Sánchez-Rojas and Gallina-Tessaro 2016).

In 2006, the monitoring of mule deer in the MBR continued, only at the San Ignacio Hill site. Up to 2018, a database of the samplings carried out during the month of March was compiled for the periods of 1996–1998, 2006–2007, and 2009–2017. Population density has been dynamic over the years, and there was no pattern in the species' behavior, even though a pattern has been described for populations in the United States of America and Canada, which decreased in the 1960s and 1990s and increased in the 1970s and 1980s (Denny 1976; Workman and Low 1976; Bergman et al. 2015). In Mapimí, 2014 was the year with the lowest number

Fig. 3.3 San Ignacio Hill in Mapimí Biosphere Reserve, Mexico. (Photo: Sonia Gallina Tessaro)

of deer per km^2, and 2006 was the year with the highest number, mean population density for the 14 years of monitoring was 2.77 deer ± 1.14 km^2 (Fig. 3.5).

The studies carried out on the population density of the mule deer make the MBR the place with the longest running study in Mexico, with the 10-year study done on a private ranch in the state of Coahuila in second place (Aguirre et al. 2016) and that done in the Sierra de La Laguna Biosphere Reserve in the state of Baja California Sur over 7 years in third place (Álvarez-Cárdenas et al. 1999). There are sites in the United States of America, such as in Arizona, where the same subspecies as that found in Mapimí occurs, and there have been long-term monitoring studies over 25 years (Anthony and Smith 1977).

Growth rate. The development of the population of mule deer in Mapimí was characterized using growth rate, which had a value of 0.94 ± 0.82, indicating that the population is stable, with years when there has been growth, since the growth rate showed a positive trend for 5 of the 14 years used in the calculations. Based on this, it can be said that the population of mule deer in Mapimí exhibits natural fluctuations over the years that are directly related to the availability of food and water. Forrester and Wittmer (2013) estimated a mean growth rate of 0.99 ± 0.04 from 48 studies of mule deer populations distributed in the United States of America and Canada. The Mapimí data coincide with this general growth pattern for the species.

Age and sex structure. Mule deer population structure in Mapimí is dominated by females by as much as 49%, followed by fawns and juveniles (31%), and lastly

Fig. 3.4 Common plants in Mapimí Biosphere Reserve, Mexico. (**a**) *Euphorbia antisyphilitica*, (**b**) *Larrea tridentata*, (**c**) *Opuntia rastrera*, (**d**) *Fouquieria splendens*, (**e**) *Prosopis glandulosa*, (**f**) *Opuntia microdasys*. (Photo: Sonia Gallina Tessaro and Luz A. Pérez-Solano)

males (20%) (Fig. 3.5). This sex ratio and age structure has been observed in the populations of cervids from different regions of its current distribution (Pac et al. 1991; Brunjes et al. 2006; Bishop et al. 2009; Bender et al. 2012; Ciuti et al. 2015).

The abundance of species that are naturally distributed in deserts usually depends on precipitation and temperature, however, for long-lived wildlife species such as cervids, this relationship is not always directly proportional, as described by Hernández-Silva (2018) for the mule deer of the San Ignacio Hill site in the MBR. To elucidate this relationship, an index of precipitation (IPPT) was estimated using the amount of annual precipitation measured at the weather station located in the

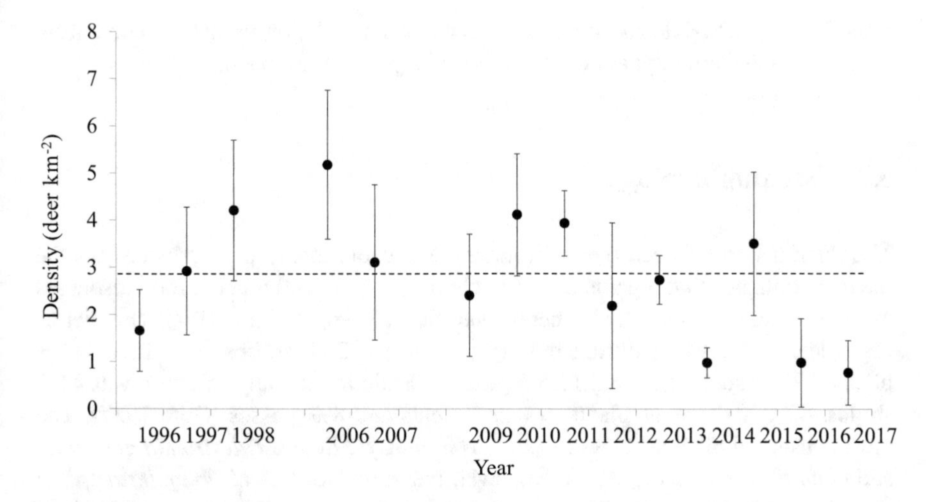

Fig. 3.5 Population density of mule deer (*Odocoileus hemionus*) over 14 years of discontinuous monitoring in the study area in Mapimí Biosphere Reserve, Durango, Mexico (dashed line: population average)

MBR. The IPPT is the relationship between the number of days with rain and the amount of rain that falls (mm), which was estimated for each year of monitoring.

This relationship exemplifies how rainfall in the desert is variable in periods of days and the amount of average annual rainfall can fall in a single day (torrential rains). This is why the distribution of rainfall in arid zones is considered random, and could affect the structure and composition of the mule deer population in the MBR by affecting variables such as water availability, temperature and humidity, all of which affect the growth of the plants that make up this species' diet (Loik et al. 2004; Xiangfei et al. 2016).

On comparing the precipitation index for the 12, 24 and 36 months prior to the month when the mule deer population was sampled in the MBR, an indirect relationship between the number of deer and the index was found for the three periods mentioned. This phenomenon implies that change in the population of mule deer is a function of how rain falls. The greatest population density occurred in the three scenarios when the rainfall events were small and more frequent (Hernández-Silva 2018). In contrast, the lowest mule deer density was recorded when IPPT was highest and this can be attributed to the torrential rains not being favorable for the MBR's natural habitat because they temporarily increase the presence of humidity and water availability in mule deer habitat.

In the arid and semiarid regions of Mexico, management should take into account precipitation patterns in the area of interest, i.e., the distribution of rainfall throughout the year. As such, estimating and analyzing the precipitation index and mule deer population density would reinforce management strategies that are carried out to improve the conservation, sustainable use, reproduction and control of this cervid. The mule deer population in the MBR has been under conservation management since 1979 (CONANP 2006) and can serve as a population model for comparing long-term population dynamics studies of this species in Mexico.

3.3 Feeding Ecology

The relationship between the environment and evolutionary processes defines the feeding strategies that organisms use to choose, acquire and process food resources in order to satisfy their requirements (Van Soest 1996; Righini 2017). The diet of mule deer in the MBR during the dry season was 35% shrubs, 37% herbaceous plants, 12.5% succulents and 11.5% grasses, while in the rainy season it was 41% shrubs, 42% herbaceous plants, 9% succulents and 8% grasses (Guth 1987). The most consumed plant species are *Euphorbia antisyphilitica*, *Tidestromia gemmata*, and *Opuntia rastrera* (Fig. 3.6). However, the inflorescences of *Fouquieria splendens*, locally known as ocotillo, represent an important nutritional resource for deer, because they contain high levels of protein (12% of dry weight), carbohydrate (67%), and 3085.98 kcal/kg of metabolizable energy; 85% are digestible nutrients and each individual plant produces around 1.5 kg of inflorescences available to the deer in the dry season (Fig. 3.7, Gallina et al. 2017). The plants consumed by mule deer provide different nutrients that meet the species' daily nutritional requirements (some plant species and their nutrients are shown in Table 3.1).

The quality of the diet includes the energy value derived from carbohydrates, lipids, and proteins, as well as the content of minerals, vitamins, and different non-nutritive elements such as lignin, cutin, suberin, silica and secondary metabolites (Robbins 1983). The nutrients are required in different quantities depending on the requirements and the physiological needs of the deer. For example, deer require a minimum of 7% of protein in the diet to maintain good metabolism by the ruminal microbiota. Their needs are higher when growing (17%), during gestation and lactation in females (18%), and during the growth of antlers in males (Ramírez-Lozano 2004).

Though controversial, it is very common to use percent fecal nitrogen (%FN) as an indicator of the crude protein content in the diet, since it correlates with the amount of protein consumed and with the digestibility of the plant species (Hodgman et al. 1996; Peripolli et al. 2011; Gil-Jiménez et al. 2015). That is, in large herbivores, FN is largely composed of metabolic nitrogen, both of bacterial and endogenous activity, and a small amount of undigested plant nitrogen (Wehausen 1995; Schwarm et al. 2009). So, it seems appropriate to make comparisons between sexes and ages to detect any differential use of resources. Fecal nitrogen evaluations have been carried out on fresh pellets, and differences between adult females and males have been detected between the dry and rainy seasons, where we observed that there is a difference in protein disposition, which is scarce in the dry season. Offspring always have a higher nitrogen content in their diet (Fig. 3.8) (De la Cruz-Morales 2017).

Fig. 3.6 The most consumed plant species by mule deer in Mapimí Biosphere Reserve, Mexico. (**a**) *Lippia graveolens*, (**b**) *Jatropha dioica*, (**c**) *Castela texana*, (**d**) *Euphorbia antisyphilitica*, (**e**) *Opuntia rastrera*, (**f**) *Agave aspérrima*. (Photo: Sonia Gallina Tessaro)

Temporal variations in %FN are due to phenological changes in the vegetation, availability of food per season, and this process is influenced by the physiological state of the deer. For example, in females, gestation and fawning require a lot of protein and minerals, while in males, preparation for the mating season requires more nutritional food resources (Bowyer and Kie 2004). In juveniles diet quality is stable throughout the year, but higher than that of adults in terms of the protein content required for their development. The %FN content for the mule deer according to the sex-age category indicates differential consumption in the amount of protein in the diet due to changes in availability during the dry and rainy seasons.

Fig. 3.7 Mule deer feeding on ocotillo inflorescences. (Photo: Luz A. Pérez-Solano)

One factor that is associated with diet quality is tannin content. Tannins influence the digestibility and absorption of protein (Hagerman and Robbins 1993; Verheyden et al. 2011). Because of this, %FN can serve as a valid index of diet quality. That is, if the intake of tannins is low, the relationship between dietary and fecal nitrogen is not altered. The same happens when the tannin content is constant over time (Verheyden et al. 2011).

3.4 Behavioral Ecology

Behavioral ecology examines how ecological constraints shape the behavior of an individual, and in turn, how individual behavior affects the dynamics of populations and communities (Sinclair et al. 2006). This approach seeks to describe the behavior of animals in terms of their adaptation to ecological conditions, including both the physical environment and the social environment (competitors, predators and parasites; Kie 1996; Davies et al. 2012). Throughout its evolutionary history, the mule deer has responded with behaviors depending on the type of habitat in which it is found, and the challenges it faces in order to survive. Some of the strategies that have been effective for the species include socialization, sexual segregation and anti-predatory strategies. For example, the degree of socialization depends on the seasonality of the region, the sex of the animal, the population and subspecies (Mackie et al. 2003). Many behavioral aspects are important for strategic conservation planning (Gosling 2003).

On the reserve, research has been carried out to estimate the selection and use of the habitat by mule deer. Some studies have found that sites with heterogeneous

Table 3.1 Bromatological analysis of plants consumed by mule deer in the Mapimí Biosphere Reserve, Durango, Mexico (modified from Cossío-Bayúgar 2015)

	Species	DM%	CP%	EE%	MM%	CF%	ADF	NDF	Lignin	Cell	HCell	Ca	P
Shrubs	*Agave asperrima*	24.49	9.09	5.04	14.28	21.04	29.44	37.29	9.31	18.27	7.85	3.80	0.37
	Euphorbia antisyphilitica	28.16	4.11	10.08	9.36	35.93	39.61	60.01	7.20	32.39	20.40	10.08	nd
	Fouquieria splendens[a]	30.51	22.02	na	6.48	na	na	Na	na	na	na	4.64	0.66
	Jatropha dioica	25.85	8.75	10.57	8.10	21.42	48.38	55.14	19.78	28.56	6.76	0.64	0.44
	Lippia graveolens	23.14	20.96	7.56	9.51	11.06	23.16	44.43	8.03	14.93	21.27	1.16	0.49
	Prosopis glandulosa[a]	56.74	17.82	5.92	4.58	26.37	35.50	65.79	17.15	13.56	30.29	1.50	0.43
	Vachellia constricta	64.75	7.43	13.67	3.51	36.47	52.66	63.36	22.02	30.04	10.70	0.48	0.29
Cactaceae	*Cylindropuntia imbricata*	22.57	6.12	5.39	35.09	12.05	33.40	53.38	13.50	19.37	19.98	10.60	0.39
	Cylindropuntia leptocaulis	41.65	9.80	13.97	4.44	15.37	33.28	38.46	17.38	15.20	5.18	5.80	0.24
	Opuntia rastrera	19.97	4.37	4.64	18.93	11.13	16.70	59.46	5.96	10.18	42.76	4.76	0.27

DM dry matter, *CP* crude protein ($N \times 6.25$), *EE* ether extract, *MM* mineral matter, *CF* crude fiber, *ADF* acid detergent fiber, *DFN* neutral detergent fiber, *Cell* cellulose, *HCell* hemicellulose, *Ca* calcium, *P* phosphorus

[a]Values from leaves

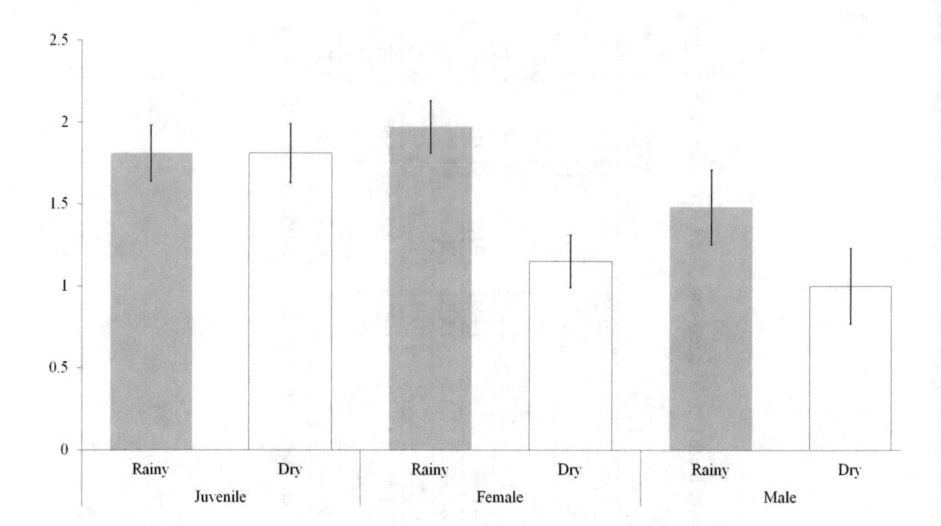

Fig. 3.8 Average percent fecal nitrogen (%FN) estimated in fresh pellets from adult male, adult female and juvenile mule deer during the rainy and dry seasons in the Mapimí Biosphere Reserve, Mexico

vegetation and terrain are preferred by deer, since this heterogeneity offers good visibility and allows the deer to quickly detect their predators (Sánchez-Rojas and Gallina 2000a). These sites are also characterized by having low plant richness (Cossío-Bayúgar 2015). While the choice and use of habitat is determined by each individual (Pérez-Solano et al. 2016), it also differs by sex (Geist 1981; Main et al. 1996; Main and Coblentz 1996; Mysterud et al. 2001; Beest et al. 2011).

Spatial ecology. The ecological interactions associated with the movement of animals can be modified by environmental factors such as landscape structure and environmental conditions, the distribution and use of resources, and even interactions with other organisms (i.e., predation), in addition to short-term reproductive and maintenance requirements (Getz and Saltz 2008; Holyoak et al. 2008; Nathan et al. 2008). Additionally, mule deer populations are spatially structured like metapopulations, so populations inhabit geographically isolated patches, often surrounded by areas of habitat not suitable for them (Sánchez-Rojas and Gallina 2006; Sánchez-Rojas and Gallina-Tessaro 2016; Wade and McDonald 2010). Throughout its distribution, the mule deer exhibits a good deal of intraspecific variation in its use of habitat resources and home ranges, even within a given population (Wallmo 1981; Marshal et al. 2006). Mule deer make different use of the characteristics of the habitat, moving seasonally between areas of favorable microclimates and resources for foraging, in order to maximize gain or minimize maintenance costs (Geist 1981; Marshal et al. 2006). Based on the deer location data from seven females and a young male mule deer, habitat use, and home range were estimated in the Mapimí Biosphere Reserve over the course of 3 years (2012–2014) by radiotelemetry monitoring (Fig. 3.9, Pérez-Solano et al. 2016, 2017).

Fig. 3.9 Mule deer group monitored by radiotelemetry over the course of 3 years in Mapimí Biosphere Reserve, Mexico. (**a**) 2012 year, (**b**) 2014 year. (Photo: Luz A. Pérez-Solano)

Habitat use. In a study of habitat use by mule deer, Pérez-Solano et al. (2016) examined the relationship between the locations recorded by radiotelemetry in both the dry and rainy seasons with topographic variables, such as distance to the nearest body of water (m), elevation (m.a.s.l.), slope of the terrain (degrees), and sun exposure (degrees), and ten types of plant associations (Fig. 3.4). During the dry season, deer were an average of 1007 m from bodies of water and during the rainy season an average of 1165 m from them. The mean slope of the terrain and elevation of the deer's locations were 2.56° and 1164 m.a.s.l. during the dry season and 1.97° and 1159 m.a.s.l. during the rainy season. Sun exposure during the dry season was 336° (Northwest) and during the rainy season, it was 338° (North), indicating that the deer search for sites that are less exposed to the thermal conditions of desert regions. The distance of a deer to the nearest body of water during the dry season suggests they are not able to endure long periods without water, though mule deer do not need to find water every day (Geist 1998; Heffelfinger et al. 2006; Marshal et al. 2006; Alcalá-Galván and Krausman 2013).

Pérez-Solano et al. (2016) found that plant associations affected the distribution of the deer throughout their habitat in both seasons, but for the group of deer in the study, there was no preference for any particular type of plant association and the observed differentiation in the use of plant associations was a function of the preferences of each individual. The greatest preference during the dry season was for the *Prosopis glandulosa*, *Hilaria* (currently *Pleuraphis*) *mutica* and *Larrea tridentata* (Pg Hm and Lt) plant association (preferred by three deer), and this association was also the one most avoided (four deer). During the rainy season, the Pg Hm and Lt association was again the most preferred by two deer and was also the most avoided during this season (three deer). Hills provide refuge from cold winds and during windy cold days in the winter and autumn, both sexes of mule deer used southern hillsides as physical protection from winds. The rest of year they were more frequently found on hillsides with a more northwesterly orientation, which are cooler (Pérez-Solano et al. 2016). Male deer exhibited a preference for a greater number of plant associations than the females did, and also for some plant associations that no females preferred, though they did use them according to their availability. These differences are attributed to individual preference rather than any difference between the sexes.

At the macro-habitat scale, the intensity of use of foraging areas increased as the density of the candelilla plant increased. In dry years, candelilla forage is the most important variable determining habitat use. In contrast, in wet years candelilla did not influence habitat use (Esparza-Carlos et al. 2011). Desert animals have behavioral and physiological adaptations to help them cope with dry conditions (Heffelfinger 2006; Hernández 2006). Habitat selection results from a combination of characteristics that vary over time and between individual preferences and physiological needs (i.e., mating and raising offspring). These individual habitat selection variables should be considered in the conservation and management programs of wildlife populations (Shields et al. 2012; Pérez-Solano et al. 2016, 2017).

Home range. Home range was estimated with the fixed Kernel method using 95% of the observations (Kernohan et al. 2001). The mean (±*SD*) home range size for the mule deer females was estimated to be 14.70 km^2 (±5.89), and for the only male that has been monitored with radio-tracking, 18.05 km^2 (Fig. 3.10, Pérez-Solano et al. 2017). These home range estimates fall within the range reported for the species, though they are at the lower end. During the dry season, the home range of the females was of 11.92 km^2 (±4.47), and this increased to 19.11 km^2 (±9.37) in the rainy season. The male has a home range of 15.29 km^2 in the dry season, which increased to 25.63 km^2 in the rainy season. Pérez-Solano et al. (2017) did not find any significant differences between seasons in the females' home range, but individual deer used different areas between seasons, and their home ranges only overlapped by a mean of 22.35% with respect to the entire area occupied each season.

Females and males make differential use of the habitat (Geist 1981; Main et al. 1996; Main and Coblentz 1996; Mysterud et al. 2001; Beest et al. 2011), and there is a segregation of males, which are located on the peripheries of the females' home ranges. The only overlap occurs at feeding sites and those where water is present. When young males turn one year old, they are expelled from the family group by the adult females. Young females remain in the family group for several years until they find and occupy a suitable home range (Geist 1981). In the study by Pérez-Solano et al. (2017), the young male deer was less than 1 year old when researchers started monitoring him and he was followed until he reached his third year. During this period, he left his family group and returned in the mating season (dry season) (Geist 1981; Mackie et al. 2003), though he mainly stayed in the same home range as the group of females.

Core areas. The core area was also estimated with the fixed Kernel method, but using 50% of the observations (Kernohan et al. 2001). The most important site within an animal's home range is the core area, which contains the greatest density of critical resources (Burt 1943; Powell 2000) and where the animal carries out at least 50% of its activities. Due to their small size and the intensive use made of these core areas, the total distances covered by animals searching for resources can be affected. The mean core area size for females estimated by Pérez-Solano et al. (2017) was 1.74 km^2 (±0.50); 1.56 km^2 (±0.32) during the dry season and 2.18 km^2

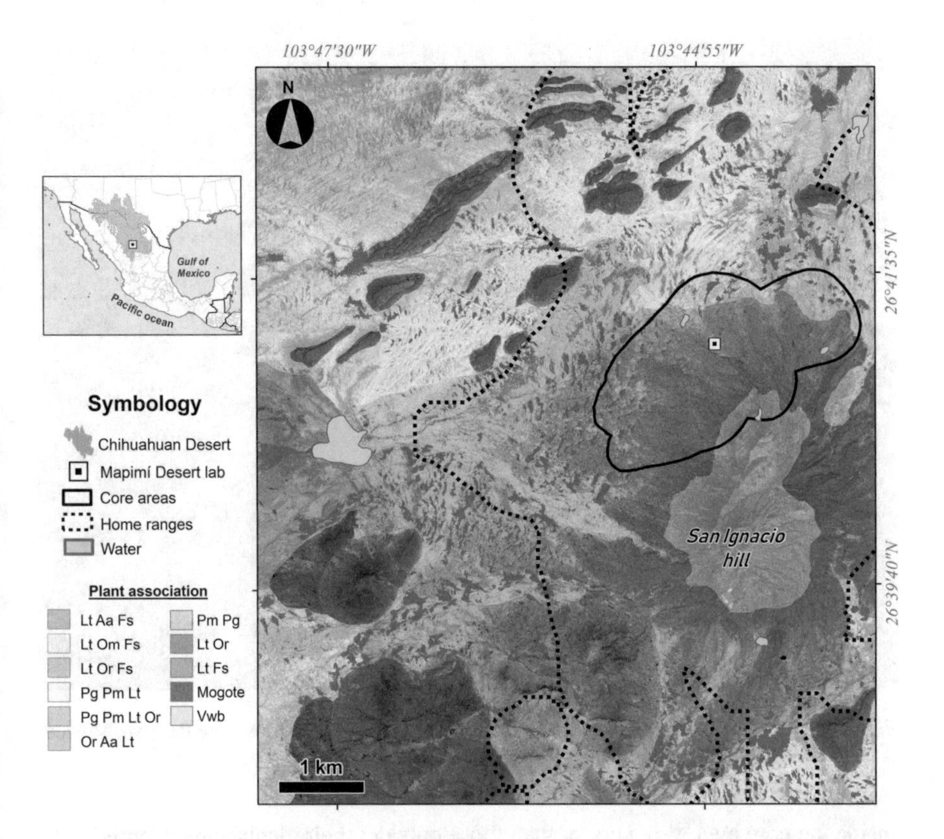

Fig. 3.10 The pooled home ranges and core areas occupied by seven female mule deer followed by radiotelemetry between 2012 and 2014. Species of plant associations: Pm = *Pleuraphis mutica*, Pg = *Prosopis glandulosa*, Lt = *Larrea tridentata*, Or = *Opuntia rastrera*, Fs = *Fouquieria splendens*, Aa = *Agave asperrima*, Om = *Opuntia microdasys*, Vwb = Vegetation associated with a dam

(±1.51) in the rainy season (Fig. 3.11). In contrast for the male, the mean core area was 3.24 km^2; 2.55 km^2 in the dry season and 4.01 km^2 in the rainy season. There were no significant differences in core area size by season, however the overlap of core areas by season was only 19.41%.

As the core activity areas are the most important sites within an animal's home range, Pérez-Solano et al. (2017) also evaluated the habitat variables related to core area selection by mule deer females and their interior movements over time (month, time of day). They used variables related to environmental conditions (temperature, relative humidity), and habitat characteristics (ten plant associations, distance to the nearest water body, slope of the terrain). The mule deer females selected the distance to the nearest body of water (<934 m at the population level and <1375 m on average for individual deer) as the most important variable in core area (Fig. 3.12), followed by two plant associations composed of *Prosopis glandulosa*, *Pleuraphis mutica*, *Larrea tridentata*, *Opuntia rastrera*, and *Fouquieria splendens*. Surprisingly, the slope of the terrain was not a selection factor. Distances covered by females

Fig. 3.11 Female mule deer monitored by radiotelemetry in Mapimí Biosphere Reserve, Mexico. (Photo: Luz A. Pérez-Solano)

inside the core area were shorter than those outside; their displacement varied with time of day and month. For example, the increase in the distance covered during June and July may be associated with their search for sites to give birth, but displacement decreased when the fawns were most vulnerable after birth (Pérez-Solano et al. 2017). Even though the majority of the critical resources are concentrated in the core areas, these sites are not sufficient to maintain the deer in the long term since many of the better quality essential resources are found outside of them. However, environmental seasonality and the physiological needs of the animals should be taken into account when assessing the core areas (Asensio et al. 2012, 2014).

Sexual segregation. Sexual segregation is defined as the differential use of space or resources between males and females, and is expressed as a pattern in the individual use of different resources, or when groups of the same sex and age carry out activities for a period of time (Main et al. 1996; Main and Coblentz 1996; Kie and Bowyer 2004; Wearmouth and Sims 2008). The quantification of sexual segregation in mule deer is complex, since it can be detected in spatial responses (e.g., selection of different areas that meet certain requirements) that are affected by time scales (Bowyer et al. 1996; Bowyer 2004; Bowyer and Kie 2004), or it can be detected in individual responses (e.g., time spent in different activities by individuals of both sexes and/or age classes) (Perez-Barberia et al. 2005; Calhim et al. 2006;

Fig. 3.12 Artificial and temporal dams in Mapimí Biosphere Reserve, Mexico. The distance to dams is the most important variable in core area selection by female mule deer. (Photo: Luz A. Pérez-Solano)

Pérez-Solano et al. 2017), or in population responses such as the formation of different sex and age groups, outside the mating season (Barboza and Bowyer 2000; Perez-Barberia and Gordon 2000; Perez-Barberia et al. 2002; Siuta and Bobek 2006). The degree of sexual segregation varies between populations of the same species and between species (Putman and Flueck 2011) and is more evident in species with strong sexual dimorphism (Ruckstuhl and Neuhaus 2002; 2005) and polygamous mating systems (Clutton-Brock et al. 1982; Alves et al. 2013).

To study sexual segregation in the mule deer of the Mapimí Biosphere Reserve, through spatial behavior and the comparison of fecal nitrogen between sexes, it was found that a scale on the order of hectares is adequate (Bowyer et al. 1996). The diffuse classification of the fecal groups, using the morphometric data of the pellets (length, width, volume and length-width ratio) differentiated the proportion of three sex-age categories (juveniles, adult females and adult males, Sánchez-Rojas et al. 2004), corresponding to a sex ratio of two females for each male. The proportion of females to juveniles was 1:0.77. The female:juvenile ratio can reflect the health status of the population and the habitat.

For this population of mule deer inhabiting San Ignacio Hill, over 11 nonconsecutive years of following their population dynamics by analyzing pellet morphology, sexual segregation was not constant across years (Sánchez-Rojas et al. 2004).

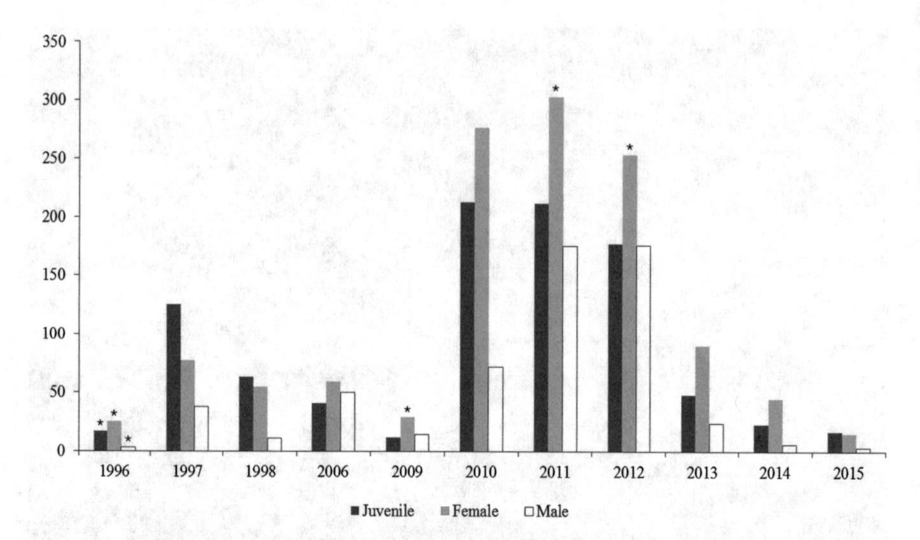

Fig. 3.13 Number of fecal groups in each of the three age and sex categories for 11 years of monitoring. The asterisk indicates that there was no independence between the categories and the space occupied by the transects

The encounter of pellets belonging to the three categories was not a random process in terms of spatial distribution and did not depend on density or on precipitation (Fig. 3.13).

Spatial segregation appears to be more greatly influenced by the spatial scale at which the phenomenon is being evaluated (Kie et al. 2002). For example, in the segregation coefficient (SC) proposed by Conradt (1998), a value of zero indicates there is no segregation, values of one indicate complete segregation, and intermediate values indicate partial segregation.

Modified for pellet count data, SC (De la Cruz-Morales 2017), is expressed as follows:

$$\mathrm{SSSC} = \sqrt{\left[1 - \frac{N-1}{X \cdot Y} \cdot \sum_{i=1}^{k} \frac{Xi \cdot Yi}{ni - 1}\right]}$$

where:

SSSC: Spatial Sexual Segregation Coefficient
X: Total number of male pellets observed
Y: Total number of female pellets observed
N: Total number of animal pellets observed ($X + Y$)
xi: Number of males in the ith type of spatial area (transect or circle)
yi: Number of females in the ith type of spatial area (transect or circle)
ni: Number of animals ($xi + yi$) in the ith type of spatial area (transect or circle)
k: Number of habitat types spatial quadrants

Analyzing the pellet counts obtained for 11 years (1996–1998, 2006, 2009–2015) for the mule deer population of the MBR, the SSSC value was 0.876, indicating segregation at a very fine scale given by the collection point. As spatial scale increases, as occurs with the transect, the value of the index decreases to 0.703, showing that the phenomenon of segregation decreases due to its dependence on scale (Kie et al. 2002).

Foraging and predation risk. All animals need to satisfy their requirements for food, water, sites for sleeping and areas for reproduction. Animals allocate their time in their search for these resources, however, much of the time they are in a state of alertness to avoid predation and avoiding predation often becomes a full-time activity (Ydenberg et al. 2007). The fear of being caught is an impetus for the animals to develop antipredator behaviors to reduce the probability of being caught, depending on predator type and habitat (Robinson et al. 2002; Caro et al. 2004). Three antipredator strategies have been identified in mule deer. (a) Modification of space use: mule deer avoid sites where they are more vulnerable to being caught by a predator. When, they are foraging, they look for habitats and microhabitats where habitat structure facilitates the detection of predator ambush and makes it possible to avoid being preyed upon (Altendorf et al. 2001; Hernández et al. 2005; Esparza-Carlos et al. 2011, 2016, 2018), while dense cover conceals yearlings from predation in bed sites (Heffelfinger 2006; Cossío-Bayúgar and Sisto-Burt 2011). (b) They modify their behavior: mule deer increase the degree of individual vigilance and reduce foraging time in dangerous areas. They can assume appropriate group sizes to increase vigilance time and reduce their vulnerability to predation (Laundré et al. 2001; Robinson et al. 2002; Caro et al. 2004; Hernández et al. 2005). (c) Avoidance over time: mule deer modify their activity patterns through time to reduce the probability of meeting with predators (Brown 1999; Robinson et al. 2002; Lomas and Bender 2007; Arias-Del-Razo et al. 2011).

In the Chihuahuan Desert, the puma (*Puma concolor*) is the only real predator of adult mule deer (Ballard et al. 2001). There are no puma diet studies in Mexican desert areas, but in other ecosystems, the main prey are ungulates: white tailed deer (36–52%) and peccary (7–11%) (Núñez et al. 2000; de la Torre and de la Riva 2009; Hernández-SaintMartín et al. 2015). In Mapimí, mule deer is the only wild ungulate, and mule deer remains are present in 5.8% of coyote and bobcat scat (Hernández et al. 2002a). These predators only kill fawns. Golden eagles can also kill fawns, but their effects on the population are minimal (Ballard et al. 2001; Heffelfinger 2006). On the Mapimí Biosphere Reserve there is no evidence that coyotes or bobcats' prey on adults; they appear to consume mule deer carrion (Hernández et al. 2002a). At a large scale, mule deer select habitat close to a body of water. In addition, for plant cover defined by height plant, the presence of bushes with a larger crown is negatively correlated with the presence of deer. Since visibility is important to reduce the risk of predation, this is how mule deer respond to predation risk (Sánchez-Rojas and Gallina 2000a). The fear of being caught affects the foraging strategy of mule deer in desert habitats (Carrera et al. 2015; Esparza-Carlos et al. 2016).

Mule deer and cattle coexistence. Within the Chihuahuan Desert, cattle ranching represents the third most important economic activity in the country (CONANP 2006). As such, the survival of many populations of mule deer has been affected by

Fig. 3.14 Cattle (*Bos taurus*) within Mapimí Biosphere Reserve

the expansion of ranching, mainly as a result of the destruction of the vegetation by the cattle (Martínez-Muñoz et al. 2003). The natural conditions of the MBR—water scarcity, low rainfall, extensive areas with halophyte pastures and matorral scrub—mean that the only economically profitable productive activity is extensive livestock farming, which is why most of the land is used for grazing in the reserve (CONANP 2006).

The study of habitat overlap and the interaction intensity of sympatric species has been reviewed using different ecological and evolutionary approaches. One of the most important explanations is the detailed study of patterns of resource use and the way species use a determined habitat (Bailey 2005). Cattle can compete with wildlife for food, but this competition varies among regions, seasons and with the type of animals involved (Kie 1996; Bolen and Robinson 2003). There are data to indicate that cattle, when managed and at a density below the carrying capacity of the environment, maintain and even improve the habitat for deer, decreasing the dry material and grasses, which in turn favors the proliferation of the herbaceous plants and shrubs that the deer consume (Kie 1996; Bolen and Robinson 2003; Findholt et al. 2004; Chaikina and Ruckstuhl 2006).

The coexistence of mule deer and cattle (Fig. 3.14, *Bos taurus*) in Mapimí was evaluated by Cossío-Bayúgar (2015) to determine the use of different areas by sampling fecal group counts. Deer used higher, steeper areas, with a high plant density and low plant richness; while cattle used flat, low areas, with low visibility or higher vertical cover all year round. Cattle used more areas with a higher horizontal cover and volume of plants (mainly mesquite), and less visibility on the upper strata (above 50 cm) associated with shaded areas, which can help them with thermoregulation (Bailey 2005; Cain III et al. 2006). In addition, vertical cover in the lower strata (0–50 cm), was found to protect calves and fawns during their first days of life by helping them maintain their body temperature and avoid attracting predators (Cossío-Bayúgar and Sisto-Burt 2011).

The results of this study may be the outcome of the plasticity of deer with respect to adapting their plant consumption to seasonal changes and food availability (Heffelfinger 2006), and also as regards the anti-predator strategy of moving around their activity area in irregular time and space patterns (Geist 1981; Robinson et al. 2002; Caro et al. 2004). There were differences in transect use for cattle and deer that were associated with seasonal changes that might be the result of exposure, influenced by San Ignacio Hill. Although some areas were used by both deer and cattle, both species used areas with different characteristics, with no competition for spatial resources or habitat use under the circumstances at Mapimí Biosphere Reserve at that time. Features such as slope, distance to water, shade and resting place availability restrict the use of certain areas for cattle, leaving some areas to be used by deer almost exclusively. This results from their differences in mobility, diet variability and selectivity, and thermoregulatory and anti-predator strategies. This information is important in the context of concern about the consequences of interactions between deer and cattle in arid environments, and because it helps establish the criteria for designating suitable grazing areas for cattle where they are less likely to interfere with mule deer. Cossío-Bayúgar (2015) recommended not increasing cattle density, and maintaining cattle in areas where they cannot interfere with deer. She also recommended that other species of herbivores such as goats or other browsing ruminants not be introduced as they may compete with the deer. Because of drought conditions, and to promote a better use of space, she recommended providing water during the dry season, for both the wildlife and the domestic animals in the area.

These zones have combinations of grassland with halophyte scrub, in which there are species of grasses such as the toboso grass (*Pleuraphis mutica*), and the alkali zacaton (*Sporobolus airoides*) and where the predominant shrub is mesquite (*Prosopis glandulosa*) (Barral and Hernández 2001; CONANP 2006), as well as those of microphyll desert scrub where shrub species such as *Larrea tridentata*, *Cassia angustifolia*, *Prosopis glandulosa*, *Celtis pallida* and *Hintonia latifolia* are found; the latter three reported to form part of the diet of bovines (Barral and Hernández 2001; Enríquez 2001; Foroughbakhch et al. 2007). Although some areas were used by both species, no competition for spatial resources or habitat was observed. Features such as slope, distance to water, shade and resting place availability restrict the use of certain areas by cattle, leaving some areas to be used by deer almost exclusively (Cossío-Bayúgar 2015).

The introduction of domestic species into a new habitat may result in resource competition with native wildlife (Bolen and Robinson 2003; Ortega 2008), particularly in arid and semiarid regions, where resources are scarce. Information about the effects of deer and cattle interactions may vary in the same habitat in relation to grazing intensity and the criteria used for their evaluation (Austin and Urness 1986; Chaikina and Ruckstuhl 2006; Findholt et al. 2004).

The interactions that occur when domestic animals and wildlife share a habitat include resource competition and, potentially, disease transmission (Craft 2015; Hassel et al. 2017). Parasites generally coexist with their natural host unless adverse environmental conditions or chronic stress in the host interferes with host immune responses (Baldomenico and Begon 2015; Craft 2015).

Deer and cattle can share different etiological agents (Aguirre et al. 1995; Cantú et al. 2007; Hoberg et al. 2008; Ortega 2008; Craft 2015; Walters and Palmer 2015). Among the endoparasites considered to have the greatest effect on mule deer are *Elaeophora schneideri*, *Setaria labiatopalillosa*, *Onchocerca cervipedis*, *Fascioloides magna*, *Haemonchus contortus*, *Ostertagia* sp., *Trichostrongylus* sp., *Nematodirus* sp., *Chabertia ovina*, *Oesophagostomum* sp., *Parelaphostrongylus* sp., *Echinococcus granulosus*, *Taenia omissa*, *Moniezia* sp. (Stubblefield et al. 1987; Hoberg et al. 2001; Jones and Pybus 2001; Pybus 2001; Myers et al. 2015), as well as several species of protozoa (i.e., *Cryptosporidium* sp., *Toxoplasma gondii*, *Capillaria* spp., *Eimeria* sp.) (Dubey and Odening 2001;Duszynski and Upton 2001; Olson and Buret 2001; Myers et al. 2015). Some of these parasites are potential pathogens and of importance in the context of public health (Foreyt 2001; Hoberg et al. 2001).

In the mule deer population of the Mapimí Biosphere Reserve, Cossío-Bayúgar et al. (2016) identified parasites belonging to the genera *Eimeria* sp., *Haemonchus* sp., *Bunostonum* sp., *Cooperia* sp., and *Trichostrongylus* sp. in cattle; and to *Eimeria* sp. in mule deer. Even though none of the parasites were common to both species, some parasites observed in cattle have been found in mule deer in other parts of their distribution. Parasite shedding increases when precipitation and temperature are higher. Rainfall and temperature appear to influence the presence of nematode parasites and therefore the degree of exposure of other animals. However, dry environments, such as that of Mapimí, decrease the survival of larvae in feces and reduce the risk of infection to other animals, particularly in direct cycle parasites.

3.5 Conclusions

Mule deer populations within the Mapimí Biosphere Reserve are protected because it is a Protected Natural Area decreed by the Mexican government in 1979. This reserve was established to preserve the region known as the Bolson de Mapimí of the Chihuahuan Desert, an ecosystem with several endemic species that have adapted to and survived these arid conditions. The study of mule deer within this protected area can be extrapolated to other populations in arid areas that are under management and sustainable use. An important aspect of the Mapimí Biosphere Reserve is that controlling the quality of the diet of wild ruminants is useful for managing their populations, and for understanding some aspects of their behavior and ecology. Studies that focus on population dynamics, behavior, spatial use of resources, as well as the study of habitat at different spatial and temporal scales, along with studies that generate knowledge about the relationship between the population and the precipitation regime are important to the management of wild species, and can help plan actions that guarantee the conservation, management and the sustainable use of this species.

Acknowledgments We are grateful to the *Comisión Nacional de Áreas Naturales Protegidas—Reserva de la Biosfera de Mapimí* and to the *Secretaría del Medio Ambiente y Recursos Naturales* for issuing the required permits (No. SEMARNAT/DGVS/00234 and 00954). The Rufford Foundation small grants program and the *Consejo Nacional de Ciencia y Tecnología*, CONACYT provided funding. We thank Rolando González-Trápaga and Francisco Herrera, along with the graduate students of the Instituto de Ecología, A. C. and of the *Universidad Autónoma del Estado de Hidalgo* for help in the field. Adriana Sandoval-Comte provided support with the Geographic Information Systems. Bianca Delfosse revised the English.

References

Aguirre AA, Hansen DE, Starkey EE et al (1995) Serologic survey of wild cervids for potential disease agents in selected national parks in the United States. Prev Vet Med 21:313–322

Aguirre BRJ, Bortoni VL, Villarreal JG (2016) Restauración y manejo del venado bura (*Odocoileus heminonus eremicus*) en el rancho Corrizalejo, San Buenaventura, Coahuila, México (años 2000–2015). In: Universidad Nacional Autónoma de México, XV Simposio sobre Venados de México, Ciudad de México, pp 65–82

Alcalá-Galván CH, Krausman P (2013) Home range and habitat use by desert mule deer in altered habitats. Calif Fish Game 99:65–79

Altendorf KB, Laundré JW, López-González CA et al (2001) Assessing effects of predation risk on foraging behavior of mule deer. J Mammal 82:430–439

Alves J, Alves da Silva A, Soares AMVM et al (2013) Sexual segregation in red deer: is social behaviour more important than habitat preferences? Anim Behav 85:501–509

Anthony R, Smith N (1977) Ecological relationships between mule deer and white-tailed deer in southeastern Arizona. Ecol Monogr 47:255–277

Arias-Del-Razo I, Hernández L, Laundré JW et al (2011) Do predator and prey foraging activity patterns match? A study of coyotes (*Canis latrans*), and lagomorphs (*Lepus californicus* and *Sylvilagus audobonii*). J Arid Environ 75:112–118

Asensio N, Lusseau CD, Schaffner M et al (2012) Spider monkeys use high-quality core areas in a tropical dry forest. J Zool 287:250–258

Asensio N, Brockelman WY, Malaivijitnond S et al (2014) White-handed Gibbon (*Hylobates lar*) core area use over a short-time scale. Biotropica 46:461–469

Austin DD, Urness PJ (1986) Effects of cattle grazing on mule deer diet and area selection. J Range Manag 39:18–21

Álvarez-Cárdenas S, Gallina S, Galina-Tessaro P et al (1999) Mule deer population dynamics in a relictual oak-pine forest in Baja California Sur, Mexico. In: Folliott PF, Ortega-Rubio A (eds) Ecology and management of forests, woodlands and shrublands in the dryland regions of the United States and México: perspectives for the 21st century. University of Arizona y Centro de Investigaciones Biológicas del Noreste, Tempe, pp 197–210

Bailey DW (2005) Identification and creation of optimum habitat conditions for livestock. Rangeland Ecol Manag 58:109–118

Baldomenico PM, Begon M (2015) Stress-host-parasite interactions: a vicious triangle? FAVE Cs Veterinarias 14:6–19

Ballard WB, Lutz D, Keegan TW et al (2001) Deer-predator relationships: a review of recent North American studies with emphasis on mule and black-tailed deer. Wildl Soc Bull 29(1):99–115

Barboza PS, Bowyer RT (2000) Sexual segregation in dimorphic deer: a new gastrocentric hypothesis. J Mammal 81(2):473–489

Barral H, Hernández L (2001) Los ecosistemas pastoreados desérticos y sus diversas formas de aprovechamiento: análisis de tres casos. In: Hernández L (ed) Historia ambiental de la ganadería en México. Instituto de Ecología A.C., Xalapa, Veracruz, México, pp 85–97

Beest FLV, Rivrud IM, Loe LE et al (2011) What determines variation in home range size across spatiotemporal scales in a large browsing herbivore? J Anim Ecol 80:771–785

Bender LC, Hoenes BD, Rodden CL (2012) Factors influencing foraging habitats of mule deer (*Odocoileus hemionus*) in the San Andres Mountains, New Mexico. Southwest Nat 57(4):370–379

Bishop CJ, White GC, Freddy DJ et al (2009) Effect of enhanced nutrition on mule deer population rate of change. Wildl Monogr 172:1–28
Bolen EG, Robinson WL (2003) Wildlife ecology and management, 5th edn. Prentice Hall, Upper Saddle River, NJ
Bowyer RT (2004) Sexual segregation in ruminants: definitions, hypotheses, and implications for conservation and management. J Mammal 85(6):1039–1052
Bowyer RT, Kie JG (2004) Effects of foraging activity on sexual segregation in mule deer. J Mammal 85(3):498–504
Bowyer RT, Kie JG, Van Ballenberghe V (1996) Sexual segregation in black-tailed deer: effects of scale. J Wildl Manag 60(1):10–17
Brown JS (1999) Vigilance, patch use and habitat selection: foraging under predation risk. Evol Ecol Res 1:49–71
Brunjes K, Ballard W, Humphrey M et al (2006) Habitat use by sympatric mule and white-tailed deer in Texas. J Wildl Manag 70(5):1351–1359
Burt WH (1943) Territoriality and home range concepts as applied to mammals. J Mammal 24:346–352
Bergman EJ, Doherty Jr.FP, White CG, Holland AA (2015) Density dependence in mule deer: a review of evidence. Wildlife Biology 21(1):18–29. https://doi.org/10.2981/wlb.00012.
Cain JW III, Krausman PR, Rosenstock SS et al (2006) Mechanisms of thermoregulation and water balance in desert ungulates. Wildl Soc Bull 34:570–581
Calhim S, Shi JB, Dunbar RIM (2006) Sexual segregation among feral goats: testing between alternative hypotheses. Anim Behav 72:31–41
Cantú AJ, Ortega A, Mosqueda J et al (2007) Immunologic and molecular identification of *Babesia bovis* and *Babesia bigemina* in freeranging white-tailed deer in northern Mexico. J Wildl Dis 43:504–507
Caro TM, Graham CM, Stoner CJ et al (2004) Adaptive significance of antipredator behaviour in artiodactyls. Anim Behav 67:205–228
Carrera R, Ballard WB, Krausman PR et al (2015) Reproduction and nutrition of desert mule deer with and without predation. Southwest Nat 60(4):285–298
Chaikina NA, Ruckstuhl KE (2006) The effect of cattle grazing on native ungulates: the good, the bad, and the ugly. Rangelands 28:8–14
Ciuti S, Jensen WF, Nielsen SE et al (2015) Predicting mule deer recruitment from climate oscillations for harvest management on the northern Great Plains. J Wildl Manag 79:1226–1238
Clutton-Brock TH, Iason GR, Albon SD et al (1982) The effects of lactation on feeding behavior and habitat use of wild red deer hinds. J Zool 198:227–236
CONANP (Comisión Nacional de Áreas Naturales Protegidas) (2006) Programa de conservación y manejo de la Reserva de la Biosfera de Mapimí, México. CONANP, Distrito Federal, México
Conradt L (1998) Measuring the degree of sexual segregation in group-living animals. J Anim Ecol 67(2):217–226
Cossío-Bayúgar A (2015) Uso del hábitat y su relación con la presencia-ausencia de parásitos en el venado bura (*Odocoileus hemionus*) de la Reserva de la Biosfera de Mapimí, México. Dissertation, Instituto de Ecología, A.C
Cossío-Bayúgar A, Sisto-Burt AM (2011) Definición y medición del bienestar animal. In: Medina-Cruz M (ed) Clínica, cirugía y producción de becerras y vaquillas lecheras. 12 Editorial AC, Ciudad de México, pp 116–128
Cossío-Bayúgar A, Romero E, Gallina S et al (2016) Variation of gastrointestinal parasites in mule deer and cattle in Mapimi Biospere Reserve, Mexico. Southwest Nat 60(2–3):180–185
Craft ME (2015) Infectious disease transmission and contact networks in wildlife and livestock. Phil Trans R Soc B 370:20140107
Davies NB, Krebs JR, West SA (2012) An introduction to behavioural ecology, 4th edn. Wiley-Blackwell, Chichester
De la Cruz-Morales NP (2017) Estudio de la segregación sexual del venado bura en la reserva de la biosfera de Mapimí, evaluando sus consecuencias ecológicas. Dissertation, Universidad Autónoma del Estado de Hidalgo

De la Torre J, de la Riva G (2009) Food habits of pumas (*Puma concolor*) in a semiarid region of Central Mexico. Mastozool Neotrop 16:211–216

Denny RN (1976) Regulations and mule deer harvest-political and biological management. In: Proceeding's: Mule deer decline in the West. Utah State University, Logan, UT, pp 87–92

Dubey JP, Odening K (2001) Toxoplasmosis and related infections. In: Samuel MW, Pybus MJ, Kocan AA (eds) Parasitic diseases of wild mammals, 2nd edn. Iowa State University Press, Ames, pp 478–519

Duszynski DW, Upton SJ (2001) *Cyclospora*, *Eimeria*, *Isospora*, and *Cryptosporidium* spp. In: Samuel MW, Pybus MJ, Kocan AA (eds) Parasitic diseases of wild mammals, 2nd edn. Iowa State University Press, Ames, pp 416–459

Eberhardt LL, Van Etten RC (1956) Evaluation of pellet group count as a deer census method. J Wildl Manag 20:70–74

Enríquez A (2001) Invasión de plantas arbustivas en los pastizales de Chihuahua. In: Hernández L (ed) Historia ambiental de la ganadería en México. Instituto de Ecología A.C., Xalapa, pp 98–107

Esparza-Carlos JP, Laundré JW, Sosa VJ (2011) Precipitation impacts on mule deer habitat use in the Chihuahuan desert of Mexico. J Arid Environ 75:1008–1015

Esparza-Carlos JP, Laundré JW, Hernández L et al (2016) Apprehension affecting foraging patterns and landscape use of mule deer in arid environments. Mamm Biol 81(6):543–550

Esparza-Carlos JP, Íñiguez-Dávalos LI, Laundré JW (2018) Microhabitat and top predator presence affects prey apprehension in a subtropical mountain forest. J Mammal 99:596–607

Findholt SL, Johnson BK, Damiran D et al (2004) Diet composition, dry matter intake, and diet overlap of mule deer, elk, and cattle. In: Transactions of the 69th North American Wildlife and Natural Resources Conference, Lawrence, Kansas, USA, pp 670–696

Foreyt WJ (2001) Veterinary parasitology reference manual, 5th edn. Blackwell Publishing, Ames, IA

Foroughbakhch R, Hernández-Piñero JL, Ramírez R et al (2007) Seasonal dynamics of leaf nutrient profile of 20 native shrubs in Northeastern Mexico. J Anim Vet Adv 6:1000–1005

Forrester TD, Wittmer HU (2013) A review of the population dynamics of mule deer and black-tailed deer Odocoileus hemionus in North America. Mammal Rev 43:292–308

Galindo-Leal C (1993) Densidades poblacionales de los venados cola blanca, cola negra y bura en Norte América. In: Medellín RA, Ceballos G (eds) Avances en el estudio de los mamíferos de México, vol 1. Asociación Mexicana de Mastozoología A.C., México, pp 371–391

Gallina S, García-Feria L, González-Trápaga R (2017) Ocotillo flowers as food resource for the mule deer during the dry season. Therya 8(2):185–188

Geist V (1981) Behavior: adaptive strategies in mule deer. In: Wallmo OC (ed) Mule and black tailed deer of North America. Wildlife Management Institute book. University of Nebraska Press, Nebraska, pp 157–223

Geist V (1998) Deer of the world: their evolution, behaviour, and ecology. Stackpole Books, Mechanicsburg

Getz WM, Saltz D (2008) A framework for generating and analyzing movement paths on ecological landscape. Proc Natl Acad Sci U S A 105:19066–19071

Gil-Jiménez E, Villamuelas M, Serrano E et al (2015) Fecal nitrogen concentration as a nutritional quality indicator for European rabbit ecological studies. PLoS One 10(4):e0125190

Gosling L (2003) Adaptive behavior and population viability. In: Festa-Bianchet M, Apollonio M (eds) Animal behavior and wildlife conservation. Island Press, Washington, pp 13–30

Guth AMCG (1987) Hábitos alimenticios del venado bura (*Odocoileus hemionus* Rafinesque 1817) en la Reserva de la Biosfera de Mapimí Dgo. Dissertation, Escuela Nacional de Estudios Profesionales Iztacala, Universidad Nacional Autónoma de México

Hagerman AE, Robbins CT (1993) Specificity of tannin-binding salivary proteins relative to diet selection by mammals. Can J Zool 71(3):628–633

Hassel JM, Begon M, Ward MJ et al (2017) Urban and disease emergence dynamics at the wildlife-livestock-human interface. Trends Ecol Evol 32:55–67

Heffelfinger J (2006) Deer of the Southwest: a complete guide to the natural history, biology, and management of southwestern mule deer and white. Texas A&M University Press, College Station, TX

Heffelfinger JR, Brewer C, Alcalá-Galván CH et al (2006) Habitat guidelines for mule deer: southwest deserts ecoregion. Mule Deer Working Group, Western Association of Fish and Wildlife Agencies, Boise, ID

Hernández HM (2006) La vida en los desiertos Mexicanos. Fondo de Cultura Económica, Ciudad de México

Hernández L, Parmenter RR, Dewitt JW et al (2002a) Coyote diets in the Chihuahuan Desert, more evidence for optimal foraging. J Arid Environ 51:613–624

Hernández L, Laundré JW, Gurung M (2005) Use of camera traps to measure predation risk in a puma-mule deer system. Wildl Soc Bull 33:353–358

Hernández L, Parmenter RR, Dewitt JW et al (2002b) Coyote diets in the Chihuahuan Desert, more evidence for optimal foraging. J Arid Environ 51:613–624

Hernández-SaintMartín AD, Rosas-Rosas OC, Palacio-Núñez J et al (2015) Food habits of jaguar and puma in a protected area and adjacent fragmented landscape of Northeastern Mexico. Nat Areas J 35:308–317

Hernández-Silva DA (2018) Manejo de fauna silvestre como herramienta en la conservación y el aprovechamiento sustentable de la biodiversidad. Dissertation, Universidad Autónoma del estado de Hidalgo

Hoberg EP, Kocan AA, Rickard LG (2001) Gastrointestinal strongyles in wild ruminants. In: Samuel MW, Pybus MJ, Kocan AA (eds) Parasitic diseases of wild mammals, 2nd edn. Iowa State University Press, Ames, pp 193–221

Hoberg EP, Polley L, Jenkins EJ et al (2008) Pathogens of domestic and free-ranging ungulates: global climate change in temperate to boreal latitudes across North America. Rev Sci Tech 27:511–528

Hodgman TP, Davitt BB, Nelson JR (1996) Monitoring mule deer diet quality and intake with fecal indices. J Range Manag 49:215–222

Holyoak M, Casagrandi R, Nathan R et al (2008) Trends and missing parts in the study of movement ecology. Proc Natl Acad Sci U S A 105:19060–19065

Jones A, Pybus MJ (2001) Taeniasis and Echinococcosis. In: Samuel MW, Pybus MJ, Kocan AA (eds) Parasitic diseases of wild mammals, 2nd edn. Iowa State University Press, Ames, pp 150–192

Kernohan BJ, Gitzen RA, Millspaugh JJ (2001) Analysis of animal space use and movement. In: Millspaugh JJ, Marzluff JM (eds) Radio tracking and animal populations. Academic, San Diego, pp 125–166

Kie JG (1996) The effects of cattle grazing on optimal foraging in mule deer (*Odocoileus hemionus*). Forest Ecol Manag 88:131–138

Kie JG, Bowyer RT (2004) Effects of foraging activity on sexual segregation in mule deer. J Mammal 85:498–504

Kie JG, Bowyer RT, Nicholson MC et al (2002) Landscape heterogeneity at differing scales: effects on spatial distribution of mule deer. Ecology 83(2):530–544

Laundré JW, Hernández L, Altendorf KB (2001) Wolves, elk, and bison: reestablishing the "landscape of fear" in Yellowstone National Park, USA. Can J Zool 79:1401–1409

Loik ME, Breshears DD, Lauenroth WK et al (2004) A multi-scale perspective of water pulses in dryland ecosystems: climatology and ecohydrology of the western USA. Oecologia 141(2):269–281

Lomas LA, Bender L (2007) Survival and cause-specific mortality of neonatal mule deer fawns, north-central New Mexico. J Wildl Manag 71(3):884–894

Mackie RJ, Kie JG, Pac DF et al (2003) Mule deer. *Odocoileus hemionus*. In: Feldhamer GA, Thompson BC, Chapman JA (eds) Wild mammals of North America. Biology, management, and conservation. Johns Hopkins University Press, Baltimore, MD, pp 889–905

Main MB, Coblentz BE (1996) Sexual segregation in Rocky Mountain mule deer. J Wildl Manag 60:497–507

Main MB, Weckerly FW, Bleich VC (1996) Sexual segregation in ungulates: new directions for research. J Mammal 77(2):449–461

Marshal JP, Bleich VC, Krausman PR et al (2006) Factors affecting habitat use and distribution of desert mule deer in an arid environment. Wildl Soc Bull 34:609–619

Martínez-Muñoz A, Hewitt DG, Valenzuela S et al (2003) Habitat and population status of desert mule deer in Mexico. Z Jagdwiss 49:14–24

Montaña C (1988) Las formaciones vegetales. In: Montaña C (ed) Estudio integrado de los recursos vegetación, suelo y agua en la reserva de la biosfera de Mapimí. Instituto de Ecología A.C., Distrito Federal, pp 167–197

Myers WL, Foreyt WJ, Talcott PA et al (2015) Serologic, trace element, and fecal parasite survey of free-ranging, female mule deer (*Odocoileus hemionus*) in eastern Washington, USA. J Wildl Dis 51(1):125–136

Mysterud A, Pérez-Barbería FJ, Gordon IJ (2001) The effect of season, sex and feeding style on home range area versus body mass scaling in temperate ruminants. Oecologia 127:30–39

Nathan R, Getz WM, Revilla E et al (2008) A movement ecology paradigm for unifying organismal movement research. Proc Natl Acad Sci U S A 105:19052–19059

Núñez R, Miller B, Lindzey F (2000) Food habits of jaguars and pumas in Jalisco, Mexico. J Zool 252:373–379

Olson ME, Buret AG (2001) *Giardia* and giardiasis. In: Samuel MW, Pybus MJ, Kocan AA (eds) Parasitic diseases of wild mammals, 2nd edn. Iowa State University Press, Ames, pp 399–415

Ortega SJA (2008) Interacciones bovinos/fauna silvestre en pastizales. In: Memorias del XI Simposio sobre venados en México Ing. Jorge G. Villareal González. Departamento de Etología, Fauna Silvestre y Animales de Laboratorio de la Facultad de Medicina Veterinaria y Zootecnia, Universidad Nacional Autónoma de México/Asociación Nacional de Ganaderos Diversificados/Consejo Estatal de Flora y Fauna de Nuevo León, A.C., Distrito Federal (Mexico City), 28–30 May 2008

Pac DF, Mackie RJ, Jorgensen HE (1991) Muler deer population organization behavior and dynamics in a northern Rocky Mountain environment. Final Report, Project W-120-R-7-18, Montana Department of Fish Wildlife and Parks

Perez-Barberia FJ, Gordon IJ (2000) Differences in body mass and oral morphology between the sexes in the Artiodactyla: evolutionary relationships with sexual segregation. Evol Ecol Res 2(5):667–684

Perez-Barberia FJ, Gordon IJ, Pagel M (2002) The origins of sexual dimorphism in body size in ungulates. Evolution 56(6):1276–1285

Perez-Barberia FJ, Robertson E, Gordon IJ (2005) Are social factors sufficient to explain sexual segregation in ungulates? Anim Behav 69:827–834

Pérez-Solano LA, Gallina-Tessaro S, Sánchez-Rojas G (2016) Individual variation in mule deer (Odocoileus hemionus) habitat and home range in the Chihuahuan Desert, Mexico. J Mammal 97:1228–1237

Pérez-Solano LA, García-Feria LM, Gallina-Tessaro S (2017) Factors affecting the selection of and displacement within core areas by female mule deer (*Odocoileus hemionus*) in the Chihuahuan Desert, Mexico. Mammal Biol 87:152–159

Peripolli V, Prates ÊR, Jardim-Barcellos JO et al (2011) Fecal nitrogen to estimate intake and digestibility in grazing ruminants. Anim Feed Sci Technol 163:17–176

Powell RA (2000) Animal home ranges and territories and home range estimators. In: Boitani L, Fuller TK (eds) Research techniques in animal ecology: controversies and consequences. Columbia University Press, New York, pp 65–110

Putman R, Flueck WT (2011) Intraespecific variation in biology and ecology of deer: magnitude and causation. Anim Prod Sci 51:277–291

Pybus MJ (2001) Liver flukes. In: Samuel MW, Pybus MJ, Kocan AA (eds) Parasitic diseases of wild mammals, 2nd edn. Iowa State University Press, Ames, pp 121–149

Ramírez-Lozano RG (2004) Nutrición del venado cola blanca. Universidad Autónoma de Nuevo León, Unión Ganadera Regional de Nuevo León, Fundación Produce Nuevo León, Monterrey, Nuevo León, México

Righini N (2017) Recent advances in primate nutritional ecology. Am J Primatol 79(4):1–5

Robbins CT (1983) Wildlife feeding and nutrition. Academic, New York

Robinson HS, Wielgus RB, Gwilliam JC (2002) Cougar predation and population growth of sympatric mule deer and white-tailed deer. Can J Zool 80:556–568

Ruckstuhl KE, Neuhaus P (2002) Sexual segregation in ungulates: a comparative test of three hypotheses. Biol Rev 77(1):77–96

Ruckstuhl KE, Neuhaus P (2005) Sexual segregation in vertebrates: ecology of the two sexes. Cambridge University Press, Cambridge

Sánchez-Rojas G (2000) Conservación y manejo del venado bura en la Reserva de la Biosfera de Mapimí. Dissertation, Instituto de Ecología A.C.

Sánchez-Rojas G, Gallina S (2000a) Factors affecting habitat use by mule deer (*Odocoileus hemionus*) in the central part of the Chihuahuan Desert, Mexico: an assessment with univariate and multivariate methods. Ethol Ecol Evol 12:405–417

Sánchez-Rojas G, Gallina S (2000b) Mule deer (*Odocoileus hemionus*) density in a landscape element of the Chihuahuan Desert, Mexico. J Arid Environ 44:357–368

Sánchez-Rojas G, Gallina S, Equihua M (2004) Pellet morphometry as tool to distinguish age and sex in the mule deer. Zoo Biol 23:139–146

Sánchez-Rojas G, Gallina S (2006) La metapoblación del venado bura en la reserva de la biósfera Mapimí, México: consideraciones para su conservación. Cuadernos de Biodiversidad 22:7–15

Sánchez-Rojas G, Gallina S (2007) Metapoblaciones el reto en la biología de la conservación: El caso del venado bura en el Bolsón de Mapimí. In: Sánchez-Rojas G, Rojas-Martínez A (eds) Tópicos en Sistemática, Biogeografía, Ecología y Conservación de Mamíferos. Universidad Autónoma del Estado de Hidalgo, Mexico, pp 115–124

Sánchez-Rojas G, Gallina-Tessaro S (2016) *Odocoileus hemionus*. The IUCN Red List of Threatened Species 2016:e.T42393A22162113. https://doi.org/10.2305/IUCN.UK.2016-1.RLTS.T42393A22162113

Schwarm A, Ortmann S, Wolf C et al (2009) More efficient mastication allows increasing intake without compromising digestibility or necessitating a larger gut: comparative feeding trials in banteng (*Bos javanicus*) and pygmy hippopotamus (*Hexaprotodon liberiensis*). Comp Biochem Physiol A Mol Integr Physiol 152:504–512

Shields AV, Larsen RT, Whiting JC (2012) Summer watering patterns of mule deer in the Great Basin Desert, USA: implications of differential use by individuals and the sexes for management of water resources. Sci World J 12:1–9

Sinclair ARE, Fryxell JM, Caughley G (2006) Wildlife ecology, conservation, and management. Blackwell Publishing, Hoboken, NJ

Siuta A, Bobek B (2006) Comparison of red deer stomachs in relation to different foraging habitats. Med Weter 62(1):32–35

Stubblefield SS, Pence DB, Warren RJ (1987) Visceral helminth communities of sympatric mule and white-tailed deer from the Davis Mountains of Texas. J Wildl Dis 23:113–120

Van Soest P (1996) Allometry and ecology of feeding behavior and digestive capacity in herbivores: a review. Zoo Biol 15:455–479

Verheyden H, Aubry L, Merlet J et al (2011) Faecal nitrogen, an index of diet quality in roe deer *Capreolus capreolus*? Wildl Biol 17(2):166–176

Wade PD, Mcdonald BK (2010) Distribution of the mule deer (*Odocoileus hemionus*) in Oklahoma: an analysis of harvest data. Proc Okla Acad Sci 90:111–116

Wallmo OC (1981) Mule and black-tailed deer distribution and habitats. In: Wallmo OC (ed) Mule and black-tailed deer of North America. University of Nebraska Press, Lincoln, pp 366–386

Walters WR, Palmer MV (2015) *Mycobacterium bovis* infection of cattle and white-tailed deer: translational research of relevance to human tuberculosis. ILAR J 56:26–43

Wearmouth VJ, Sims DW (2008) Sexual segregation in marine fish, reptiles, birds and mammals: behaviour patterns, mechanisms and conservation implications. Adv Mar Biol 54:107–170

Wehausen JD (1995) Fecal measures of diet quality in wild and domestic ruminants. J Wildl Manag 59:816–823

Workman GW, Low JB (1976) Mule deer decline in the west: a symposium. Utah State University and Utah Agricultural Experiment Station, Logan, UT

Xiangfei Y, Tonghui Z, Xueyong Z et al (2016) Effects of rainfall patterns on annual plants in Horqin Sandy Land, Inner Mongolia of China. J Arid Land 8(3):389–398

Ydenberg RC, Brown JS, Stephens DW (2007) Foraging: an overview. In: Stephens DW, Brown JS, Ydenberg RC (eds) Foraging: behavior and ecology. University of Chicago Press, Chicago, pp 1–28

Chapter 4
Distribution and Abundance of White-Tailed Deer at Regional and Landscape Scales at Tehuacán-Cuicatlán Biosphere Reserve, Mexico

Salvador Mandujano, Odalis Morteo-Montiel, Carlos Yáñez-Arenas, Michelle Ramos-Robles, Ariana Barrera-Salazar, Eva López-Tello, Pablo Ramirez-Barajas, Concepción López-Téllez, and Adriana Sandoval-Comte

Abstract In México, the white-tailed *Odocoileus virginianus* inhabits a wide type of habitats including the tropical dry forest. In particular, in the Tehuacán-Cuicatlán Biosphere Reserve (TCBR) this species has a wide distribution and it is important for hunting subsistence. In this chapter we analyzed the distribution and abundance of this cervid at regional and landscape scales in the TCBR and buffer zone (ca. 700,000 ha). The subspecies best represented in the reserve would be *O. v. mexicanus*; *O. v. oaxacensis* and *O. v. toltecus* but it is need a specific genetics studies on this topic. Using ecological niche and habitat suitability modeling, it is suggested that 92% of the surface of this site presents favorable conditions for the presence of this deer. Mean population density estimation is 2.3 deer/km^2, which is lower compared to other regions of the country dominated by tropical dry forests. However, considering the large surface of habitat, the total abundance it is significant. White-tailed deer mortality it is affect by natural predators as *Felis concolor*, *Canis latrans* and *Lynx rufus,* and illegal hunters and feral dogs. In TCBR, this deer species coexist with *Pecari tajacu* and *Mazama temama*, and with domestic such as *Capra*

The original version of this chapter was revised. The correction to this chapter is available at https://doi.org/10.1007/978-3-030-28868-6_19

S. Mandujano (✉) · O. Morteo-Montiel · C. Yáñez-Arenas · M. Ramos-Robles
P. Ramirez-Barajas · A. Sandoval-Comte
Red de Biología y Conservación de Vertebrados, Instituto de Ecología A.C., Xalapa, Veracruz, Mexico
e-mail: salvador.mandujano@inecol.mx

A. Barrera-Salazar · C. López-Téllez
Facultad de Biología, Benemérita Universidad Autónoma de Puebla, Puebla, Mexico

E. López-Tello
Instituto de Neuroetología, Universidad Veracruzana, Xalapa, Veracruz, Mexico

S. Gallina-Tessaro (ed.), *Ecology and Conservation of Tropical Ungulates in Latin America*, https://doi.org/10.1007/978-3-030-28868-6_4

hircus, *Bos taurus* and *Equus*. This species it is not in danger situation in the TCBR and could be sustainably used in management wildlife units (UMAs), but it is urgent to diminish the illegal hunt.

4.1 Introduction

In Mexico, the white-tailed deer *Odocoileus virginianus* (Zimmermann, 1780) is found throughout the territory except the Baja California peninsula, part of northern Sonora, and areas of the Chihuahuan desert (Mandujano et al. 2014). Its high reproductive, behavioral, and ecological plasticity, together with the modification of the habitat by human activities, are factors that have allowed this species to expand its geographic range. Consequently, this deer lives in a great variety of types of plant communities. For example, in Mexico it inhabits temperate forests of pine-oak, xerophilous shrubs, tropical forests such as dry and rainy forests (Ortega-Santos et al. 2014; Mandujano et al. 2014).

The white-tailed deer is a highly prized species throughout México to supplement the consumption of animal protein, for trade, craftsmanship, and recreation, and has been part of the cosmogony and rites of various indigenous cultures (Mandujano and Rico-Gray 1991; Montiel-Ortega et al. 1999; González-Marín et al. 2003; Naranjo et al. 2004). Also, the white-tailed deer is one of the main species managed mainly in the north of the country, where it has been demonstrated that it can be a form of economically profitable exploitation (Villarreal 1999), and with wide opportunities in the tropical zones of the center and southeast of the country (Villarreal-Espino 2006; Mandujano 2016). This species of deer is one of those used in the Units for Conservation, Management and Use Sustainability of Wildlife (UMA) (Ortega-Santos et al. 2014), and studied in Natural Protected Areas (Gallina et al. 2007). The white-tailed deer is one of the most studied species in the world, mainly in the United States and Canada (Gallina et al. 2010; Hewitt 2011). In contrast, in Latin America this deer species has been less studied, despite its ecological, economic and cultural importance (Weber and Gonzalez 2003). Of the four deer species that inhabit Mexico, the white-tailed deer is the most studied (Mandujano 2004).

In the Tehuacán-Cuicatlán Biosphere Reserve inhabited wild ungulates such as the white-tailed deer, collared peccary *Pecari tajacu* and red brocket deer *Mazama temama,* and domestic ungulates such as goats (*Capra hircus)*, cattle (*Bos taurus*), and horses and donkeys (*Equus* spp.) (Fig. 4.1; Ortiz-García and Mandujano 2011; Ortíz-García et al. 2012; Pérez-Solano et al. 2012; Yañez-Arenas et al. 2012, 2014; Pérez-Solano and Mandujano 2013; Ramos-Robles et al. 2013; Barrera-Salazar et al. 2015; Yañez-Arenas and Mandujano 2015; Mandujano et al. 2016a, b; Pérez-Solano et al. 2016; Pérez-Solano and Mandujano 2018). The objective of this chapter is to integrate data of different topics to analyze the distribution and abundance of white-tailed deer in the Tehuacán-Cuicatlán Biosphere Reserve.

Fig. 4.1 Wild and domestic ungulates inhabiting in the Tehuacán-Cuicatlán Biosphere Reserve (TCBR). From the upper left to the lower right image: white-tailed deer, collared peccary, red brocket deer, goats, donkeys, and cattle

4.2 Regional Scale Analysis

4.2.1 Study Region

The Tehuacán-Cuicatlán Biosphere Reserve (TCBR) is part of the Sierra Madre del Sur and occupies the northwestern zone of the sub-province of the Meseta of Oaxaca, within the region of La Cañada; it is located in the extreme southeast of the state of Puebla and northeastern Oaxaca, Mexico, between 17 °39′ to 18° 53′ N and 96° 55′ to 97°44′ W (Fig. 4.2). The area of this reserve is 490,187 ha, and its altitude varies from 600 to 2950 m above sea level; average annual temperature varies between 18 and 24.5 °C (Mandujano et al. 2016a). The annual average of precipitation in the valley region varies from 250 to 500 mm, and occurs mainly from May to October,

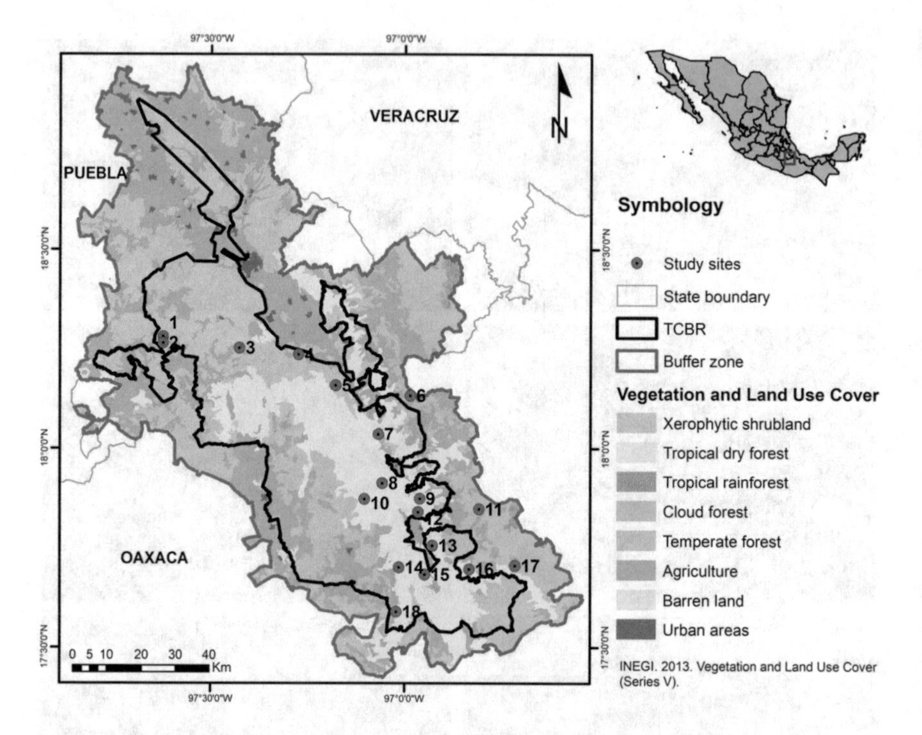

Fig. 4.2 Geographical location of the Tehuacán-Cuicatlán Biosphere Reserve in the states of Puebla and Oaxaca, Mexico. The main types of vegetation are presented according to the National Forest Inventory Series III). The numbers correspond to the sampled localities where estimates of deer density were obtained (see Table 4.1)

with greater rainfall between June and September. The Tehuacán-Cuicatlán region contains about 10% of the flora of Mexico and more than 400 species of vertebrates (Dávila et al. 2002). However, its diversity and floristic endemism have monopolized most research efforts, leaving aside knowledge of animal populations and/or communities such as mammals. The main types of vegetation and land use in the region that includes part of the National Protected Areas and the priority terrestrial region of the Tehuacán-Cuicatlán Valley are (Fig. 4.3): the tropical dry forest (29% of the reserve's territory); land dedicated to agriculture, raising livestock and forest exploitation (22%); the forest of oak and pine (21%); the desert scrub with predominance of thorny shrubs and an important presence of cacti (10%); the crassicaule scrub with vegetation dominated by large cacti (8%); and other types of vegetation (10%) (Dávila et al. 2002).

4.2.2 Subspecies Distribution

In Mexico inhabits 14 subspecies of white-tailed deer (Mandujano et al. 2010). Three maps of the species distribution were obtained for the TCBR (Fig. 4.4). According to Kellogg (1956), *O. v. oaxacensis* is the principal subspecies inhabited the TCBR;

Fig. 4.3 Example of the main types of vegetation in the TCBR: Xerophilous scrub, tropical dry forest, and temperate forest

according to Hall (1981) it would be *O. v. toltecus*; while according to Villarreal (1999), the subspecies best represented in the reserve would be *O. v. mexicanus*; *O. v. oaxacensis* and *O. v. toltecus* would be in the southeast part and in the Sierra de Juarez, respectively. Based on these data, there is no certainty of the identity and geographical limits of the subspecies that inhabit this reserve. Studies in Mexico show that morphometric (Logan-López et al. 2006) and genetic data (de la Rosa-Reyna et al. 2012) allow to determine the genetic diversity and differentiation of white-tailed deer subspecies. In the TCBR it is important to define the geographical limits of possible subspecies or variations in morphotypes, because it is relevant both in terms of conservation of the genetic diversity of the species, as well as its possible management in UMA. In this sense, although there are no geographic barriers that could geographically separate different populations, the altitudinal and longitudinal differences of this reserve originate a gradient with climatic, topographic, vegetation types and other conditions. The white-tailed deer is a very plastic species and adaptable to different environmental conditions of habitat, so variations in size, color, and antler shapes could be expected in this environmental gradient. Some data on male antler measures in the region show a lot of variation (Villarreal-Espino 2006). Genetic data obtained in different parts of the TCBR will allow better elucidate this problem. At the level of management in UMA, the topic is relevant because if hunting permits are granted at the species level, the management plans require that the name of the white-tailed deer subspecies be entered.

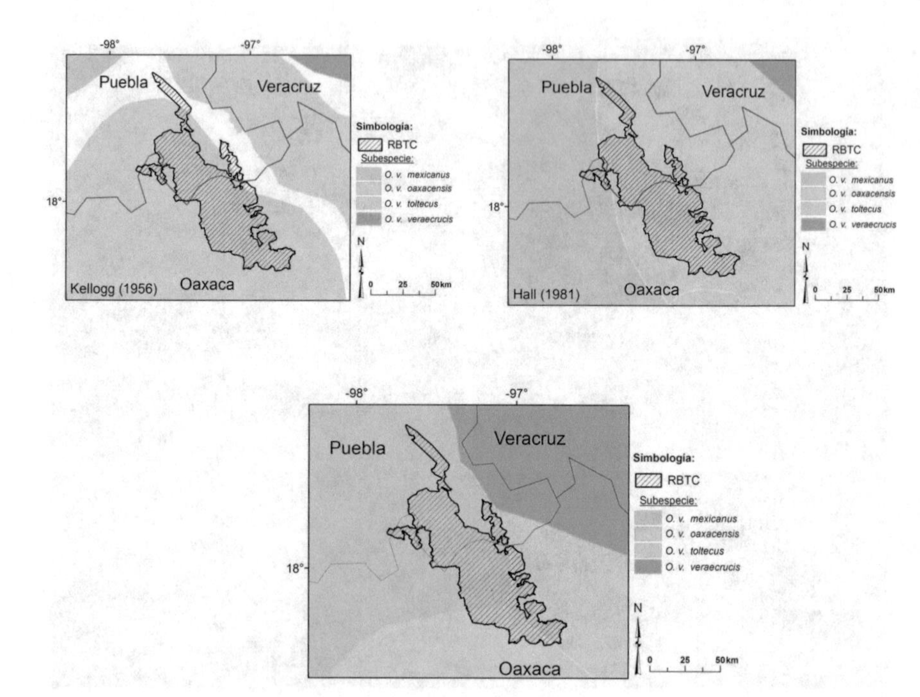

Fig. 4.4 Geographical distribution maps of white-tailed deer subspecies according to Kellogg (1956), Hall (1981) and Villarreal (1999)

4.2.3 Potential Distribution and Abundance

Using ecological niche modeling, it is suggested that 92% of the surface of this site presents favorable conditions for the presence of this deer (Fig. 4.5). The most important variables associated with this distribution are: the distance to human locations, the slope of the land, the total annual precipitation, the maximum temperature of the warmest month and the isothermally (Ortíz-García et al. 2012). On the other hand, the sites where the white-tailed deer finds suitable conditions of habitat as food in the quantity and quality adequate to cover their nutritional needs for the maintenance, growth and reproduction of the individuals. In addition, adequate coverage which allows them to protect themselves from extreme environmental conditions, such as sunstroke; it also allows them to hide from the presence of natural predators in the area such as the puma *Felis concolor*, coyote *Canis latrans*, bobcat *Lynx rufus*, and feral dogs, and illegal hunters (Fig. 4.6). Additionally, in these places the deer find rivers, streams and other permanent sources of water to meet their needs; and other special requirements such as salt sites, where they can supplement their requirements for some specific nutrients. A notable aspect of the TCBR is that although local densities tend to be low (see Table 4.1), the amount of

Table 4.1 Population density estimation (mean ± SD) of white-tailed deer in different locations in the Tehuacán-Cuicatlán Biosphere Reserve

Key	Site	Density (ind/km^2)
1	San Juan Raya	3.0 (0.5)
2	San Sebastián Frontera	0.8 (0.3)
3	San Francisco Xochiltepec	0.7 (0.3)
4	San José Miahuatlán	2.9 (0.9)
5	San Gabriel Casa Blanca	3.7 (0.8)
6	Zaragoza	0.8 (0.2)
7	San Juan de los Cués	2.0 (0.3)
8	Santa María Tecomavaca	1.1 (0.3)
9	Santiago Quiotepec	0.3 (0.1)
10	Santa María Ixcatlán	5.0 (1.6)
11	Concepción Pápalo	0.4 (0.2)
12	San Juan Bautista Cuicatlán	3.1 (0.8)
13	San Pedro Chicozapotes	0.7 (0.4)
14	San Pedro Jaltepetongo	3.7 (1.0)
15	San José del Chilar	2.9 (0.5)
16	San Juan Tepeuxila	3.1 (0.9)
17	San Juan Teponaxtla	3.8 (0.7)
18	Santa María Almoloyas	3.9 (0.8)

Key represents the geographic location (see Fig. 4.2)

continuous habitat is very extensive and in good condition (Yañez-Arenas et al. 2012, 2014). Predicted density categories (low, medium, and high) was produced for the TCBR (Fig. 4.5). These data clearly suggest that the TCBR can potentially maintain an important population of white-tailed deer.

4.2.4 Habitat Suitability

Models of habitat suitability index (HSI) have been used in the management of various species of wildlife among which is the white-tailed deer (Delfín-Alfonso et al. 2009). Maps produced through this process can identify areas at different suitability levels; they have been utilized in many studies to provide a synoptic view of habitat suitability for specific species as well as assess suitability habitats for species assemblages. The habitat model developed in this study considers five environmental variables (vegetation type, mean annual temperature, slope, aspect, and water sources) and two variables of human pressure (proximity to human villages and dirt roads). The scale resolution of the HIS was 1 ha. HSI results indicate that of the total area, 15.2%, 37.6% and 47.1% was classified as low, medium and higher suitability habitat, respectively. However, when considering human activities, the high quality habitat classification decrease 47%. The HSI identified areas of high quality for

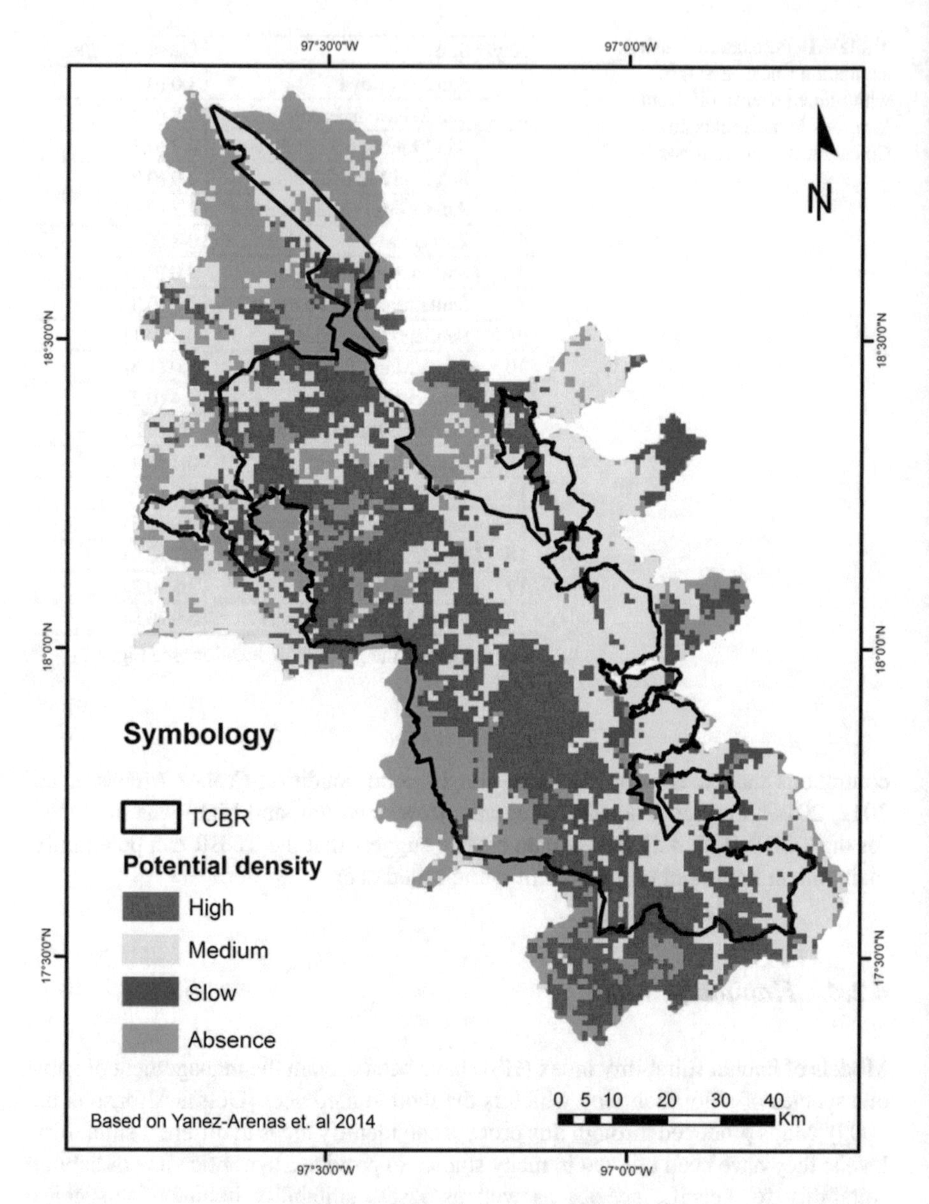

Fig. 4.5 Map of the potential distribution and density of white-tailed deer in the reserve and surrounding areas using ecological niche modeling

conservation and management opportunities (Fig. 4.7). These results suggest applying this model HIS with variations in other UMA and sites with similar ecological conditions as the TCBR.

Fig. 4.6 Main predatory species of white-tailed deer: cougar, coyote, bobcat and dogs

4.3 Landscape Scale Analysis

4.3.1 Population Densities

Wildlife management requires reliable biological information to make proper decisions that affect populations and habitats using data of distribution, abundance, and habitat resources availability/use. Based on the count of fecal groups in 136 transects and using the PELLET program (Mandujano 2014), the estimates of the density of white-tailed deer varied between 0.2 to 4.2 ind/km^2 depending on the locality (Table 4.1). The general average for the RBTC is estimated at 2.3 deer/km^2, which is lower compared to other regions of the country dominated by tropical dry forests.

4.3.2 Spatial Explicit Autocorrelation

Spatial autocorrelation of an ecological response occurs when nearby locations in general have more similar values than distant locations, due to the relationship between distance and biological processes such as speciation, extinction, dispersion,

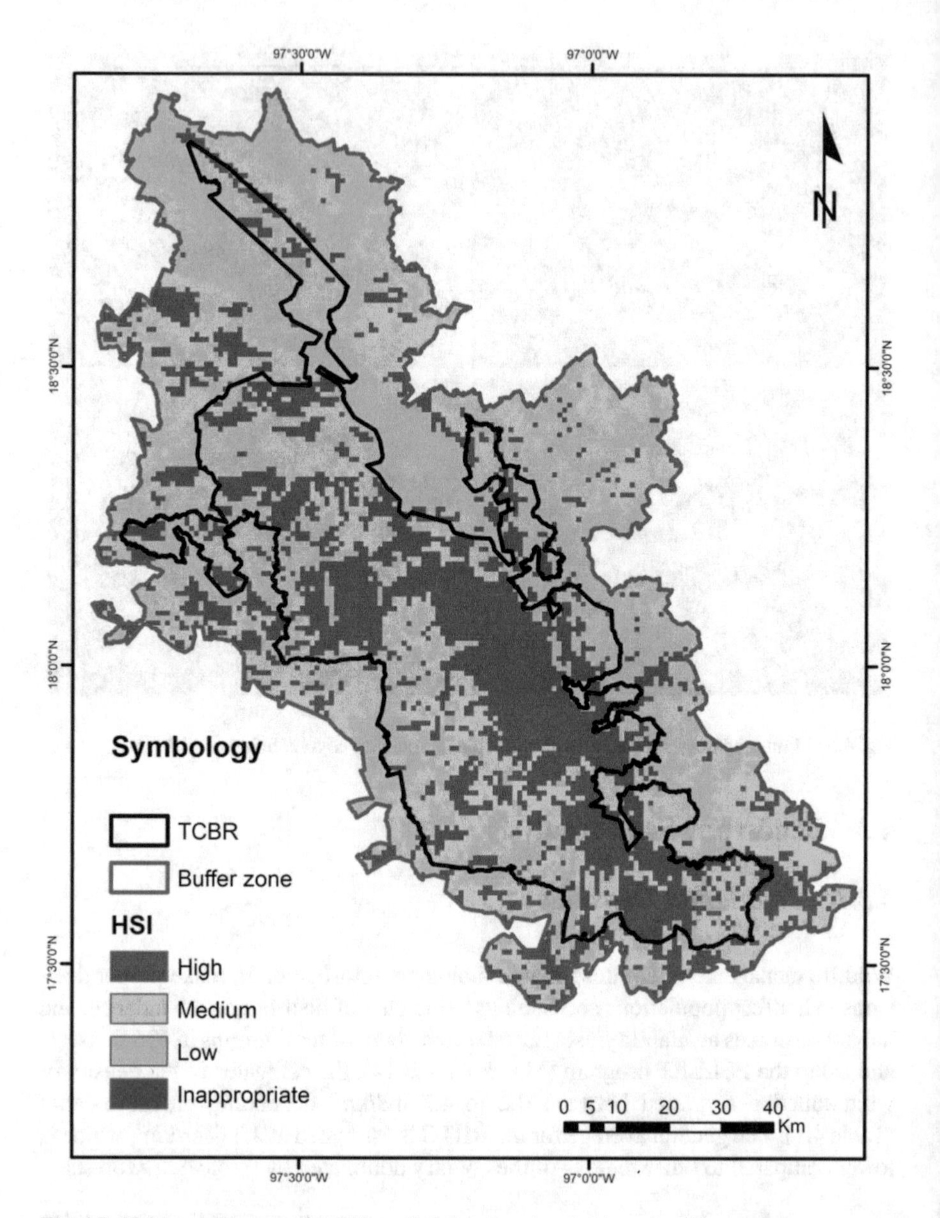

Fig. 4.7 Habitat quality map for white-tailed deer based on habitat suitability modeling

or species interactions (Yañez-Arenas and Mandujano 2015). According to the spatially explicit regression methods, only two variables were related with white-tailed deer: annual mean temperature and seasonality in precipitation (Fig. 4.8). Analysis of the residuals through correlograms in the ordinary regression models consistently presented a positive spatial autocorrelation at short distances. This could mainly be due to the absence in the analysis of spatially structured explicative variables that reflect biological processes in the deer populations, such as dispersion movements

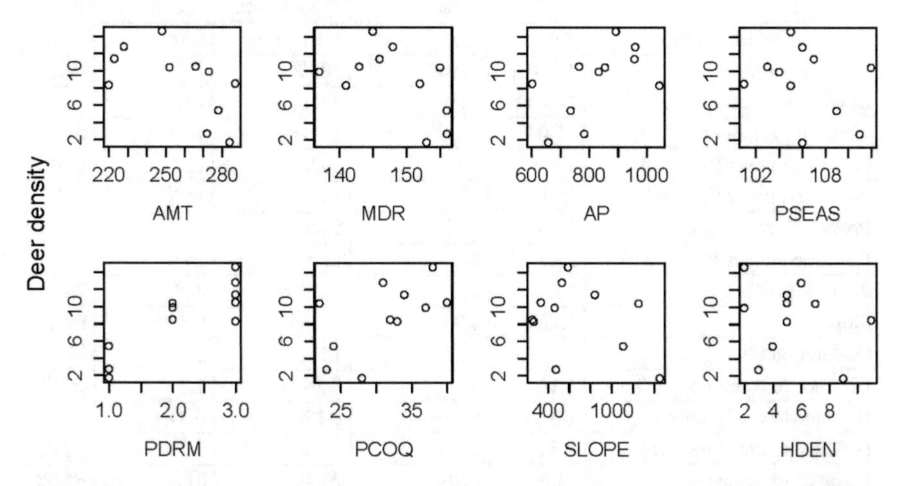

Fig. 4.8 Relationship between white-tailed deer density (ind/km^2) and environmental variables in TCBR. Abbreviations: *AMT* annual mean temperature, *MDR* mean diurnal range, *AP* annual precipitation, *PSEAS* precipitation seasonality, *PDRM* precipitation in the driest month of the year, *PCOQ* precipitation in the coldest quarter of the year, *SLOPE* slope, and *HDEN* human population density

of individuals, interaction with other species (predators and food), anthropogenic effects, and other demographic factors. The spatial explicit models eliminated, or considerably diminished, the effects of the spatial autocorrelation.

4.3.3 Deer-Habitat Relationships

Regression analysis showed that the habitat variables that had a negative relationship with density were basal area and cattle presence (Ramos-Robles et al. 2013). On the other hand, distance from roads, and from the town with the highest population density, were both positively related to deer density (Table 4.2). Ordination of transects based on habitat and human attributes defined three distinct groups: Casa Blanca-Tecomavaca, Chilar, and Quiotepec (Fig. 4.3). The first component was positively associated with tree density, aspect, distance to rivers and presence of livestock, and negatively associated with the distance from the largest human community. Meanwhile, the second component was associated positively with cover variables, and negatively with distance to the nearest road. In the second ordination of transects according to tree composition, it was found that Quiotepec and Chilar had similar plant composition; in contrast, Tecomavaca and Casa Blanca showed different plant associations. In sites with higher white-tailed deer density, the main plant species were *Lantana camara*, *Vallesia glabra*, *Bunchosia biocellata*, *Amphipterygiuum adstringens*, and *Bursera aptera*, while sites with lower deer density were associated with *Neobuxbaumia* sp., *Bursera schlectendalii*, and *Acacia* sp. Our results suggest that the studied locations have different habitat conditions for white-tailed deer. In general, Casa Blanca and Tecomavaca were more similar,

Table 4.2 Habitat description of four sites in the study area

Variables	Chilar	Casa Blanca	Tecomavaca	Quiotepec	P
Deer density (deer/km^2)	2.9a	2.6a	1.1b	0.1c	<0.05
Basal area (m^2)	173.6 a	219 a	188.8 a	299 b	0.02
Density of trees (ind/100 m^2)	18.4 ab	14 a	3.8 c	21.8 b	<0.05
Understory richness	15.4 b	12 a	17.3 bc	14.9 abc	0.01
Protection cover 0–50 (%)	7.8	9.8	10.9	11.1	0.97
Protection cover 51–100 (%)	15.1	11.4	19.3	15.7	0.76
Protection cover 101–150 (%)	19.9	17.5	21.2	18.7	0.96
Protection cover 151–200 (%)	29.7	24	28.4	26.0	0.93
Altitude (msl)	937.7 b	879.5 ab	825.7 a	849.9 a	0.03
Slope (°)	10.2 b	4.7 a	8.8 ab	8.6 b	0.04
Orientation (°)	98.8 a	68.5 a	124 a	271 b	<0.05
Distance to water bodies (km)	1.4	0.6	0.3	1.2	<0.05
Distance to roads (km)	1.5 b	6.5 a	1.1 b	0.3 c	<0.05
Distance to locations (km)	2.5 b	6.8 a	5.9 a	3.7 c	<0.05
Distance to human (km)	51.3	13.6	24.4	31	<0.05
Livestock (feces per transect)	1.1 a	0 a	1 a	5 b	<0.05
Number of plant families	27	19	31	34	
Number of plant species	123	119	138	96	
Lantana camara		103.4			
Bursera sp.[a]		57.7			
Bursera linanoe[a]		48.1			
Agonandra sp.		47.8			
Randia thurberi[a]		40.3			
Neobuxbaumia tetetzo				554.0	
Bursera schlectendalii[a]				313.7	
Mimosa sp.				311.1	
Parkinsonia proecox				207.4	
Bursera sp.[a]				147.6	
Vallesia glabra			52.0		
Acacia pringlei			43.8		
Bursera sp.[a]			39.7		
			31.8		
Parkinsonia proecox			28.8		
Randia thurberi[a]	40.1				
Bunchosia biocellata	39.1				
Amphipterygium adstringens	37.8				
Bursera aptera[a]	35.8				
Lippia alba	32.4				
Bursera sp.[a]	32.4				
Opuntia sp.[a]	30.6				
Cyrtocarpa procera	30.3				
Asclepias curassavica	30.2				
Stenocereus stellatus					

Values with different letters are statistically different by ANOVA test. The most important plant species are presented with regard to their relative importance value for each location.

[a]Lists of the plants that are eaten by white-tailed deer during the dry season (according Vasquez et al. 2016)

[b]Values with different letters are statistically different by ANOVA test

while Chilar and Quiotepec were the most contrasting locations. However, this ordination of locations was not consistent with deer densities. Significantly higher densities were found in sites with more suitable habitat conditions, while lower densities were found in sites with a higher incidence of human activity. In addition, the multivariate analysis showed that even if some habitat and human variables were associated with deer abundance, the explained variance was low, suggesting that other factors we did not measure could be affecting deer density. Similarly, some of the habitat characteristics, such as vegetation structure and land and human pressure, were significantly different among locations, which could influence the variance in density values among them.

4.3.4 Site Probability Occupation

The optimum models with minor AIC value in single-season models, considered habitat covariate as the principal variable to explain occupancy by white-tailed deer (Table 4.3). During the wet season the occupancy was higher in Scrub-TDF than TDF, and during the dry season: Scrub-TDF and TDF. The detection probability

Table 4.3 Summary of single-season occupancy (psi) and probability of detection (P) models in two sites (Casa Blanca and Chicozapotes) in TCBR

Season/covariables	AIC	ΔAIC	AICw	ML	#Par	−2 × LL
Wet season						
psi(habitat), P(•)	165.4	0	0.92	1	2	161.4
psi(•), P(•)	170.8	5.4	0.06	0.07	2	166.8
psi(habitat), P(survey)	175.2	9.7	0.01	0.01	13	149.2
psi(habitat), P(survey + habitat)	175.7	10.3	0.005	0.01	14	147.7
psi(•), P(survey + habitat)	176.3	10.9	0.004	0.004	14	148.3
psi(•), P(survey)	180.6	15.2	0.001	0.001	13	154.6
psi(habitat), P(habitat)	190.4	24.9	0	0	2	186.4
psi(•), P(habitat)	197.0	31.6	0	0	2	193.0
Dry season						
psi(habitat), P(survey)	116.3	0	0.37	1	13	90.3
psi(habitat), P(•)	117.1	0.8	0.25	0.68	2	113.1
psi(habitat), P(survey + habitat)	118.1	1.7	0.16	0.42	14	90.1
psi(•), P(survey)	119.2	2.9	0.09	0.24	13	93.2
psi(•), P(survey + habitat)	119.6	3.2	0.07	0.20	14	91.6
psi(•), P(•)	120.1	3.8	0.06	0.15	2	116.1
psi(habitat), P(habitat)	132.9	16.6	0.0001	0.0003	2	128.9
psi(•), P(habitat)	136.2	19.9	0	0	2	132.2

Constant (•), *AIC* Akaike information criterion, *ΔAIC* relative difference between the AIC of the second model and the top-ranked model, *AICw* weight of the model, a measure of the support of each model, *ML* (*Model-likelihood*) ratio between the weight of each model and the top ranked model, *#Par* the number of parameters estimated for each model, *−2 × LL* −2LogLikelihood estimated was considered constant in wet season, while in dry season was influenced by survey. Our results show of both single-season and multiple-season models suggest that occupancy was higher in the Scrub-TDF than TDF habitats and did not differ between seasons. Previously, Ramos-Robles et al. (2013) estimated the population densities of white-tailed deer in the study region and found them low (< 6 ind/km^2). These results are according to single-season models where covariates habitat have most influence in the first-ranked models. Detection probability values of the white-tailed deer were relatively low, over the entire area of study, while occupancy values were relatively mid and high. Given the design of the present study, it was expected that deer detection probability would vary temporally and between habitats (MacKenzie et al. 2006). Our results also highlight the importance of considering the detection probability in order to estimate occupancy, to correct interpretation of spatial and temporal changes in abundance.

4.4 Implications for Conservation and Sustainable Use

White-tailed deer represents an important resource for the human communities of TCBR (Fig. 4.9). Subsistence hunting of white-tailed deer is a practice that has been carried out in the region for a long time ago. The number of deer and their contribution in kilos of meat per year in the region or in some local community has not been quantified. It is known that in some parts of the country, this type of hunting and this deer provides significant protein in communities. In addition, this species is one of the most appreciated by sports hunters throughout the country. Unfortunately, in the TCBR hunters of different hunting clubs and/or individually, hunt in an irregular and uncontrolled manner, being in some cases an activity that reduces populations of white-tailed deer. Consequently, the execution of projects on conservation and use of white-tailed deer represents a viable economic, social, ecological and cultural option for the management and sustainable use of its natural resources and its rational use (Mandujano 2007; Mandujano et al. 2010; Escalante and Martínez-Meyer 2013; López-Téllez et al. 2016). This could allow generating sources of employment, thus avoiding migration and enabling new generations to take an interest in caring for their natural resources.

In general, the trade in meat and deer derivatives are prohibited by law, if this is done outside the UMAs that are currently the only legal way to be able to use the deer and other resources, under a previous series of requirements. That is, if a site is not registered as an UMA, any hunting activity technically and legally could be considered illegal, although the subsistence hunting scheme allows for some considerations. However, defining whether the current hunt is subsistence or not is complicated. Besides this hunting, in the community is done in a similar way to other localities in the region, sporadically they face the problem of sports hunters who arrive at their lands without any authorization and pertinent permission. Considering the limitations of UMA as an independent unit of management and their small size in those located in the center and southeast of the country, Mandujano and González-Zamora (2009) suggest a network of regional systems where small UMA are connected to

Fig. 4.9 Monitoring program and workshops with local people as a basis for the conservation and sustainable use of white-tailed deer in the TCBR. Upper right photo show the first legally hunted white-tailed deer trophy male in the Casa Blanca Wildlife Conservation and Sustainable Management Unit (UMA)

large ANP. These systems at the landscape level could allow deer dispersal movements and maintain viable minimum populations. To create this system of networks for conservation and management at the regional or landscape level, two recent ecological models can be applied: archipelago-type reserves and metapopulation models of the source-sink type (Naranjo and Bodmer 2007).

Acknowledgements This chapter was benefitted from the economic support of CONACYT by the projects: CB-2009-01-130702 and CB-2015-01-256549. We thanks the Red de Biología y Conservación de Vertebrados of Instituto de Ecología, A. C., and many people who assisted during the field and laboratory work.

References

Barrera-Salazar A, Mandujano S, Villarreal Espino-Barros OA, Jiménez-García D (2015) Classification of vegetation types in the habitat of white-tailed deer in a location of the Tehuacán-Cuicatlán Biosphere Reserve, Mexico. Trop Conserv Sci 8:547–563

Dávila P, Arizmendi MC, Valiente-Banuet A, Villaseñor JL, Casas A, Lira R (2002) Biological diversity in the Tehuacán-Cuicatlán Valley. Mexico Biodivers Conserv 11:421–442

De La Rosa-Reyna XF, Calderón RD, Parra GM, Sifuentes AM, Deyoung RW, García de León FJ, Arellano W (2012) Genetic diversity and structure among subspecies of white-tailed deer in Mexico. J Mammal 93:1158–1168

Delfín-Alfonso C, Gallina S, López–González CA (2009) Evaluación del hábitat del venado cola blanca utilizando modelos espaciales y sus implicaciones para el manejo en el centro de Veracruz, México. Trop Conserv Sci 2:215–228

Escalante T, Martínez-Meyer E (2013) Ecological niche modelling and wildlife management units (UMAs): an application to deer in Campeche, Mexico. Trop Subtrop Agroecosyst 16:183–191

Gallina S, Mandujano S, Delfín-Alfonso C (2007) Importancia de las Áreas Naturales Protegidas para conservar y generar conocimiento biológico de las especies de venados en México. In: Halffter G, Guevara S, Melic A (eds), Hacia una cultura de conservación de la diversidad, Monografías del 3er. Milenio Vol. 6. Sociedad Entomológica Aragonesa, Comisión Nacional para el uso y conocimiento de la biodiversidad, Comisión Nacional de Áreas Naturales Protegidas, Instituto de Ecología AC y MAB-UNESCO

Gallina G, Mandujano S, Bello-Gutiérrez J, López-Fernández H, Weber M (2010) White-tailed deer *Odocoileus virginianus* (Zimmermann, 1780). In: Duarte JMB, González S (eds) Neotropical cervidology: biology and medicine of Latin American deer. Jaboticabal, Brazil: Funep and Gland, Switzerland: IUCN, pp 101–118

González-Marín RM, Montes E, Santos J (2003) Caracterización de las unidades de manejo para la conservación, manejo y aprovechamiento sustentable de la fauna silvestre en Yucatán, México. Trop Subtrop Agroecosyst 2:13–21

Hall ER (1981) The mammals of North America. Wiley, New York

Hewitt D (2011) Biology and management of white-tailed deer. CRC Press/Taylor & Francis Group, Boca Raton, FL

Kellogg R (1956) What and where are the whitetails? In: Taylor WP (ed) The deer of North America. The Stackpole Company, Harrisburg

Logan-López K, Cienfuegos-Rivas E, Sánchez FC, Mendoza G, Sifuentes AM, Tarango-Arámbula LA (2006) Caracterización morfométrica de cuatro subespecies de venado cola blanca (*Odocoileus virginianus*) en la zona noreste de México. Revista Científica XVI(1):14–22, Universidad del Zulia, Maracaibo, Venezuela

López-Téllez C, Barrera-Salazar A, Ramírez-Vera B, Chávez-Herrera S, Mandujano S, Salazar-Torres JM (2016) Primera experiencia de la cacería deportiva del venado cola blanca en UMA extensiva en la RBTC. In: Mandujano S (ed) Venado cola blanca en Oaxaca: potencial, conservación, manejo y monitoreo. Instituto de Ecología AC y Comisión Nacional para el Conocimiento de la Biodiversidad, Xalapa, Veracruz

MacKenzie DI, Nichols JD, Royle JA, Pollock KH, Bailey LL, Hines JE (2006) Occupancy estimation and modeling. Academic Press, Burlington, MA

Mandujano S (2004) Análisis bibliográfico de los estudios de venados en México. Acta Zool Mex (ns) 20:211–251

Mandujano S (2007) Carrying capacity and potential production of ungulates for human use in a Mexican tropical dry forest. Biotropica 39:519–524

Mandujano S (2014) PELLET: an Excel®-based procedure for estimating deer population density using the pellet-group counting method. Trop Conserv Sci 7:308–325

Mandujano S (2016) Venado cola blanca en Oaxaca: potencial, conservación, manejo y monitoreo. Instituto de Ecología AC y Comisión Nacional para el Conocimiento de la Biodiversidad, Xalapa, Veracruz

Mandujano S, González-Zamora A (2009) Evaluation of natural conservation areas, and wildlife management units to support minimum viable populations of white-tailed deer in Mexico. Trop Conserv Sci 2:237–250

Mandujano S, Rico-Gray V (1991) Hunting, use, and knowledge of the biology of the white-tailed deer, *Odocoileus virginianus* (Hays), by the maya of central Yucatan, Mexico. J Ethnobiol 11:175–183

Mandujano S, Delfín-Alfonso CA, Gallina S (2010) Comparison of geographic distribution models of white-tailed deer *Odocoileus virginianus* (Zimmermann, 1780) subspecies in Mexico: biological and management implications. Therya 1:41–68

Mandujano S, Gallina S, Ortega-Santos JA (2014) Venado cola blanca en México. In: Valdez R, Ortega-Santos JA (eds) Ecología y manejo de fauna silvestre en México. Biblioteca básica de agricultura, Colegio de Postgraduados, Universidad Autónoma de Chapingo, Guadalajara, Jalisco

Mandujano S, Barrera-Salazar A, Yañez-Arenas CA, Ramos-Robles MI, López-Tello E, Gallina S, Villarreal-Espino Barros OA, Vergara-Castrejón A, Morteo-Montiel O (2016a) Venado cola blanca en la región de la Cañada de la Reserva de Biosfera Tehuacán-Cuicatlán de Oaxaca. In: Mandujano S (ed) Venado cola blanca en Oaxaca: potencial, conservación, manejo y monitoreo. Instituto de Ecología AC y Comisión Nacional para el Conocimiento de la Biodiversidad, Xalapa, Veracruz

Mandujano S, López-Téllez C, Barrera-Salazar A, Romero-Castañón S, Ramírez-Vera B, López-Tello E, Yañez-Arenas CA, Castillo-Correo JC (2016b) UMA extensiva de venado cola blanca en San Gabriel Casa Blanca, Reserva de Biosfera Tehuacán-Cuicatlán: perspectivas social y ecológica. In: Mandujano S (ed) Venado cola blanca en Oaxaca: potencial, conservación, manejo y monitoreo. Instituto de Ecología AC y Comisión Nacional para el Conocimiento de la Biodiversidad, Xalapa, Veracruz

Montiel-Ortega S, Arias LM, Dickinson F (1999) La cacería tradicional en el norte de Yucatán: una práctica comunitaria. Rev Geog Agríc 29:43–51

Naranjo E, Bodmer RE (2007) Source-sink systems and conservation of hunted ungulates in the Lacandon forest, Mexico. Biol Conserv 138:412–420

Naranjo E, Guerra MM, Bodmer RE, Bolaños JE (2004) Subsistence hunting by three ethnic groups of the Lacandon forest, México. J Ethnobiol 24:233–253

Ortega-Santos JA, Villarreal GJG, Mandujano S, Gallina S, Weber M, Clemente SF, Valdez R (2014) Retos y estrategias de conservación y aprovechamiento de la fauna en México. In: Valdez R, Ortega-Santos JA (eds) Ecología y manejo de fauna silvestre en México. Biblioteca básica de agricultura, Colegio de Postgraduados, Universidad Autónoma de Chapingo

Ortiz-García AI, Mandujano S (2011) Evaluación de la calidad del hábitat para el pecarí de collar en una Reserva de Biosfera de México. Suiform Soundings 11:14–27

Ortíz-García AI, Ramos-Robles MI, Pérez-Solano LA, Mandujano S (2012) Distribución potencial de los ungulados silvestres en la Reserva de Biosfera de Tehuacán-Cuicatlán, México. Therya 3:333–348

Pérez-Solano LA, Mandujano S (2013) Distribution and loss of potential habitat of the Central American red brocket deer (*Mazama temama*) in the Sierra Madre Oriental, Mexico. IUCN Deer Specialist Group Newslett 25:11–17

Pérez-Solano LA, Mandujano S (2018) Radiotelemetría GPS: aplicación en el monitoreo del ganado caprino en la Reserva de la Biosfera de Tehuacán-Cuicatlán, Oaxaca, México. Agroproductividad 11:63–69

Pérez-Solano LA, Mandujano S, Contreras-Moreno F, Salazar-Torres JM (2012) Primeros registros del temazate rojo *Mazama temama* en áreas aledañas a la Reserva de la Biosfera de Tehuacán-Cuicatlán, México. Rev Mex Biodivers 83:875–878

Pérez-Solano LA, Hidalgo-Mihart MG, Mandujano S (2016) Preliminary study of habitat preferences of red brocket deer (*Mazama temama*) in a mountainous region of central Mexico. Therya 7:197–203

Ramos-Robles MI, Gallina S, Mandujano S (2013) Habitat and human factors associated with white-tailed deer density in the tropical dry forest of Tehuacán-Cuicatlán Biosphere Reserve, Mexico. Trop Conserv Sci 6:70–86

Vasquez Y, Tarango L, López-Pérez E, Herrera J, Mendoza G, Mandujano (2016) Variation in the diet composition of the white tailed deer (Odocoileus virginianus) in the Tehuacán-Cuicatlán Biosphere Reserve. Rev Chapingo S Cien Fores Amb 22:57–68

Villarreal J (1999) Venado Cola Blanca: Manejo y aprovechamiento cinegético. Unión Ganadera Regional de Nuevo León, Monterrey

Villarreal-Espino O (2006) El venado cola blanca en la mixteca poblana: Conceptos y métodos para su conservación y manejo. Fundación Produce Puebla AC, Puebla

Weber M, Gonzalez S (2003) Latin American deer diversity and conservation: a review of status and distribution. Ecoscience 10:443–454

Yañez-Arenas CA, Mandujano S (2015) Evaluating the relationship between white-tailed deer and environmental conditions using spatially autocorrelated data in tropical dry forests of central Mexico. Trop Conserv Sci 8:1126–1139

Yañez-Arenas CA, Martínez-Meyer E, Mandujano S, Rojas-Soto O (2012) Modelling geographic patterns of population density of the white-tailed deer in central Mexico by implementing ecological niche theory. Oikos 121:2081–2089

Yañez-Arenas CA, Mandujano S, Martínez-Meyer E (2014) Predicting the density and abundance of white-tailed deer based on ecological niche theory. IUCN Deer Specialist Group Newslett 26:20–30

Chapter 5
Ungulates of Calakmul

Rafael Reyna-Hurtado and Khiavett Sanchez-Pinzón

Abstract Calakmul region located in the south of Yucatan peninsula in southern Mexico is an amazing site that contains a vast diversity of ecosystems and some of the largest extensions of tropical forests in Mexico. These forests still host six species of ungulates, including the largest and last survivor of the Neotropical megafauna, the Central American tapir, and one of the rarest ungulate species in Mexico, the white-lipped peccary, as well as another peccary species (collared peccary), the white-tailed deer and two species of brocket deer. All these species face serious conservation threats, as they are some of the most preferred prey species for of subsistence hunters and/or require large extensions of habitat in good conservation status to fulfill its basic survival needs for survival. This chapter is an attempt to summarize what is currently known on the ungulate species of the Calakmul region and we end by pointing out gaps in the existing information, information that is missing and is absolutely necessary to apply conservation and management plans of these highly interesting and endangered species.

5.1 Introduction

Calakmul region is an extraordinary region in Mexico that embraces a vast diversity of habitats and wildlife mixed with rich cultural traditions and a historical legacy as the core of ancient and modern Mayan civilizations. The Yucatan peninsula and Central America have served as a bridge between wildlife communities of the North and South America biota. The wildlife community in this region is therefore, a product of these geological and climatic events and is composed of elements of the Neoartic and Neotropical realm creating a unique wildlife community (Arroyo-Cabrales and Alvarez 2003). In the Calakmul region we also find the largest protected tropical forest of Mexico, the Calakmul Biosphere Reserve with 7238.5 km^2

R. Reyna-Hurtado (✉) · K. Sanchez-Pinzón
Department of Biodiversity Conservation, El Colegio de la Frontera Sur, Unidad Campeche, Campeche, Campeche, Mexico
e-mail: rreyna@ecosur.mx

S. Gallina-Tessaro (ed.), *Ecology and Conservation of Tropical Ungulates in Latin America*, https://doi.org/10.1007/978-3-030-28868-6_5

that together with the Maya Biosphere in Guatemala and some reserves in Belize conform the Maya forest, a trinational forest that after the Amazon forest is the largest tropical forest in America. Tapirs, jaguars, pumas, herds of peccaries still roam the forest understory, while spider monkeys and groups of howler monkeys find their way above the forest canopy. Currently, the region fosters some of the largest remaining tracks of tropical forest in good state of conservation in Mexico and its value for the protection of large Neotropical wildlife is becoming more evident as other areas of the country have higher rates of deforestation.

Six species of ungulates inhabit the Calakmul region; one is the last representative of the Perissodactyla order, the Central American Tapir (*Tapirus bairdii*) and the largest of all Neotropical terrestrial mammals. The Artiodactyla order is also represented with two species of peccaries, the white-lipped peccary (*Tayassu pecari*) and the collared peccary (*Pecari tajacu*) and three species of deer, the white-tailed deer (*Odocoileus virginianus*), the grey brocket deer (*Mazama pandora*), and the Central American red brocket deer (*Mazama temama*). These ungulate species are main target species among subsistence hunters because they are source of high-quality animal protein among many rural families in the area (Escamilla et al. 2000; Weber 2000; Reyna-Hurtado and Tanner 2007). All these species need also large areas of habitat to maintain viable populations and some of them require large extensions of space as they perform long distance movements to fulfill their basic needs (Reyna-Hurtado et al. 2009, Fig. 5.1).

Fig. 5.1 A view from above of the Calakmul Pyramid in the ancient Maya city of Calakmul, Campeche State. (Photos: Rafael Reyna-Hurtado)

In this section we explore current knowledge on these six species of ungulates, their conservation status and we end the chapter suggesting some future directions in what we consider it is critical to know to protect and conserve these amazing species in the Calakmul region.

5.2 Central American or Baird's Tapir (*Tapirus bairdii*)

The common names in the Yucatan peninsula include "Danta," "Anteburro," or "Tizimín." The Central American tapir is the only representative of the order Perissodactyla in Mexico. Baird tapir is a massive round animal that weights up to 300 kg, reaches up to 1.3 m and have a protuberance in the nose that give them a particular shape (Emmons and Feer 1990). Baird's tapirs prefer well-conserved forests with available water and use low-flooded forest while avoiding dry deciduous forest, however, it can visit secondary/perturbed forest to some degree if the species is not hunted (Reyna-Hurtado and Tanner 2005). In fact, in a study using tracks as indicator of relative abundance, tapirs were three times more abundant in communal forests (0.42 tracks/km) contiguous to the Calakmul Biosphere Reserve, than within the interior of the protected area (0.03 tracks/km) (Reyna-Hurtado and Tanner 2007). This research showed that communal forests might have better forest quality. Thus, the higher abundance of water outside the Calakmul protected area (O'Farrill et al. 2014) and the fact that tapirs are not the favorite prey of subsistence hunters, in this region might explain why the population persists and even grow more abundant than the one in the nearby protected area (Reyna-Hurtado and Tanner 2007). In the Calakmul Region, some arguments for not killing tapirs include the excessive weight and consequently work for meat disposal and a potential waste of meat, meat taste is not palatable, and a general respect for an animal rarely seen by hunters (RRH, SC, GO, pers. comm. villagers, Ejido Nuevo Becal, Ejido 20 de Noviembre, Calakmul, Campeche).

In the Calakmul Region, tapirs depend on water all year round visiting repeatedly water bodies locally known as "aguadas." Since 2008 in a monitoring effort of "aguadas" using camera traps at the interior of the Calakmul Biosphere Reserve, reported that tapir was one of the most frequent species photographed (Reyna-Hurtado et al. 2010), with an abundance of 37.57 individuals/1000 night traps (Pérez-Córtez et al. 2012). A recent study suggests that tapirs also can persist well in dry areas such as the north of the Calakmul Biosphere Reserve (Carrillo et al. 2002). In the only study in Mexico that followed a tapir for several years in the Calakmul Biosphere Reserve it was reported that one individual moved in a polygon of at least 23.9 km^2 of size but that was projected that it may moved in an almost 40 km^2 polygon (Reyna-Hurtado et al. 2016).

Sanchez-Pinzón et al. (in preparation) investigated the effect of moon light on tapir activity on the Calakmul Biosphere Reserve and in one adjacent community forest where there are human activities including hunting. They found that tapirs are shyer in the community forest and are less active in the nights with moon light in the community forest than in the protected area (Fig. 5.2).

Fig. 5.2 Central American tapir (*Tapirus bairdii*) in a pond in Calakmul Biosphere Reserve, Campeche, Mexico. (Photo: Rafael Reyna-Hurtado and Khiavett Sanchez-Pinzon)

In the Yucatan Peninsula, Baird's tapirs are the only known species that can successfully disperse large seeds of tree species such as the zapote tree (*Manilkara zapota;* O'Farrill et al. 2006). In an in-depth study of the relationship of tapirs and zapote trees, O'Farrill et al. (2012) found that seeds that pass through the digestive track of tapir are able to germinate in a similar rate than seeds that did not pass through the tapir's digestive system; therefore, facilitating long distances dispersal of zapote tree (O'Farrill et al. 2012). Despite the present of what we believe are stable populations of Baird's tapir in the Yucatan Peninsula, the long-distance movement patterns of this species might be under threat given current rates of habitat fragmentation due to an increase in human activities (road infrastructure and agriculture) and changes in water resource availability due to climate change (O'Farrill et al. 2013). The disruption of the movement patterns of this species can have detrimental and cascading effects on other species and on the functional role of this important long-distance seed disperser of large seeds putting this species at risk of being functionally extinct (O'Farrill et al. 2013).

The Maya forest that comprises the forest of the Greater Calakmul Region and the forest in the Peten area in Guatemala may harbor the largest tapir population in its whole distribution range (Naranjo 2009). Although this population may be the largest, several threats exist. For example, in the forest that surrounds the Calakmul Biosphere Reserve there has been an increase in conflicts with villagers due to tapir's crop raiding (M. Sanvicente, R. Reyna-Hurtado, S. Calme obs. pers.). In a recent study carried out in four communities of the Calakmul region Mac Gregor et al. (in preparation) found that tapirs are causing damages to crops in crop fields that are far away from villages and that include beans among the crops.

Despite an increase in tapir research in the Yucatan Peninsula in recent years, we believe that basic research of tapir in the Yucatan Peninsula must verse over areas most frequented by tapirs and therefore areas that need to be conserved to maintain a stable and connected populations. Research on tapir's dispersal capacities (movement) and connectivity among populations must be a priority due to current habitat encroachment. We need to know the status of the population under different human-induced and ecological conditions. Therefore population estimates (relative abundance, density) are very important information to inform management and conservation plans and initiatives.

5.3 White-Lipped Peccary (*Tayassu pecari*)

White-lipped peccary is known in the Calakmul region as "Senso," "Jabalín," "Hauilla," and "Kitam" in Mayan language (Reyna-Hurtado et al. 2014a). The white-lipped peccary belongs to the order Artiodactyla and is one of the two peccary species of the Tayassuidae family (known as the pigs of the new world) that exists in Mexico. White-lipped peccaries are pig-like animals with long black hairs in all the body except for a white area under the cheeks and the lips that give them its particular name. With more than a meter in length, 60 cm height and weighting up to 40 kg, white-lips are larger than the other species, the collared peccaries, and have longer legs (Emmons and Feer 1990). White-lipped peccary is a social animal that lives in groups up to 300 individuals with anecdotical reports of 700 and even a thousand individuals in a single group (R. Bodmer pers. comm.; Fragoso 2004; Mayer and Wetzel 1987, Fig. 5.3).

White-lipped peccary is among the favorite prey species for subsistence and sport hunters in the area and it is estimated that its historical distribution range in Mexico has been reduced in an 87% in the last 50 years (Altrichter et al. 2012; Reyna-Hurtado et al. 2017). In terms of the population estimates, in a long-term study in the Calakmul region it has been found that relative abundance of this species is three times larger in the protected area than in the surrounded communal forest (Reyna-Hurtado and Tanner 2007) with a density estimation of 0.42 individuals/km^2 for the protected area. It has been documented also that the species have been eliminated completely from some areas outside the Calakmul reserve and only survive in few numbers in some communal forest (Reyna-Hurtado 2009). White-lipped peccaries are easily hunted in the dry season in the few remaining water bodies ("aguadas") and it has been documented that subsistence hunting combined with sport hunting can eradicate a whole group in a single dry season (Reyna-Hurtado et al. 2010).

Fig. 5.3 White-lipped peccaries (*Tayassu pecari*) in Calakmul Biosphere Reserve, Campeche, Mexico. (Photo: Rafael Reyna-Hurtado)

White-lipped peccaries are specialist animals that prefer humid-tall and well-conserved forest with available water (Sowls 1997). In the Calakmul region it has been found that white-lipped peccary use more than expected the medium semi-perennial forest and the low-flooded forest while avoiding the dry semi-deciduous forest (Reyna-Hurtado and Tanner 2005; Reyna-Hurtado et al. 2009; Briceño-Méndez et al. 2014). This species is an assiduous visitor of water bodies ("aguadas") during the dry season of every year and its foraging movements are strongly influenced by water availability in time and space, to the degree that groups of this species have been classified as "central place foragers" with the central place being the "aguadas" (Reyna-Hurtado et al. 2012).

In a long-term study carried in Mexico about its movement patterns, Reyna-Hurtado et al. (2009) documented that groups of this species move in areas larger than 100 km^2 and that the movement patterns are strongly influenced by water and preferred forest availability (Reyna-Hurtado et al. 2009). There is empirical evidence that groups of this species can travel in coordinate way and may visit places that they remember, inferring that spatial memory can play a big role in its movements (Reyna-Hurtado and Tanner 2007). White-lipped peccaries feed on fruits in an 80% and invertebrates are also important part of its diet (Pérez-Cortéz and Reyna-Hurtado 2008), these items are found in greater quantities in the medium semi-perennial forest and in the low-flooded forest of the Yucatan Peninsula (Reyna-Hurtado et al. 2009). Moreira-Ramírez et al. (2018) followed three groups of white-lipped peccaries over the Calakmul Biosphere Reserve and one group outside the reserve and found larger home ranges 140 km^2 in the group living in the hunted site as this group always keeps moving. Authors conclude that this group was behaving in that way due to hunting pressure.

The forest of the Calakmul region hold one of the last and the largest population of white-lipped peccary in Mexico after the big reduction of its distribution range in the last 50 years (Altrichter et al. 2012; Reyna-Hurtado et al. 2017). For example, it was documented an estimate of 0.43 individuals/km^2 which may translate into approximately 1500 individuals in the southern part of the Calakmul Biosphere Reserve (assuming the forest maintain similar conditions) (Reyna-Hurtado et al. 2010). These findings contrast with research carried in communal forest where this species was very rare and its abundance very low in comparison with the protected area (Reyna-Hurtado 2009). These results raise the possibility of this species being isolated in the few protected areas and will become extinct in the near future in the communal forest where it still persists. This is not the best scenario for the conservation of this endangered species of wildlife. In addition, the predicted reduction in rainfall in the center of the peninsula due to climate change (Magrin et al. 2007) will greatly affect the species persistent in the dry areas of the Calakmul region and may force the species to migrate to the communal forest where there is more availability of standing water. In South America diseases are believed to be one potential cause of observed' large declines of this species (Fragoso 1997). In the Calakmul region we have no information on this topic, except for some evidence of skin diseases observed recently in the Maya forest (Reyna-Hurtado et al. 2014b).

It will be fundamental to know relative abundance and density when possible because these are very important information for conservation of this endangered species as we do not know the degree of hunting impact, of seasonal changes, and of diseases on this species. Documenting movement patterns and the possible use of spatial memory could be an exciting research that will contribute to understand the long-distance movements this species perform and the decision process in which these movements occur, as well as the ecological conditions that trigger them.

5.4 Collared Peccary (*Pecari tajacu*)

In the Calakmul region collared peccary are locally know as: "Puerco de monte," "Puerco cinchado," "Coche de monte," and "Kitam" in Maya language. The collared peccary belongs to the order Artiodactyla and is one of the two peccary species of the Tayassuidae family (known as the pigs of the new world) that exists in Mexico. Collared peccaries are pig-like animals with grey-yellowish hair in all the body except for a white collar that goes around the shoulder area and that gives them its particular name. Collared peccary is significant smaller than white-lipped peccary and its length ranges from 90 to 100 cm approximately and they weight between 15 and 28 kg with a more rounded head and shorter legs than white-lipped peccary (Emmons and Feer 1990). Collared peccaries are also social animals with groups from 2 to 50 animals but less cohesive than groups of white-lipped peccaries. Individuals usually remain close each other but they can travel separated for several periods of time, especially if they are in danger of predation when they run in

Fig. 5.4 Collared peccary (*Pecari tajacu*) in Calakmul Biosphere Reserve, Campeche, Mexico. (Photo: Rafael Reyna-Hurtado–Khiavett Sanchez-Pinzón)

different directions, a behavioral strategy different than the stand and defend of the white-lipped peccary (Sowls 1997, Fig. 5.4).

The collared peccary in the Calakmul region is find in tropical semi-perennial forest with some degree of tolerance to perturbed areas. For example, there are reports of collared peccaries in highly perturbed areas as the secondary habitat ("acahuales") surrounding communities in the Calakmul (R. Reyna-Hurtado, personal observation), and collared peccary is a common animal in the Calakmul Biosphere Reserve and in the communal forest that surround it (Reyna-Hurtado and Tanner 2007) despite being the most hunted animal in some communities (*ejidos* as they are named in Mexico, Weber 2000; Reyna-Hurtado and Tanner 2007).

There are not major studies of this species in the Yucatan peninsula that have been carried to this date. Reyna-Hurtado and Tanner (2005, 2007) reported on habitat use and population abundance as part of a study focus on the whole ungulate community. They found that collared peccary have similar abundance in a set of hunted areas than in the Calakmul protected area, this finding is surprising given that some reports point the collared peccary as the most hunted animal in some communities (Escamilla et al. 2000; Weber 2000). Reyna-Hurtado and Tanner (2005) also found that collared peccaries are habitat generalists but prefer tall humid habitats (medium sub-perennial forest) when available, and avoid in certain degree the dry forest. Briceño-Mendes et al. (2016) in a comparative study between the two species of peccaries found that collared peccary have an average of 6 individuals per groups with a range of 1 to 44 for a total of 85 groups observed. The same authors found a similar relative abundance between a hunted area and the protected area in the Calakmul region using track´ counts in linear transects (Briceño-Mendes et al. 2016). In the Calakmul region there are not major studies so it is urgent to know the status of the population and how it is affected by hunting and other human activities in the non-protected forest. Determine group sizes and fission–fusion social behavior would be an interesting research in this species, especially the comparison of protected versus non-protected areas. Finally, movement patterns and home range are essential information to elaborate management plans of this species as we will know the area needed to maintain a viable population in a given site.

5.5 White-Tailed Deer (*Odocoileus virginianus*)

White-tailed deer is known as "venado" or "venado real" in Calakmul region or "Quej" in Mayan language. The white-tailed deer is the largest of the three species of deer Mexico with a length of 1.11–2.22 m and between 30 and 50 kg weight. The color is brownish with grey in some areas, especially the head (Emmons and Feer 1990). White-tailed deer is a very important species for the rural communities of Yucatan peninsula and despite being one of the most studied animals in the world, white-tailed deer have been ignored by science in the Yucatan peninsula with only a handful of studies in diverse topics (Fig. 5.5).

In a 2 years study in the Calakmul region Weber (2005) found that white-tailed deer are browser consumers and that 70% of its diet comprised leaves and stems all year around, a high contrast with the more fruit-eating brocket deer (*Mazama* spp.). Also, white-tailed deer was the species of deer with the more diverse diet of the three deer species of the Yucatan peninsula. Weber (2005) also found a lower density with 0.021 individuals per square kilometer and relative abundance did not varies from a hunted community than in the Calakmul Biosphere Reserve in the Campeche state. Weber (2005) estimated the white-tailed deer as being more common than brocket deer in agriculture and secondary forested areas. In addition this author found that white-tailed deer was the most hunted species of the three deer species in the Calakmul region and is the one providing the highest biomass of all species hunted in terms of kilograms of meat (Weber 2000).

Reyna-Hurtado and Tanner (2007) found also that relative abundance of this species was higher in hunted/perturbed areas than in the Calakmul Biosphere Reserve and that white-tailed deer was a common species in secondary forest and prefer low-flooded forest in the hunted areas (Reyna-Hurtado and Tanner 2005). Contrasting with this, in a more recent study, Ramírez (2016) using camera traps deployed in ponds, she found that white-tailed deer was more abundant in the Calakmul Biosphere Reserve than in a community forest. This result may be a consequence of white-tailed deer visiting ponds to feed on herbaceous species that grow there. These findings highlight the persistence of this species in highly perturbed areas and the potential as prey species for a well-organized management plan

Fig. 5.5 White-tailed deer (*Odocoileus virginianus*) in Calakmul Biosphere Reserve, Campeche, Mexico. (Photo: Rafael Reyna-Hurtado)

that take care of the conservation of the populations. Important information needed for this species would be to know how this species is able to move through and survive in highly perturbed areas and avoid in the possible hunter encounter is a very interesting topic of study. White-tailed deer is the most appreciated species for subsistence hunters and is the subject of several cultural traditions among the Mayan hunters, therefore, studying the role in the traditional cultures as the Maya culture is also a very interesting topic that could promote good hunting practices.

5.6 Brocket Deer Species (*Mazama pandora* and *Mazama temama*)

Brocket deer are known as "venados cabritos," "cabritos," and "Chack Yuk" for the red brocket deer and "Sac Yuk" for the gray brocket deer in Mayan language. Brocket deer species includes two species in the Calakmul region, the gray brocket deer (which was formerly classified as *M. gouzabira* but recently was renamed *M. pandora*, Medellin and Aranda 1998) and the red brocket deer, formerly classified as *M. americana* (Weber 2005) but who has been renamed as *M. temama*, based on genetic and morphometric criteria (Bello-Gutiérrez et al. 2010). Brocket deer are

Fig. 5.6 Red brocket deer (*Mazama temama*) in Calakmul Biosphere Reserve, Campeche, Mexico. (Photo: Marco Briceno Mendez–Rafael Reyna-Hurtado)

small deer with 90–120 cm in length and no more than 20 kg in weight. *M. pandora* is slightly heavier than *M. temama*, and the weight of 21 individuals of *M. pandora* from Calakmul Biosphere Reserve averaged 17.5 kg for females and 20.5 per males (Weber and Medellin 2010), while *M. temama* averaged 16.3 kg for adult males hunted in the Calakmul region (Weber 2014). Brocket deer are also very important species for subsistence hunters, especially in the south areas of the Yucatan peninsula around the Calakmul Biosphere Reserve (Escamilla et al. 2000; Weber 2000) and the red brocket deer have been subjected to sport hunting in the last years in some communities where sport hunting is allowed under the UMA (Units for Wildlife Management and Conservation) scheme (Weber et al. 2006, R. Reyna-Hurtado pers. obs., Fig. 5.6).

Brocket deer are small, shy animals that live in well conserved forest rarely seen in perturbed areas (Reyna-Hurtado and Tanner 2005). These features make them difficult to study, in an effort to capture them to attach radiotelemetry devices, a total of 2 years of efforts prove to be very difficult to capture very few animals (Weber and Reyna-Hurtado unpublished data). The home range for the gray brocket deer was preliminarily estimated in less than 50 ha for a single animal during a 3-month follow up (Weber and Reyna-Hurtado unpublished data).

Weber (2005) in an in-depth study in feeding habits found that the gray brocket deer was a generalist in terms of the species that feed on while the red brocket deer was a specialist with more fruits consumed than the gray brocket deer and the

Fig. 5.7 Grey brocket deer (*Mazama pandora*) in Calakmul Biosphere Reserve, Campeche, Mexico. (Photo: Rafael Reyna-Hurtado–Khiavett Sanchez-Pinzón)

white-tailed deer. This author found that gray brocket deer switch from frugivorous to browsers during the year while red brocket deer specialized in fruits all year around (Weber 2014, Fig. 5.7).

Population estimates of brocket deer are scarce and when available is a mix of the two species given the almost no possibility of differentiate the tracks of one species of the other. In the Calakmul Biosphere Reserve, Reyna-Hurtado and Tanner (2007) found that brocket deer signs (again the combination of tracks of the two species) were the most abundant of all ungulate species and that track relative abundance did not vary between a set of hunted sites and the Calakmul protected area. The same authors also found that tracks of brocket deer were the most abundant tracks in dry forest of the protected area and that dry forest were used more than expected for these species, while in the hunted sites the low-flooded forest was used more than expected (Reyna-Hurtado and Tanner 2005). Using camera traps Ramírez (2016) found that *M. temama* was more abundant in a community forest than in Calakmul Biosphere Reserve and the opposite occurred for *M. pandora* that was more abundant in the protected area than in the community forest. The vegetation types of both sites can explain these findings. In Calakmul Biosphere Reserve the dry forest are the favorite habitat of *M. pandora* while the more humid community forests are favorite of *M. temama* (Bello-Gutiérrez et al. 2010).

Some estimates of density using observations in transects are between 0.90 and 1.5 deer/km^2 for both species of *Mazama* for the Calakmul region (Weber 2005).

Information that is highly needed for these two species includes the impact of hunting activities and deforestation in the population of the two species and the home range size and movement patterns for both species would be exciting research with conservation implications.

5.7 Conclusions and Future Research Directions

Calakmul region is a land that still conserves large tracts of forests in a good conservation status; these forests are the hope for the conservation of large Neotropical fauna that need large amount of habitat to maintain viable populations. In this chapter we have reviewed what is currently known about ecological aspects of the six tropical ungulates that live there. Although there are important and substantially advances in the knowledge of ecological aspects of some of these species, there is still a lot to do in terms of research with conservation goals. One aspect that was raised in almost all species was the need to investigate the degree that humans are impacting populations in almost all these species. All kind of human activities are taking place every day in the forest, timber extraction, fires, forest fragmentation, hunting, pollution as well as tourism, and we lack of information on how these activities are impacting wild populations of large fauna. For example, it was found that the synergic effect of subsistence hunting associated with sport hunting are having a bigger impact in an endangered species with the almost complete elimination of groups of white-lipped peccary in some communal forests surrounding the Calakmul Biosphere Reserve (Reyna-Hurtado et al. 2010). Also, sometimes, the human impact can be subtle and not evident at first look but evident when research focuses on behavior, as it was found that tapirs are shier in forest with humans than in the protected area as the activity level of them during the nights with moon light differs between these two areas (Sanchez-Pinzón et al. in preparation).

Another research aspect that was highlighted as priority was movement ecology. The fact that many of these species have large home ranges and high dispersal abilities present unique opportunities to design research aimed to know the extent of their movements, of their home range and define, or refine what is known about, habitat preferences. In addition to these topics that are exciting research, they also provide important information to known the extent of an area to be protected if we want to conserve a viable population of a species. We also need to know how animals disperse and what characteristics a potential corridor may have to function as a real corridor.

In summary, the Calakmul region is an amazing area that still holds populations of the tropical ungulates in a good conservation status. We must assure that human communities living in the area have all basic needs without depleting forest resources and the species that live on them. This challenge is becoming urgent to attend as forests are being transformed every day. The task is one that needs to be tackled by the entire society; therefore, academia, NGOs, the rural societies and the governmental institutions need to work closely to address these challenges.

Acknowledgements We acknowledge the editors of the book for the invitation to write this chapter. We are grateful to our own institution, El Colegio de la Frontera Sur, Unidad Campeche for support while writing the chapter.

References

Altrichter M, Taber A, Beck H, Reyna-Hurtado R, Keuroghlian A (2012) Range-wide declines of a key Neotropical ecosystem architect, the near threatened white-lipped peccary Tayassu pecari. Oryx 46:87–98

Arroyo-Cabrales J, Alvarez T (2003) A preliminary report of the late quaternary mammal fauna from Lotún cave, Yucatán, Mexico. In: Schubert BW et al (eds) Ice Age cave faunas of North America. Indiana University Press and Denver Museum of Nature and Science, Bloomington, pp 262–272

Bello-Gutiérrez J, Reyna-Hurtado R, Jorge W (2010) Central American red brocket deer *Mazama temama* (Kerr 1792). In: Duarte JMB, González S (eds) Neotropical cervidology: biology and medicine of Latin American deer. Funep/IUCN, Jaboticabal/Gland, pp 166–171

Briceño-Mendes M, Naranjo E, Altrichter M, Reyna-Hurtado R (2016) Responses of two sympatric species of peccaries (*Tayassu pecari* and *Pecari tajacu*) to hunting in Calakmul, Mexico. Trop Conserv Sci 9:1–11. https://doi.org/10.1177/1940082916667331

Briceño-Méndez M, Reyna-Hurtado R, Calme S, García-Gil G (2014) Preferencias de hábitat y abundancia relativa de *Tayassu pecari* en un área con cacería en la región de Calakmul, Campeche, México. Rev Mex Biodivers 85:242–250

Carrillo E, Saenz JC, Fuller TK (2002) Movements and activities of white-lipped peccaries in Corcovado. Biol Conserv 108:317–324

Emmons LH, Feer F (1990) Neotropical rainforest mammals. A field guide. The University of Chicago Press, Chicago

Escamilla A, Sanvicente M, Sosa M, Galindo-Leal C (2000) Habitat mosaic, wildlife availability, and hunting in the tropical forest of Calakmul, México. Conserv Biol 14:1592–1601

Fragoso JM (1997) Desapariciones locales del baquiro labiado (*Tayassu pecari*) en la ¿Amazonía: migración, sobre-cosecha o epidemia? In: Fang TG, Bodmer R, Aquino R, Valqui M (eds) Manejo de fauna silvestre en la Amazonía. UNAP, University of Florida UNDP/GEF e Instituto de Ecologia, La Paz, pp 309–312

Fragoso JMV (2004) A long-term study of white-lipped peccary (*Tayassu pecari*) population fluctuation in northern Amazonia. In: Silvius K, Bodmer RE, Fragoso JMV (eds) People in nature. Wildlife Conservation in South and Central America. Columbia University Press, New York, pp 286–296

Mac Gregor I, Reyna-Hurtado R, Molina D (in preparation) Baird's tapir: predicting patterns of crop damage around Calakmul Biosphere Reserve, Campeche, Mexico

Magrin G, Gay García C, Cruz-Choque D, Gímenez JC, Moreno AR, Nagy GJ, Nobre C, Villamizar A (2007) In: Parry ML, Canziani OF, Palutikof JP, van der Linden PJ, Hanson CE (eds) Latin América. Climate change 2007: impacts, adaptability and vulnerability. Contribution of working group II to the fourth assessment report of the intergovernmental panel on climate change. Cambridge University Press, Cambridge, pp 581–615

Mayer JJ, Wetzel RM (1987) Tayassu pecari. Mamm Species 293:1–7

Medellin R, Aranda M (1998) The taxonomic status of the Yucatan brown brocket *Mazama pandora* (Mammalia: Cervidae). Proc Biol Soc Wash 111:1–14

Moreira-Ramírez J, Reyna-Hurtado R, Hidalgo-Mihart M, Naranjo EJ, Ribeiro MC, García-Anleu R, McNab R, Radachowsky J, Mérida M, Briceño-Méndez M, Ponce-Santizo G (2018) White-lipped peccary home-range size in the Maya Forest of Guatemala and México. In: Reyna-Hurtado R, Chapman CA (eds) Movement ecology of neotropical forest mammals. Springer, Switzerland, pp 21–37

Naranjo E (2009) Ecology and conservation of Baird's tapir in Mexico. Trop Conserv Sci 2:140–158
O'Farrill G, Calmé S, Sengupta R, Gonzalez A (2012) Effective dispersal of large seeds by Baird's tapir: a large-scale field experiment. J Trop Ecol 28:119–122
O'Farrill G, Campos-Arceiz A, Galetti M (2013) Frugivory and seed dispersal by tapirs: an insight on their ecological role. J Integr Zool 8(1):4–17
O'Farrill G, Gauthier-Schampaert K, Rayfield B, Bodin Ö, Calmé S, Gonzalez A, Sengupta R (2014) The potential connectivity of waterhole networks and the effectiveness of a protected area under various drought scenarios. PLoS One 9(5):e95049. https://doi.org/10.1371/journal.pone.0095049
O'Farrill G, Calmé S, Gonzalez A (2006) Manilkara zapota: a new record of a species dispersed by tapirs. Tapir conservation, newsletter of the IUCN/SSC tapir specialist. Group 15:32–35
Pérez-Cortéz S, Reyna-Hurtado R (2008) La dieta de los pecaríes (Pecari tajacu y Tayassu pecari) en la región de Calakmul, Campeche, México. Rev Mex Mastozool 12:17–42
Pérez-Cortez S, Enriquez PL, Sima-Panti D, Reyna-Hurtado R, Naranjo E (2012) Influencia de la disponibilidad de agua en la presencia y abundancia de Tapirus bairdii en la selva de Calakmul, Campeche, Mexico. Rev Mex Biodivers 83:753–756
Ramírez L (2016) Abundancia relativa y patrones de actividad por venados en dos sitios de la región de Calakmul, Campeche, México. Tesis licenciatura, Universidad Autónoma de Campeche, San Francisco de Campeche, México
Reyna-Hurtado R (2009) Conservation status of the white-lipped peccary (Tayassu pecari) outside the Calakmul Biosphere Reserve in Campeche, Mexico: a synthesis. Trop Conserv Sci 2:159–172
Reyna-Hurtado R, Tanner G (2005) Habitat preferences of an ungulate community in Calakmul forest, Campeche. Mexico Biotropica 37:676–685
Reyna-Hurtado R, Tanner GW (2007) Ungulate relative abundance in hunted and non-hunted sites in Calakmul Forest (Southern Mexico). Biodivers Conserv 16:743–757
Reyna-Hurtado R, Rojas-Flores E, Tanner GW (2009) Home range and habitat preferences of white-lipped peccary groups (*Tayassu pecari*) in a seasonal tropical forest of the Yucatan Peninsula, Mexico. J Mammal 90:1199–1209
Reyna-Hurtado R, Naranjo E, Chapman CA, Tanner GW (2010) Hunting and the conservation of a social ungulate: the white-lipped peccary *Tayassu pecari* in the Calakmul, Mexico. Oryx 44:88–96
Reyna-Hurtado R, Chapman CA, Calmé S, Pedersen E (2012) Searching in heterogeneous environments: foraging strategies in the white-lipped peccary (Tayassu pecari). J Mammal 93:124–133
Reyna-Hurtado R, March I, Naranjo EJ, Mandujano S (2014a) Pecaríes en México. In: Valdez R, Ortega JA (eds) Ecología y manejo de fauna silvestre en México. Colegio de Postgraduados y New Mexico State University, Texcoco, pp 353–375
Reyna-Hurtado R, Moreira J, Briceño M, Sanvicente M, McNab R, García R, Mérida M, Sandoval E, Ponce G, Hyeroba D (2014b) White-lipped peccaries with skin problems in the Maya Forest. Suiform Soundings 13:28
Reyna-Hurtado R, Sanvicente-López M, Pérez-Flores J, Carrillo-Reyna N, Calmé S (2016) Insights into the multiannual home range of a Baird's tapir (*Tapirus bairdii*) in the Maya Forest. Therya 7(2):271–276
Reyna-Hurtado R et al (2017) White-lipped peccary in Mesoamerica: status, threats and conservation actions. Suiform Soundings 15:31–35
Sanchez-Pinzón K, Reyna-Hurtado R, Meyer N (in preparation) Moon light and the activity patterns of *Tapirus bairdii* in the Calakmul region, Southern Mexico. Therya
Sowls LK (1997) Javelinas and the other peccaries: their biology, management and use, 2nd edn. Texas A&M University Press, College Station
Weber M (2000) Effects of hunting on tropical deer populations in southeastern México. MSc Thesis, Royal Veterinary College, University of London
Weber M (2005) Ecology and conservation of sympatric tropical deer populations in the Greater Calakmul Region, south-eastern, Mexico. Dissertation, School of Biological and Biomedical Sciences, University of Durham

Weber M (2014) Temazates y venados cola blanca tropicales. In: Valdez R, Ortega JA (eds) Ecología y manejo de fauna silvestre en México. Biblioteca básica de agricultura, Colegio de Posgraduados, Texcoco, pp 435–466
Weber M, Medellin RA (2010) Yucatan brown brocket deer Mazama pandora (Merriam 1901). In: Barbanti-Duarte JM, Gonzalez S (eds) Neotropical cervidology: biology and medicine of Latin American deer. FUNEP/IUCN, Jaboticabal, pp 166–170
Weber M, García-Marmolejo G, Reyna-Hurtado R (2006) The tragedy of the commons Mexican style: a critique to the Mexican UMAs concept as applied to wildlife management and use in South-Eastern Mexico. Wildl Soc Bull 34:1480–1488

Chapter 6
Ecology and Conservation of Ungulates in the Lacandon Forest, Mexico

Eduardo J. Naranjo

Abstract This study presents an assessment of the variations and trends of ungulate populations presence, abundance, and uses over a decade (2001–2012) in Montes Azules Biosphere Reserve (MABR) and in a sample of neighboring communities of the Lacandon Forest, Mexico. During two periods of study (1998–2001 and 2010–2012), abundances (distance sampling and camera-trapping) and uses (interviews and participant observation) of five ungulate species (Baird's tapir, *Tapirus bairdii*; white-lipped peccary, *Tayassu pecari*; collared peccary, *Pecari tajacu*; red-brocket deer, *Mazama temama*; and white-tailed deer, *Odocoileus virginianus*) were assessed in four sites within MABR and in eight surrounding communities. Abundances based on direct sightings and camera-trapping were considerably higher in slightly hunted sites inside MABR than in the sorrounding communities for Baird's tapir and the white-lipped peccary, but not for the other species. There were variations in the abundances (based on direct sightings) of all species between 2001 and 2012: Baird's tapir declined (−59.7%), while the white-tailed deer (421%), white-lipped peccary (64%), collared peccary (37%), and red-brocket deer (22%) increased their abundances. These variations may be related to regional changes in the landscape in sinergy with social and cultural shifts in the communities of the Lacandon Forest. The collared peccary had the highest mean harvest rate (0.61 individuals/km^2/year) in the study area, followed by the red-brocket deer (0.36), white-tailed deer (0.29), white-lipped peccary (0.07), and Baird's tapir (0.02). Significantly higher harvest rates of ungulates were detected in 2012 compared to those estimated in 2001, probably due to a higher availability of access roads and firearms for hunters in 2012, combined with demographic growth and prevalent poverty of households in the communities of the study area.

E. J. Naranjo (✉)
El Colegio de la Frontera Sur, San Cristobal de Las Casas, Chiapas, Mexico
e-mail: enaranjo@ecosur.mx

S. Gallina-Tessaro (ed.), *Ecology and Conservation of Tropical Ungulates in Latin America*, https://doi.org/10.1007/978-3-030-28868-6_6

6.1 Introduction

Neotropical ungulates (orders Artiodactyla and Perissodactyla) constitute important natural resources providing ecological services and representing income and food sources for residents of southeast Mexico's tropical forests (Naranjo et al. 2010). However, these ecosystems are being rapidly transformed into farming and grazing areas, which in sinergy with unsustainable hunting are threatening many vulnerable wildlife populations, including some ungulate species such as Baird's tapir (*Tapirus bairdii*), and the white-lipped peccary (*Tayassu pecari*), which are endangered in Mexico (SEMARNAT 2010). Other ungulate species such as the collared peccary (*Pecari tajacu*), red brocket deer (*Mazama temama*), and white-tailed deer (*Odocoileus virginianus*) are known to be more tolerant and in some cases, even benefited from moderate hunting and habitat fragmentation (Naranjo and Bodmer 2007; Tejeda et al. 2009). Wildlife conservation and sustainable use requires basic information on the biology and ecology of focal populations (Caughley and Sinclair 1994; Ojasti and Dallmeier 2000).

Understanding the interactions between wildlife populations, their habitats, and human societies as users, is essential to develop viable strategies for their conservation and sustainable use (Robinson and Bodmer 1999; Bodmer and Pezo 2001). Previous studies have shown that uncontrolled hunting and habitat loss in general have negative effects on the distribution and abundance of many wildlife populations in tropical forests (Robinson and Bennett 2000). However, there is a wide variety of patterns in wildlife use and habitat transformation driving diverse impacts on populations, depending on the cultural and socioeconomic features of the users, and on particularities of the physical environment (Robinson and Redford 1991; Robinson and Bennett 2000). Unfortunately, this kind of information is still scarce and disperse in southeastern Mexico (Cuarón 2000; Escamilla et al. 2000; Jorgenson 2000; Naranjo et al. 2009).

The Lacandon Forest (Spanish: *Selva Lacandona*) of Chiapas, Mexico, is a relevant region for conservation of biological and cultural diversity, and it harbors complex environmental and social problems (Vásquez and Ramos 1992; De Vos 2002). There are numerous human settlements in the Lacandon Forest, most of them small and inhabited by indigenous and mestizo groups of diverse origins, whose primary productive activities are farming, cattle ranching, harvest of non-timber forest products, subsistence hunting, and fishing (INE 2000). A growing number of local assessments of wildlife abundance and use have been done in the Lacandon Forest (e.g., March 1987; Naranjo et al. 2004a, b; Naranjo and Bodmer 2007). Nevertheless, this work constitutes the first long-term effort to evaluate trends in ungulate populations presence and abundance resulting from harvesting and habitat transformation. The author conducted surveys on the presence, abundance, and uses of ungulates and other wildlife species in Montes Azules Biosphere Reserve (MABR) and selected neighboring communities of the Lacandon Forest in two periods: 1998–2001 and 2010–2012. The purpose of this study was to assess the variations and trends of ungulate populations presence, abundance, and uses over a decade (2001–2012) in MABR and in a sample of neighboring communities of the Lacandon Forest, Mexico. It was expected that (1) the abundance of vulnerable ungulates (Baird's tapir and white-lipped peccary) would be lower in community territories than inside MABR, while abundance of the other ungulate species (collared peccary, red-brocket deer, and white-tailed deer) would be similar between sites; (2) the most hunted ungulate species would be the most abundant in the communities surveyed; and (3) vulnerable species would show negative trends in their abundances, whereas the remainder ungulates would not have significant changes between 2001 and 2012.

6.2 Methods

The Lacandon Forest of Mexico constitutes the southwestern sector of the Great Maya Forest, and it is one of the most important tracts of rainforest remaining in the country (Vásquez and Ramos 1992). The area is located in the northeastern portion of the state of Chiapas (16°05′–17°15′ N, 90°30′–91°30′ W), and it is delimited by the Guatemalan border on the east, north, and south, and by the Chiapas Highlands on the west. The Lacandon Forest originally was covered by over a million hectares of tropical rainforest, of which less than a half remain today (Cuarón 1997; Soto-Pinto et al. 2012). Its eight protected areas comprise around 4000 km^2; INE 2000; Fig. 6.1), and they are considered among the top priorities for conservation in Mesoamerica because of their high biodiversity, hydrological richness, and connectivity with the rainforests of northern Guatemala and the southern portion of the Yucatan Peninsula (CONABIO 1998). The area probably harbors some of the largest Mexican populations of precious hardwood trees and large vertebrate species harvested by both indigenous and mestizo residents (Vásquez and Ramos 1992; Medellín 1994; Naranjo 2002). Therefore, many government agencies and academic institutions are interested in promoting sustainable development in the region.

During the two periods of study (1998–2001 and 2010–2012), abundances and uses of ungulate species were assessed in four sites within MABR and in eight surrounding communities (Bethel, Galacia, Lacanjá-Chansayab, Loma Bonita, Metzabok, Nueva Palestina, Playón de la Gloria, and Reforma Agraria; Fig. 6.1). A total of 14 line transects (1–5 km each) were walked about every 2 months in the study area. Presence and relative abundance of each ungulate species were evaluated with the help of previously trained local guides and graduate students. Two different techniques were used to assess presence and abundance. The first consisted of repeated counts of individuals and their tracks along line transects walked slowly (average speed: 1.5 km/h) during the first hours of daylight (usually 6:00 to 10:00 h), and before dusk (16:00–18:00 h). Abundance indices (individuals and tracks per km walked) were calculated from the frequencies of individuals and tracks observed in each transect (Naranjo 2002; Naranjo et al. 2004a).

The second technique was photo-trapping using camera-traps (Cuddeback E2, Attack, and Moultrie D-55). Photo-trapping was only applied in the second period

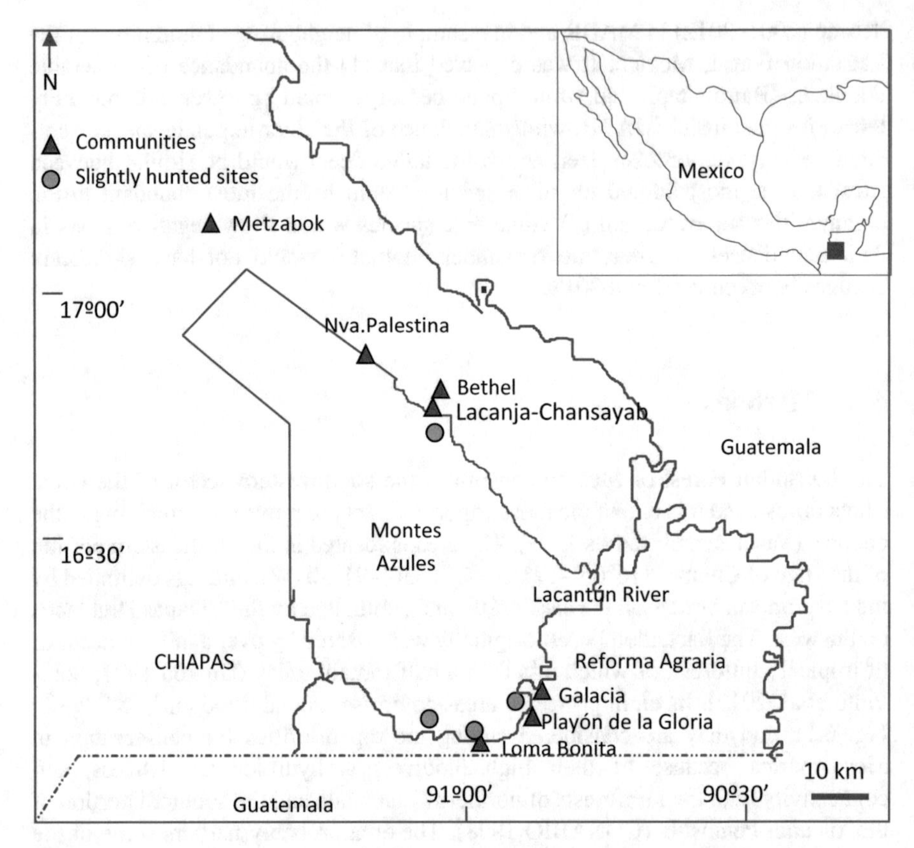

Fig. 6.1 Study area and sampling sites in the Lacandon Forest, Chiapas, Mexico. Red triangles show the communities visited. Green circles indicate study sites selected within Montes Azules Biosphere Reserve (MABR)

(2010–2012) in each site and transect. Cameras were deployed in a 3 × 3 km grid containing nine stations per site that were 1 km apart from each other (Chavez et al. 2007). Cameras were active between 2 and 3 months per sampling site, including parts of both dry and rainy seasons. Photographic rates were used to estimate abundance indices (records per 1000 camera-days) for all five ungulate species in each study site. Ungulate abundances were compared among species and study sites, and between seasons and periods of study using standard statistical procedures (Sokal and Rohlf 1995).

Uses of ungulates by residents of the study area were documented through participant observation and interviews with subsistence hunters and their families. Users were questioned about their hunting preferences (e.g., hunting methods, frequencies, seasons, and sites) for each species (Naranjo 2002). Interviewees were men and women over 15 years of age, which were randomly selected in each community. Participant observations and interviews allowed to obtain annual harvest rates (individuals hunted/km^2) for each species (Naranjo et al. 2004b).

Hunting frequencies and harvest rates were compared between periods of study using either Student's t tests or Wilcoxon's sign tests (Sokal and Rohlf 1995). Additionally, Spearman's correlation analyses were applied to detect potential associations between population abundances and their correspondent harvest rates in the communities of study.

6.3 Results and Discussion

6.3.1 Population Abundances

A total of 2642 records (730 direct sightings, 72 photocaptures, and 1840 tracks) of five ungulate species were obtained along 2486 km of line transects walked and an effort of 1817 camera-days applied in the study area during the two periods analyzed. Based on direct sightings, the most abundant ungulates recorded per 100 km walked were the white-lipped peccary (average 2001–2012: 18.1), and the collared peccary (6.9). These two species were followed by the red-brocket deer (1.1), Baird's tapir (0.6), and white-tailed deer (0.3). Recorded tracks per 100 km walked were more frequent for Baird's tapir (average 2001–2011: 21.8), collared peccary (16.7), red-brocket deer (11.8), white-lipped peccary (5.9), and white-tailed deer (3.1) (Table 6.1). Photocaptures per 1000 camera-days were higher for the collared peccary (average MABR-communities: 15.1), red-brocket deer (9.1), Baird's tapir (4.7), white-lipped peccary (2.8), and white-tailed deer (1.2; Table 6.2).

Abundances of all species combined based on direct sightings and tracks during the second period of study (2010–2012) were not significantly different between conserved sites within MABR and the surrounding communities (direct sightings:

Table 6.1 Numbres of individuals (*N*) and relative abundances (Ab) of ungulates recorded through direct sightings and tracks in Montes Azules Biosphere Reserve (MABR) and neighboring communities (Ejidos) in the Lacandon Forest, Mexico

	Tracks (footprints)				Direct sightings			
Species	MABR (*N*)	Ejidos (*N*)	AbMABR (*N*/100 km)	AbEjidos (*N*/100 km)	MABR (*N*)	Ejidos (*N*)	AbMABR (*N*/100 km)	AbEjidos (*N*/100 km)
Tapirus bairdii	116	16	36.3	6.2	2	0	0.6	0
Pecari tajacu	33	32	10.3	12.4	29	17	9.1	6.6
Tayassu pecari	34	0	10.6	0	130	0	40.6	0
Mazama temama	52	21	16.3	8.1	5	2	1.6	0.8
Odocoileus virginianus	3	32	0.9	12.4	1	2	0.3	0.8
Total	238	101	74.3	39.1	167	21	52.2	8.2

Table 6.2 Numbers of photo-captures and mean relative abundance (AbRel) of ungulates recorded through camera-trapping in Montes Azules Biosphere Reserve (MABR) and neighboring communities (Ejidos) in the Lacandon Forest, Mexico

Species	MABR (*N* photos)	Ejidos (*N* photos)	AbRel MABR (*N*/1000 camera-days)	AbRel Ejidos (*N*/1000 camera-days)
Tapirus bairdii	11	1	7.6	1.7
Pecari tajacu	29	2	20.0	10.2
Tayassu pecari	8	0	5.5	0
Mazama temama	14	5	9.7	8.5
Odocoileus virginianus	1	1	0.7	1.7
Total	63	9	43.5	22.1

$U = 33.0$; $n = 5$; $P = 0.31$; tracks: $t = 1.12$; $n = 5$; $P = 0.30$), where persistent hunting was present and the landscape had been transformed into a matrix of farming and grazing areas encompassing forest fragments of different sizes and conditions. However, abundances (both from direct sightings and camera-trapping) were considerably higher in slightly hunted sites inside MABR than in the surrounding communities for endangered and vulnerable species (i.e., Baird's tapir and white-lipped peccary), but not for the other species ($P > 0.05$), as expected at the beginning of the study.

No significant differences in the abundances of the five ungulate species were detected between the two periods of study (direct sightings: $W = -9.0$; $n = 5$; $P = 0.31$; tracks: $W = -3.0$; $n = 5$; $P = 0.81$). However, at the species level there were noticeable variations in their abundances based on direct sightings between 2001 and 2012. A decline was found for Baird's tapir (−59.7%). In contrast, the white-tailed deer showed a sharp increase (421%) in its abundance in the same lapse. More modest increases were found for the white-lipped peccary (64%), the collared peccary (37%), and the red-brocket deer (22%; Table 6.3). Track abundances went upwards within periods, particularly for the white-tailed deer (3682%). There were slight increases in track abundances of the red-brocket deer (16%) and Baird's tapir (11%), while it stayed stable for the white-lipped peccary, and even decreased for the collared peccary (−48%).

The variations of ungulate abundances between the two periods of study may be related to regional changes in the landscape in sinergy with social and cultural shifts in the communities of the Lacandon Forest, including those around MABR (Naranjo et al. 2004a, 2015; Amador-Alcalá et al. 2013; Naranjo 2018). Nevertheless, it is important to acknowledge that such variations may be also partially due to differences in the skills of collaborators participating in field samplings of each study period. Yet the shifts detected in the abundances of species such as the white-tailed deer, Baird's tapir, and both peccary species, may have derived from a complex series of factors interacting during the decade compared in the study, resulting in environmental changes (e.g., habitat transformation, climatic change), and social dynamics (e.g., human population growth, modifications in farming and hunting practices), among others (Soto-Pinto et al. 2012; Naranjo et al. 2015; Naranjo 2018).

Table 6.3 Ungulates observed (*N*) and abundance indices (AbN and AbTracks) estimated from direct observations and tracks in line transects of the Lacandon Forest, Chiapas, Mexico in 1998–2001 (Naranjo 2002), and in 2010–2012

Species	*N*		Tracks		AbN/100 km		AbTracks/100 km	
	2012	2001	2012	2001	2012	2001	2012	2001
Tapirus bairdii	2	22	132	520	0.35	0.89	22.85	20.81
Pecari tajacu	46	147	65	554	7.96	5.87	11.25	22.17
Tayassu pecari	130	345	34	149	22.5	13.78	5.89	5.97
Mazama temama	7	25	73	274	1.21	1	12.64	10.95
Odocoileus virginianus	3	3	35	4	0.52	0.1	6.05	0.16
Total	188	542	339	1501	32.5	21.6	60.1	58.7

Environmental and social changes may influence the distribution and abundance of wildlife populations in different ways, favoring generalist species, negatively affecting vulnerable ones, and maintaining a dynamic stability on many others. It was expected that vulnerable ungulates species such as Baird's tapir and the white-lipped peccary, declined in numbers between 2001 and 2012. This was true for the first, but not for the second. Four potential explanations of this situation are: (1) The above cited differences in the observation skills of field assistants between periods; (2) An active maintenance of the forest cover in some of the communities included in the study (e.g., Lacanjá, Chansayab, Metzabok, and Reforma Agraria); (3) A decline in hunting pressure on ungulate species in the study area; and (4) A temporary increase in density of ungulate populations remaining in forest fragments within community territories around MABR.

The temporary increase of density resulting from forest fragmentation would be consistent with the expected response of generalist species such as the white-tailed deer and the collared peccary, which are tolerant to habitat transformation, hunting, and human activity in general (Halls 1984; Sowls 1997). Similarly, points 2 and 3 (maintenance of forest cover and decline of hunting pressure), could be related with the growing number of conservation initiatives promoted and applied by government agencies, nongovernment organizations, and academic institutions during the last 15 years in the Lacandon region (Carabias et al. 2016). Some of those initiatives consist of programs such as the implementation of payments for environmental services, the organization of multiple environmental education workshops, and subsidies for community-based conservation and management of biodiversity.

6.3.2 Uses of Ungulate Species

During the second period of the study, 277 interviews applied to residents of three of the communities included in the survey allowed to record 345 hunting events related to the use of ungulate species (Table 6.4). The collared peccary (*N* = 139 hunting events), was the most frequently cited ungulate in the interviews, followed

Table 6.4 Ungulate species and average numbers of individuals hunted (*N*) in the communities of Bethel, Lacanjá-Chansayab, and Nueva Palestina, Lacandon Forest, Mexico

Species	Common name	Bethel *N*	Lacanjá-Chansayab *N*	Nueva Palestina *N*	Total *N*
Tayassu pecari	White-lipped peccary	11	5	12	28
Pecari tajacu	Collared peccary	67	8	64	139
Odocoileus virginianus	White-tailed deer	51	8	30	89
Mazama temama	Red-brocket deer	55	7	19	81
Tapirus bairdii	Baird's tapir	2	3	3	8
Total		186	31	128	345

Data from interviews applied in 2010–2012

by the white-tailed deer (*N* = 89), red-brocket deer (*N* = 81), white-lipped peccary (*N* = 28), and Baird's tapir (*N* = 8; Table 6.4). The five ungulate species were primarily hunted to obtain meat for human consumption, which was cited by 30% of the interviewees (Table 6.5). Other hunting purposes recorded were crop damage control (21% of interviewees), handicraft making (7%), companionship (pets, 7%), clothing (hides; 4%), medicine (3%), and trade (1%).

All five species were considered important traditional meat sources by local hunters of the study area. Peccaries and deer species were regarded as high quality prey in the communities visited. However, tapirs were not usually hunted because (1) the taste of their meat was not considered as good as that of deer and peccaries; and (2) their densities were usually very low, and consequently they were hard to find during hunting trips. White-lipped peccaries were also difficult to find in the study area, but they still were considered preferred prey by residents of the Lacandon Forest. Collared peccaries, and occasionally, red-brocket deer and tapirs were hunted to mitigate damage on corn and beans fields. Skulls of peccaries and deer antlers were sometimes used as trophies to adorn houses, while pelts of those ungulates were useful to make domestic chairs, carpets, and drums. Deer antlers, peccary canines, and hooves of all of them could be sold as decorative items. Newborns of all ungulate species (including tapirs), were occasionally kept as pets in the hunters' households (Table 6.5). Hunters of the Lacandon Forest usually employed firearms (rifles 0.22 caliber, and shotguns 0.16 caliber (83% of interviewees), machetes (48%), and trained dogs (17%) for taking ungulate prey. The most common hunting methods observed in the study area were: active search with trained dogs (68%), waitings (on elevated points) at strategic sites (63%), and nocturnal "flashlighting" (38%). The most frequently used hunting sites were corn crops and their surroundings (89%), secondary vegetation plots (*acahuales*; 40%), and forest fragments (7.2%).

Among the ungulates present in the study area, the collared peccary was recipient of the highest mean harvest rate for the two periods (0.61 individuals/km^2/year), followed by the red-brocket deer (0.36), white-tailed deer (0.29), white-lipped

Table 6.5 Main uses and parts used of ungulate species in three indigenous communities of the Lacandon Forest, Mexico

Species	Bethel uses	Bethel parts used	Lacanjá Chansayab uses	Lacanjá Chansayab parts used	Nueva Palestina uses	Nueva Palestina parts used
Tayassu pecari	F, H, D	Me, Ca	F	Me	F	Me
Pecari tajacu	F, D, H, P	Me, Sk, Ca	F, H	Me, Ca	F	Me
Odocoileus virginianus	F, P, O	Me, Sk, An, Ho	F, O	Me, An, Sk	F, O, M	Me, Sk, An
Mazama temama	F	Me	F, H	Me, Sk	F, M, O, C	Me, Sk, An
Tapirus bairdii	F	Me	F	Me	F, C	Me, In

Uses: *F* food, *Dc* damage control, *P* pelt, *O* ornamental (trophies), *H* handicraft, *C* companionship (pet), *M* medicine. Parts used: *Me* meat, *In* individual, *Sk* skin, *An* antlers, *Ho* hooves, *Ca* canines

Table 6.6 Annual harvest rates (HR: individuals hunted/km^2/year) of ungulates in communities of the Lacandon Forest, Mexico (2001 versus 2012)

Species	HR 2001[a]	HR 2012	Mean HR 2001–2012
Pecari tajacu	0.42	0.8	0.61
Tayassu pecari	0.04	0.1	0.07
Odocoileus virginianus	0.08	0.5	0.29
Mazama temama	0.22	0.5	0.36
Tapirus bairdii	0.01	0.03	0.02
Total	0.77	1.93	1.35

[a]Naranjo et al. (2004b)

peccary (0.07), and Baird's tapir (0.02; Table 6.6). Significantly higher harvest rates of ungulates ($t = 3.97$, df = 4, $p = 0.002$) were detected in 2012 compared to those estimated in 2001 (Naranjo 2002; Naranjo et al. 2004b). The increase in harvest rates may be due to a higher availability of access roads and firearms for hunters in 2012, combined with demographic growth and prevalent poverty of households in the communities of the study area. An alternative explanation is the possibility that interviewers collaborating in the 2012 survey were more successful than those of 2001 in gaining confidence with interviewees, which would have produced more accurate records of animals hunted in the visited villages.

This study provides quantitative information on the trends of abundance and use of ungulate populations in the Lacandon Forest, which constitutes one of the most important reservoirs of biodiversity in Mexico. This information will be helpful to support conservation and sustainable management strategies for ungulate species in the study area, some of them endangered. This project was focused on improving our knowledge on animal species relevant for the nutrition and economy of inhabitants of rural communities living in poverty in southeastern Mexico.

Acknowledgements The author is grateful to the people of the Lacandon Forest, who facilitated permits and greatly helped in the fieldwork of this study. Graduate and undergraduate students of El Colegio de la Frontera Sur (ECOSUR) and Universidad de Ciencias y Artes de Chiapas (UNICACH) were enthusiastic collaborators, especially Jorge Bolaños, Fredy Falconi, Michelle Guerra, and Evelin Amador. The National Commission for Protected Areas (CONANP) kindly granted permits for working in Montes Azules Biosphere Reserve. Mexico's Council for Science and Technology (CONACYT) and ECOSUR provided financial and logistic support during the two periods of study.

References

Amador-Alcalá S, Naranjo EJ, Jiménez-Ferrer JG (2013) Wildlife predation on livestock and poultry: implications for predator conservation in the rainforest of south-east Mexico. Oryx 47:243–250

Bodmer RE, Pezo E (2001) Rural development and sustainable wildlife use in Peru. Conserv Biol 15:1163–1170

Carabias J, De la Maza J, Cadena R (eds) (2016) Conservación y desarrollo sustentable en la Selva Lacandona: 25 años de actividades y experiencia. Natura y Ecosistemas Mexicanos, México

Caughley G, Sinclair ARE (1994) Wildlife ecology and management. Blackwell Science, Oxford

Chavez C, Ceballos G, Medellín R, Zarza H (2007) Primer censo nacional del jaguar. In: Ceballos G, Chávez C, List R, Zarza H (eds) Conservación y manejo del jaguar en México: estudios de caso y perspectivas. CONABIO/UNAM, México, pp 133–141

Comisión Nacional para el Conocimiento y Uso de la Biodiversidad (CONABIO) (1998) La biodiversidad de México: un estudio de país. CONABIO, México

Cuarón AD (1997) Land-cover changes and mamal conservation in Mesoamerica. PhD dissertation, University of Cambridge, Cambridge

Cuarón AD (2000) Effects of land-cover changes on mammals in a Neotropical region: a modelling approach. Conserv Biol 14:1676–1692

De Vos J (2002) Una tierra para sembrar sueños: historia reciente de la Selva Lacandona 1950–2000. CIESAS-Fondo de Cultura Económica, México DF

Escamilla A, Sanvicente M, Sosa M, Galindo C (2000) Habitat mosaic, wildlife availability, and hunting in the tropical forest of Calakmul, Mexico. Conserv Biol 14:1592–1601

Halls LK (ed) (1984) White-tailed deer: ecology and management. Stackpole Books, Harrisbugh

Instituto Nacional de Ecología (INE) (2000) Programa de manejo, Reserva de la Biósfera Montes Azules, México. Secretaría de Medio Ambiente, Recursos Naturales y Pesca, México

Jorgenson JP (2000) Wildlife conservation and game harvest by Maya hunters in Quintana Roo, Mexico. In: Robinson JG, Bennett EL (eds) Hunting for sustainability in tropical forests. Columbia University Press, New York, pp 251–266

March IJ (1987) Los Lacandones de México y su relación con los mamíferos silvestres: un estudio etnozoológico. Biotica 12:43–56

Medellín RA (1994) Mammal diversity and conservation in the Selva Lacandona, Chiapas, Mexico. Conserv Biol 8:780–799

Naranjo EJ (2002) Population ecology and conservation of ungulates in the Lacandon Forest, Mexico. PhD dissertation, University of Florida, Gainesville

Naranjo EJ (2018) Baird's tapir ecology and conservation in Mexico revisited. Trop Conserv Sci 11:1–4

Naranjo EJ, Bodmer RE (2007) Source-sink systems of hunted ungulates in the Lacandon Forest, Mexico. Biol Conserv 138:412–420

Naranjo EJ, Bolaños JE, Guerra MM, Bodmer RE (2004a) Hunting sustainability of ungulate populations in the Lacandon Forest, Mexico. In: Silvius KM, Bodmer RE, Fragoso JMV (eds)

People in nature: wildlife conservation in South and Central America. Columbia University Press, New York, pp 324–343

Naranjo EJ, Guerra MM, Bodmer RE, Bolaños JE (2004b) Subsistence hunting by three ethnic groups of the Lacandon Forest, Mexico. J Ethnobiol 24:233–253

Naranjo EJ, Dirzo R, López-Acosta JC, Rendón-von Osten J, Reuter A, Sosa-Nishizaki O (2009) Impacto de los factores antropogénicos de afectación directa a las poblaciones silvestres de flora y fauna. In: CONABIO (ed) Capital natural de México, Estado de conservación y tendencias de cambio, vol 2. CONABIO, México, pp 247–276

Naranjo EJ, López-Acosta JC, Dirzo R (2010) La cacería en México. Biodiversitas 91:6–10

Naranjo EJ, Amador SA, Falconi FA, Reyna-Hurtado RA (2015) Distribución, abundancia y amenazas a las poblaciones de tapir centroamericano (Tapirus bairdii) y pecarí de labios blancos (Tayassu pecari) en México. Therya 6:227–249

Ojasti J, Dallmeier J (2000) Manejo de fauna silvestre Neotropical. Smithsonian Institution/Man and Biosphere Program, Washington

Robinson JG, Bennett EL (eds) (2000) Hunting for sustainability in tropical forests. Columbia University Press, New York

Robinson JG, Bodmer RE (1999) Towards wildlife management in tropical forests. J Wildl Manage 63:1–13

Robinson JG, Redford KH (eds) (1991) Neotropical wildlife use and conservation. University of Chicago Press, Chicago

Secretaría de Medio Ambiente y Recursos Naturales (SEMARNAT) (2010) Norma Oficial Mexicana NOM-059-SEMARNAT-2010. Protección ambiental, especies nativas de flora y fauna silvestres de México, categorías de riesgo y especificaciones para su inclusión, exclusión o cambio, y lista de especies en riesgo. Diario Oficial de la Federación, 30 de diciembre de 2010:1-78

Sokal RR, Rohlf FJ (1995) Biometry. WH Freeman, New York

Soto-Pinto L, Castillo-Santiago MA, Jiménez-Ferrer G (2012) Agroforestry systems and local institutional development for preventing deforestation in Chiapas, Mexico. In: Moutinho P (ed) Deforestation around the world. InTech, Rijeka, pp 333–350

Sowls LK (1997) Javelinas and other peccaries: their biology, management, and use. Texas A&M University Press, College Station

Tejeda C, Naranjo EJ, Cuarón AD, Perales H, Cruz-Burguete JL (2009) Habitat use of wild ungulates in fragmented landscapes of the Lacandon Forest, southern Mexico. Mammalia 73:211–219

Vásquez MA, Ramos M (eds) (1992) Reserva de la Biósfera Montes Azules, Selva Lacandona: Investigación para su conservación. Publicaciones Especiales Ecosfera, San Cristóbal de Las Casas

Chapter 7
Ungulates of Costa Rica

Marco A. Ramírez-Vargas and Lilliana M. Piedra-Castro

Abstract Costa Rica's 51,100 km^2 terrestrial territory represents only the 0.03% of the world's land area. Despite being a small country, Costa Rica harbors approximately 5% of the world's known biodiversity, this makes it one of the world's 20 countries with greater species diversity. About 2.5 MY ago, Costa Rica and Panamá served as an intercontinental bridge between North and South America. This land bridge favored a two-way movement of biota between continents. This closure was of special significance for ungulates in the Neotropics. With the isthmus establishment, the Cervidae, Tayassuidae and Tapiridae families (nonexistent in South America until the closure) gained an access route to the new continent, thus initializing the colonization and diversification of Southern ungulates in America. Nowadays five species inhabit Costa Rica: the Central American red brocket (*Mazama temama*), the white-tailed deer (*Odocoileus virginianus*), the white-collared peccary (*Pecari tajacu*), the white-lipped peccary (*Tayassu pecari*) and the Baird's tapir (*Tapirus bairdii*). In order to establish the actual ecological and biological state of the knowledge of the different species inhabiting Costa Rica, we reviewed the literature published between 2008 and 2018. On general terms, most research on ungulates was focused on the Caribbean low-humid lands, establishing relative population indexes during short time periods. Worth saying, many studies reporting ungulate abundances were actually jaguar monitoring studies. Also, a big deal of variation between sampling efforts was noted. No long-term population monitoring of any ungulate species was found. Regarding species, the white-collared peccary and the Baird's tapir are the most studied species recently. On the other hand, the Central American red brocket is a data-deficient species. The white-lipped peccary represents a serious conservation concern due to its low population size, as a result of historical illegal hunting and habitat loss. Regarding habitat analysis, in qualitative terms, there is no change in the species range within the country. The white-collared peccary, the Baird's tapir, and the white-tailed deer are the most registered ungulates. Few information about habitat suitability and its relationship with population viability or biological connectivity and its impacts (for example

M. A. Ramírez-Vargas · L. M. Piedra-Castro (✉)
Natural Resources and Wildlife Management Lab, National University, Heredia, Costa Rica
e-mail: lilliana.piedra.castro@una.ac.cr

S. Gallina-Tessaro (ed.), *Ecology and Conservation of Tropical Ungulates in Latin America*, https://doi.org/10.1007/978-3-030-28868-6_7

roadkill) has been accomplished recently. Recent ecological and behavioral research is focused on the analysis of interspecific interactions of the white-collared peccary in low humid forests on the one hand and the social behavior of highland Baird's tapirs in the other. There is no recent study focusing on basic ecological aspects of the white-tailed deer, the Central American red brocket or white-lipped peccary. On veterinarian terms, research on wild ungulates is a relatively new area, and has focused mainly on parasitology. Many important research have been done in the country, improving the general knowledge of the species. However, in order to ensure ungulate conservation in Costa Rica, more research and actions such as long-term monitoring for establishing population trends, studies of ecological and genetic connectivity between populations, local communities' involvement in ungulate monitoring, and turning ungulates into key species within different wildlife areas should become a priority.

7.1 Introduction

Costa Rica is a 51,100 km^2 large country representing only 0.03% of the world's land area. Despite being a small country, it harbors approximately 5% of the world's known biodiversity, becoming one of the world's 20 countries with greater species diversity (SINAC 2014). This great biotic richness is due, in part, to the fact that the country served as an intercontinental bridge between North and South America. This land bridge (the Panamá Isthmus) favored a two-way movement of the biota between masses known as "The Great American Biotic Exchange" (GABI) about 2.5 MY ago (Stehli and Webb 1985; Webb 2006). Costa Rica and Panama formation was of special significance for ungulates in the Neotropics. With the isthmus establishment, the Cervidae, Tayassuidae, and Tapiridae families (nonexistent in South America until the closure) gained an access route to the new continent, thus initializing the colonization and diversification of ungulates in southern Central America and South America (Ruiz-García et al. 2007). Five species of ungulates are known to inhabit the country:

Central American red brocket (Cervidae: *Mazama temama*): locally known as "cabro de monte." It has been reported in low and humid lands of the Caribbean, Central and South Pacific slope, from sea level to approximately 2300 m above sea level (masl). According to the National System of Conservation Areas (SINAC) resolution (R-SINAC-CONAC-092-2017) the Central American red brocket has reduced or threatened populations in Costa Rica.

White-tailed deer (Cervidae: *Odocoileus virginianus*): locally known as "venado cola blanca." It inhabits the country from 0 to 1300 masl, in a variety of environments such as wooded paddocks, pastures, scrubs (where it is sometimes mixed with domestic ungulates), as well as dry to humid secondary forests. The species was introduced in Coco's Island (Costa Rican Pacific). The species was declared as Costa Rica's national wildlife symbol in 1995. White-tailed deer does not exhibit threatened or endangered populations within the country (R-SINAC-CONAC-092-2017).

White-collared peccary (Tayassuidae: *Pecari tajacu*): locally known as "saíno". Inhabits throughout the country, from 0 to 3000 masl, in dry forests, gallery forests, moist forests and mature secondary forests, as in crop and pastures areas. According to the R-SINAC-CONAC-092-2017 resolution, the species has reduced or threatened populations in Costa Rica.

White-lipped peccary (Tayassuidae: *Tayassu pecari*): locally known as "cariblanco" or "chancho de monte." Early records of the species placed it from 0 to 1900 masl; but recent records expand its distribution up to 3000 masl, in both Caribbean and Pacific slopes. It prefers mature or slightly disturbed moist forests. According to the R-SINAC-CONAC-092-2017 resolution, the species has reduced or threatened populations in Costa Rica.

Baird's tapir (Tapiridae: *Tapirus bairdii*): locally known as "danta". It can be found from moist to dry mature and secondary forests, as well as in bamboo (*Chusquea* spp.) areas, from sea level to 3800 masl on both slopes. According to R-SINAC-CONAC-092-2017 resolution, the species has reduced or threatened populations in Costa Rica.

The following chapter aims to collect recent information (between 2008 and 2018) on wild ungulates generated in Costa Rica. Topics such as ecology, biology, distribution and conservation of the five species are included in order to update the state of the knowledge of these species in the country, thus establishing investigation priorities. The information here presented was generated through consultation and review of research papers, from different databases and information sources such as ISI Web of Science, Binabitrop Central America, SINABI Costa Rica, government technical reports, and personal interviews of park rangers.

Given the fact that the great majority of research carried out in the country is within the limits of protected wild areas, it is worth mentioning briefly how these areas are organized in Costa Rica. The National Conservation Areas System (SINAC) was established in 1995, belonging to the Environment and Energy Ministry (MINAE). The country was then divided into 11 large Conservation Areas (CA) (Fig. 7.1), which are land units managed by a sustainable development strategy. Within the CAs, the Protected Wildlife Areas (PWA) are located. These areas, being representative of different life zones (Holdridge 1967) (Fig. 7.2), are defined by their natural, cultural, and/or socioeconomic importance, complying several conservation and management objectives (La Gaceta 2012). These areas are classified as: Forest Reserves (FR), Protective Areas (PA), National Parks (NP), Biological Reserves (BR), National Wildlife Refuges (NWR), Wetlands (W), Natural Monuments (NM), Marine Reserves (MR), and Marine Management Areas (MMA) (SINAC 2018).

7.2 Population Analysis

According to national regulations, with the exception of the white-tailed deer (declared national wildlife symbol since 1995), all of the Costa Rica's wild ungulates are threatened. As it is common throughout its distribution, ungulates have

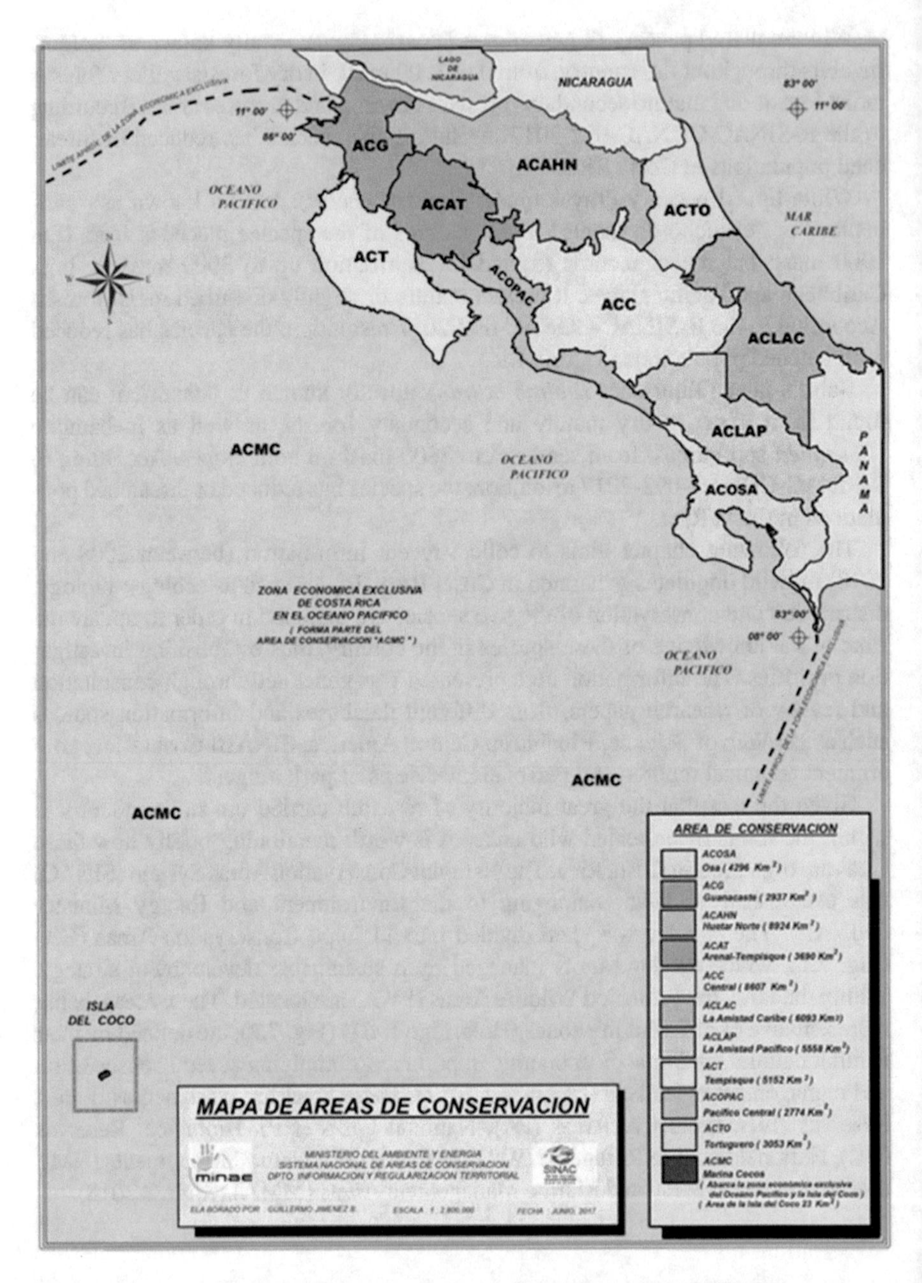

Fig. 7.1 Division of Costa Rica into Conservation Areas (CA). Source: SINAC 2018

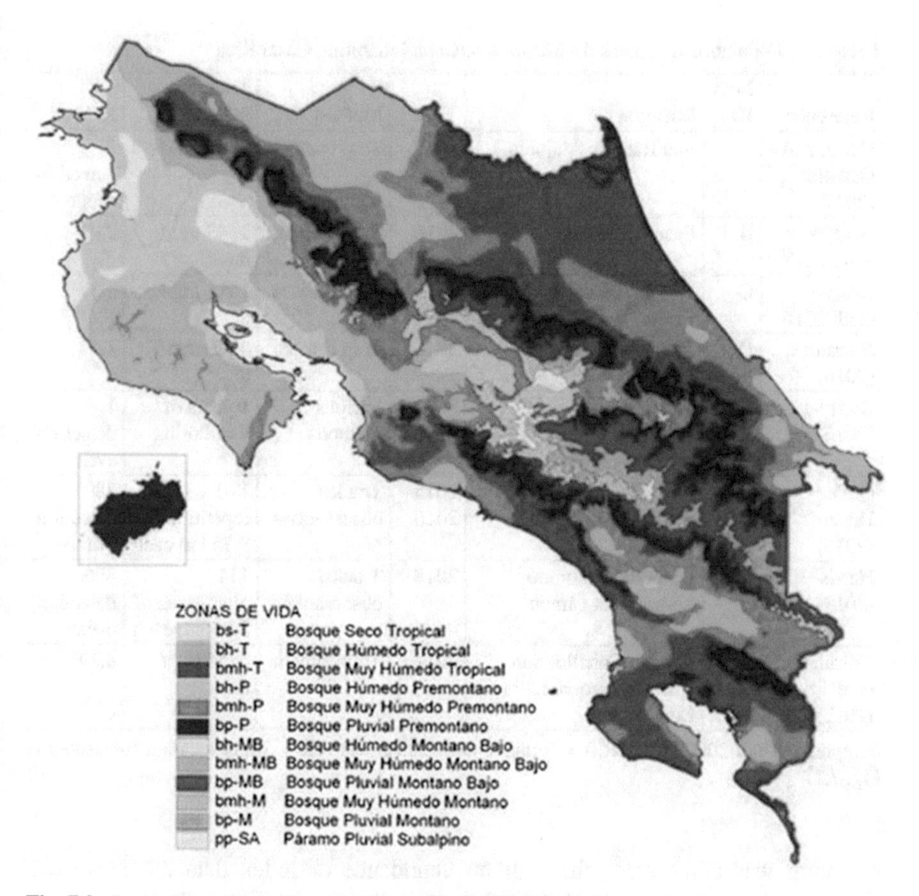

Fig. 7.2 Costa Rica's life zones according to Holdridge (1967). Source: Bolaños et al. 2005

been considered objective species, both for sport hunting and consumption. Both hunting and the marked habitat loss that occurred within the country between 1960s and 1990s, are factors that contributed with the current state of threat of the populations.

For the Central American red brocket, seven studies included population data (Table 7.1). Greater research effort was concentrated in the Caribbean region, using trap cameras as sampling method (Fig. 7.3). In only four cases, the authors estimated relative abundance of the species. Only one study made a time tracing (6 years), however a population estimate is not presented.

Regarding the white-tailed deer, two studies presented population data for the species (Table 7.2) between 2008 and 2018. Records are from the Caribbean, an area where the species has always been considered rare (Table 7.2).

Regarding the white-lipped peccary two studies included population index (Table 7.3). The largest sampling effort was concentrated in the North Caribbean using camera traps as a sampling method. Only in one of the cases several years of

Table 7.1 Population estimates for *Mazama temama* inhabiting Costa Rica

Reference	Map ID	Location	Year	Method	Trapping effort	R.A
Cartín and Carrillo (2017)	A	San Ramón, Alajuela.	2017	Trap camera	720 trap/days	3 detection events
Arroyo et al. 2017	B	Pacuare, Limón	2016	Trap camera	1643 trap/ days	2.01
Arroyo et al. 2016	C	Barra del Colorado, Limón	2016	Trap camera	1611 trap/ days	1.37
Barrantes (2016)	D	Maquenque, Heredia-Alajuela	2016	Trap camera	4370 trap/ days	1.37
Brett (2016)	E	Jalova, Limón	2010–2016	Various methods	6 years of monitoring	1 detection event
Quilez-Huezo (2016)	F	Barra del Colorado-Tortuguero, Limón	2015–2016	Tracks/ observations	170 repetitions of 3.55 km each	380 detection events
Harris (2018)	G	Barra del Colorado-Tortuguero, Limón	2018	Tracks/ observations	111 repetitions of 3.55 km each	396 detection events
Corrales et al. (2012)	L	Braulio Carrillo, San José-Cartago-Heredia-Limón	2011	Trap camera	3755 trap/ days	8.79

Sampled period: 2008–2018. R.A = Relative abundance; Map ID = Study location (presented in Fig. 7.3)

sampling was completed, although no abundance or index data are presented. However, this is the only species that has a population viability analysis.

By simulating scenarios incorporating both biological (i.e. age of sexual maturity) and ecological (i.e. density-dependence reproduction) variables, Rivera (2014) determined the effect of hunter-gold miners on the white-lipped peccary population that inhabits the Corcovado NP. If a 250 hunter-gold miners pressure is maintained in the NP, white-lipped population has a 40% probability of extinction in a five-year period and a 99% chance of the extinction within a 10 year period. Results highlighted the importance of implementing strict actions in the Corcovado Park, as it is one of the last country's areas that still conserves an important white-lipped peccary population.

The white-collared peccary is the species with the highest number of abundance estimates, (11 studies) (Table 7.4). The largest sampling effort is concentrated in the north Caribbean, using trap cameras. Even though several studies collected population data, only two (Quilez-Huezo 2016 and Harris 2018, both in Barra del Colorado-Tortuguero) can be used for a preliminary analysis of population trends. For the 2015–2016 period 261 detection events were reported while 187 detections were recorded in 2018. It could be estimated that a population decrease of the species has occurred in the Barra del Colorado area.

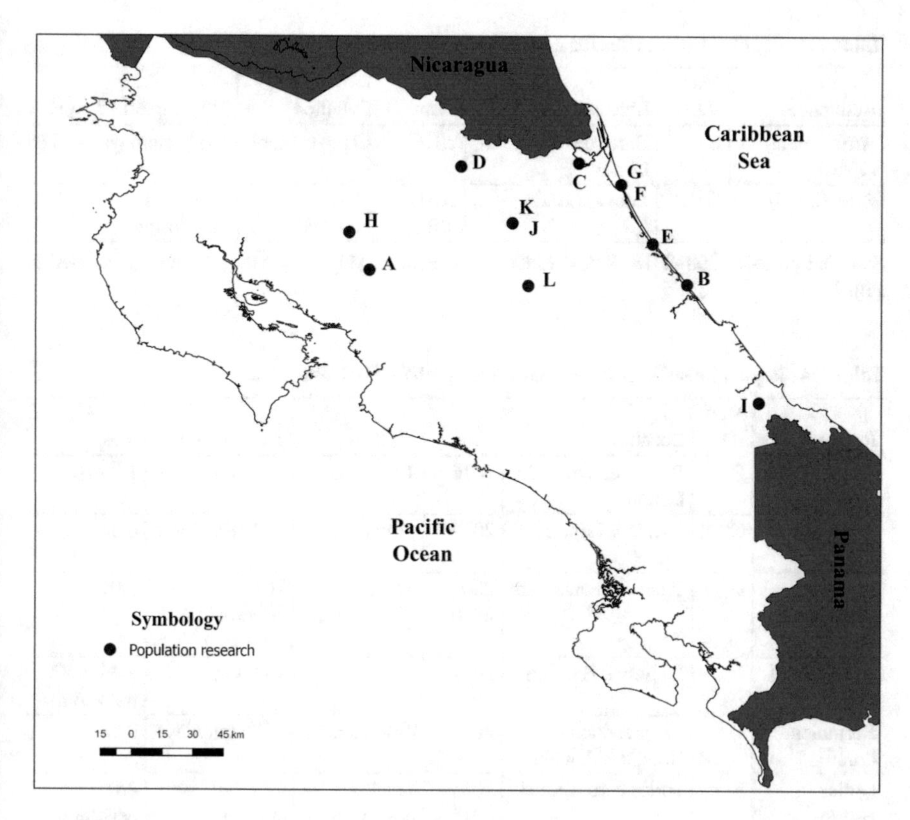

Fig. 7.3 Localities where ungulate population estimates have been provided. Sampled period 2008–2018

Table 7.2 Population estimates for *Odocoileus virginianus* inhabiting Costa Rica

Reference	Map ID	Location	Year	Method	Trapping effort	R.A
Barrantes (2016)	D	Maquenque, Heredia-Alajuela	2016	Trap camera	4370 trap/days	2.74
Brett (2016)	E	Jalova, Limón	2010–2016	Various methods	6 years of monitoring	NR

Sampled period: 2008–2018. R.A = Relative abundance; Map ID = Study location (presented in Fig.7.3)

For the Baird's tapir, eight studies included some kind of population estimate (Table 7.5). The greatest research effort was in the Huetar Norte and Caribbean region, using trap cameras as the main sampling method. In none of the cases a time follow-up was accomplished. It is noteworthy that for many years, general research on Baird's tapir was carried out only in lowland areas. Until 2006, González-Maya et al. (2012) provided the first estimate of tapir density in highlands. Authors pointed

Table 7.3 Population estimates for *Tayassu pecari* inhabiting Costa Rica

Reference	Map ID	Location	Year	Method	Trapping effort	R.A
Arroyo et al. (2016)	B	Barra del Colorado, Limón	2016	Trap camera	1611 trap/days	0.68
Brett (2016)	E	Jalova, Limón	2010–2016	Various methods	6 years of monitoring	NR

Sampled period: 2008–2018. R.A = Relative abundance; Map ID = Study location presented in Fig. 7.1

Table 7.4 Population estimates for *Pecari tajacu* inhabiting Costa Rica

Reference	Map ID	Location	Year	Method	Trapping effort	R.A
Arroyo et al. (2016)	B	Barra del Colorado, Limón	2016	Trap camera	1611 trap/days	1.24
Arroyo et al. (2016)	C	Pacuare, Limón	2016	Trap camera	1643 trap/days	0.30
Arévalo et al. (2015)	H	Tilarán, Guanacaste	2009–2010	Tracks	101 trail surveys (303 km)	0.01
Arroyo et al. 2013	J	La Selva, Heredia	2011	Tracks	13.2 km	5.34–6.82 tracks/km
Barrantes (2016)	D	Maquenque, Heredia-Alajuela	2016	Trap camera	4370 trap/days	0.68
Quilez-Huezo (2016)	F	Tortuguero, Limón	2015–2016	Tracks/observations	170 trail surveys (3.55 km each)	261 detection events
Romero et al. (2013)	K	La Selva, Heredia	2005–2007	Observations	348 survey/days	19.05–65.92 ind/km^2
Cartín and Carrillo (2017)	A	San Ramón, Alajuela.	2017	Trap camera	720 trap/days	19 detection events
Brett (2016)	E	Jalova, Limón	2010–2016	Various methods	6 years of monitoring	1 detection event
Harris (2018)	G	Barra del Colorado-Tortuguero, Limón	2018	Tracks/observations	111 (trail surveys) 3.55 km each	187 detection events
Corrales et al. (2012)	L	Braulio Carrillo, San José-Cartago-Heredia-Limón	2011	Trap camera	3755 trap/days	29.83

Sampled period: 2008–2018. R.A = Relative abundance; Map ID = Study location (presented in Fig. 7.3)

out that tapir density there was high when compared with data from lowland areas. This suggests that country's highlands are tapir havens that maintain greater ecological integrity, providing resources to sustain larger populations.

Table 7.5 Population estimates for *Tapirus bairdii* inhabiting Costa Rica

Reference	Map ID	Location	Year	Method	Trapping effort	R.A	A.A
Barrantes (2016)	D	Maquenque, Heredia-Alajuela	2016	Trap camera	4370 trap/days	19.22	NR
González-Maya et al. (2012)	I	Valle del Silencio, Limón	2006	Trap camera	540 trap/days	2.93 ind/km^2	17–36
Chassot et al. (2009)	J	San Juan-La Selva, Heredia-Alajuela	2008	SIG	–	0.53–1.60 ind /km^2	69–208
Arroyo et al. (2016)	B	Barra del Colorado, Limón	2016	Trap camera	1611 trap/days	1.18	NR
Quilez-Huezo (2016)	F	Tortuguero, Limón	2015–2016	Tracks/observations	170 repetitions of 3.55 km each	21 detection events	NR
Brett (2016)	E	Jalova, Limón	2010–2016	Various methods	6 years of monitoring	–	NR
Harris (2018)	G	Barra del Colorado-Tortuguero, Limón	2018	Tracks/observations	111 trail surveys (3.55 km each)	60 detection events	NR
Schanck et al. (2017)	M	Corcovado, Puntarenas	2000–2015	Ecological modeling	–	0.81 ind/km^2	
Corrales et al. (2012)	L	Braulio Carrillo, San José-Cartago-Heredia-Limón	2011	Trap camera	3755 trap/days	19.71	NR

Sampled period: 2008–2018. R.A = Relative abundance; A.A = Absolute Abundance. Map ID = Study location (presented in Fig. 7.3)

7.3 Distribution and Habitat Requirements

Historically, establishment of species distribution patterns are based on individual collection for later entry into museum collections. However, especially for endangered species, photographic records taken by the public have been used for the generation of distribution analysis. This information can be accessed through different virtual platforms (e.g. Global Biodiversity Information Facility: https://www.gbif.org/). People involvement in sampling efforts can complement and even increase the information generated by professional researchers, which in Latin America usually have important limitations for data generation. Citizen action is particularly useful in Costa Rica, being a country with an important ecotourism activity, in which hundreds of generally well trained naturalist guides participate in the sighting and identification of wildlife. From information generated jointly by citizens and researchers and accessed by us through the GBIF platform, different aspects of the distribution of ungulate species in the country during the last 10 years are revealed.

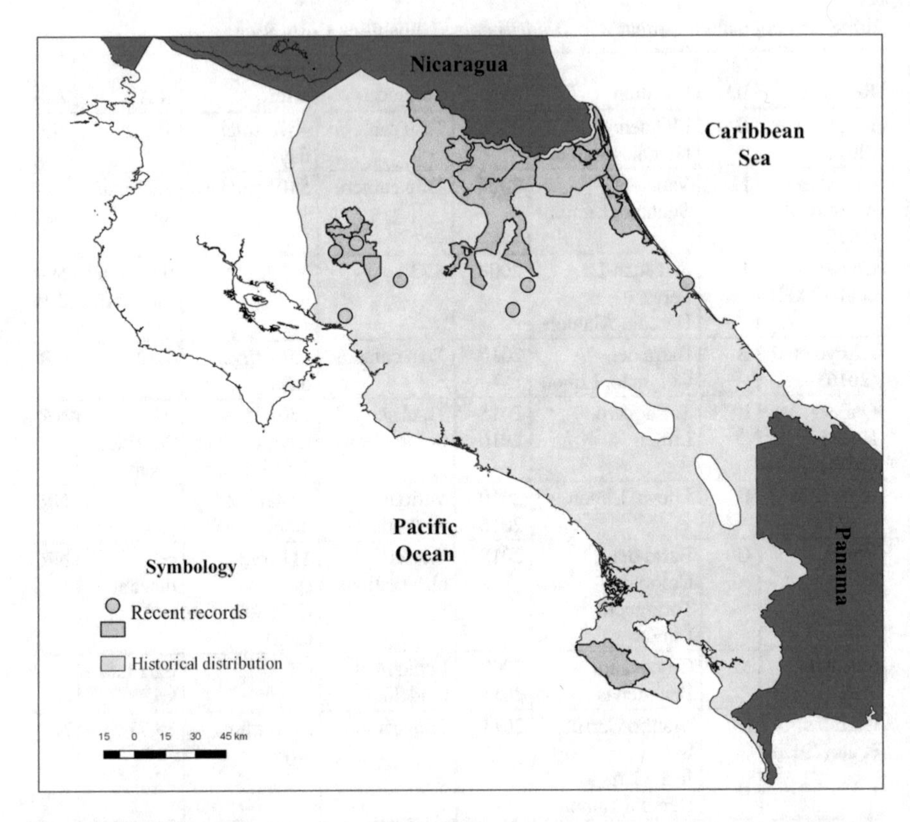

Fig. 7.4 Historical distribution (Carrillo et al. 1999) and recent records (2008–2018) of *Mazama temama* in Costa Rica. Records obtained from bibliographic data and the Global Biodiversity Information Facility (https://www.gbif.org/). Polygons area protected wildlife areas where the species has been recorded but an exact location has not been provided

In general terms, all records were obtained within the limits of the species historical distributions, therefore is not possible to ensure that there has been a change, at least in qualitative terms, in species distribution within the country. Most of the records are white-collared peccary, followed by tapir and white-tailed deer. It is noteworthy the little record of Central American red brocket and the white-lipped peccary were obtained.

For the red brocket, the few records obtained are from the Tilaran and the Central Volcanic mountain range, as well as the North Caribbean and the Osa Peninsula (Fig. 7.4); while no records were obtained in the rest of the country. In Costa Rica the red brocket is characteristic of well-preserved humid forest, so it can be a bioindicator species, although no habitat analysis for the species has been developed.

Unlike the red brocket, the white-tailed deer is a frequently seen species, especially along lowland areas of the North and Central Pacific (Fig. 7.5). Although there are few records in the North and South Caribbean, these areas are considered potential but rare habitat for the species, so the reduced sighting here is not surprising. In habitat terms, there is no recent research for this subject.

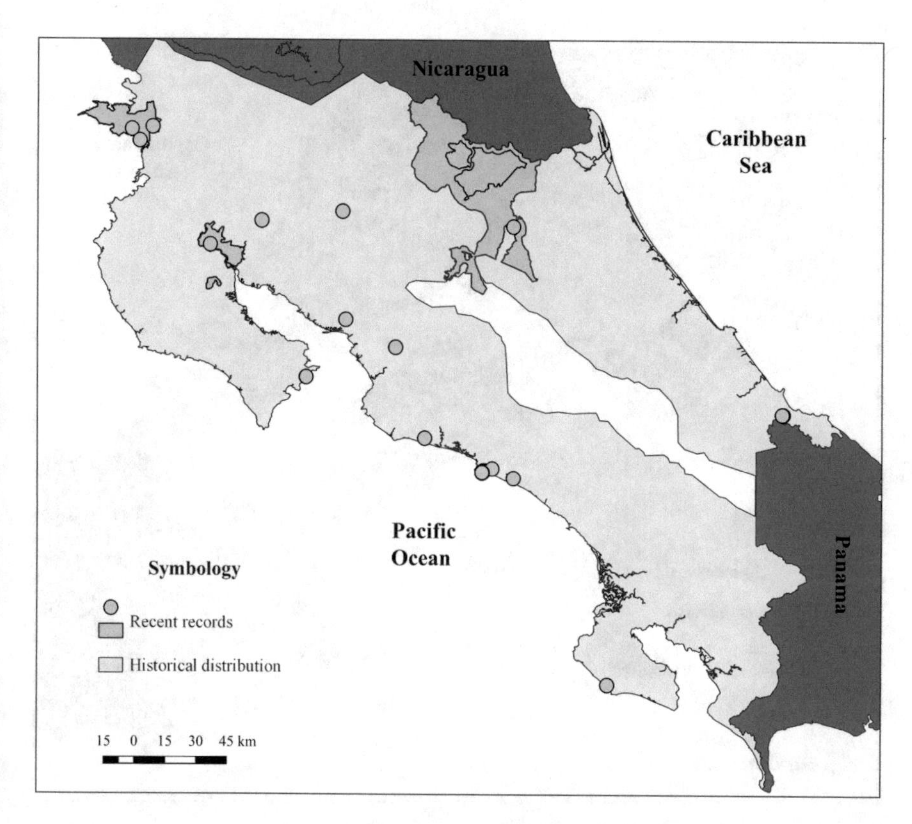

Fig. 7.5 Historical distribution (Carrillo et al. 1999) and recent records (2008–2018) of *Odocoileus virginianus* in Costa Rica. Records obtained from bibliographic data and the Global Biodiversity Information Facility (https://www.gbif.org/). Polygons are protected wildlife areas where the species has been recorded but an exact location has not been provided

The white-lipped peccary has had few sighting in recent years (Fig. 7.6). Recent records are based on the North Caribbean, especially in the San Juan-La Selva biological corridor and the Tortuguero NP, as well as in the Corcovado NP (south Pacific). Its scarce registry is a reflection of the decimated populations. Regarding its state, Altrichter et al. (2011) pointed out that in Costa Rica the species disappeared from 89% of its original distribution.

Given the species precarious situation, it is important to highlight recent records. In 2016, white-lipped pecari was rediscovered in the Rincón de la Vieja NP (where it was considered extinct) and in the Caño Negro NWR (North Caribbean) (Leonardi et al. 2010; Guerrero and Morazán 2016). While the first record of the species was obtained in the PN Barbilla (Talamanca mountain range) during 2017 (Esquivel-Cambronero et al. 2017). Those records may be evidence of the species recovery in certain areas; however, in the absence of more study efforts, it is a premature affirmation.

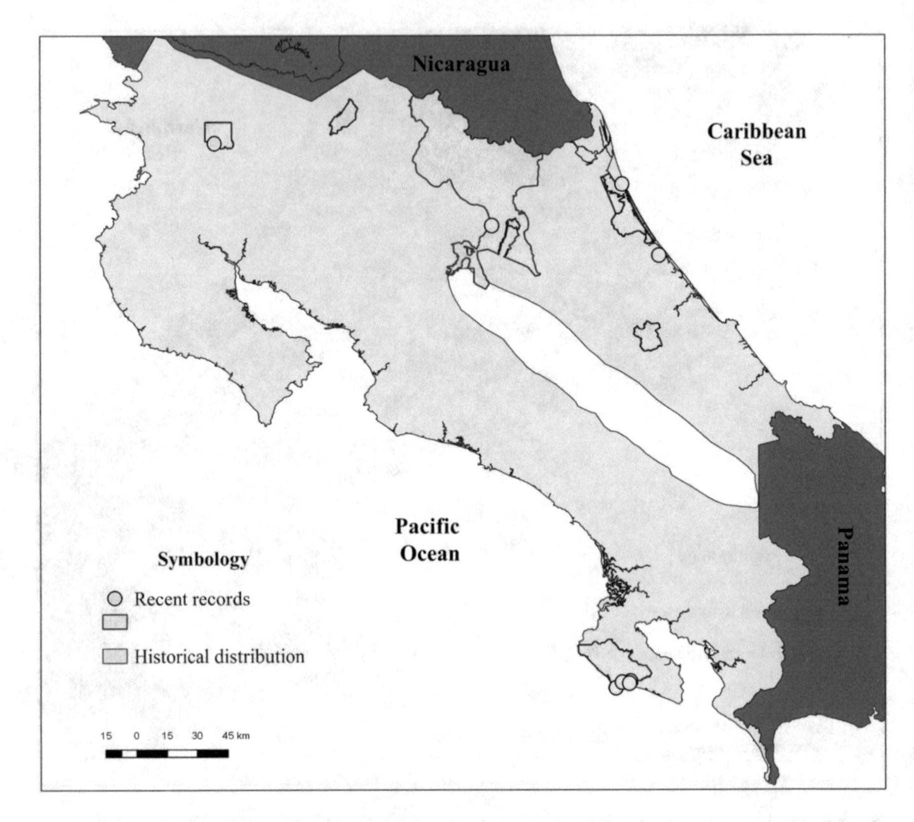

Fig. 7.6 Historical distribution (Carrillo et al. 1999) and recent records (2008–2018) of *Tayassu pecari* in Costa Rica. Records obtained from bibliographic data and the Global Biodiversity Information Facility (GBIF, https://www.gbif.org/). Polygons are protected wildlife areas where the species has been recorded but an exact location has not been provided

Contrary to what happens with the white-lipped peccary, the white-collared pecari is frequently recorded in the country (Fig. 7.7). In areas such as the Braulio Carrillo NP and the La Selva Biological Station the species is even considered abundant. However, there is few records in areas such as the Nicoya peninsula (north Pacific) and the south Caribbean. The white-collared pecari is considered a more adaptable species than the white-lipped peccary, which makes it a more resilient species. However, hunting pressure and habitat transformation, makes it rare in certain parts of the country, especially non-protected areas. No recent information on habitat availability was obtained.

Baird's tapir has considerable amount of records in certain areas, such as the Central Volcanic mountain range and the North Caribbean (especially the San Juan-La Selva Biological Corridor) and the Osa Peninsula. However, there is significant lack of information about the species in the North and Central Pacific and Southern Caribbean regions (Fig. 7.8). Regarding habitat availability, Schank et al. (2015) determined that 66% of the country can be classified as Baird's tapir

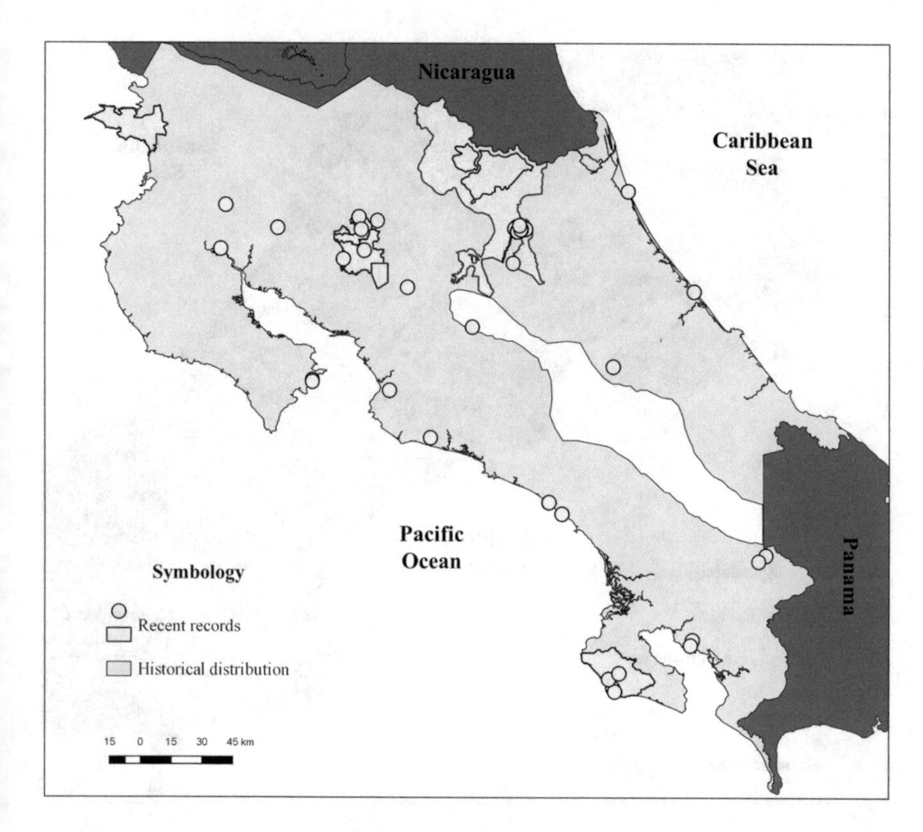

Fig. 7.7 Historical distribution (Carrillo et al. 1999) and recent records (2008–2018) of *Pecari tajacu* in Costa Rica. Records obtained from bibliographic data and the Global Biodiversity Information Facility (GBIF, https://www.gbif.org/). Polygons are protected wildlife areas where the species has been recorded but an exact location has not been provided

potential habitat, however only 35% is in protected wildlife areas. Although the model used has limitations, it is an important starting point for the management of conservation units for Baird's tapir in Costa Rica.

7.4 Habitat Fragmentation and Connectivity

Although Costa Rica has considerable forest cover (52% of the country) (Fonseca et al. 2010; Canet 2015) heavy deforestation during the 1960s and 1970s as well as a growing road network (7759.5 km in a 51,100 km^2 country) has caused heavy forest fragmentation which jeopardizes species survival. Currently, different initiatives have been designed trying to connect habitats and wildlife populations. Through the National Program of Biological Corridors, at least 44 corridors have been established. However, generally there are no studies evaluating functional connectivity of this areas.

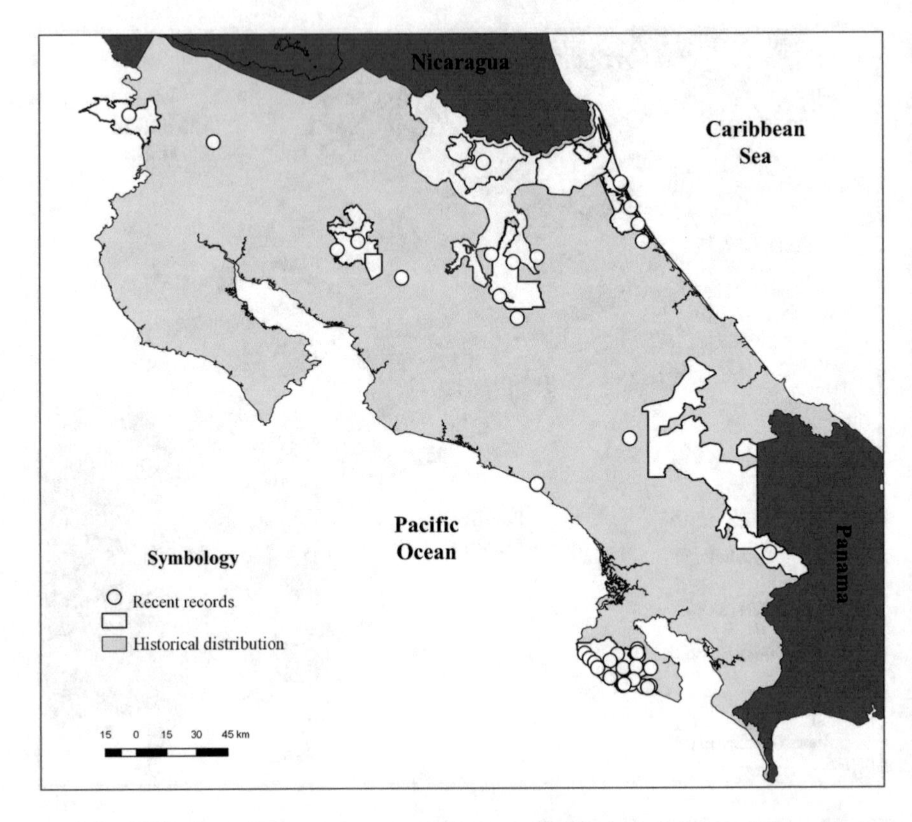

Fig. 7.8 Historical distribution (Carrillo et al. 1999) and recent records (2008–2018) of *Tapirus bairdii* in Costa Rica. Records obtained from bibliographic data and the Global Biodiversity Information Facility (GBIF, https://www.gbif.org/). Polygons are protected wildlife areas where the species has been recorded but an exact location has not been provided

Chassot et al. (2009) using the Baird's tapir as model, highlighted that the country's Northern forest swamps are key on maintaining the Nicaragua–Costa Rica connectivity within the Mesoamerican Biological Corridor. However, small-sized forest fragments and habitat degradation seriously threatens corridor functionality. For this area, Baird's tapir potential habitat is 130,022ha (130 km^2). However, many of the areas designated as viable habitat, are small forest fragments which do not meet the minimum needs for tapirs regarding their home range.

Subsequently, Cove et al. (2013) between 2009 and 2011 analyzed which habitat conditions influence Baird's tapir use of the corridor the most. The main explanatory variables was adjacency to main protected areas (core zones) and native and/or forest cover (even secondary). Although Baird's tapir presence is rare, it seems to be more resistant and more capable of using the available patches into the fragmented corridor than previously thought. Although there are adjacent sites with Baird's tapir favorable environmental conditions, if they are small in size (<75 h) they may not be large enough to support the presence of tapirs, only if they are connected to larger areas. The importance of the Indio Maíz Reserve (Nicaragua) as a population source of tapirs for the north of CR was also highlighted, which indicates the importance of bilateral conservation action.

In Costa Rica, many protected wildlife areas, such as the Braulio Carrillo, Los Quetzales, and Carara NPs, are somehow crossed by roads. Road network increase threatens long-term survival of ungulates and other mammals, not only by population isolation and direct killing, but also affecting plant composition, forest cover, food quality and availability, and microclimatic regimes (Watson 2005; Carvajal and Díaz 2016; Mónge-Nájera 2018). Evaluations carried in the country, placed the Baird's tapir as the most affected ungulate species regarding road kills. From 2010–2017, 23 killings occurred in the Cerro de la Muerte área (Los Quetzales NP) (Sancho 2017).

Because road killing affects species conservation, recently a "Roads and Wildlife Commission" araised in the country. The commission brought together academy, nongovernmental organizations, state institutions and society, in order to plan actions that minimize road impacts. Among the most important actions promoted by the commission are the road ecology symposium (2013), radio podcasts, lectures and workshops, traffic signal installations, and an environmental guide for road design. Though these actions are of special importance, more research is required in order to better asses how roads affect ungulate populations in the country.

7.5 Ecology and Behavior

The ecological research around wild ungulates in Costa Rica has focused on the analysis of interspecific interactions of the collared peccary and on the social behavior of the tapir. There is no recent study focusing on basic ecological aspects of the white-tailed deer, Central American red brocket, or white-lipped peccary.

The white-collared pecari is an important herbivore, feeding on a variety of plants while forming herds. Kuprewicz (2010) and Ávalos et al. (2016), indicated that because peccaries are voracious seeds and plant predators, the herds have negative effects on the survival and establishment of important rain forest plants such as *Astrocaryum alatum*, *Iriartea deltoidea*, *Socratea exorrhiza*, *Dipteryx panamensis* and *Mucuna holtonii*, as demonstrated in the La Selva RB. These studies showed that chemical and physical defenses of seeds are not enough against the white-collared peccary, so a significant number of seeds get destroyed, while the consumption of the emerging roots (such as the ones of *S. exorrhiza*) decreases the size of the cone, which especially affects the establishment of medium-sized palms.

Modification of forest structure by white-collared peccary has implications on faunistic diversity such as birds, amphibians and reptiles. Michel et al. (2014, 2015) analyzed the effect of different *P. tajacu* densities on liana density and coverage which are used as foraging and nesting elements for different rainforest understory bird species: *Epinecrophylla fulviventris, Microrhopias quixensis* and *Terenotriccus erythrurus*. The study took place in the Bartola Refuge (Nicaragua),

La Selva and Tirimbina BS's (Costa Rica), and Barro Colorado, Gigante, and Limbo areas (Panamá). White-collared peccary density in all sites explained up to 92% of the in situ variation of the average density and coverage of forest lianas, thus negatively affecting bird abundance (although the total area of forest and precipitation also had negative effects on the abundance of birds).

Reider et al. (2013), also in the La Selva BS, showed that white-collared peccary increase affects amphibian and reptile populations. Given that understory herpetofauna depends on leaf litter as habitat, trampling by peccaries modifies microhabitat dynamics. Evidence showed that more (16%) amphibian and reptile abundance (mostly juveniles) were found in areas where peccaries were not excluded. Peccary trampling not only creates small puddles, but by soil mixing and nutrients adding (through urine and manure) peccaries could influence leaf litter decomposition. Decomposition feeds the diverse community of invertebrates, which in turn constitute the largest source of prey for amphibians and reptiles.

For the Baird's tapir new research was developed in the country, understanding aspects of its behavioral ecology, which is poorly known. Gomez-Hoyos et al. (2018) developed the first study describing tapir vocalizations on the wild. In the Talamanca mountain range, they determined that tapirs have two types of calls: one is composed of two notes (average 0.625 ± 0.069 and 0.333 ± 0.080 s respectively) with a dominant frequency of 4.940 ± 248.3 Hz, while the second one is composed of a single note (1.121 ± 0.063 s) with a dominant frequency of 6.471 ± 704.1 Hz. Interactions between individuals are probably related to resource availability while vocalizations probably reflect communication between pairs (male-female) or to avoid agonistic behavior.

7.6 Ungulate Veterinary Research

Veterinary research on wild ungulates has been poorly developed and has focused mainly on tapirs. Kidney and Berrocal (2008) described, for the first time, equine sarcoids (non-metastatic fibroblastic cutaneous neoplasia) in two captive tapirs in Costa Rica. The first case was a 2-year-old male who had an active growing mass on the inner surface of the left ear for 7 months. The second case was a 3.6 year old female with a similar injury. Both tapirs were born in captivity and were closely related, sharing grandfather. A sarcoid was also diagnosed in a resident horse living in the same facility. Through PCR and ISH analysis, bovine papillomavirus DNA was detected in the sarcoid tissue of both tapirs, so it is believed that the cause of equine sarcoids was viral. Although BPV transmission in equine sarcoids remains unclear, circumstantial evidence suggests that flies are involved in the pathogenesis and epidemiology of equine sarcoids. BPV infection source could not be confirmed; however, cattle present may have played an important role. Since sarcoid was diagnosed in a horse, all three animals could have contracted BPV infection from the same source. Given the kinship between individuals (same grandfather) there may also be a genetic susceptibility.

Alvarado et al. (2018) analyzed the presence of gastrointestinal parasites in wild tapirs and its relation with environmental variables in the northwest region of the Talamanca mountain range. Eight parasite taxa were recorded: Ascarididae and Anoplocephalidae families; *Kiluluma* sp. (reported for the first time in the Central American tapir), *Strongylus vulgaris* (also reported in domestic animals), *Balantidium coli* (reported as zoonotic); *Tziminema unachi* (recently described), *Buisonella tapiri*, and *Blepharocorys cardionucleata* (commensal protozoa). Results showed that forest type and cover are the best explanatory variables for egg, larva, and cyst composition within sites (latrines), while livestock-farm distance could also be a determinant. More studies are needed in order to understand how habitat disturbance could affect cross-transmission of parasitic diseases among domestic animals, humans and wildlife.

Given current antibiotic resistance related to agriculture, aquaculture, and domestic–wildlife interaction, Rojas-Jiménez et al. (2018) analyzed antibiotic sensitivity profiles in commensal strains of *Escherichia coli* obtained from wild and captive Baird's tapir feces from the Talamanca mountain range highlands. Results suggested that intestinal microbiota of wild tapirs remained isolated from the selective pressure of the antibiotics, possibly due to low influence of anthropogenic activity. However samples collected from the captive individual, showed a multiresistance profile. This situation shows the importance of better antibiotic management in within zoos and rescue centers, as well as in areas around wildlife protected areas.

7.7 General Conclusions

After analyzing the main stream bibliography related to ungulate research in Costa Rica, a series of highlights and conclusions rise:

Research efforts are focused on the low wetlands of the Caribbean, while northern dry forests, central pacific humid forests, and highlands have received less attention.

There is no mid-to-long term population monitoring of any ungulate species in the country, thus population trends establishment is limited. Just one research program accomplished 6 years of monitoring, but focused on establishing species presence rather than population estimates.

There is a lack of research regarding habitat suitability and functional connectivity of biological corridors.

While Baird's tapir and the white-collared peccary are subject of important research, limited information regarding the Central American red brocket, the white-tailed deer and the white-lipped peccary is available.

Most studies that reported ungulate abundance indexes were actually jaguar population analysis.

Most monitoring efforts have employed trap cameras, however, there is a big deal of variation in sampling efforts between studies, thus making comparisons difficult. In addition, trap cameras are limited for individual identification.

Road and disease ecology are newly developing research areas in the country.

Research and action priorities are as follows:

Species population trends.
Disease ecology.
Genetics and reproductive ecology.
Road ecology.
Population connectivity (source–sink and metapopulation dynamics).
Wildlife captivity management.
Continue with the national forest increase in order to maintain and expand habitats and connectivity.
Involvement of local communities in ungulate population management outside protected areas.
Incorporate ungulate species in National System of Conservation Areas research priorities.

Acknowledgements We thank the Natural Resources and Wildlife Management Lab (LARNAVISI) of the School of Biological Sciences at the Universidad Nacional of Costa Rica for financial support granted for this research. Special thanks to Dr. Sonia Gallina Tessaro of the Instituto de Ecología, A.C (INECOL) for her support during the investigation and writing process; to Vanessa Morales and Maikol Castillo of LARNAVISI for their support in map designing; to the scientific managers of the different Costa Rican Conservation Wildlife Areas in providing information about research performed in their areas; and to Esteban Brenes of Ñai Conservation for information sharing.

References

Altrichter M, Taber A, Beck H, Reyna-Hurtado R, Lizarraga L, Keuroghlian A, Sanderson EW (2011) Range-wide declines of a key Neotropical ecosystem architect, the near threatened white-lipped peccary *Tayassu pecari*. Oryx 46(1):87–98. https://doi.org/10.1017/S0030605311000421

Alvarado R, Jiménez M, Jiménez A, Retamosa M (2018) Presencia de parásitos gastrointestinales en la danta centroamericana (*Tapirus bairdii*) y la relación de la composición de parásitos en sus letrinas con variables ambientales en la región noroeste de la Cordillera de Talamanca, Costa Rica. Tesis Maestría. Instituto de Conservación y Manejo de Vida Silvestre. Universidad Nacional de Costa Rica

Arévalo JE, Méndez Y, Roberts M, Alvarado G, Vargas S (2015) Monitoring species of mammals using track collection by rangers in the Tilarán mountain range, Costa Rica. Cuadernos de Investigación UNED 7(2):249–257

Arroyo-Arce S, Thomson I, Fernández C, Salom-Pérez R (2017) Relative abundance and activity patterns of terrestrial mammals in Pacuare nature reserve, Costa Rica. Cuadernos de Investigación UNED 9(1):15–21. https://doi.org/10.1901/jeab.1981.36-101

Arroyo-Arce S, Thomson I, Salom-Pérez R (2016) Relative abundance and activity patterns of terrestrial mammalian species in Barra del Colorado wildlife refuge, Costa Rica. Cuadernos de Investigación UNED 8(2):131–137. http://www.scielo.sa.cr/pdf/cinn/v8n2/1659-4266-cinn-8-02-00131.pdf

Arroyo-Arce S, Berrondo L, Canto Y, Carrillo N, Gómez-Carrillo N et al (2013) Uso de dos tipos de bosque por saínos (*Pecari tajacu*) en estación "La Selva", Costa Rica. Rev Cult Cient:32–39

Avalos G, Cambronero M, Vargas O (2016) Quantification of browsing damage to the stilt root cone of *Socratea exorrhiza* (Arecaceae) by Collared Peccaries (*Pecari tajacu*, Artiodactyla: Tayassuidae) at La Selva, Costa Rica. Brenesia 85

Barrantes M (2016) Evaluación inicial del Refugio Nacional de Vida Silvestre Mixto Maquenque a partir de la Unidad de Conservación del Jaguar Cerro Silva-Indio-Maíz-Tortuguero. Tesis Licenciatura. Escuela de Ciencas Biologicas. Universidad Nacional de Costa Rica

Bolaños R; Watson V, Tosi J (2005) Mapa ecológico de Costa Rica (Zonas de Vida), según el sistema de clasificación de zonas de vida del mundo de LR Holdridge), Escala 1:750 000. Centro Científico Tropical, San José, Costa Rica

Brett M (2016) Análisis a seis años del inventario de especies y recolección de datos de biodiversidad en el área de Jalova. Parque Nacional Tortuguero, Costa Rica

Canet G (2015) Recuperación de la cobertura forestal en Costa Rica, logro de la sociedad costarricense. Ambientico 253:17–22. ISSN 1409-214X

Carrillo E, Wong G, Sáenz J (1999) Mamíferos de Costa Rica. INBio, Santo Domingo. Heredia, p 248

Cartín M, Carrillo E (2017) Estado poblacional de mamíferos terrestres en dos áreas protegidas de la región central occidental de Costa Rica. Rev Biol Trop 65(2):493–503. https://doi.org/10.15517/rbt.v65i2.24418

Carvajal V, Díaz F (2016) Registro de mamíferos silvestres atropellados y hábitat asociados en el cantón de la Fortuna, San Carlos, Costa Rica. Biocenosis 30(1–2):49–58

Chassot O, Monge G, Jiménez V (2009) Evaluación del hábitat potencial para la danta centroamericana (*Tapirus bairdii*) en el corredor biológico San Juan- La Selva, Costa Rica. Rev Geog Am Centr 42:97–112

Corrales D, Salom R, Carazo J, Araya D (2012) Evaluación inicial de la unidad de conservación del jaguar (Panthera onca) Parque Nacional Braulio Carrillo, Costa Rica

Cove MV, Pardo Vargas LE, de la Cruz JC, Spínola RM, Jackson VL, Saenz JC, Chassot O (2013) Factors influencing the occurrence of the endangered Baird's tapir *Tapirus bairdii*: potential flagship species for a costa Rican biological corridor. Oryx. https://doi.org/10.1017/S0030605313000070

Esquivel-Cambronero A, Sáenz-Bolaños C, Montalvo V, Alfaro-Alvarado LD, Carrillo E (2017) First records of *Tayassu pecari* (Artiodactyla : Tayassuidae) in the Barbilla National Park. Suiform Soundings, Newsletter of the IUCN/SSC Wild Pig, Peccary, and Hippo Specialist Groups 15(2):28–30

Fonseca W, Cháves H, Alice F, Rey JM (2010) Cambios en la cobertura del suelo y áreas prioritarias para la restauración forestal en el Caribe de Costa Rica. Rec Nat Amb 59-60:99–107

Gómez-Hoyos DA, Escobar-Lasso S, Brenes-Mora E, Schipper J, González-Maya JF (2018) Interaction behavior and vocalization of the Baird's tapir *Tapirus bairdii* from Talamanca, Costa Rica. Neotrop Biol Conserv 13(1):17–23. https://doi.org/10.4013/nbc.2018.131.03

Gonzalez-Maya JF, Schipper J, Polidoro B, Hoepker A, Zarrate-Charry D, Belant JL (2012) Baird's tapir density in high elevation forests of the Talamanca region of Costa Rica. Integr Zool:381–388. https://doi.org/10.1111/j.1749

Guerrero S, Morazan F (2016) Redescubrimiento de *Tayassu pecari* (Artiodactyla : Tayassuidae) en el Refugio Nacional de Vida Silvestre Mixto Caño Negro, Costa Rica. Cuadernos de Investigación UNED 8(2):225–229. https://doi.org/10.22458/urj.v8i2.1565

Harris A (2018) Estudio de grandes mamíferos en la Estación Biológica Caño Palma y alrededores. RNVS Barra del Colorado) y en el Parque Nacional, Tortuguero

Holdridge LR (1967) Life zone ecology. Tropical Science Center, San José, Costa Rica

Kidney B, Berrocal A (2008) Sarcoids in two captive tapirs (*Tapirus bairdii*): clinical, pathological and molecular study. Vet Dermatol 19(6):380–384. https://doi.org/10.1111/j.1365-3164.2008.00698.x

Kuprewicz EK (2010) The effects of large terrestrial mammals on seed fates, hoarding, and seedling survival in a Costa Rican rain forest. University of Miami

La Gaceta (2012). Reglamento a la Ley de Biodiversidad, Decreto Ejecutivo No. 34433-MINAE. 10 de diciembre 2012

Leonardi ML, Amit R, Watson R, Gordillo E, Carrillo E (2010) Presencia de *Tayassu pecari* (Artiodactyla: Tayassuidae) en Parque Nacional Rincón de la Vieja, Guanacaste, Costa Rica. Brenesia 73:146–147

Michel NL, Carson WP, Sherry TW (2015) Do collared peccaries negatively impact understory insectivorous rain forest birds indirectly via lianas and vines? Biotropica 47(6):745–757. https://doi.org/10.1111/btp.12261

Michel NL, Sherry TW, Carson WP (2014) The omnivorous collared peccary negates an insectivore–generated trophic cascade in costa Rican wet tropical forest understorey. J Trop Ecol 30:1–11

Monge-Nájera J (2018) Road kills in tropical ecosystems: a review with recommendations for mitigation and for new research. 66 (2): 722–738. doi:https://doi.org/10.15517/RBT.V66I2.33404

Quilez-Hueso I (2016) Estudio de grandes mamíferos en la Estación Biológica Caño Palma y alrededores. RNVS Barra del Colorado) y en el Parque Nacional, Tortuguero

Reider KE, Carson WP, Donnelly MA (2013) Effects of collared peccary (*Pecari tajacu*) exclusion on leaf litter amphibians and reptiles in a Neotropical wet forest, Costa Rica. Biol Conserv 163:90–98. https://doi.org/10.1016/j.biocon.2012.12.015

Rivera CJ (2014) Facing the 2013 gold rush: a population viability analysis for the endangered white-lipped peccary (*Tayassu pecari*) in Corcovado National Park, Costa Rica. Nat Resour 05:1007–1019. https://doi.org/10.1016/j.jtemb.2014.07.020

Rojas-Jimenez J, Arguedas R, Barquero-Calvo E, Alcázar-García P (2018) Análisis del perfil de sensibilidad a los antibióticos en aislamientos de *Escherichia coli* obtenidos a partir de heces de *Tapirus bairdii* de vida libre, y su relación con la actividad antropogénica, en zonas altas de la Cordillera de Talamanca, Costa Rica. Escuela de Medicina Veterinaria. Universidad Nacional de Costa Rica

Romero A, O'Neill BJ, Timm RM, Gerow KG, McClearn D (2013) Group dynamics, behavior, and current and historical abundance of peccaries in Costa Rica's Caribbean lowlands. J Mammal 94(4):771–791. https://doi.org/10.1644/12-MAMM-A-266.1

Ruiz-García M, Randi E, Martínez-Agüero M, Alvarez D (2007) Relaciones filogenéticas entre géneros de ciervos neotropicales (Artiodactyla : Cervidae) mediante secuenciación de ADN mitocondrial y marcadores microsatelitales. Rev Biol Trop 55(2):723–741

Sancho M (2017) Lanzan campaña para prevenir atropello de dantas. https://www.crhoy.com/ambiente/lanzan-campana-para-prevenir-atropello-de-dantas

Schank C, Mendoza E, Vettorazzi MJG, Cove MV, Jordan CA, O'Farrill G, Meyer N, Lizcano DJ, Estrada N, Poot C, Leonardo R (2015) Integrating current range-wide occurrence data with species distribution models to map the potential distribution of Baird's tapir. Tapir conservation, newsletter of the IUCN /SSC tapir specialist Group 24(33):15–30. https://doi.org/10.5281/zenodo.23417

Schank CJ, Cove MV, Kelly MJ, Mendoza E, O'Farrill G, Reyna-Hurtado R, Miller JA (2017) Using a novel model approach to assess the distribution and conservation status of the endangered Baird's tapir. Divers Distrib 23(12):1459–1471. https://doi.org/10.1111/ddi.12631

SINAC (2014) V Informe nacional al convenio sobre la diversidad biológica, Costa Rica. GEF-PNUD, San José, Costa Rica, p 192

SINAC (2018) Sistema nacional de áreas de conservación. http://www.sinac.go.cr/ES/Paginas/default.aspx

Stehli FG, Webb SD (eds) (1985) The great American biotic interchange. Plenum Press, New York

Watson ML (2005) Habitat fragmentation and the effects of roads on wildlife and habitats. New Mexico Department of game and fish, New Mexico, p 18

Webb D (2006) The great American biotic interchange: patterns and processes. Ann Missouri Bot Gard 93(2):245–257

Chapter 8
The Ungulates of Honduras

Fausto Antonio Elvir Valle and Héctor Orlando Portillo Reyes

Abstract The distribution and ecology of the five species of ungulates that occur in Honduras, *Tapirus bairdii* (Danto or Tapir) of the Tapiridae family, *Pecari tajacu* (Quequeo, Chancho de Monte) and *Tayassu pecari* (Jagüilla) of the Tayassuidae family, and *Odocoileus virginianus* (White-tailed Deer) and *Mazama temama* (Tilopo, Güisisil) from the Cervidae family, are presented. The history of the registration and monitoring of terrestrial mammals in Honduras is briefly narrated. Registration maps and potential distribution of the species, generated by the MaxEnt Program using presence records as a database for the analysis are presented. Spatial analysis is done through the use of Geographic Information Systems (GIS) that allow for the digitalization and measurement of distribution polygons. The analysis is complemented by the use of several layers that allow for ecological relationships, distribution percentages and threats. The distribution and ecology of the ungulates in Honduras is wide with particular characteristics of each species; they are found in buffer zones and core areas of protected areas and within private protected areas. The region of the Honduran Moskitia in the eastern part of Honduras records the presence of the five species indicating that ungulates share the same territories especially in areas of low human presence; this territory is considered as the genetic bank for these and other wild species. The threats faced by ungulates in general are the same, loss of vegetation, poaching, change of land use, and traffic. Genetic constitution is unknown, except for *Tapirus bairdii* where there is some phylogeny work. The future of ungulates in Honduras is uncertain, so it is recommended that governments and private organizations pay adequate attention by promoting conservation programs, scientific research, and laws that protect species and their areas of occurrence.

F. A. Elvir Valle (✉) · H. O. Portillo Reyes
Research and Biological Monitoring, Science Foundation for the Study and Conservation of Biodiversity (INCEBIO) Fundación en Ciencias para el Estudio y Conservación de la Biodiversidad, Tegucigalpa, Francisco Morazán, Honduras

S. Gallina-Tessaro (ed.), *Ecology and Conservation of Tropical Ungulates in Latin America*, https://doi.org/10.1007/978-3-030-28868-6_8

8.1 Introduction

The first records of mammals for Honduras come from William Wells's trips in 1857, where he tells his encounters with mammals and writes it in his book "Explorations and Adventures in Honduras." Gaumer (1886), Townsend (1887–1888), Wittkugel (1891), Perry (1892), and Reed (1901) also make their contribution to the registration of these species.

On May 29, 1942, George Goodwin publishes "Mammals of Honduras" that appears in the Bulletin of the American Museum of Natural History of New York, in this document they describe, classify and locate places where species were collected, the work of The field in the collection of the samples was by Cecil Underwood from 1932 to 1939 in the departments of Francisco Morazán, Olancho, La Paz, Lempira, and Copán. In the 1970s, Professor Ibrahim Gamero Idiáquez publishes his book "Mamíferos de mi Tierra" (Mammals of my Land) in which he made a general mention of the mammals of Honduras. Years later, Dr. Noé Pineda Portillo compiles information from geographers and historians from 1932 to 1939 who visited Honduras and described fauna seen in their travel, his book is titled "Researchers of the Honduran Biography" and publishes it in the year of 2005.

For the year of 1998 Leonel Marineros and Francisco Martínez publish their book "Field Guide of the Mammals of Honduras" where they describe records collected by personal (students and professors) of the School of Biology in their field trips through the country and shows distribution maps of different mammal of Honduras including the records of George Goodwin. In the year 2002–2005, the Biodiversity Program in Priority Areas PROBAP-AFE-COHDEFOR-BM was implemented in the country, this project generated information on large and medium mammal records as well as birds in approximately 24 Protected Areas. In the first months of the year 2006 camera traps were installed for the first time in the Pico Bonito National Park in the department of Atlántida and the first photographic records of fauna in the wild were obtained in Honduras.

To estimate the potential distribution, we used the MaxEnt Program (Maximum Entropy). The MaxEnt is a multipurpose program, based on a statistical approach called maximum entropy, which allows making predictions, using data on the presence or occurrence of the distribution potential of a species (Phillips et al. 2006; Phillips and Dudík 2008). Modeling programs for the distribution of species such as MaxEnt (Philips et al. 2006) allow us to approximate the total range of distribution, thus being a practical tool to identify the areas in which a species is probable to occur (Scheldeman and Van Zonneveld 2011). Models, work as a valuable tool to determine species distribution (Morales 2012). For this analysis, current data available in different platforms, in the form of technical reports, scientific articles, and camera trap works carried out in the country by native authors and in collaboration with other researchers, were used. 180 data were used for the Tapir or Danto (*T. bairdii*), 42 data for Quequeo (*P. tajacu*), 16 data for the jagüilla (*T. pecari*), 97 data for the white-tailed Deer (*O. virginianus*) and 32 records for the Tilopo (*M. temama*) and 19 bioclimatic variables (temperatures, humidity, precipitation), taken from the Worldclim database (http://www.worldclim.org/bioclim).

To evaluate the performance of the model, the value of the area under the curve (AUC) was considered, which gives a value, closer to one, greater sensitivity to the test (Moisen et al. 2006). The probabilistic distribution values are between 0 and 1 used to generate the models with the environmental requirements; this distribution represented in an output map that uses the scale of colors that indicate this probability. The values between 0.62 and 1 indicate optimal conditions for the distribution of the species. Values between 0.38 and 0.62 indicate intermediate conditions and values from 0 to 0.38 indicate unfavorable conditions for potential distribution, based on presence correlations and bioclimatic conditions (Phillips et al. 2006; Portillo and Elvir 2016).

The Moskitia area in the eastern part of Honduras, maintain the presence of the five species, which was used as a reference to estimate the minimum distance of registration of species that is 200 m and that indicates that ungulates share same territories especially in areas with adequate forest cover. In other areas similar distances are estimated although not with the presence of the species that include this analysis. For the elaboration of the maps, digitization of polygons and distance estimation, the Arc View 3.3 and Quantum GIS 3.6 Programs were used; the change land use map (2014), Census map (2001), and Protected Areas layers were used.

8.2 Species

Tapirus bairdii (Tapir or Danto).
Pecari tajacu (Quequeo, Chancho de Monte).
Tayassu pecari (Jagüilla).
Odocoileus virginianus (Whitetail Deer).
Mazama temama (Venado Tilopo, Güisisil).

They are distributed in a particular way in protected and unprotected natural areas; all are included with some status of conservation in national and international lists.

8.2.1 Tapirus bairdii *(Tapir, Danto)*

The Central American tapir, also known as tapir in Honduras (*Tapirus bairdii*), is the largest tapir of the three of the tropical forest of the Americas (Medici et al. 2005; Portillo et al. 2016, Fig. 8.1). In the territories of the ethnic groups is known with different names according to its language, Tilba (Miskitu), Pánh-ka (Tawahka), Dandei (Garífuna), and Chajú (Pech) (Marineros and Martínez, 1998).

8.2.1.1 Distribution

This herbivore belongs to the order Perissodactyla and was originally distributed, almost continuously, from southern Mexico to northern Colombia and Ecuador ((Alston 1882; Matola et al. 1997; Portillo et al. 2016). In 2016 (Portillo et al.) using

Fig. 8.1 Danto, camera trap photograph in the Reserve of the Man Biosphere of Río Plátano, Aukaben area, courtesy of Wildres Rodríguez and Marcio Martínez, Regional Institute of Forest Conservation, Río Platano

the MaxEnt Program modeled the potential distribution of the Tapir for Honduras, based on the model obtained, a map was prepared that predicts an area of potential distribution of the tapir in Honduras of 19,752 km^2, which represents 17.55% of the model predicts an area of 2416 km^2 in the cloud forests of the Honduran Caribbean, which represents 12.23% of the total area of potential distribution for the tapir (Fig. 8.2). Partially, the model shows in the mountainous center of the country, (the departments of Olancho, Yoro and El Paraiso) an area of 1426 km^2, representing 7. 22%, of the total potential territory of the tapir. For lowland tropical forests, in the region of the Honduran Moskitia, the model indicates an area of distribution of 15,910.33 km^2, representing 80.55% of its territory. Based on the distribution model, it can be determined that the range of the tapir distribution in Honduras is within the limits of the following protected areas: in the Caribbean; the Cusuco National Park, the Merendón Reserve Zone, the Wildlife Refuge de Texiguat, Pico Bonito National Park, Nombre de Dios National Park, and The Sierra de Río Tinto National Park; In the mountainous central part: The Pech Anthropological and Forestry Reserve (RAFT), El Carbon Mountain, The Botaderos National Park and The La Muralla Wildlife Refuge; in the Moskitia: in the Reserve of the Man and Biosphere of the Río Plátano, The Tawahka Asagni Biosphere Reserve, The Patuca National Park, and the Territorial Council of the Indigenous Federation of the Mokorón and

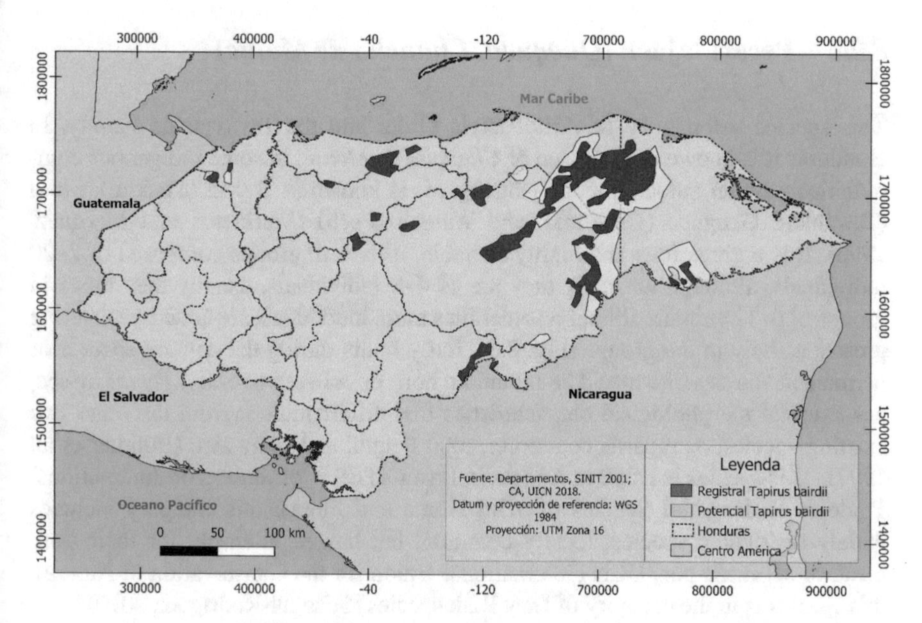

Fig. 8.2 Current and potential distribution of the danto in Honduras, showing the departments where the species occurs

Segovia Zone (FINZMOS). The protected areas mentioned above have an extension of 21,020 km^2 (Fig. 8.1).

However, it can be determined that the current distribution of the species is smaller, because part of the area includes portions of pine forests, mixed pine-oak forests and agriculture areas, that do not represent a suitable habitat for the species, they correspond to habitats of low suitability (Schank et al. 2015). Performing the spaces analysis for the tapir and selecting only the values that are between 0.62 and 1, the potential distribution area for the tapir is reduced to 4354 km^2, which represents 3.87% of the national territory and 22% of the area with intermediate and optimal conditions. The potential area of the tapir territory represents approximately 17.55% of the national territory. The Caribbean and central areas of Honduras are fragmented and isolated, with agricultural landscapes that include crops of African palm, pineapples, sugar cane and bananas, among others, which have become permanent and continuous barriers for the tapir (Schank et al. 2015). The model indicates that the potential distribution area of the tapir is included in 21,020.79 km^2 of protected areas, with some category of legal protection. However, the category and legal status do not guarantee the conservation of the species, which is in a critical situation, reaching a high probability of extinction in places such as the Cusuco National Park, the Texiguat Refuge Wildlife, the Pico Bonito National Park, and the Sierra del Río Tinto National Park (McCann et al. 2012). Records of this species in protected areas are restricted to high altitude sites, mostly in the cloud forest, as a consequence of hunting pressures and habitat loss (Portillo et al. 2016).

8.2.2 Pecari tajacu *(Quequeo, Chancho de Monte)*

This species belongs to the Artiodactyla Order and the Tayassuidae Family. In Honduras it is known as *Quequeo* or *Chancho de Monte*; in some indigenous communities with an autochthonous language, it is known as Buksa (Miskitu), Siw-í (Tawahka), Guegueu (Garifuna), and Wareka (Pech) (Marineros and Martínez, 1998). It is a gregarious and highly sociable, it lives in groups composed of 2–20 individuals, although normally they are of 4–8 individuals, usually they move in groups of 6–12 animals although sometimes more individuals are added, registering greater activity in the groups (Fig. 8.3). Early hours during the day are spent near permanent sources of water. The mountain horn or collared peccary, *Pecari tajacu*, has external morphological characteristics that differentiate it from the other two existing species, as regards coat color, total length, and body size (Bodmer et al. 1997). This species is listed in Appendix II of CITES (Convention on International Trade in Endangered Species of Wild Fauna and Flora); this category includes widely distributed species, locally common, but hunted intensely for their skin, meat, or sport. As per IUCN (International Union for the Conservation of Nature), this species is in the category of Low Risk species (Sabogal-Rodríguez 2010).

Fig. 8.3 Quequeo in La Tigra National Park, camera trap photograph courtesy of Portillo et al. (2006), GIBH-BALAM y AMITIGRA

8.2.2.1 Distribution

The historical, current, and potential distribution was estimated using the MaxEnt Program with 42 registry data and 19 environmental variables. It was estimated an area of registry occupation of 143, 903,745 ha and a potential distribution area of 184,083.51 ha with suitable conditions for the expansion of the species (Fig. 8.4). The area of potential and registry occupation occurs in the departments of Copán, Ocotepeque, Lempira, Intibucá, La Paz, Santa Bárbara, Cortés, Atlántida, Colón, Yoro, Comayagua, Francisco Morazán, El Paraíso, Olancho, and Gracias a Dios; they include 116 municipalities in these departments. 95.23% (40 data) are within protected areas, and 4.76% (2 data) are outside protected areas, but very close to the area.

The protected areas in which its presence is registered in the north zone of the country are Cusuco National Park, Lancetilla Botanical Garden, and Sierra de Río Tinto National Park.

In the west it is registered in Cerro Azul Copán National Park, Trifinio-Montecristo National Park, Güisayote Biological Reserve, El Pital Biological Reserve, Cacique Lempira Biosphere Reserve, Lord of the Mountains, Opalaca Biological Reserve, Guajiquiro Biological Reserve, and Santa Barbara national park.

In the middle of the country it is registered in Montaña Nacional de Comayagua, Hierba Buena Biological Reserve, La Tigra National Park, El Chile Biological Reserve, El Uyuca Biological Reserve, and Monserrat Biological Reserve.

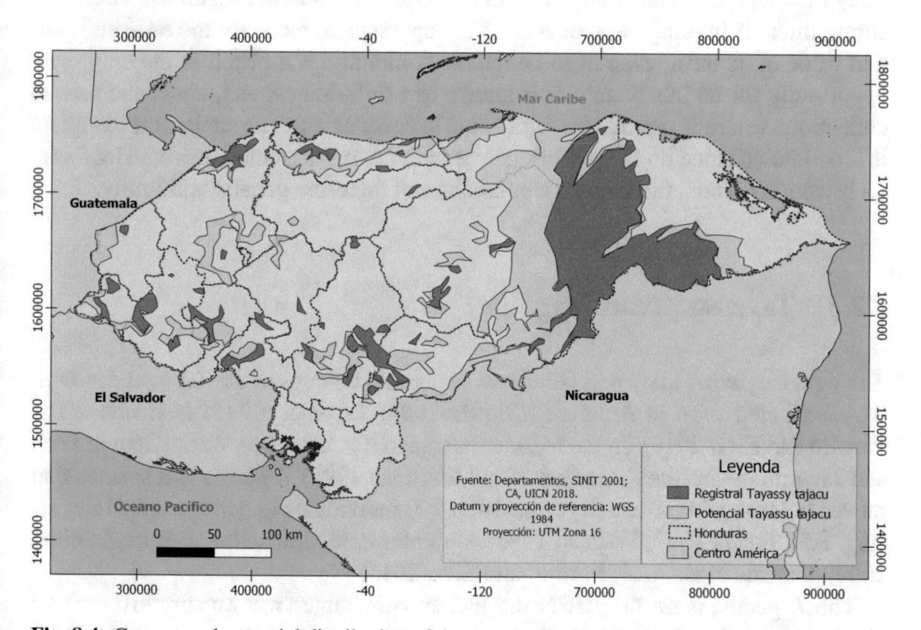

Fig. 8.4 Current and potential distribution of the quequeo in Honduras, showing the departments where the species occurs

In the east of Honduras it is registered in the Sierra de Agalta National Park, the Reserve of the Man and Biosphere of the Río Plátano, the Tawahka Asagni Biosphere Reserve, and the Rus-Rus Biological Reserve and the Mokorón Forest Reserve that belong to the Territorial Council of the Indigenous Federation of the Mokorón and Segovia area (FINZMOS). The following land uses are identified in the distribution area: pastures and crops, dense coniferous forest, mixed forest, humid broad-leaved forest, deciduous broad-leaved forest, sparse conifer forest.

8.2.2.2 Current Situation

In recent monitoring works, with camera traps conducted in the Caribbean of Honduras, there are no records of the species or are very scarce, the reason could be the large coverage area of intensive agricultural crops of pineapple, bananas and African palm, as well as touristic infrastructure.

In the Honduran Moskitia and certain areas of the West, troops of up to 50 individuals are reported, according to Marineros and Martínez (1998). In the country there are localities where they are not captured because they are considered to be aggressive, since they invade milpa (corn crops) surrounding their domains causing damage to the crops of subsistence. The strongest pressures affecting the species are the loss of habitat due to changes in land use, deforestation and hunting. This species is closely related to water courses so deforestation and loss of coverage directly affects their routine activities by limiting their access to water and food sources. They also use the same trails with their closest cousins the jagüillas. The other strong threat is hunting because it is a very appreciated piece for the taste and protein value of its meat, even in some remote communities it becomes the only meat opportunity for its inhabitants, it is known as exhibition centers, zoos, and private collections where it reproduces with some success as a poultry animal. Phylogeny is little known since no DNA work has been done, it is assumed that the Honduran La Moskitia harbors the largest populations and therefore genetic variability.

8.2.3 Tayassu pecari *(Jagüilla)*

The *Tayassu pecari* known in Honduras as jagüilla belongs to the Tayassuidae family, distributed only in America ((Slowls 1984; Taber et al. 1994; Grubb 2005; Portillo and Elvir 2016). In the Miskitu language it is known as Wari, Kitan in Pech and Jáwuria in Garífuna (Marineros and Martínez 1998) *T. pecari* is a species that moves in large groups, usually from 10 to 300 individuals in dense tropical forests (Fig. 8.5, Slowls 1997; Fragoso 1998; Altrichter et al. 2001; Reyna-Hurtado et al. 2009; Almeida et al. 2013; Portillo and Elvir 2016).

The *T. pecari* is on the IUCN red list, as vulnerable (www.uicnredlist.org), in Honduras it is placed on the list of concerns according to resolution gg-dapv-003-98 AFE/COHDEFOR, (SERNA, 2008). For Honduras, studies of mammals have

Fig. 8.5 Two Jagüillas, camera trap photograph in Warunta Mountains and Bodega, courtesy Héctor Portillo and Jonathan Hernández, 2008

become especially important with the use of camera traps, which have contributed significantly to the biodiversity of large and medium terrestrial mammals, recording species that are difficult to observe (Portillo and Elvir 2013).

8.2.3.1 Distribution

The current records of jagüilla for Honduras come from the field guide of the wild mammals of Honduras of Marineros and Martínez (1998) and photocaptures of studies for jaguars and other mammals that have been realized in the regions of the Middle, Caribbean and Moskitia of Honduras, which scarcely recorded jagüillas by camera traps (Portillo and Elvir 2013). The potential distribution of the jagüilla was modeled using the MAXENT Program (Fig. 8.6). The model predicts a territorial extension of 6126 km^2 that corresponds to 5.5% of the territory of the country. The potential area for this species is located in three main sites: Protected area of the Reserve of the Man and the Biosphere of Río Plátano (RHBRP), with approximately 70% of the prediction of the model (4288 Km^2) by 20% (1225 Km^2) in the indigenous territories of Rus-Rus, Mocorón, and Warunta and 10% (613 Km^2) in the Biosphere of the Tawahka Asagni Reserve, represented to the greatest extension in the department of Gracias a Dios in the broadleaved forest with approximately 95% of the territory and 5% between the departments of Colón and Olancho, this based on the forest coverage map of 2014 (Portillo and Elvir 2016).

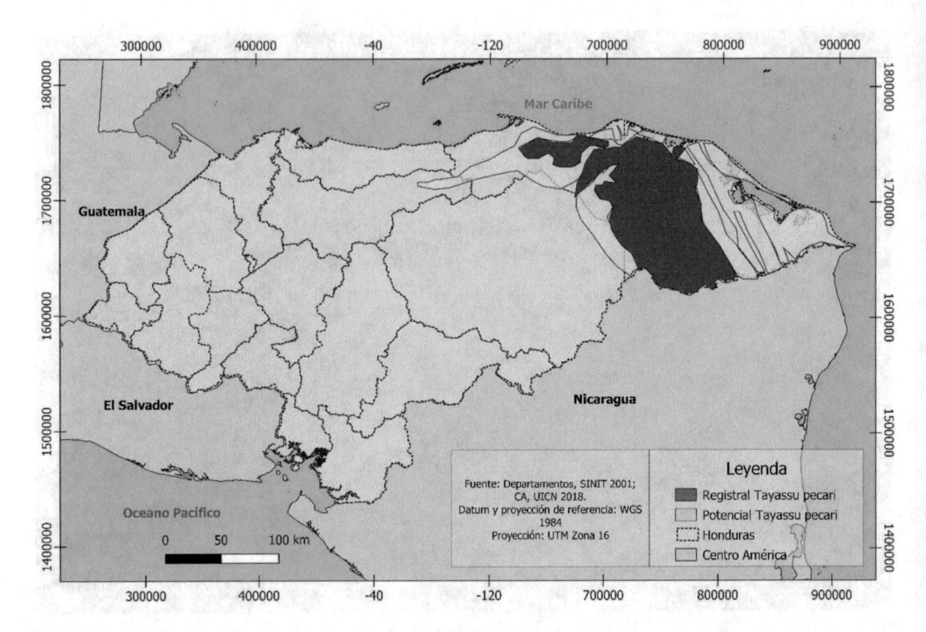

Fig. 8.6 Current and potential distribution of the jagüilla in Honduras, showing the departments where the species occurs

The ecosystems present in the potential distribution area, are the following: Broad-leaved forest in karstic hills, broad-leaved lowland forests, submontane broad-leaved forests, and alluvial broadleaf forests (Mejía and House 2002). The river network represented (large and short rivers of great flow) 816 linear km, among the navigable rivers we have: Rus-Rus, Mocorón, Plátano, Pao, Patuca, Wampú, and Sikre rivers, which drain into the Caribbean Sea. It is assumed that one of the historical habitats for the jagüilla in Honduras was the tropical humid forest (bh-t), which had a territorial extension of 32,504 km^2 distributed throughout the Moskitia region, the Caribbean and the Northwest of Honduras (Holdrige 1971). As a result the potential distribution modeling of *T. pecari* of 6126 km^2 was generated, if the historical distribution is taken as the reference point, loss represent the 26,378 km^2 which has been dramatically reduced in 81.2% of the potential historical habitat for the jagüilla, and the current area would represent 18.8%. The Moskitia region is the only region where the presence of jagüilla has been recorded for Honduras.

8.2.3.2 Current Situation

According to Portillo and Elvir (2016) Honduras does not have records of *T. pecari* that show its multitemporal distribution between the historical and the contemporary. For the group of CSE / IUCN peccary specialists (2008) based on monitoring criteria for local resources monitoring (AFE/COHDEFOR/PROBAP 2005), they propose the distribution area for *T. pecari* of 0.8% of the national territory, which represents 896 km^2, assuming in general terms its distribution for Honduras. Castañeda (2009) reports historical data for three indigenous communities in the Honduran Moskitia in Auratá, Rus, and Mokorón; it is reported that in the 1950s the effort for hunting jagüilla was 15 min, in the 1970s an effort of capture lasted from 1 h to 1 day, in the 1990s it was worth 2 days of effort, and at the time of study hunting a jagüilla required more than 3 days of effort.

As reported by Castañeda as an interesting fact in the indigenous community of Rus-Rus, near the Coco River or Segovia, in December of 1999, a herd of jagüillas suddenly entered into the village, between 90 and 100 individuals, surprising the community in an unusual event. In the same way it happened in the community of Auratá, near the Karataska lagoon for the year 2006 (Castañeda 2009). Regarding the size of the herds, hunters mentioned observing in the 1950s up to 1970s, 100–150 individuals in a herd in the 1990s 40–80 individuals (Castañeda 2009). For the year 2008, in the area known as Tapalwás, camera traps were installed with a sampling effort of 2400 camera nights, and only 27 individuals were recorded in 13 independent photographs (Portillo and Hernández 2011); however, there was an encounter with a herd of approximately 150–200 individuals according to the Miskito indigenous guides.

According to Almeida et al. (2013), *T. pecari* functions ecologically as an important seed disperser and as a prey species for predators such as the jaguar. It is susceptible to human presence especially when it is pressured by hunting, loss and fragmentation of habitat, as well as domestic diseases (Sowls 1984; Bodmer et al. 1997; Fragoso 1998; Altrichter and Boaglio 2004; Beck 2004; Reyna-Hurtado and Tanner 2005, 2007; Reyna-Hurtado et al. 2009; Moreira-Ramírez et al. 2016; Portillo and Elvir 2016).

It is urgent to take conservation measures for the region, given its ecological role as a disperser of seeds and prey species for large predators.

8.2.4 **Odocoileus virginianus** *(White-Tailed Deer, Venado Cola Blanca)*

This species belong to the Order Artiodactyla and the family Cervidae, it is an animal that in adult stage is large and imposing. The white-tailed deer was declared the "National Mammal," through Executive Decree 36–93. It is called in Spanish Venado Cacaste, Venado Ramudo in the indigenous Lenca area, Sula in the Miskitu language, Sánah-pih in Tawahka, Ichá in Pech, and Usari in Garífuna (Marineros and Martínez 1998, Fig. 8.7).

It is a ruminant mammal one of the most adaptable in the world, besides being considered as one of the flagship species in wildlife management (Beltrand and Díaz de la Vega 2010; Portillo et al. 2015a). It is a versatile species that lives from the lowlands to mountain systems above 3000 m altitude. The availability of food, water, forest, climatic conditions and the presence of predators and competitors,

Fig. 8.7 White-tailed deer, camera trap photograph, courtesy of Fausto Elvir INCEBIO, Francis Hernández and Martha Moreno UNACIFOR, 2012

influence the size of the populations of this species (Galindo-Leal and Weber 1994; Villareal and Treviño 1995; Gallina et al. 1998).

The genus *Odocoileus* is native to the American continent where 38 subspecies are recognized (Portillo et al. 2015a, b). In Honduras the white-tailed deer is one of the most persecuted prey by humans, as well as by natural predators, among which the felines and canids especially "jaguar" (*Panthera onca*)(Portillo and Elvir 2015), "puma" (*Puma concolor*), "ocelot" (*Leopardus pardalis*), and "coyote" (*Canis latrans*). All those species in the structure of biotic communities accomplish a complex ecological function (Marineros and Martínez 1998), but under strong pressure for the loss of habitat and the increase of hunting (Secaira 2013; Portillo et al. 2015a, b).

8.2.4.1 Distribution

The distribution of whitetail deer in Honduras was analyzed using 97 records, data was collected per sighting, photos and videos of camera traps. The MAXENT Program was used to model the potential distribution using 19 environmental variables. It is estimated a potential distribution of 436,372.37 ha, 51.57% of the data (51) are located outside protected areas and 47.42% (46) within the protected areas (Fig. 8.8).

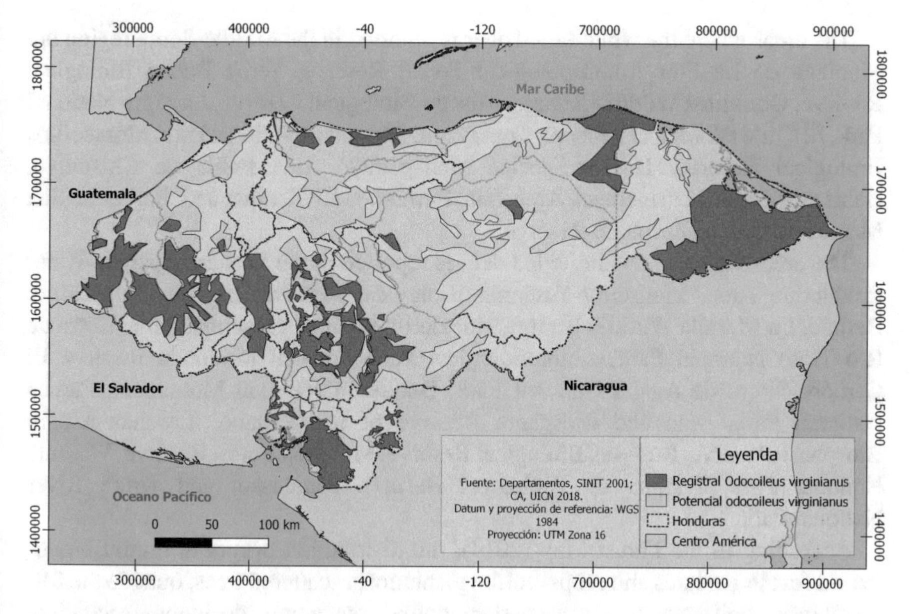

Fig. 8.8 Current and potential distribution of the white-tailed deer in Honduras, showing the departments where the species occurs

The distribution by registers shows that the white-tailed deer is spread out in the departments of Ocotepeque, Copán, Santa Bárbara, Cortés, Atlántida, Colón, Gracias a Dios, Olancho, Yoro, Francisco Morazán, Comayagua, La Paz, Intibucá, Lempira, Valle, Choluteca, El Paraíso, and all municipalities with the exception of those located south of Olancho, Cortés, Yoro, Santa Bárbara, Francisco, Morazán, and Ocotepeque, departments of Lempira, Intibucá, and La Paz on the southern border within the Republic of El Salvador, except the insular department of Islas de la Bahía where wild populations are not reported, although there are introduced individuals.

The protected areas where the white-tailed deer is reported in the western region are Montecristo-Trifinio National Park, Santa Bárbara Mountain National Park, Puca Wildlife Refuge, Cerro Azul National Park, Güisayote Biological Reserve, Pacayita Volcano Biological Reserve, Refugio de Erapuca Wildlife, Celaque Señor de la Montaña Biosphere Reserve, Congolón National Park, Piedra Parada Coyocutena, Opalaca Biological Reserve, Montaña Verde Wildlife Refuge, Mixcure Wildlife Refuge, Sabanetas Biological Reserve, El Jilguero Water Production Area, and Guajiquiro Biological Reserve.

The areas where the white-tailed deer is reported in the north and Caribbean region are Multiple Use Area, Yojoa Lake, Cerro Azul Meámbar National Park, El Cajón Resource Reserve, Mico Quemado Ecological Reserve Zone, Yoro Mountain National Park, Pijol Peak National Park, San Agua Production Area Pedro Sula-Cofradía-Naco, Cusuco National Park, Omoa-Cuyamel National Park, Jeannette Kawas National Park, Punta Izopo National Park, Texíguat Wildlife Refuge, Barras de Cuero and Salado Wildlife Refuge, Pico Bonito National Park, National Park Name of God, Capiro National Park, and Calentura.

The areas where the white-tailed deer is reported in the Middle-South region are Montaña de La Flor Anthropological Forest Reserve, Yerba Buena Biological Reserve, Corralitos Wildlife Refuge, Misoco Biological Reserve, La Tigra National Park, El Chile Biological Reserve, Comayagua Mountain National Park, Montecillos Biological Reserve, Habitat-Species Management Area Bahía de Chismuyo, Habitat-Species Management Area San Bernardo, El Jicarito and Barbería, San Marcos de Colón Biosphere Reserve.

The areas where the white-tailed deer is reported in the East are Apagüíz Water Production Area, Monserrat-Yuscarán Biological Reserve, El Armado Wildlife Refuge, La Muralla Wildlife Refuge, Botaderos Mountain National Park, Sierra de Río Tinto National Park, Anthropological Pech Forestal Mountain Reserve El Carbon, Sierra de Agalta National Park, Boquerón National Monument, Patuca National Park, Man and Biosphere Reserve of Río Plátano, Tawahka-Asagni Biosphere Reserve, Rus-Rus Biological Reserve, Mokorón Forest Reserve, Warunta National Park, Laguna de Biological Reserve Karataska, and Kruta River National Park.

According to the Forest Map (2014), the distribution occurs in humid broadleaved forest, pastures and crops, coffee plantations, scattered trees, outside the forest, dense coniferous forest, secondary humid vegetation, deciduous secondary vegetation, deciduous broadleaved forest, riparian forest, other deciduous forest, sparse coniferous forest and savannas. In the case of land use called pastures and crops, it has been verified at field that it is actually tropical and subtropical dry forest in different strata and a quite threatened ecosystem.

The results of the analysis extend the area of distribution to almost the entire country but differentiating sites with greater potential, some of these sites include the Honduran Caribbean but in the last 5 years there have been no records or are very scarce, there are photographic records of the dry forest in a sector of the municipality of Olanchito with human presence, dogs, cattle, and roads, where the animals are thin and with signs of bodily harm. In the rest of the country, photographic records reveal healthy animals.

The species faces strong hunting pressure from residents of large cities that are organized in "Sports Hunting Clubs," move to the places of occurrence and pursue the animal with trained dogs and high-caliber weapons. It is always alert to evade hunters who consider it a hunting trophy. It reproduces well under conditions of captivity so it is common to see it in zoos, exhibition centers and private collections, in some countries of America have much experience in the management in semicautivity in ranches or ejidos as is the case of Mexico. Phylogeny studies have not been done but it is assumed that there should be some genetic variety given its wide distribution and expansion capacity.

According to Portillo et al. (2015a) it is one of the species with hunting potential that can benefit conservation and community development as other countries as Mexico. The white-tailed deer in Honduras is one of the main sources of protein for rural communities and indigenous groups, as well an important species in the food chain, structured by predators and preys (ICADE 2007; Portillo et al. 2015a).

8.2.5 Mazama temama *(Tilopo, Güisisil)*

The Tilopo deer or Güisisil is one of the two species of Cervidae that occur in Honduras.Other common names are Snapuka (Miskitu), Sánah-pauni (Tawahka), Icha Pawá (Pech), and Usari (Garífuna), (Marineros and Martínez 1998).

The Tilopo inhabits preferably well-conserved areas (Mandujano 2004); the red temazate deer, kid deer, or tylope (*Mazama temama*) from Central America (Medellín et al. 1988; Villarreal et al. 2009) is a small Neotropical cervid. Its continental distribution extends from Mexico to Central America (Coates-Estrada and Estrada 1986; Geist 1998; Weber and Gonzáles 2003; Villarreal et al. 2013, Fig. 8.9). The temazates are little known species; their study is of particular interest (Mandujano 2004). The lack of red temazate information is a serious problem that prevents the establishment of its risk of extinction, for this reason the International Union for Conservation of Nature (IUCN) categorizes it as a deficient of data (DD) (Bello et al. 2010; Ramírez and Hernández 2012). Visual identification requires experience, knowledge and training. Those sites might not be the current one, in areas such as the Honduran Caribbean the hunting is strong and in recent monitoring work with camera traps none of the two species of cervids is registered for Honduras.

Fig. 8.9 Tilopo deer in the Community of Las Marías, Reserve of the Man and Biosphere of Río Plátano, camera snapshot camera courtesy of Proyecto ECOSISTEMAS, Jonathan Hernández and Napoleón Morazán, 2012

8.2.5.1 Distribution

The historical and contemporary distribution of the Tilopo deer in Honduras was analyzed using 32 records, 28 data (87.5%) comes from protected areas and 4 data (12.5%) outside but adjacent to them, a registry distribution of 148, 979,757 ha was estimated. and a potential of 176, 506,029 ha (Fig. 8.10).

The records of the species reflects that the Tilopo is distributed in 12 departments of the country, Copán, Ocotepeque, Lempira, Cortés, Intibucá, La Paz, Comayagua, Francisco Morazán, Colón, Olancho, El Paraíso, and Gracias a Dios and 19 municipalities belonging to these departments. he registry distribution reflects that the Tilopo is distributed in 12 departments of the country, Copán, Ocotepeque, Lempira, Cortés, Intibucá, La Paz, Comayagua, Francisco Morazán, Colón, Olancho, El Paraíso, and Gracias a Dios and 19 municipalities belonging to these departments. The distribution in these sites may not be the current especially of the Honduran Caribbean where hunting is strong. In recent monitoring work with camera traps, none of the two species of cervids were recorded for Honduras.

It is registered in the following protected areas: Biosphere Reserve Cacique Lempira Lord of the Mountain, Puca Wildlife Refuge, Patuca National Park, Merendón Water Production Zone, ZPA El Jilguero, El Chile Biological Reserve, RVS La Muralla, Agalta NP, PN Patuca, Tawahka Biosphere Reserve, Man and Biosphere Reserve of Río Plátano, and Rus-Rus Biological Reserve.

The estimated distribution identifies two sites that can be considered as hotspots in the department of Olancho in the protected areas of the Tawahka-Asagni

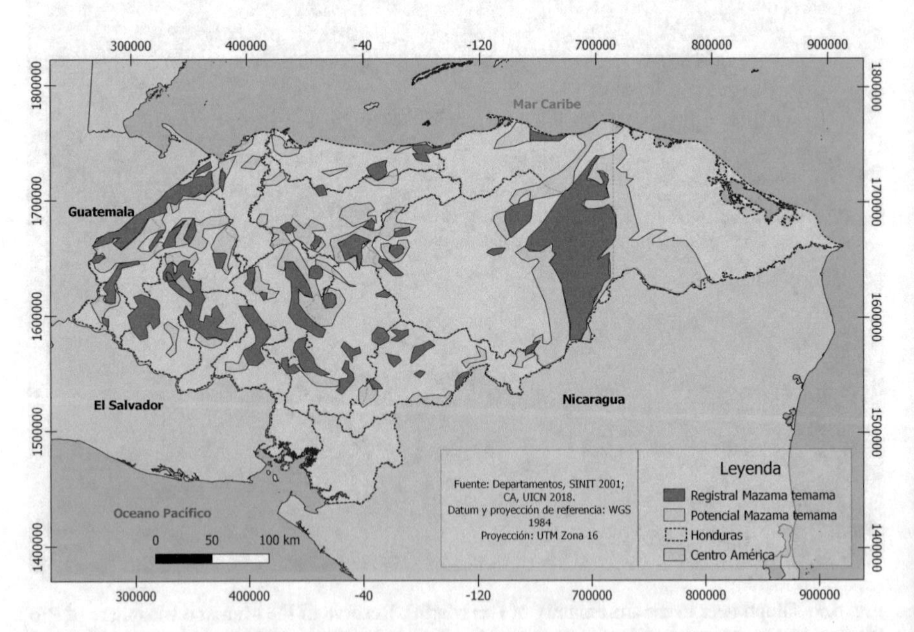

Fig. 8.10 Registry distribution and potential of the tilopo deer in Honduras, shows the departments where the species occurs

Biosphere Reserve, Patuca National Park and Sierra de Agalta National Park, a part of the Gracias a Dios department in the Reserve of the Man and Biosphere of Río Plátano and the department of Colón. The species is registered in the west in the departments of Lempira, Intibucá, and La Paz including the protected areas of Güisayote Biological Reserve, Erapuca Wildlife Reserve, Pacayita Volcano Biological Reserve, Cacique Lempira Biosphere Reserve Lord of the Mountains, Puca Wildlife Reserve, Montaña Verde Wildlife Reserve, Opalaca Biological Reserve, Mixcure Wildlife Reserve, El Chiflador Wildlife Reserves, Guajiquiro, Montecillos. The program predicts suitable conditions for the species in Comayagua Mountain National Park, Cusuco National Park, Motagua River Barras, Merendón Water Production Area, Pico Bonito National Park, La Muralla National Park, Misoco Biological Reserve, Reserve Biological El Chile, Yoro Mountain National Park, and La Tigra National Park.

Even though the distribution per records and the potential distribution predict adequate conditions for the species in some areas, at present these areas are heavily intervened and the actual distribution could not be assessed; the two species of cervids that occur in Honduras are victims of intensive hunting for the value of their meat, although people do it illegally and there is no legislation that punishes the crime. The hunting of the cervids is done without discrimination of sex, size, or species. In the East of Honduras in the department of Olancho in the city of Catacamas, in the sector known as Piedras Blancas, the captivity of a juvenile male in a private residence was reported; the owner bought it from a farmer who had said it came from the Sierra National Park of Agalta close to the city. In another case, the captivity of five individuals, an adult male and a juvenile and three females, was reported in the department of Lempira in western Honduras in the area known as Mejocote; the animals shared the enclosure with poultry and were fed with fruits of season and vegetables; the owners were asked about their origin and they confirmed the animals came from the Montaña Verde Wildlife Reserve, the north of the departments of Lempira and Intibucá. As in the majority of the countries of distribution of the species, in Honduras too little is known about the Tilopo; in some sectors the antlers of hunted animals are kept as a trophy. In general, very little is known of the Cervidae family; Portillo et al. (2015a, b) have carried out specific research works, but it is necessary to know more about their ecology and phylogeny.

Acknowledgements To David Mejía for his valuable cooperation so that this chapter would become a reality.

References

AFE/COHDEFOR/PROBAP (2005) Estudio multitemporal del Parque nacional Patuca, la Reserva de la Biósfera Tawahka y el Área Propuesta de Rus Rus. Informe Final de Consultoría, p 75

Almeida Jácomo AT, Malzoni Furtado M, Kayo Kashivakura C, Marinho-Filho J, Sollmann R (2013) White-lipped peccari home-range size in a protected area and farmland in the central Brazilean grassland. J Mammal 94:137145

Alston E (1882) Mammalia: Biologia Centrali-Americana: Contributions to the knowledge of the fauna and flora of México and Central América. In: Godman FD, Salvin O (eds) London 1918. http://www.sil.si.edu/digitalcollections/bca/explore.cfm

Altrichter M, Carrillo E, Saenz J, Fuller T (2001) White-lipped peccary (tayassu pecari, Artiodactyla: Tayassuidae) diet and fruit availability in a Costa Rican rain forest. Revista de Biología tropical 49:1183–1192

Altrichter M, Boaglio G (2004) Distribution and relative abundance of peccaries in the argentine Chaco: associations with human factors. Biol Conserv 116:217–225

Beck H (2004) Seed predation and dispersal by peccaries throughout the neotropics and its consequences: a review and synthesis. Department of Biology, University of Miami, Coral Gables 6:77–115

Bello-Gutiérrez J, Reyna-Hurtado R, Wilham J (2010) Central American red brocket deer *Mazama temama* (Kerr, 1992). In: Duarte JMB, González S (eds) Neotropical cervidology: Biology and medicine of Latin American deer. Jaboticabal, Brazil: Funep and Gland, Switzerland: IUCN, pp 166–171

Beltrand C, Díaz de la Vega AD (2010) Estimación de la densidad poblacional del venado cola blanca Texano (Odocoileus virginianus texanus), Introducido en la UMA "Ejido de Amanalco", Estado de México. Ciencia Ergo Sum 17:154–158

Bodmer R, Aquino R, Puertas P, Reyes C, Fang T, Gottdenker N (1997) Manejo y uso sustentable de pecaríes en la Amazonía Peruana. Occ Pap UICN/SSC No. 18. UICN-Sur, Quito, Ecuador y Secretaría CITES, Ginebra, Suiza

Castañeda F (2009) Datos preliminares sobre la distribución del jaguar (Panthera onca), el estado de sus especies presa, y el conflicto felinos - ganadería en la Moskitia Hondureña. WCS, p 14

Coates-Estrada R, Estrada A (1986) Manual de identificación de campo de los mamíferos de "Los Tuxtlas". UMAM, México

Fragoso J (1998) Home range and movement patterns of white-lipped peccary (*Tayassu pecari*) herds in the northern brazilian Amazon. Biotropica 30:458–469

Galindo-Leal C, Weber M (1994) Translocation of deer subspecies: reproductive implications. Wildl Soc Bull 22:117–120

Gallina S, Pérez-Arteaga A, Mandujano S (1998) Patrones de actividad del venado cola blanca (Odocoileus virginianus texanus) en un matorral xerófilo de México. Bol Soc Biol Concepción (Chile) 69:221–228

Geist V (1998) Deer of the world, their evolution, behavior and ecology. Stackpole Books, USA

Goodwin G (1942) Mammals of Honduras. Bull Am Mus Nat Hist 79:107–195

Grubb P (2005) Family Tayassuidae. In: Wilson D, Reeder D (eds) Mammal species of the world, 3rd edn. The Johns Hopkins University Press, Baltimore, pp 643–644

Instituto de Conservación Forestal (2014) Mapa Forestal

Instituto para la Cooperación y Desarrollo (ICADE) (2007) Línea base de la fauna silvestre en la Reserva de la Biosfera Tawahka Asagni y la parte sur de la Reserva de la Biosfera del Río Plátano. Reporte final. Instituto para la Cooperación y Desarrollo-Centro Agronómico Tropical de Investigación y Enseñanza, Honduras, P 97

Mandujano S (2004) Análisis bibliográfico de los estudios de venados en México. Acta Zoologica Mexicana (n.s.) 20:211–251

Marineros L, Martínez F (1998) Guía de campo de los mamíferos de Honduras. INADES, Tegucigalpa, p 374

Matola S, Cuarón D, Rubio-Torgler H (1997) Status and action plan of Baird's tapir (*Tapirus bairdii*). In: Brooks DM, Bodmer RE, Matola S (eds) Tapirs: Status survey and conservation action plan, pp 29–45

McCann N, Wheeler M, Coles T, Bruford M (2012) Rapid ongoing decline of Baird's tapir in Cusuco National Park, Honduras. Integr Zool 7:420–428

Medellín R, Gardner A, Aranda M (1988) The taxonomic status of the Brown brocket, *Mazama pandora* (Mammalia: Cervidae). Proc Biol Soc Washington 11(1):1–14

Medici E, Carrillo L, Montenegro O, Miller P, Carbonell F, Chassot O, Cruz-Aldán E, García M, Estrada-Andino N, Shoemaker A, Mendoza A (2005) Baird's tapir (Tapirus bairdii)

Conservation workshop population and habitat viability assessment (phva). Belize, Central America, August 15–19

Mejía T, House P (2002) Mapa de ecosistemas vegetales de Honduras. Manual de Consultas AFE/COHDEFOR. Proyecto PARA, Tegucigalpa, p 37

Moisen G, Freeman E, Blackard J, Frescino T, Nicklaus E, Edwards T Jr (2006) Predicting tree species presence and basal area in Utah. A comparison of stochastic gradient boosting generalized additive models and tree-based methods. Ecol Model 199:102–117

Morales N (2012) Modelos de distribución de especies: software Maxent y sus aplicaciones en conservación. Rev Conserv Ambient 2:1–5

Moreira-Ramírez J, Reyna-Hurtado R, Hidalgo-Mihart M, Naranjo E, Riveiro C, Garcia-Arleu R, Merida M, Ponce-Santizo G (2016) Importancia de las aguadas para el pecari de labios blancos (Tayassu pecari) en la selva Maya, Guatemala. Therya 7:1–14

Phillips S, Dudík M (2008) Modeling of species distributions with Maxent: new extensions and a comprehensive evaluation. Ecography 31:161–175

Phillips S, Anderson R, Schapire R (2006) Modelling distribution and abundance with presence only-data. J Appl Ecol 43:405–412

Portillo H, Elvir F (2016) Distribución potencial de la jagüilla (Tayassu pecari) en Honduras. Rev Mex Mastozool Né 6(1):15–23

Portillo H, Elvir F (2015) Registros y distribución potencial del jaguar (Panthera onca) en Honduras. Rev Mex Mastozool Né 5(2):55–65

Portillo H, Elvir F, Martínez M (2016) Distribución, ecología y estado actual del tapir (Tapirus bairdii) en Honduras. Rev Mex Mastozool Né 6(2):50–56

Portillo H, Elvir F, Hernández J, Leiva F, Flores M, Martínez I, Vega H (2015a) INCEBIO. Rev Mesoamericana 19(2):23–30

Portillo H, Elvir F, Hernández J, Leiva F, Flores M, Martínez I, Vega H (2015b) Datos Preliminares de la Densidad Poblacional del Venado Cola Blanca (Odocoileus virginianus) en la Zona Núcleo del Parque Nacional La Tigra, Honduras. Mesoamericana 19(2). Honduras.

Portillo H, Elvir F (2013) Composición, estructura y diversidad de los mamíferos terrestres grandes y medianos en 16 áreas protegidas en Honduras, usando fotocapturas como evidencia de registro. Mesoamericana 17:15–31

Portillo-Reyes H, Hernández J (2011) Densidad del jaguar (Panthera onca) en Honduras: primer estudio con trampas cámara en La Mosquitia Hondureña. Rev Latinoamericana Conserv 2:45–50

Ramírez-Bravo OE, Hernández Santín L (2012) Nuevos registros del temazate rojo (Mammalia: Artiodactyla: Cervidae: Mazama temama) en el Estado de Puebla, México. Acta Zool Mex 28(2):487–490

Reyna-Hurtado R, Tanner G (2005) Habitat Preferences of Ungulates in Hunted and Nonhunted Areas in the Calakmul Forest, Campeche, Mexico. Biotropica 37:676–685

Reyna-Hurtado R, Tanner GW (2007) Ungulate relative abundance in hunted and non-hunted sites in Calakmul Forest (Southern Mexico). Biodivers Conserv 16:743–757

Reyna-Hurtado R, Rojas-Flores E, Tanner G (2009) Home range and habitat preferences of white-lipped peccaries (*Tayassu pecari*) in Calakmul, Campeche, México. J Mammal 90:1199–1209

Sabogal-Rodríguez S (2010) Filogeografía y conservación del pecari de collar (Pecari tajacu) en cuatro departamentos de Colombia. Tesis Maestría en Biología, Universidad Nacional de Colombia, Facultad de Ciencias, Departamento de Biología, p 119

Schank C, Mendoza E, García Vettorazzi M, Cove M, Jordan C, O'Farrill G, Meyer N, Lizcano D, Estrada N, Poot C, Leonardo R (2015) Integrating current range-wide occurrence data with species distribution models to map the potential distribution of Baird's Tapir. Tapir Conservation, the newsletter of the UICN/SSC Tapir Specialist Group 24 (33)

Scheldeman X, van Zonneveld M (2011) Manual de capacitación en análisis espacial de diversidad y distribución de plantas. Bioversity International, Roma, Italia. 186 .ISBN 978-92-9043-908-0

Secretaria de Recursos Naturales y Ambiente (serna) (2008) Especies de Preocupación especial en Honduras, Tegucigalpa, Honduras

Secaira E (2013) Análisis y síntesis de los planes de conservación elaborados para 10 áreas protegidas de Honduras: basados en análisis de amenazas, situación y del impacto del cambio climático y definición de metas y estrategias del Proyecto Pro-Parque. ICF y USAID Pro-Parque, p 57

Slowls L (1984) The peccaries. Universidad de Arizona Press, Tucson, p 57

Slowls L (1997) Javelinas and other peccaries: their biology, management, and use. The Texas A&M University Press, College Station

Taber A, Doncaster C, Neris N, Colman F (1994) Ranging behaviour and activity patterns of two sympatric peccaries, *Catagonus wagneri* and *Tayassu tajacu*, in the Paraguayan Chaco. Mammalia 58:61–72

Villareal O, Hernández H, Franco F, García F, Utrera F (2013) Densidad poblacional del temazate rojo (Mazama temama) en dos sierras del Estado de Puebla, México. Rev Colombiana Cienc Anim 5(1):24–35

Villarreal J, Treviño A (1995) Estimación de las poblaciones silvestres de venado cola blanca Texano (Odocoileus virginianus texanus) del noreste de México. XIII Simposio sobre Fauna Silvestre. Universidad Nacional Autónoma de México y Universidad de Colima, Colima, México

Villarreal-EB FOA, Hernández F, Camacho J, Mendoza J, Campos G, Cortés L (2009) Plan de manejo para el venado temazate rojo (Mazama temama) para la Sierra Madre Oriental. In: Franco Hernández J, Villarreal O, Camacho JC (eds) Producción animal y desarrollo sustentable en rumiantes. México, BUAP, pp 77–104

Weber M, González S (2003) Latin american deer diversity and conservation: a review of atatus and distribution. Ecoscience 10(4):443–454

Chapter 9
Tropical Ungulates of Colombia

Olga L. Montenegro, Hugo F. López-Arévalo, Catherine Mora-Beltrán, Diego J. Lizcano, Hernán Serrano, Elizabeth Mesa, and Alejandra Bonilla-Sánchez

Abstract Twelve species of ungulates are present in Colombia, represented by three species of tapirs, two of peccaries and seven species of deer. A synthesis about the evolutionary and recent history for each of the group is presented, highlighting recent paleontological information for northern South America. Genetic and molecular studies indicate that deer present the greatest uncertainty regarding the number of species and their phylogenetic relationships. By species richness, the Caribbean region, with eight species, is the most diverse, followed by the Andean region with seven, the Orinoquia and Amazonia with six species and finally the Pacific region with four ungulate species. All the species present a greater proportion of their distribution outside protected areas, being two species of Andean ungulates, the mountain tapir (*Tapirus pinchaque*) and the rabbit deer (*Pudu mephistophiles*) the only ones reaching 20% of their distribution inside protected areas. Seven species are found in one of the threatened categories of IUCN, in which all species of tapirs are included. Finally, information about hippopotamus (*Hippopotamus amphibius*) and feral pig (*Sus scrofa*), alien species with wild populations in Colombia, is presented.

9.1 Evolutionary and Recent History

Most Information about fossil ungulates from South America is based on records from the southern part of the continent, mainly from Patagonia. There, ancient ungulates are known from the early Cenozoic, reaching their greatest diversification

O. L. Montenegro (✉) · H. F. López-Arévalo · C. Mora-Beltrán · H. Serrano · E. Mesa
Universidad Nacional de Colombia, grupo en Conservación y Manejo de Vida Silvestre, Bogotá, Cundinamarca, Colombia
e-mail: olmontenegrod@unal.edu.co

D. J. Lizcano
The Nature Conservancy, Northern Andes and Southern Central America (NASCA) conservation program, Bogota, DC, Colombia

A. Bonilla-Sánchez
Universidad de Antioquia, Grupo Mastozoología, Medellín, Antioquia, Colombia

S. Gallina-Tessaro (ed.), *Ecology and Conservation of Tropical Ungulates in Latin America*, https://doi.org/10.1007/978-3-030-28868-6_9

around 50 million years ago (Eocene) with five orders and 40 genera, with a subsequent reduction to the three orders, Astrophoteria, Litopterna and Notoungulata, in addition to the 50% reduction of genera (Horovitz 2012).

In Colombia, the fossil record of the Miocene, between 16 and 22 million years ago, comes mainly from La Venta site, in the department of Huila. However, recent paleontological findings from the Colombian Guajira, the Cocinetas basin and the Falcon basin in Venezuela, evidence at this site the presence of ancient native South American ungulates. Besides genera of wide distribution in South America, other ungulates restricted to low latitudes of South America such as *Hiolacotehrium* and *Huilaterium* were found, and a new species of Astrapotheria is described as *Hiolacotehrium miyou* (Carrillo et al. 2018). These authors suggest that variation between fossil fauna from northern and southern South America is related to climate differences.

Whit the Panama isthmus formation, the Great American Biotic Exchange (GIBA) accelerated, allowing entry of a larger number of groups including North American ungulates. Nearly 1 million years prior to the final closure of the isthmus, there are confirmed records of peccaries (Tayassuidae) and llamas (Camelidae). The presence of tapirs and peccaries in South America prior to the Pliocene, must be considered as doubtful until now. The arrival of other ungulates such as horses (Equidae) occurred 2.6–2.4 Mya, while deer (Cervidae), and new camelid species arrived 1.8 Mya (Chávez Hoffmeister 2016). Fossil record of tapirs is large, but it does not completely agree with molecular clock evidence (Norman and Ashley 2000). The best timing estimation of tapir entrance to South America indicate a late Miocene arrival (MacFadden and Higgins 2004).

There is some archaeological information about native ungulates in Colombia. Near Bogotá, for example, bone remains from the Pleistocene and Holocene were found at the archeological site Aguazuque located at 2600 m above sea level, where environmental conditions are ideal for preservations of ancient cultural evidence (Correal Urrego 1990). Besides two groups of mastodons, bone remains include American horses (*Equus amerhipuus lasallei*), and deer (*Odocoileus virginianus*). The presence of megafauna remains in some other studied Pleistocene archaeological sites, such as Tibitó 1 and Tequendama, suggest their posterior extinction. Disappearance of megafauna and American horse, during the tardiglacial, were related to both ecological and human factors, as such an increase in forests and a reduction of grasslands during the Holocene as well as extermination by hunters (Correal Urrego 1990).

Once established occupations of hunter-gatherers and planters who settled in unflooded lands, such as Aguazuque (Bogotá plateau), 7000–2200 b.p, it is remarkable the amount of deer and guinea pigs (genus *Cavia*) remains, as well as a large diversity and abundance of bone tools such as perforators, engravers, needles, punches, blades, burnishers, and knives made from deer's shoulder blade (Martínez-Polanco et al. 2019; Peña-León and Rincón-Rodríguez 2019; Rincón-Rodríguez 2019). Martínez-Polanco et al. (2015), examining those deer remains of Aguazuque dating from 2725 ± 35 b.p to 5025 ± 40 b.p did not find size changes over time in deer bones, that might suggest overexploitation, and habitat changes were proposed as a factor related to deer disappearance.

Current native ungulate species in Colombia include several species of deer, peccaries and tapirs. Also, there are several domestic species which were introduced by Europeans for breeding purposes about 500 years ago. Those species are mainly pigs, horses, cows, sheep and goats (Ramírez-Chaves et al. 2011). Pigs (*Sus scrofa*) were introduced to New Granada (now Colombia) in 1536, by Spanish soldiers led by Sebastián de Belalcázar (Ramírez-Chaves et al. 2011). Today there are several feral pig populations, which have been recently studied (López-Arévalo et al. 2018). Also, water buffaloes (*Bubalus bubalis*) were introduced first to help oil palm cultivation and, more recently, for meat trading purposes. In addition, during the 1980 decade, several African species were illegally introduced for recreational purposes. Such introductions were linked to the culture of drug trafficking and involved large ungulates such as African hippos, *Hippopotamus amphibius*, which since the 1990s have a free population whose control have been attempted by capture, sterilization and fencing strategies (Cornare 2018).

In the following sections, we review the wild ungulate species present in Colombia, their distribution and advances in their knowledge during the last decade. Finally, research and conservation priorities are proposed. Information on feral pigs and hippos is included, given the implications of their existence as wild populations.

9.2 Current State of Knowledge

This synthesis about each ungulate group (tapirs, peccaries and deer) living in Colombia, includes distribution by biogeographic regions, genetic and molecular aspects, ecology, use, management and conservation issues. Also, information about two exotic ungulates exhibiting wild populations in Colombia (hippopotamus and feral pigs) is included.

9.2.1 Species Account

We recognize 12 native ungulate species occurring in Colombia, including three species of tapirs, two species of peccaries and seven species of deer (Table 9.1).

9.2.2 Tapirs

For long time three species of tapirs living in Colombia had been recognized *Tapirus terrestris*, *T. pinchaque* and *T. bairdii* (Hershkovitz 1954). In 2013 a new species of tapir, *Tapirus kabomani*, was described based on morphological, morphometric and molecular data (Cozzuol et al. 2013). This new species would be distributed in Brazil and Southern Colombia (department of Amazon) in sympatry

Table 9.1 Native ungulate species living in Colombia, their distribution by biogeographic regions and their IUCN threat category

Species	IUCN category	Andes	Caribbean	Pacific	Orinoquia	Amazonia
PERISSODACTYLA						
Tapirus bairdii (Gill, 1865)	EN		X	X		
Tapirus pinchaque (Roulin, 1829)	EN	X				
Tapirus terrestris (Linnaeus, 1758)	VU		X		X	X
ARTIODACTYLA						
Pecari tajacu (Linnaeus, 1758)	LC	X	X	X	X	X
Tayassu pecari (Link, 1795)	VU	X	X	X	X	X
Mazama americana (Erxleben, 1777)	DD	X	X		X	X
Mazama nemorivaga (Cuvier, 1817)	LC				X	X
Mazama rufina (Pucheran, 1851)	VU	X				
Mazama sanctaemartae J. A. Allen 1915	NE		X			
Mazama temama (Kerr, 1792)	DD		X	X		
Odocoileus virginianus (Zimmermann, 1780)	LC	X	X	Extinct	X	X
Pudu mephistophiles (de Winton, 1896)	VU	X				

with *T. terrestris*. However, a debate on the validity of *T. kabomani* has been in place (Cozzuol et al. 2014; Ruiz-García et al. 2015a; Ruiz-García et al. 2016; Voss et al. 2014). Given this uncertainty and lack of further information, we decided not to include *T. kabomani* here until more information resolving the topic is available.

Although Groves and Grubb (2011) use genus *Tapirella* for the Central American tapir, with no further explanation, we use here the genus *Tapirus* for the three species living in Colombia. Concerning subspecies, we included in *Tapirus terrestris* two subspecies, *T. t. colombianus* and *T. t. terrestris* following Hershkovitz (1954). The first one inhabits the Caribbean region and the other one the Orinoquia and Amazonia regions.

Divergence time for the ancestors of *T. bairdii* and the group *T. terrestris* + *T. pinchaque* has been estimated in about 18–20.4 Mya using mitochondrial sequences (COII) or from 15 to 16.5 Mya (12S rRNA) (Norman and Ashley 2000). Also, this same author estimated a divergence time of the ancestors of *T. terrestris* and *T. pinchaque* to be from 2.5–2.7 to 1.5–1.6 Mya. However, Ruiz-García et al. (2012) estimated a divergence time between these linages (*T. bairdii and T. terrestris-T. pinchaque*) from 9.6 to 10.9 Mya. Also, by using sequences of Cyt-b and Bayesian trees Ruiz-García et al. (2012) found that the ancestors of *T. terrestris* and *T. pinchaque* diverged around 3.9 Mya and the divergence between the most frequent haplotype of *T. terrestris* and the main haplotype of *T. pinchaque* using the median of the union network is about 1.55 ± 0.32 Mya.

The Baird's tapir (*Tapirus bairdii*) is the largest tapir species in the Neotropics (Hershkovitz 1954). Its common name is *macho de monte* or *danto* in the Chocó region of Colombia. Additionally, there are nine names in indigenous languages applied to this species in Colombia (Rodríguez-Mahecha et al. 1995).

Ruiz-Garcia et al. (2012, 2015a) found that *T. bairdii* belongs to a different molecular group that *T. terrestris* + *T. pinchaque* and proposed that the first one could be more related to the fossil tapirs from North America. Also, Ruiz-García et al. (2015a) found a very low mitochondrial genetic diversity in *T. bairdii*, compared with other Neotropical tapirs and suggested a potential bottleneck and/or the gene drift in this species probably due to both natural and human factors.

The Baird's tapir is the only tapir species registered both at high and low altitudes (Naranjo and Vaughan 2000). The Baird's tapir habitat includes tropical and montane forests between 0 and 3000 masl (González-Maya and Schipper 2009). Historically distributed throughout Mexico, Central America and the northern region of South America, this species current distribution in Colombia is restricted to a few forest patches. Most of the Baird's tapir habitat has been cleared for logging, agriculture and extensive cattle ranching (García et al. 2016). Despite its limited distribution in Colombia, *T. bairdii* uses a wide variety of climates, including the extremely wet Chocó region, where Baird's tapir has been recorded in Los Katios National Park (Mejia-Correa and Diaz-Martinez 2014), to the dry forests and natural savannas of the upper Sinu River in the Caribbean region, where Hershkovitz registered the species in the first half of the twentieth century, sympatric to *Tapirus terrestris colombianus* (Hershkovitz 1954). However, Baird's tapirs have not been recently recorded in the upper Sinu River (Solari et al. 2013). In the Bahia Solano region, Baird's tapir was reported as locally extinct in the second half of twentieth century due to overhunting (Ulloa et al. 2004), even though substantial forest cover remained. The southernmost record of Baird's tapir in Colombia is from Nariño department, on the Pacific slopes of the Andes in southern Colombia, where Baird's tapirs were reported as declining in the 1980s due to overhunting (Orejuela 1992). The most recent distribution analysis suggests the species to be restricted to the Chocó region in Colombia (Schank et al. 2015).

Baird's tapir ecology has been little studied in Colombia. An analysis of Baird's tapir diet has identified 28 plant species used as food in Katios National Park (Mejia-Correa and Diaz-Martinez 2014). The density of Baird's tapir has been estimated in 1.02 individuals/km^2 using capture recapture methods assisted by camera trapping and visual transects in Katios National Park (Mejia-Correa and Diaz-Martinez 2014). Most current estimates of tapir population size based on a Poisson point process models, which allows for predictions of population size, show that total predicted abundance based on spatial covariates modeling, is higher than expert opinions about the species. Local density estimates from the Poisson point process model were like available independent assessments in different regions (Schank et al. 2017).

Baird's tapirs are hunted for meat, pelts and body parts, which are used in traditional folk medicine. This occurs although harming, killing or selling tapirs or its parts is illegal in Colombia (Minambiente 2017a). A study of hunting among Embera-Katíos indigenous communities recorded tapir extractions (Racero-Casarrubia et al. 2008), and although those authors do not mention *T. bairdii*, it is very likely that it will also be used due to its sympatric distribution with *T. colombianus* in the area (Hershkovitz 1954). Also, previous hunting studies carried out in the Pacific region had no records of *T. bairdii* uptakes (e.g., Hernández 1995; Rubio-Torgler 1997; Castiblanco 2002), suggesting very low populations or absence of this species in parts of that region since several decades ago.

A Population Viability Analysis (PVA) carried out for the Baird's tapir populations in Central America, demonstrated that the species is extremely sensitive to threats that affect adult reproductive success (Medici et al. 2005), highlighting the impact of habitat loss and hunting on the long-term survival of the species. In terms of conservation planning, the species is included in the National Plan for the Conservation of the *Tapirus* genus in Colombia, (Montenegro 2005), and it is currently listed as Endangered (EN) on the 2016 IUCN Red List of Threatened Species. In Colombia, *T. bairdii* is found in two protected areas (Los Katíos and Las Orquídeas National Parks), which represent 4.4% of its area of occurrence in the country (Fig. 9.3).

Mountain tapir *Tapirus pinchaque* is the smallest species of the *Tapirus* genus. However, it is the largest mammal in the Andes. This species went unnoticed by science until the nineteenth century, when it was discovered by French naturalist François Désiré Roulin in 1829, with a skull from the Sumapaz *páramo* and another one from the Quindío *páramo* in Colombia. However, the original publication describing this species was elaborated by Cuvier (1829). The name *pinchaque* or *panchique* was applied to a mythical animal, reincarnation of a *cacique* (Indian leader) of the south of the country, near Popayán, and indigenous Guambinos, Puracés, Moguez, Coconucos, or Polindras adjudged it to mountain tapir. In other regions of the country this species is known as *danta lanuda* (wooly tapir), *danta conga*, or *danta apizarrada* (Rodríguez-Mahecha et al. 1995).

According to Ruiz-García et al. (2016), mountain tapir populations of Colombia and Ecuador experienced a high historical gene flow, which is reflected in their low genetic differentiation, found from the information coming from the sequencing of 15 mitochondrial genes. Although the Colombian population presents a slightly higher nucleotide diversity, these authors did not detect any molecular subspecies or significantly different evolutionary units for *T. pinchaque*. These two countries contain more than 98% of the current mountain tapirs.

The southern limit of *Tapirus pinchaque* distribution is northern Peru (Lizcano and Sissa 2003). In Ecuador Mountain tapir is distributed mainly in the western flank of the Andes, in the Amazonian slope (Ortega-Andrade et al. 2015). In Colombia it is found from the southern Andes in the Amazonian slope (Noguera-Urbano et al. 2014) along the central mountain range northward up to Los Nevados

National Park (Cavelier et al. 2011; Lizcano et al. 2002) (Figs. 9.1b and 9.2i). The recent record of a skull collected in 1911 in the Frontino highlands in the western mountain range (Arias Alzate et al. 2010) opens possibilities to think that the mountain tapir could have a historical distribution that included the Western Cordillera, showing a need for a biogeographical re-evaluation of Mountain tapir distribution.

Mountain tapir uses Andean forests and páramos located between 2000 and 4500 m of altitude. Although tapir occasionally travel through the snow, it uses secondary Andean forests more frequently than the mature forests or *páramos* (Lizcano and Cavelier 2004a; Cavelier et al. 2011). Tapir is an herbivore found more on the foliar side of the spectrum browser-grazer (Bodmer 1990), therefore its dependents on tree and shrubs of forests rather than on *páramo* grasslands. Tapir's diet includes more than 1 hundred species, among which stand out *Chusquea* spp., *Lupinus* spp., *Gunnera manicata*, *Oxalis* spp., and ferns (Downer 2001). Many of the consumed species contain toxic compounds or elements that reduce digestibility, which is why tapirs frequent salt licks to help detoxification and supplementation of its diet (Acosta et al. 1996; Lizcano and Cavelier 2004b).

Although mountain tapir is mainly a solitary animal, males can follow females, forming temporary pairs during the mating season. Tapir is mainly crepuscular, with a bimodal pattern of activity, with peaks in dawn and dusk hours. Activity pattern is also negatively correlated with temperature, being more active in the less warm hours of the day (Lizcano and Cavelier 2000; Lizcano and Cavelier 2004a). Although mountain tapir is a large and strong animal, it can become a prey of pumas and Andean bears, as it has been recorded in the Central Andes of Colombia (Rodríguez et al. 2014).

In Colombia, densities of 1 tapir/550 ha and a relative abundance of 37 tapirs/100 cameras-night have been estimated in the Ucumari regional park. Three individual tapirs were identified from camera trap records in an area of 3.5 km^2 in Purace National Park (Abud et al. 2012).

Previous studies from the 2000s estimated that only 13% of mountain tapir habitat was protected within the Colombian National System of Protected Natural Areas (SNANP) (Lizcano et al. 2002). However, creation of new protected areas since the 2000s has increased up to 23% the percentage protected tapir (Fig. 9.3). However, potential current and future distribution under climate change scenarios has been estimated. Potential area of occurrence was about 28,000 km^2 with a predicted reduction of 38.11% due to habitat loss and about 35–47% due to climate change effects. This prediction took into consideration effects of habitat loss, ecosystem availability and the role of current National System of protected Areas (SNANP). However, the synergistic effect of both phenomena could represent a greater risk in the short term with an estimated reduction of about 55–65% of the potential distribution (Lizcano et al. 2015).

In terms of conservation planning, the National Plan for the Conservation of the *Tapirus* genus in Colombia, highlights the need for delimiting distribution areas and ecosystems availability for mountain tapir, in order to obtain a better evaluation of the main threats and to identify priority conservation units (Lizcano et al. 2005; Montenegro 2005). *T. pinchaque* is categorized as an endangered (EN) in both the red book of mammals of Colombia (Rodríguez-Mahecha et al. 2006), and in the

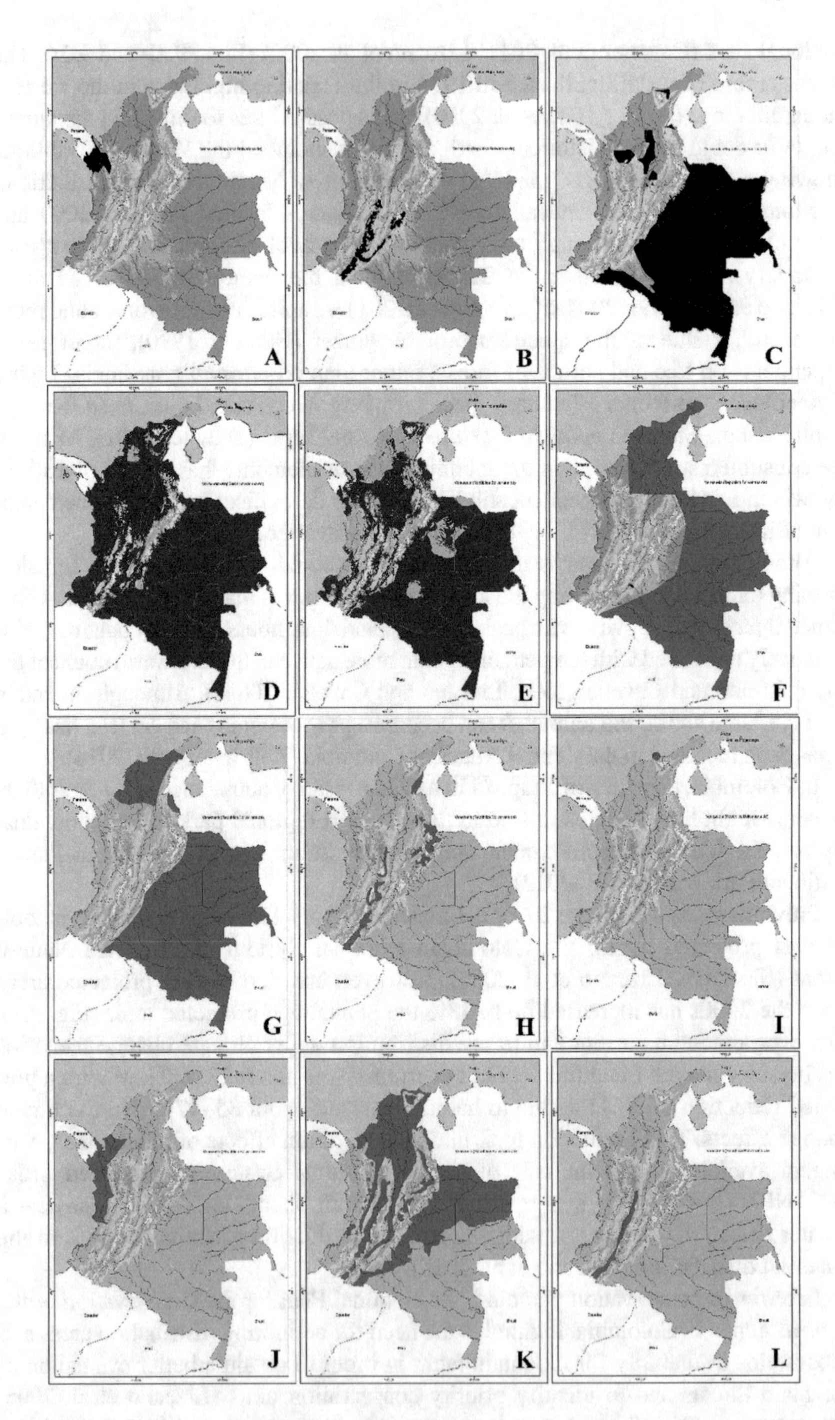

Fig. 9.1 Distribution maps of ungulate species present in Colombia. (**a**) *Tapirus bairdii.* (**b**) *Tapirus pinchaque.* (**c**) *Tapirus terrestris.* (**d**) *Pecari tajacu.* (**e**) *Tayassu pecari.* (**f**) *Mazama americana.* (**g**) *Mazama nemorivaga.* (**h**) *Mazama rufina.* (**i**) *Mazama sanctaemartae.* (**j**) *Mazama temama.* (**k**) *Odocoileus virginianus.* (**l**) *Pudu mephistophiles*

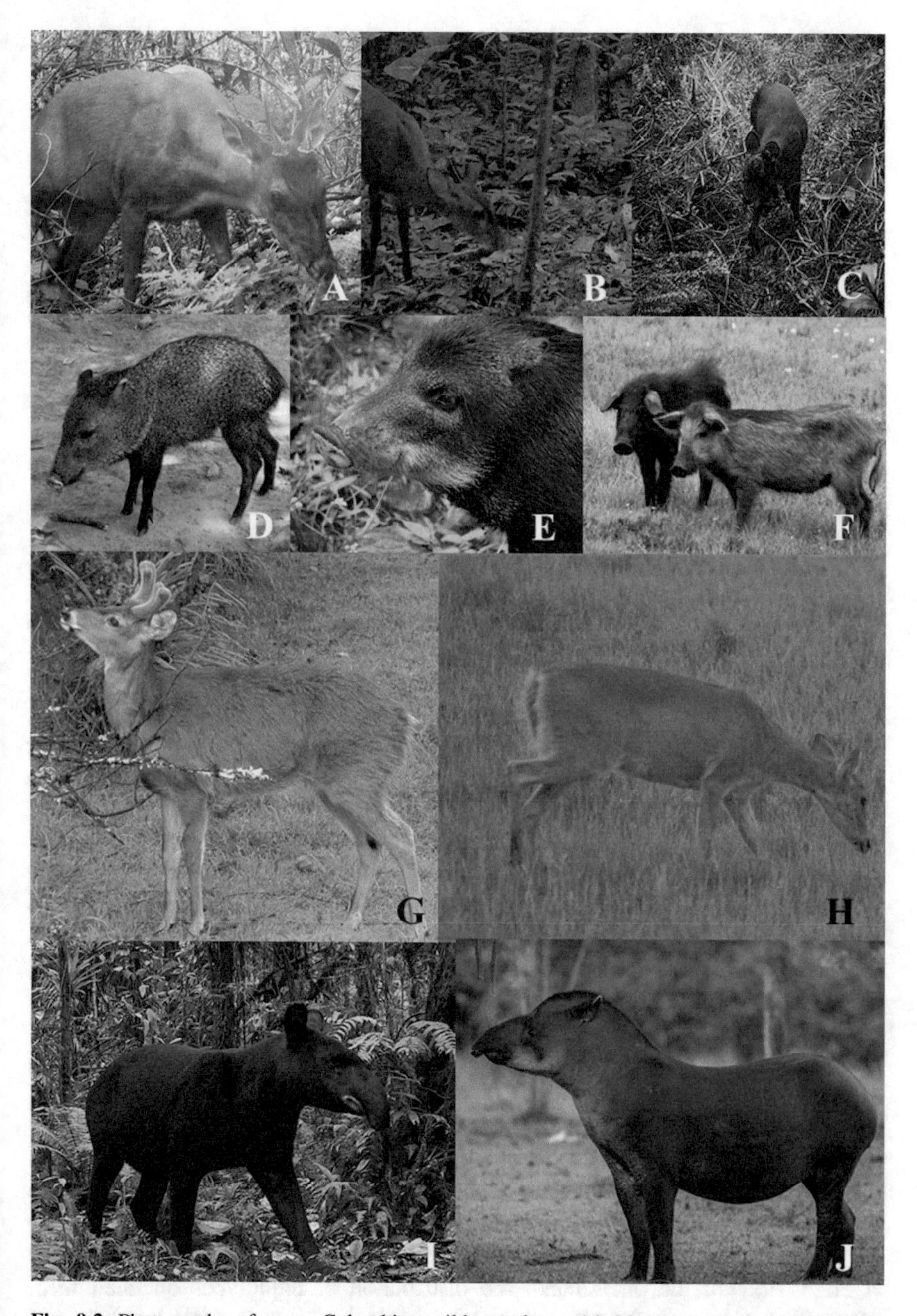

Fig. 9.2 Photographs of some Colombian wild ungulates. (**a**) *Mazama americana* (Instituto Amazónico de Investigaciones científicas SINCHI). (**b**) *Mazama nemorivaga* (Abelardo Rodríguez). (**c**) *Mazama rufina* (Abelardo Rodríguez). (**d**) *Pecari tajacu* (Olga L. Montenegro). (**e**) *Tayassu pecari* (Olga L. Montenegro). (**f**) *Sus scrofa,* asilvestrado (Grupo en Conservación y Manejo de Vida Silvestre). (**g**) *Odocoileus virginianus goudotii* (Olga L. Montenegro). (**h**) *Odocoileus virginianus apurensis* (Grupo en Conservación y Manejo de Vida Silvestre). (**i**) *Tapirus pinchaque* (Diego J. Lizcano). (**j**) *Tapirus terrestris* (Diego J. Lizcano)

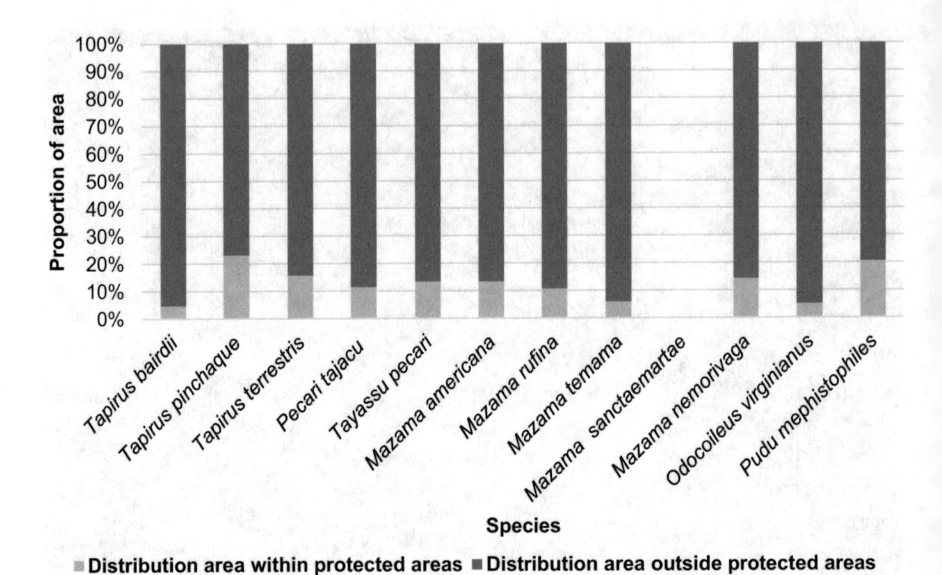

Fig. 9.3 Proportion of the distribution of Colombian ungulate species in National Natural Parks defined according to the Sistema Nacional de Áreas Protegidas (SINAP)

official list of threatened Colombian species (Minambiente 2017a). Additionally, it is listed in appendix II of CITES. At the regional level, tapir faces threats such as hunting in its southern range. Other threats are habitat deterioration from agricultural advance and infrastructure development in northern Colombian Andes. *T. pinchaque* is present in most of the protected areas of the Andes in Colombia (Table 9.1). Although there is a national conservation plan for this species in Colombia (Montenegro 2005), at the regional level tapir conservation initiatives are just beginning to be outlined with departmental plans and by national parks authorities. This species has been recognized within the species used by Andean indigenous communities, and in some of them it is part of the species targeted for hunting monitoring.

Lowland tapir *T. terrestris*, receives different common names such as danta, Colombian tapir, Amazon tapir or cuanta or *ranta* in the Sierra Nevada de Santa Marta. In Southern Colombia (department of Cauca) lowland tapir is known as gran bestia (great beast). Also, there are more than 45 names in indigenous dialects for lowland tapir throughout the national territory (Rodríguez-Mahecha et al. 1995).

A study of the evolutionary history and phylogeography of *T. terrestris* indicates its South American origin and its rapid expansion, since the first fossil record from the late Pleistocene was found in the southern limits of its distribution (de Thoisy et al. 2010). On the other hand, the distribution of haplotypes obtained from sequencing of two overlapping fragments of cytochrome b, indicate low levels of genetic structuring at the continental scale (de Thoisy et al. 2010). Recently, sequences of the Cyt-b gene from samples of large geographical distribution of lowland tapirs (Ruiz-García et al. 2015a) confirms that the greatest genetic diversity of this species is found in the northwestern part of the Amazon. These authors also point out the presence of six haplogroups within *T. terrestris* of which five are present in Colombia (Ruiz-García et al. 2015a).

Colombian lowland tapir (*Tapirus terrestris colombianus*) inhabits continental ecosystems of the Caribbean region in Colombia (Constantino 2005) down lowlands of northern central cordillera of Colombia (Arias Alzate et al. 2009) and it is separated from *T. t. terrestris* by the Andean cordilleras. *T. t. colombianus* was described by Hershkovitz (1954) based on the skin and skull of a young adult male, from the site called *El salado*, on the eastern flank of the Sierra Nevada de Santa Marta, department of Magdalena. The other subspecies inhabiting Colombia, *Tapirus terrestris*, has a greater distribution in the country, inhabiting the Orinoquia and Amazonia regions (Figs. 9.1c and 9.2j), and extending to the neighboring countries Venezuela, Brazil, Peru, and Ecuador (Montenegro 2005). In Colombian Orinoquia region, many new records of lowland tapir have been available from Meta, Guaviare and Vichada departments, thanks to the increased use of camera trapping (Díaz et al. 2016). In Vichada, along the Bita river basin, lowland tapir has been registered as abundant (Mosquera-Guerra et al. 2018).

From direct monitoring of a subadult female of *T. t. colombinus* in northern Antioquia, Sábato et al. (2008) provided natural history data and recorded that her main activity was resting mainly in shelters and, in a lesser extent, in water bodies. He found also, that she foraged on at least 22 plant species. In La Guajira department, Kaston et al. (2008) registered *T. t. colombianus* in five sites of Dibulla municipality based on camera trapping and skull fragments. These authors suggested that this population is the northernmost of its distribution and is isolated from other populations. More recently, a monitored individual in Sierra Nevada of Santa Marta was found using preserved or recovering forest mixed with patches of secondary vegetation where its activity was the most frequent (González-D et al. 2014).

In Amacayacu National Park (Colombian Amazon), from May 2009 to September 2010 Cabrera (2015) simultaneously studied a total of nine salt licks, each in areas with different hunting pressures. Most visits were of solitary individuals. The common visit pattern was several entries during 1 or 2 days followed by 5 ± 6.4 days of absence and reappearance for another couple of days, followed by periods of total absence of 20 ± 6.5 days.

Two studies on lowland tapir movements have been published. The first one deals with comparison of lowland tapir home range estimations in the middle Caquetá river basin, based with two types of information, store-on-board (SoB) and satellite-transmitted (IT) data sets (Cabrera et al. 2016). Home range estimations for a young male and adult female were obtained using different analytical methods. Male home ranges estimates ranged from 834 ha to 1798.7 ha (with 95% data set) and from 81 ha to 311.7 ha (with 50% data set). For the adult female, home range estimates varied from 189.5 ha to 914.9 ha (95% data set) and from 9.1 to 293.2 ha (50% data set). Differences in those estimations come from the analytical procedures tested in the study. Part of the above data was used in the second study (González-D et al. 2017), where authors were able to differentiate between lowland tapir foraging behaviors and exploring behaviors. Foraging behaviors were those of long trips and selectively use of resources, and exploring behaviors were short displacements with a reduced use of resource areas (González-D et al. 2017).

At the same region Vélez et al. (2017) determined species and plant components (leaves, fruits) utilized by lowland tapirs in different seasons and habitats, in relation to changes in food availability. They also found that tapirs browsed on least 50 plants species and consumed fruits from at least 18 species, according to seasonal availability. Concerning population estimates, Gómez et al. (2016) found a mean relative abundance index of 0.08 in Tuparro National Park and 0.06 in Puinawai Natural Park. In the Tiquie river (Osorno et al. 2014) estimated a lowland tapir density of 0.37 individuals/km^2.

Regarding use, lowland tapir has been repeatedly identified within the top five most hunted species in terms of amount of biomass extracted for subsistence and trade, by Amazonian communities (Wallace et al. 2012; van Vliet et al. 2014). In most of the Amazonia region tapir hunting occurs mainly at salt licks (Lozano 2004; Sarmiento 2007; Cabrera 2015). In the transition between Amazonia and Orinoquia region, an evaluation of hunting practiced by Sikuani communities of Matavén indigenous reserve, in savanna areas, suggests a decline of tapir populations, due to increase in human population, the presence of new settlers and with them the establishment of illegal coca crops (Plata Rangel 2005). In Puinawai National Natural Reserve (Guainía department), villagers from *Punta Pava* reported tapirs to be most desired and valuable meat (Guzmán 2005). In the same reserve, lowland tapir, along with peccaries, contributed the largest quantity of consumed biomass between 2005 and 2009 by a *Puinave* community (Tafur-Guarin 2010).

In the Tiquie river (Vaupés department) the *Tuyuca* and *Bará* ethnic groups living in the Puerto Loro and Bellavista communities showed a lowland tapir extraction rate of 8.639 Kg/Km^2 (Osorno et al. 2014). Harvest rates gets up to 0.093 individuals/consumer∗year, which is lower than averages reported in other Amazonian sites. As in other ethnic groups, lowland tapirs have an important symbolism in these ethnic group's cosmology (Osorno et al. 2014).

Maldonado (2010), estimated tapir extraction to be between 21.3 individuals/km^2 in San Martín de Amacayacu and 4 individuals/km^2 in Mocagua, in southern Colombian Amazon. Ungulate population declines seem to be occurring near some indigenous communities. For example, *Ticuna* and *Yagua* ethnic groups in southern Colombian Amazon indicate that for hunting large animals, such as deer, tapirs, and peccaries, long trips into the jungle are needed, which may require from 4 to 10 h by boat, and several days of searching (Sandrin et al. 2016). Hunting occurs in mainly in high water periods. However, not everyone at the community consume tapir meat due to cultural restrictions especially for pregnant women (Sandrin et al. 2016).

As mentioned above, salt licks are most used sites for tapir hunting, and a negative effect on tapir visits to those sites could be expected from this practice. However, a 2-year study in sites with different hunting pressure, in the Amacayacu National Natural Park, did not evidence significant changes in tapir visits to salt licks. Apparently, the free spaces resulting from the harvest of individual tapirs are quickly occupied, either by local subadults or adults coming from other areas of the park, preventing the effects of hunting within the PNN Amacayacu to be significant (Cabrera 2015).

Concerning conservation status, lowland tapir is listed globally as vulnerable (VU) (Naveda et al. 2008). In Colombia *T. t. colombianus* was categorized as critically endangered (CR) in the red book of Colombian mammals (Rodríguez-Mahecha et al. 2006) because of its small range and few populations (Arias Alzate et al. 2009). A national conservation program for all tapir species in Colombia was published by the Ministry of Environment in 2005 (Montenegro 2005). A plan for implementation of conservation measures for *T.t. colombianus* was proposed for La Guajira department (Corpoguajira and Fundación Nativa 2007).

For lowland tapir in Amazonia and Orinoquia regions, several areas were identified as conservation priorities more than a decade ago (Rodríguez 2004). More recently, a regional conservation program for lowland tapir populations in Casanare, Arauca y Vichada (Orinoquia region) was proposed. Currently, there are in place conservation efforts that include consolidation of private reserves both in the Bita river (Vichada) and the lower Casanare river (Arauca department) (Alviz and Pérez 2015).

The lowland tapir is protected in 20 national parks in Colombia (Fig. 9.3). However, to establish effective conservation strategies it is necessary to understand the close relationship between tapirs, the habitats that they use and the use of people, to understand the effects of hunting, and to better understand their key habitat requirements.

9.2.3 Peccaries

In Colombia there are two representatives of the Tayassuidae family, collared peccary *Pecari tajacu* and white-lipped peccary *Tayassu pecari*. Collared peccary *Pecari tajacu* receives different names in Colombia, such as *chácharo* and *saíno* in Orinoquia, *tatabro* in the central part of the country, *cerrillo* in the Amazon and *báquiro* in the border with Venezuela (Montenegro et al. 2008). In addition, *P. tajacu* receives more than 20 names in indigenous dialects throughout the national territory (Rodríguez-Mahecha et al. 1995).

Collared peccary is found in the five biogeographical regions of Colombia (Table 9.1). A niche model using MaxEnt (Montenegro et al. 2008; Montenegro et al. unpublished data) predicts its wide distribution in Colombia, except for the arid areas of the northern department of La Guajira, the highest areas of the Sierra Nevada de Santa Martha and the three mountain ranges (Figs. 9.1d and 9.2d). There is field verification of *P. tajacu* populations in 94 municipalities of 24 departments of the country (Montenegro et al. 2008). In the Córdoba department collared peccary populations have been recorded in 10 municipalities (Humanez-López et al. 2016). Potential distribution of *P. tajacu* in Colombia might reach approximately 1,021,462 km^2, about 89% of the country's continental territory. However, according to IDEAM (2015), about 34% of the Colombian terrestrial ecosystems have been transformed, so the distribution of the collared peccary would currently be close to 674,000 km^2 or less, because transformation processes continue.

There are some studies of phylogenetics, phylogeography, and population genetics of collared peccary in Colombia. Góngora et al. (2006) found that collared peccaries in this country are paraphyletic and that two clades (North and South) can be recognized. Sabogal (2010) found similar results, with a further subdivision of the South clade into two subclades (South 1 and 2). At the genetic-population level, Sabogal (2010) found high population structure for three regions: Andean, Caribbean and Orinoco. Given the agreement of the phylogenetic, phylogeographic, and population genetics results, three management units for this species in Colombia are proposed, as well as a possible hybridization phenomenon between lineages of *P. tajacu* in the southeastern department of Antioquia (Sabogal 2010).

Pecari tajacu has been recorded in several environments in Colombia, such as dry forest ecosystems and swamp complex in the District of Cartagena de Indias (Tinoco-Sotomayor 2018), the very humid forests of Chocó (Palacios-Mosquera et al. 2008), oligotrophic forests of the Guiana Shield (Gómez and Montenegro 2012), ecotones between forest and savannahs dominated by *Mauritia flexuosa* palms in the Orinoquia (Trujillo and Mosquera-Guerra 2016) and in Igapó forest in the Amazon (Acevedo-Quintero and Zamora-Abrego 2016). Studies on the ecology of *Mauritia* palm trees at Calderón River (Amazonia) found that collared peccary did not consume or scatter seeds of this palm (Acevedo-Quintero and Zamora-Abrego 2016) but ate its seedlings, affecting palm survival (González-B 2016).

Regarding population abundance, estimations from the National Park El Tuparro (PNNT) and the National Natural Reserve Puinawai (RNP), were 0.24 individuals/km and 0.05 individuals/km respectively (Gómez and Montenegro 2012). Abundance differences, also analyzed with occupancy models, seem to be associated with variation in landscape configuration between the two areas (Gómez et al. 2016). These abundances are low compared to those of other Neotropical forests, and resemble those of areas with high hunting pressure, fragmented forests and dry environments. Low collared peccary abundances are also recorded at the Tiquie River (Vaupés department), with estimations of 0.159 and 0.665 individuals/km^2 at Bellavista and Puerto Loro sites respectively (Osorno et al. 2014). However, PNNT, RNP and Tiquié River are in the Colombian Guiana, a biogeographic region characterized by low productivity soils (Morán 1997), where the fauna exhibits natural low densities. In the department of Amazonas an index of relative abundance of 0.53 was estimated for the collared peccary, in an unprotected area subject to selective logging and with low hunting pressure in the Calderón River (Payán and Escudero-Páez 2015).

Peccaries are used as source of protein by rural, indigenous and black communities in all regions of the country, with emphasis on the Pacific, Amazonian and Orinoco regions. Hunting is mainly of subsistence, although there may be a small local trade. In the Puinawai National Natural Reserve, the natives of Zancudo community (Puinaves and Curripacos) consume this species and harvesting models show this extraction as sustainable (Tafur-Guarin 2010).

Communities of Puinave, Curripaco, Piapoco, and Sicuane ethnic groups living at the Inírida River basin (Guainía department), use the two species of peccaries for subsistence and commercial hunting (Ferrer-Pérez et al. 2009). In the Tuparro biosphere reserve (Vichada department), most hunted biomass come from the two species of peccaries. However, *P. tajacu* tends to be more consumed (Martínez-Salas et al. 2016). In the Tiquie River (Vaupés), collared peccaries account for only 4.71% of consumed biomass due to a great diversification that these communities make in their hunting practices (Osorno et al. 2014). Both *T. pecari* and *P. tajacu* are the most hunted species at the Amazonian Tikuna communities of San Martín de Amacayacu and Mocagua villages (Bonilla-Riveros 2014).

Use of *P. tajacu* at the Ayapel swamp complex (Caribbean region) include consumption, some trade and keeping young animals as pets (David et al. 2017). Also, at the Pacific region, some communities of the middle Atrato River occasionally hunt peccaries, mainly as a tradition rather than a subsistence practice, since their main source of protein are domestic animals (Palacios-Mosquera et al. 2008).

In Colombia, several areas of swine production are close to peccary habitat. Technified swine farms are located mainly at the Andean region where contact between domestic pigs and peccaries is prevented. In the Orinoco region, in contrast, pig production occurs mainly in non-technified farms where pigs roam freely. Such a production system facilitates the occurrence of feral pig populations and their contact with peccaries (Montenegro et al. 2008; López-Arévalo et al. 2018). Serological surveys carried out in zones where *P. tajacu* and feral pigs (*S. scrofa*) coexist have shown several shared pathogens (viruses and bacteria) such as porcine circovirus, vesicular stomatitis (subtypes of New Jersey and Indiana), and *Leptospira interrogans* serovars bratislava, grippotyphosa, icterohaemorrhagiae, and pomona (Montenegro et al. 2018).

Globally *P. tajacu* is considered a species of Low Concern (LC) because of its wide range and ecological flexibility (Góngora et al. 2011). In Colombia, there is not a risk assessment for this species. However, its international trade is regulated by CITES Appendix II. Collared peccary is found in 47 out of 59 Colombian protected areas. The extension of such areas covers about 10% of collared peccary distribution in the country (Fig. 9.3). At regional scale, this species shows threats such as hunting (Humanez-López et al. 2016) and habitat loss, a growing situation and that might endanger some local populations.

White-lipped peccary *Tayassu pecari* (Fig. 9.2e) is known in Colombia as *manao* in the Andean, Caribbean and Pacific regions, *cajuche* or *cafuche* in the Orinoquia and *puerco de monte* in the Amazon. This species also receives at least 23 names Colombian indigenous languages (Rodríguez-Mahecha et al. 1995).

The distribution of the white-lipped peccary in Colombia has been a bit confusing. Several authors show lowlands of the whole country up to 1800 m altitude as white-lipped peccary habitat (Emmons 1999; Alberico et al. 2000; Morales-Jiménez et al. 2004; Solari et al. 2013); while others exclude areas of La Guajira and north-eastern Colombian Orinoquia as suitable habitat for this species (Mayer and Wetzel 1987; Eisenberg 1989; March 1993).

During a multinational workshop in 2005, a distribution map was proposed maintaining white-lipped peccary naturally absent in the northern east of the Colombian Orinoquia and extinct in the entire Caribbean region, the inter-Andean valleys, foothills of the eastern mountain range, northern part of Meta department and the west of Caquetá department, given deforestation at those areas (Taber et al. 2008). Keuroghlian et al. (2013) shows this same distribution, since the topic had not been updated. It is currently known that some populations at the Caribbean region still exist, specifically in the interior and buffer zone of Paramillo National Park in the southern Córdoba department (Humanez-López et al. 2016). Also, a white-lipped population was found at Puerto Rondón, in northeastern Orinoquia region where tropical savannas dominate (Montenegro et al. 2018). However, these records come from riparian forests and not from savannas. This species has also been recorded at the Serranía de La Lindosa, in San José del Guaviare (Montenegro et al. 2017). Other white-lipped populations were registered in 48 municipalities of 15 departments of Colombia (Montenegro et al. 2008). The potential area of occurrence of this species in Colombia, estimated by niche modeling, would be 824,690 km^2 which is the 72% of the continental area of the country (Fig. 9.1e). However, due to the transformation of about 34% of the continental territory of Colombia, an occurrence area of 544,295 km^2 is suggested.

Rodríguez (2014) and Ruiz-García et al. (2015b) examined the genetic structure and molecular phylogeography of *T. pecari*, sequencing the mitochondrial control region (D-loop) and genotyping with three microsatellites *T. pecari* individuals representing the four morphological subspecies considered by Groves and Grubb (1993) in northwestern South America (*Tayassu pecari spiradens*, *T. p. aquatoris*, *T. p. pecari*, and *T. p. albirostris*). These authors found that the molecular results do not correlate with the morphological subspecies described so far by *T. pecari* in northwestern South America. Neither did they detect evidence of notable historical demographic changes in the population in northwestern South America.

In the central region of Chocó, white-lipped peccary was found using different types of habitat, such as remnants of natural vegetation, secondary forests and cultivated areas (Palacios-Mosquera et al. 2010). Camera trapping has shown this species using edges of *Mauritia* palm areas in five localities of the Colombian Orinoquia (Trujillo and Mosquera-Guerra 2016) and in flooded forest in the Amazonian (Acevedo-Quintero and Zamora-Abrego 2016). In the Calderón River basin (Amazon) *T. pecari* was the main consumer of *Mauritia flexuosa* palm eating up to 45.3% of fallen fruits (Acevedo-Quintero and Zamora-Abrego 2016). Most palm fruits were partially destroyed and left under their parent trees, increasing seed mortality. This result shows the white-lipped peccary as the main *Mauritia* seed predator at this study site (González-B 2016).

In riparian forests along the Bita river (Vichada) white-lipped peccary showed an early morning activity (Mosquera-Guerra et al. 2018). At this study site, records of this species were more abundant during the rainy season. Herd size has seldom been registered in Colombia. In areas of Colombian Guiana, herds of 20 individuals have been observed (Gómez et al. 2016).

There are few populations estimates for white-lipped peccary in Colombia. In the central Choco region, in natural and secondary forest remnants, relative abundance of 3.0 individuals/km (confidence interval of 2.7–4.0) have been reported (Palacios-Mosquera et al. 2010). In the Bita river (Vichada) a white-lipped peccary relative abundance index of 0.176 individuals/km ($n = 1360$) was estimated, being the most abundant species compared to other medium and large mammals of the region (Mosquera-Guerra et al. 2018). Lower relative abundance of white-lipper peccary was found in two protected areas of Colombian Guiana (PNN Tuparro and RNN Puinawai) using occupancy models. The estimated mean relative abundance was 0.02 in the two areas and a positive relationship was found between the abundance of this species and the percentage of flooded forest in the landscape (Gómez et al. 2016).

Like collared peccary, the white-lipped peccary has a relationship with the feral pigs. In a study of pathogens shared with feral pigs, an individual of white-lipped was positive for *Leptospira interrogans* serovar bratislava and for vesicular stomatitis virus, New Jersey strain, both pathogens shared also with collar pecari (Montenegro et al. 2018).

White-lipped peccary is also one of the main hunted species. In the Tuparro Biosphere Reserve, a Piaroa Indian community harvested 0.375 individuals which provided 9.38 kg of biomass consumed per person per year (Martínez-Salas et al. 2016). For the Tikuna Indians inhabiting Martín de Amacayacu (Amazonia) white-lipped peccary is one of the top ten used species, and its capture occurs mainly in the low water season (Bonilla-Riveros 2014). In contrast, in Tiquie river (Vaupés), white-lipped peccary contributed only to 4.68% of the annual biomass consumed by the Bellavista and Puerto Loro communities (Osorno et al. 2014). This is because a low animal abundance makes villagers to focus in a large diversity of preys (Osorno et al. 2014).

Globally, white-lipped peccary has been categorized as a Vulnerable (VU) species (Keuroghlian et al. 2013). However, in Colombia, this species has no yet been included neither in the Colombian red book of mammals (Rodríguez-Mahecha et al. 2006), or the officially threatened species list (Minambiente 2017a), because of lack of a risk analysis. Currently, white-lipped peccary inhabits 46 out of 59 protected areas, where about 10% of its distribution is included (Fig. 9.3). Probably the most vulnerable white-lipped peccary populations in Colombia are those located at forest remnants the Caribbean region, except for those inhabiting the Paramillo National Park in southern Córdoba (Humanez-López et al. 2016), and the few populations of northeastern Orinoquia (municipality of Puerto Rondón), where there are no protected areas.

9.2.4 Deer

Of wild ungulates in Colombia, deer present the greatest uncertainty regarding the number of species and their phylogenetic relationships. The need for a thorough review of the taxonomy and phylogeny of Neotropical cervids has been identified by several authors and a review of the current status can be found in Gutiérrez et al. (2015) and Escobedo-Morales et al. (2016). Based on phylogenetic analyzes covering a wide geographical coverage, these authors propose the absence of monophyly in most genera of the *Odocoileini* tribe and suggest that definition of new genera and species are needed inside this group. In these studies, Colombian material has been used, although there are very few specimens from certain localities such as *M. temama* of Magdalena, DNA from *O. virginianus* from the central zone of the country, among others, and mitogenome from *M. gouazoubira* (Ruiz-García et al. 2007; Caparroz et al. 2015; Gutiérrez et al. 2015; Escobedo-Morales et al. 2016). In this chapter we propose the presence of seven deer species in Colombia (Table 9.1).

For the *Mazama* genus, we recognize five species in Colombia, including *Mazama americana* (Erxleben, 1777), *Mazama rufina* (Pucheran, 1851), *Mazama temama* (Kerr, 1792), as red brocket group and *Mazama sanctaemartae*, J.A. Allen 1915 and *Mazama nemorivaga* (Cuvier, 1817) as a gray brocket group. We excluded species for which we did not find solid evidence of being in Colombia. Ramírez-Chaves et al. (2016) did not include *Mazama americana* in their latest list Colombian mammals and instead they listed *M. zamora* and *M. zetta* as valid species in Colombia, with no further explanations. However, we keep *Mazama americana* until more information is available. In Allen's description of *M. a. zamora*, (Allen 1915) only one specimen from southeastern Ecuador is mentioned, and although Cabrera (1960) indicates this subspecies to occur in southeastern Colombia, he did not mention a solid evidence of such distribution. Conversely, *M. a. zetta* was described based on ten specimens from Medellín, Antioquia (Thomas 1913) and later Allen (1915) reported it also in Antioquia. Here, we included those records in Antioquia as *M. americana* until the taxonomy of this group is better understood. In red brocket group also, we include *Mazama temama* a species presents in Colombia, at the Caribbean and Pacific regions based on the distribution shown by Bello et al. (2016). This species is also listed by Ramírez-Chaves et al. (2016) as present in Colombia. For Andean dwarf red broket deer we included only *M. rufina*, and excluded *Mazama bricenii* because it was found to be a synonymous of *M. rufina,* based on molecular information (Gutiérrez et al. 2015).

About gray brocket deer group, recently *M. gouazoubira* was excluded from Colombia (Black-Decima and Vogliotti 2016) and the subspecies present in the country remain as species, *M. sanctaemartae* and *M. nemorivaga*, as proposed by Allen (1915). *M. nemorivaga* is currently accepted as a valid species for both the IUCN listing (Rossi and Duarte 2016) and for The Mammal Diversity Database (Burgin et al. 2018) as gray brocket deer of Colombian Amazonia. We decided to keep *M. sanctaemartae* as a full species following Allen (1915) there are twelve specimens as a support that allows separating at the species level the records of the town of Bonda, Santa Marta, Magdalena and the work of Sarria Perea (2012) where he performs a phylogenetic analysis of some *Mazama* for Colombia, it is concluded that the Magdalena specimens have a high support. Additionally, Gutiérrez et al. (2015) present phylogenetic support for the deer of that locality. Finally, we exclude *M. murelia* due to the little morphological support, since it was described by Allen (1915) with a single juvenile specimen from the department of Caquetá.

Concerning *Mazama* species distribution across the biogeographical regions of Colombia, there are three species at the Andean and Caribbean regions and two species at the Orinoquia and Amazonia regions (Table 9.1). In the Pacific region, we propose the presence *M. temama* only (Fig. 9.1j). We made this decision, by examining voucher specimens from both Chocó and Valle del Cauca departments which are kept at the "Alberto Cadena García" mammalian collection of the Institute of Natural Sciences of the National University of Colombia. Those specimens are labeled as *M. americana* and *Mazama* sp. and require further work. Population status and better understanding of distribution is needed for deer at this region.

For distribution maps of *Mazama* species in Colombia we considered information from Varela et al. (2010) for *M. Americana* (Figs. 9.1f and 9.2a), Lizcano et al. (2010); Lizcano and Alvarez (2016) for *M. rufina* (Figs. 9.1h and 9.2c), Rossi and Duarte (2016) for *M. nemorivaga* (Figs. 9.1g and 9.2b) and Bello et al. (2016) for *M. temama* (Fig. 9.1j), adding the proposal in this chapter. For *M. sanctaemartae* we used those records in Magdalena from Allen (1915) (Fig. 9.1i).

Concerning the ecology of *Mazama* species in Colombia, a few studies addressing occupancy, activity patterns and habitat use are available. Occupancy models were used to estimate abundance of *M. americana* and *M. nemorivaga* in two protected areas in the Colombian Guiana (Gómez et al. 2016). In the Tamá National Natural Park activity pattern of *M. rufina* was studied, finding this species to be mainly nocturnal (Caceres-Martínez et al. 2016). Habitat use by *M. rufina* has been studied in the Mamapacha massif, in Boyacá (Sarria Perea and Vargas Munar unpublished data) and in a eucalyptus reforested area in Caldas department (Ramírez-Mejía and Sánchez 2016). No ecological studies were found for *M. sanctaemartae* and *M. temama* in Colombia.

Mazama species are commonly used by communities for consumption, pelt procurement or as pets. Those uses have been reported in all regions of the country, especially for *M. temama* and *M. americana* at the Caribbean and Pacific regions (Cuesta-Ríos et al. 2007; Racero-Casarrubia et al. 2008; de la Ossa and de la Ossa 2011; Racero-Casarrubia and González-Maya 2014; Tinoco-Sotomayor et al. 2016; Asprilla Perea and Díaz Puente 2018). Also *M. rufina* is used in the Andean region (Ramírez 2007; Parra-Colorado et al. 2014) and *M. nemorivaga* in Orinoquia and Amazonia (Plata Rangel 2005).

Globally, *M. rufina* is categorized as Vulnerable (VU) due to population decline associated with habitat loss, which in Colombia is estimated to be at least 50% of its range (Lizcano and Alvarez 2016). Management and conservation plans are available for *M. rufina* in Quindío (Mantilla-Meluk et al. 2017), and Cundinamarca departments (CAR 2016). Those documents include biological and aspects, regional contexts and main threats. *M. nemorivaga* is of Low Concern (LC) because of its wide distribution which includes several protected areas. Risk assessments for *M. americana* and *M. temama* are difficult due to their taxonomic uncertainty (Bello et al. 2016; Duarte and Vogliotti 2016), and are currently categorized as Data Deficient (DD) species. Finally, *M. sanctaemartae* is not evaluate for the IUCN because it is not included in the species list. None of the *Mazama* species living in Colombia is registered in any of the CITES appendices or in the red book of mammals of Colombia (Rodríguez-Mahecha et al. 2006).

Distribution of *M. americana* and *M. nemorivaga* coincide with 24 out of 59 protected areas in Colombia. *M. rufina* is found in eight protected areas in the Andean region and *M. temama* could be in five protected areas of the Pacific and Caribbean regions. *M. sanctaemartae* is not currently in any of the protected area because it is only known for one locality. Overall, less than 20% of the range of *Mazama* species in Colombia is included in some protected area (Fig. 9.3), which might not be insufficient for their conservation, especially for *M. temama* where only 5.9% of its range is protected.

White-tailed deer in Colombia has different common names, according with ecosystem or region where it is found (Table 9.1). Those names are *venado de páramo*, *venado sabanero*, *venado llanero*. Also some local names refer to antler characteristics, such as *venado de cornamenta*, *venado reinoso*, *venado de racimo*, *venado carameludo o caramerudo*, *venado de ramazón*, *venado cachiliso*, *venado cachiforrado*, *venado cachienvainado*. Other names refer to deer physical appearance, such as *venado blanco*, *venado grande*, *venado pelón*, or simply *ciervo*, *cierva*, *ciervita* or *venado*. Rodríguez-Mahecha et al. (1995), also report 24 names in indigenous languages, which may reflect variation in forms and ages.

Taxonomy and phylogeny of American cervids based on molecular analyses, have raised a debate on the identity and distribution of the species through their range, and need for an exhaustive taxonomic review of the Neotropical deer have been pointed out (Escobedo-Morales et al. 2016; Gutiérrez et al. 2015). For Neotropical *Odocoileus* there are different interpretations both at the generic level and at the species level. Molina and Molinari (1999) suggested that Venezuelan and other Neotropical white-tailed deer do not belong to *O. virginianus* and proposed that those *Odocoileus* deer living in the Venezuelan Andes are *O. lasiotis* and those living in lowlands in Venezuela are *O. cariacou*, O. *margaritae* and *O. curassavicus*. On the other hand, Moscarella et al. (2003), through a phylogeographic study, determined moderate levels of genetic polymorphism in the three Venezuelan subspecies, suggested that those deer could be considered as different evolutionary entities. However, they clarified that there is not enough information to raise those subspecies to full species, as suggested by Molina and Molinari (1999).

The taxonomy of white-tailed deer in Colombia is even more complicated, since very few Colombian samples have been included in the above studies. For this reason and until taxonomy of white-tailed deer in Colombia is better understood, here, we maintain the name *O. virginianus*, as suggested by López-Arévalo et al. (2019). Also, we included four subspecies, *O. v. goudotii*, *O. v. ustus*, *O. v. apurensis* and *O. v. tropicalis*. The first two subspecies refer to the gray forms living on the Andean highlands (*páramos*). *O. v. goudotii* (Fig. 9.2g) include populations of most Andean region and *O. v. ustus* include those populations of southern Andes, in the border with Ecuador. The other two subspecies refer to the ocher-yellowish forms with *O. v. apurensis* (Fig. 9.2h) for those white-tailed deer living in the eastern savannas and northern Colombian Amazon and *O. v. tropicalis* inhabiting in the pacific coast. This taxonomy differs from Solari et al. (2013) who list three names (*O. goudotti* and *O. cariacou* and *O. ustus*) as full species in the whole country.

Other studies of white-tailed deer in Colombia have addressed different topics. A review of 91 documents, written from 1912 to 2017 (Mateus-Gutiérrez et al. 2019) indicates a diversity of topics, including biology (24.2%), species generalities (22%), ecology (20.9%), archaeozoology (13.2%), taxonomy and distribution (9.9%), management and conservation (8.8%) and phylogenetic relationships with other Neotropical cervids (1.1%). Many of those studies are unpublished undergraduate and graduate thesis. Most reviewed documents were written during the 2000–2010 decade.

Captive white-tailed deer populations in different Colombian zoos is increasing and their study has provided information on reproductive aspects, gestation, behavior in captivity, diet (Mateus-Gutiérrez and López-Arévalo 2019), movement and habitat use (Camargo-Sanabria et al. 2019), as well as differences in the antler cycles between deer coming from the high zones and lowlands. Also, variation in fur color between those deer (gray in high altitude deer and ochre yellowish in lowland deer) and even crosses between the two forms were detected (Guzmán-Lenis and López-Arévalo 2019).

Ecological aspects of white-tailed deer natural populations have been addressed in the Andean and Orinoco regions for *O. v. goudotii* and *O. v. apurensis* respectively. In the Andean region, the largest number of studies has been conducted in the Chingaza National Natural Park (CNNP) looking at white-tailed deer abundance and habitat use. The first density estimations, using feces counts, were 3.3 white-tailed deer per km^2 (Ramos 1995). Almost a decade latter (Mateus Gutiérrez 2014) estimated densities of 11.5 deer/km^2 and 15 deer/km^2, in two sites of CNNP, based on pellet counts and an in situ estimated defecation rate of 23.26 fecal groups/individual/day. A population dynamics model using estimates and sightings during the last 20 years suggested an exponential growth of deer population in CNNP and a current occupation of 31% (238 km^2) (Rodríguez-Castellanos 2016). The most recent density estimations show variations from 4.6 deer/km^2 to 8.9 deer/km^2 between two sites in CNNP with an inverse relationship between density and distance to water bodies (Gómez 2017). Habitat data indicates that white-tailed deer in CNNP use at least ten different cover types including both native and exotic vegetation (González-Zárate et al. 2005).

Another white-tailed deer density estimate in the Eastern Cordillera was 2.7 deer/km^2 in the Siscunsi *páramo* in Boyacá department (Rayo Avendaño 2017). In Santander department, Alarcón (2009) found densities from 2.5 deer/km^2 to 6.8 deer/km^2 across different cover types including *páramo* natural vegetation, mixed forests and managed pastures. The highest abundance was found at managed pastures. At this study, age structure was 7:2:1 (fawns: juveniles: adults).

In the Orinoquia, region there are density estimations of white-tailed deer *O. v. apurensis* at two sites in Casanare department (Pérez-Moreno et al. 2019). Those estimates varied from 13.3 deer/km^2 to 43.3 deer/km^2. Six cover types at those sites were identified, with dominance of scattered and dense grasslands and savannas. Deer showed preference for scattered grasslands. In another study Rojas (2010) found that deer preferred savanna over forest, even though the latter had a higher habitat quality index.

In Colombia, as in all its range, white-tailed deer has been a species of high hunting value both in the past (Correal and Van Der Hammen 1977; Peña-León and Pinto 1996; Martínez-Polanco et al. 2015) and in the present (Ojasti and Dallmeier 2000; Blanco-Estupiñán and Zabala-Cristancho 2005). In the Andean area, some communities perceive white-tailed deer as a crop pest. Some villagers use adult deer as a meat source and fawns as pets. Also, deer pelts are sometimes exhibited in some local religious ceremonies or their skulls and feet are kept as home decorations (Blanco-Estupiñán and Zabala-Cristancho 2005; Parra-Colorado et al. 2014). In the Orinoquia region, besides the same uses, some local artists make musical instruments with deer pelts (Plata Rangel 2005). In the Caribbean region, although deer meat and skin are used, white-tailed deer is a rare prey (Racero-Casarrubia and González-Maya 2014; Tinoco-Sotomayor et al. 2016).

Regarding conservation status, *O. v. tropicalis* is considered Critically Endangered (CR) in Colombia (López-Arévalo and González-Hernández 2006) whereas *O. v. apurensis* is of Low Concern (LC). Subspecies *O. v. goudotii* y *O. v. ustus* have Deficient Data (DD) and no risk assessments are available for them. None of these subspecies is listed any CITES appendices. White-tailed deer is recorded in 30 out of 59 Colombian protected areas, which account for about 10% of its range in the country (Fig. 9.3). Conservation challenges for this species are mainly outside protected areas because of the ongoing transformation of natural areas.

The common name for *Pudu mephistophiles* in Colombia is *venado conejo* (rabbit deer). This is the only *Pudu* species registered in Colombia so far (Ramírez-Chaves et al. 2016) and it is the least known deer species in Colombia. Distribution of this species in Colombia is partially known (Fig. 9.11). There are nine voucher specimens of *P. mephistophiles* in the mammal collection of University of Cauca (Rivas-Pava et al. 2007) and seven more in the Natural Science Institute (National University of Colombia) and Alexander von Humboldt Institute. Those specimens come from departments of Cauca, Risaralda, Boyacá and Valle del Cauca. Some years ago, rabbit deer was registered in the Munchique National Park (Cauca), on the western slope of the Western Cordillera, being the first report of this species at this protected area (Mejía-Correa and Díaz-Martínez 2009).

Heckeberg et al. (2016), included genus *Pudu* in their deer phylogenetics study finding that *P. mephistophiles* is not related to *Pudu puda*, the other species of the genus, and suggested that *Pudu* is polyphyletic leaving this genus in an uncertain situation respect to other the cervid tribes. These findings highlight a need for further studies including more samples, especially from Colombia, since the above analysis used only data from Ecuador.

Ecology of rabbit deer is poorly known in Colombia. A study on *Puma concolor*'s diet in the Puracé National Natural Park, found the rabbit deer to be the most important prey of this cat and suggest an important relationship between these two species (Hernández-Guzmán et al. 2011). Some rural communities at Quindío department use rabbit deer as food or as a pet. (Parra-Colorado et al. 2014). Currently, a master's thesis is being conducted at the University of Cauca, looking at rabbit deer's distribution and its conservation status in Colombia (Gómez unpublished data).

In terms of conservation, rabbit deer is categorized globally as Vulnerable (VU) because of decline of their small and fragmented subpopulations through the high northern Andes (Barrio and Tirira 2008). In Colombia, the red book of mammals (Rodríguez-Mahecha et al. 2006), listed *P. mephistophiles* as Near Threatened (NT), but no recent assessments are published so far. The official current list of threatened species in Colombia does not include this species either (Minambiente 2017a). This deer species is listed in Appendix II of CITES.

Rabbit deer is present in 8 out of 59 protected areas, covering 20.5% of its range in Colombia (Fig. 9.3). At local scale, rabbit deer is present in 74.5% of the Departmental System of Protected Areas (SIDAP) of Quindío (Gómez-Hoyos et al. 2014).

9.3 Exotic Ungulates

9.3.1 Hippopotamus (Hippopotamus amphibius)

Hippopotamus (*Hippopotamus amphibius*) is native to Africa, where it is currently a Vulnerable species (VU), mainly due to habitat loss and uncontrolled hunting (Lewison and Pluhaček 2017). In early 1980s, drug trafficker Pablo Escobar introduced six hippos from the United States to his *Hacienda Nápoles*, a property located 249 km from Bogotá (Valderrama 2012). After Escobar's death in 1993, the land was expropriated by the Colombian State and much of the exotic fauna of this site was remitted to different zoos in Colombia. However, hippos remained at *Hacienda Nápoles*, where its population began to increase. By 2012, the regional environmental agency estimated that 35 hippos remained, both in the *Hacienda Nápoles* and outside, because some animals escaped from confinement (Cornare 2012).

During the last few years free hippos, mainly young males, have been registered more than 200 km far from their initial point of introduction. Dispersal of these animals occurred both northward up to Cimitarra, Santander, and southwestward down to Sonsón, Antioquia (Cornare 2014). Several press releases continue to record sightings in different sectors of the Magdalena River basin. *Hacienda Nápoles* is currently a theme park and houses the largest hippopotamus population outside Africa (El Espectador 2016).

Hippopotamus in Colombia has shown an alarming invasive potential. Factors associated with this potential include its rapid population growth, high survival of reproductive adults, and longevity, (40–45 years in the wild and 60 in captivity) (Weigl 2005). Other contributing factors are absence of hippo predators in Colombia, forest conversion into pastures and other open areas for livestock and agriculture and abundance of water bodies along the middle basin of the Magdalena River.

Control measures have been difficult to implement. In 2009, regional environmental authorities tried a controlled hunting, resulting in an adult male's death, a fact that generated a strong debate and protests by civil society. A legal action (*Tutela* 05001-23-33-000-2013-00604-01) against the regional agency Cornare, prevented the execution of more animals, limiting the action of environmental authorities. African experts consulted by the Colombian government proposed controlling hippo reproduction through castration or sacrifice of free males and construction of safe enclosures to house them (Semana 2009). The Institute of Natural Sciences of the National University of Colombia, in 2014, warned about the urgent need of eradicating this species from Colombia given the biological, economic and social implications (López-Arévalo and Montenegro 2015). In 2016, two adult males were successfully sterilized by a team of veterinarians from CES University (Cornare 2014). However, the high costs of these procedures have limited this strategy.

Relocation has been another proposed strategy. A successful transfer of a young hippo (3–6 years old) from Puerto Triunfo Municipality (Antioquia) to a zoo in Cundinamarca was carried out recently, as well as fencing of a management area near *Hacienda Nápoles* (Cornare 2018) . Remission of hippos to their original range in Africa was another proposed alternative but it was not accepted, not only because of its prohibitive costs, but due to high risk of introducing parasites and pathogens that might have been acquired by hippos in Colombia (Monsalve and Ramírez 2018).

Although hippos are usually herbivores, it is known that they can eat meat (Dudley et al. 2016). In Colombia attacks on calves have been reported and there is fear among fishermen due to the risk of encounters with hippos in rivers and lagoons in the municipality of Puerto Berrio, in the department of Antioquia (Valderrama 2012). Although this species is the most lethal large terrestrial mammal in the world, killing about 500 people per year in Africa (BBC 2016), in Colombia, the general public has a different perception, because to date there have been no deadly attacks on humans. However, there is that potential, because people have become accustomed to seeing them, even in urban areas. It is even reported that there are complaints about the sale of juveniles as an exotic pet (Monsalve and Ramírez 2018).

The potential negative effects of a wild hippopotamus population in the Middle Magdalena include transformation of water bodies by opening channels, overconsumption of surrounding vegetation, water nitrification by dung and urine deposition, potential disease transmission and impacts on local biodiversity because of absence of native ecological equivalents (Valderrama 2012).

Simulations of possible population trajectories based on the initial number of introduced individuals and using features of their life history, as well as recruitment, growth and survival rates available for African populations, predict a continuous growth of this population (López-Arévalo and Montenegro 2015). This growth can be even faster considering the absence of natural predators in Colombia, which would increase their survival. Other scenarios examined with these simulations show that adult sterilization would have a slow effect on controlling this population and that an increase of adult mortality by direct control is the measure with the fastest effect in reducing hippo's rate of population growth.

In conclusion, *Hippopotamus amphibius* is an exotic ungulate species at an establishment phase in Colombia. There are ecological, economic and social implications that justify an eradication program of the hippo wild population. However, such a program requires strong coordination among regional and national environmental authorities, armed forces and judicial entities, accompanied by an appropriate awareness campaign to citizens, to address the problem.

9.3.2 *Feral Pigs (*Sus scrofa*)*

In Latin America *Sus scrofa* comes from Iberian pigs taken initially to Haiti by Columbus in 1493 (Benítez 2001). At present, there is a great variety of phenotypes of Creole pigs from various breeds brought during the Spanish conquest. In Colombia, three creole races have been identified whose local names are *zungo, sampedreño* and *casco de mula (mule-hoofed)* (Barrera et al. 2007); which are of slight dimensions and have smaller litters than commercial breeds. Additionally, two breeds of creole pigs from the Pacific region have different morphometric traits and minor sexual dimorphism (Arredondo 2013). Such a breed seems to be different from those described by Barrera et al. (2007).

In several regions free domestic pigs have formed feral populations. Research on feral pigs started during last century (Rosell et al. 2001; Wolf and Conover 2003), being currently the most studied feral mammal species in the world. Those studies have addressed population estimations (Nores et al. 2000; Saunders 1995) and demographics (e.g., Fernández-Llario and Carranza 2000; Focardi et al. 2008). Likewise, there is a great deal of research focused on management and control as an invasive species (Barrera et al. 2007; Hanson 2006; Oslinger et al. 2006), as well as on their effects on ecosystems (Wolf and Conover 2003).

In Colombia, *Sus scrofa* is one of the 62 exotic mammals that has managed to establish feral populations in the territory mainly in the Orinoquia region (Casanare and Arauca departments) (Fig. 9.2f), given local management practices that involve release of individuals for latter capture for consumption (Ramírez-Chaves et al. 2011).

Currently, feral pig populations are known in Orinoquia region, specifically in the flooded savannas of the departments of Arauca and Casanare (Baptiste et al. 2010; Guerrero and Montenegro 2012; López-Arévalo et al. 2018; Montenegro et al. 2009; Universidad Nacional de Colombia and Corporinoquia 2015). Likewise, there are indications of at least one feral population in the Pacific region (Baptiste et al. 2010).

Recently, a study was carried out in the Orinoquia floodplains, including the departments of Arauca, Casanare and northwestern Meta to assess feral pig population status, available habitat, and home range (López-Arévalo et al. 2018). This study identified from 95 to 114 feral pig population nuclei along the inspection of 2306.5 km of roads and paths through the study region. Density estimations in five sites (two in Arauca and three in Casanare) ranged from 7 pigs/km^2 to 44.2 pigs/km^2, with a mean value of 32 pigs/km^2. Suitable habitat for feral pigs accounted for 56.3% of the study area. Estimated mean home range was 1.33 ± 0.93 km^2 (ranges 0.23–4.87 km^2) and maximum daily movement was 0.66 ± 0.46 km. Pig displacement was longer at night (0.59 ± 0.49 km) than during the day (0.31 ± 0.24 km).

The feral pigs are of great importance in the Orinoquia culture, being a significant component of the regional economy and gastronomy. Three management types were identified according to the freedom pigs have. The first one is pig rearing under confinement, the second is a semi-tamed breeding in which individuals live autonomously feeding on their own but maintaining regular interactions with humans, and the third one where pigs do not have contact with humans, but they are kept inside properties and might be hunted when they are adults.

9.4 Final Remarks

Much of the reviewed studies on ungulates in Colombia are undergraduate and graduate theses. Unfortunately, most of this information seldom gets to a scientific publication. Despite this, university repositories currently give visibility to such information. Examples of these repositories are those of National University of Colombia (bdigital.unal.edu.co) Javeriana Pontifical University, (repository.javeriana.edu.co) and Francisco José de Caldas District University (repository.udistrital.edu.co), among others. However, it is necessary to encourage grey literature to reach formal publication.

Ungulate records, studies and monitoring have been increasing because of popularity of camera trapping (Díaz et al. 2016; López-Arévalo et al. 2018), both inside protected areas (Portal Fototrampeo-Parks Natural Nationals of Colombia) and outside them. Some new records of rare and elusive species, such as tapirs or rabbit deer, for example, were available only from camera trapping in various sites.

On the other hand, new political scenarios in Colombia have an effect in wild areas. The signing of the Peace Accords with the FARC_EP guerrilla has allowed access to new areas previously unexplored, both through teaching practices (Delgadillo-Ordoñez et al. 2017), as well as scientific expeditions. For example, Colombia BIO is an ongoing initiative started on 2016 that have supported around 20 expeditions until 2018, with participation of scientific institutions and universities (Colciencias 2019). Such expeditions have provided new species records including ungulates. Unfortunately, cessation of hostilities in different territories has allowed the increase of deforestation which by 2017, had at least eight active cores (and, SMBYC 2017), which causes habitat loss and an increase of hunting, both factors affecting ungulates. This situation poses the challenge of making peace as an opportunity to strengthen environmental protection and to plan sustainable rural development in Colombia (Morales 2017).

Finally, increasing ungulate knowledge and conservation as well as preserving biological and cultural diversity, depends on the joint action of educational, management and research institutions. A strong collaborate work is needed in order to achieve the objectives, policies, and commitments that as a country we have proposed in different instruments (CDB 2010; Minambiente 2017b).

References

Abud M, Duque S, Calero H, Valderrama S (2012) Abundance of mountain tapir in Puracé national park, Colombia. Conservation Leadership Programme final report. Samanea Foundation, Colombia

Acevedo-Quintero JF, Zamora-Abrego JG (2016) Mamíferos medianos y grandes asociados a un cananguchal de la Amazonia colombiana. In: Lasso CA, Colonnello G, Moraes RM (eds) Morichales, cananguchales y otros palmares inundables de Suramérica. Parte II: Colombia, Venezuela, Brasil, Perú, Bolivia, Paraguay, Uruguay y Argentina. Serie Editorial Recursos Hidrobiológicos y Pesqueros Continentales de Colombia. Instituto de Investigación de Recursos Biológicos Alexander von Humboldt (IAvH), Bogotá, pp 221–239

Acosta H, Cavelier J, Londoño S (1996) Aportes al conocimiento de la biología de la danta de montaña, *Tapirus pinchaque*, en los Andes Centrales de Colombia. Biotropica 28:258–266

Alarcón SM (2009) Estado poblacional del venado vola blanca (*Odocoileus virginianus*) en la Vereda Molinos, municipio de Soatá, Boyacá. Implementando una estrategia para el uso, manejo y conservación de la especie. Tesis pregrado, Universidad Distrital Francisco José de Caldas, Bogotá

Alberico MA, Cadena J, Hernández-Camacho MY (2000) Mamíferos (Sinapsia: Theria) de Colombia. Biota Colomb 1:43–75

Allen JA (1915) Notes on American deer of the genus Mazama. Bull Am Mus Nat Hist 34, article 18

Alviz A, Pérez AK (2015) Plan para la conservación de la danta de tierras bajas (*Tapirus terrestris*) en los departamentos de Casanare, Arauca y Vichada. Corporinoquia – Fundación Orinoquia Biodiversa, Yopal

Arias Alzate A, Downer CC, Delgado-V CA, Sánchez-Londoño JD (2010) Un registro de tapir de montaña (*Tapirus Pinchaque*) en el norte de la Cordillera Occidental de Colombia. Mastozool Neotrop 17(1):111–116

Arias Alzate A, Palacio Vieira JA, Muñoz-Durán J (2009) Nuevos registros de distribución y oferta de hábitat de la danta colombiana (*Tapirus terrestris colombianus*) en las tierras bajas del norte de la Cordillera Central (Colombia). Mastozool Neotrop 16(1):19–25

Arredondo JV (2013) Caracterización de los sistemas de producción tradicional, morfología y diversidad genética del cerdo criollo de la Región Pacífica colombiana. Tesis doctoral, Universidad Nacional de Colombia, Palmira

Asprilla Perea J, Díaz Puente J (2018) Traditional use of wild edible food in rural territories within tropical forest zones: A case study from the northwestern Colombia. In: New trends and issues Proceedings on Humanities and Social Sciences, 10th World Conference on Educational Sciences (WCES-2018), 01/02/2018–03/02/2018, Prague, Czech Republic

Baptiste M, Castaño N, Lasso C, Cárdenas D, Gutiérrez F, Gil D (eds) (2010) Análisis de riesgo y propuesta de categorización de especies introducidas para Colombia. Instituto de Investigación de Recursos Biológicos Alexander von Humboldt (IAvH), Bogotá

Barrera G, Martínez R, Ortegón Y, Ortiz A, Moreno F, Velásquez H, Abuabara Y (2007) Cerdos Criollos Colombianos: Caracterización racial, productiva y genética. Corporación colombiana de investigación agropecuaria. Corpoica, Bogotá

Barrio J, Tirira D (2008) *Pudu mephistophiles*. The IUCN red list of threatened species 2008. https://doi.org/10.2305/IUCN.UK.2008.RLTS.T18847A8647714. Accessed 15 Jan 2019

BBC (2016) What are the world's deadliest animals? BBC-News. https://www.bbc.com/news/world-36320744. Accessed 11 Feb 2019

Bello J, Reyna R, Schipper J (2016) *Mazama temama*. The IUCN red list of threatened species 2016. https://doi.org/10.2305/IUCN.UK.2016-2.RLTS.T136290A22164644. Accessed 15 Jan 2019

Benítez W (2001) Los cerdos criollos de América Latina. Estudio FAO Producción y Sanidad Animal 148:13–36

Black-Decima PA, Vogliotti A (2016) *Mazama gouazoubira*. The IUCN red list of threatened species 2016. https://doi.org/10.2305/IUCN.UK.2016-2.RLTS.T29620A22154584. Accessed 15 Jan 2019

Blanco-Estupiñán L, Zabala-Cristancho A (2005) Recopilación del conocimiento local sobre el venado cola blanca (*Odocoileus virginianus*) como base inicial para su conservación en la zona amortiguadora del Parque Nacional Natural Pisba, en los municipios de Tasco y Socha. Tesis de grado, Escuela de Ciencias Biológicas, Universidad Pedagógica y Tecnológica, Tunja

Bodmer RE (1990) Ungulate frugivores and the browser-grazer continuum. Oikos 57:319–325

Bonilla-Riveros TA (2014) Usos, prácticas e ideologías socioculturales de la cacería de dos comunidades Tikuna, ubicadas en el sur de la Amazonía colombiana. Tesis maestría, Universidad Nacional de Colombia, Sede Amazonía, Leticia

Burgin CJ, Colella JP, Kahn PL, Upham NS (2018) How many species of mammals are there? J Mammal 99:1–11

Cabrera A (1960) Catalogo de los mamiferos de América del Sur. Rev Mus Argentino Cienc Nats "Bernardino Rivadavia". Fortschr Zool 4:309–732

Cabrera J (2015) Una historia de dos ciudades: cacería y conservación por fuera de áreas protegidas. El caso de la danta (Tapirus terrestris) en el Parque Nacional Natural Amacayacu, Colombia. Capítulo 5. In: Payán E, Lasso CA, Castaño-Uribe C (eds) I. Conservación de grandes vertebrados en áreas no protegidas de Colombia, Venezuela y Brasil. Serie Editorial Fauna Silvestre Neotropical. Instituto de Investigación de Recursos Biológicos Alexander von Humboldt (IAvH), Bogotá, p 99–114

Cabrera JA, Molina E, Gonzalez T, Armenteras D (2016) ¿Funciona el plan B? Estimación de rango de hogar a partir de datos almacenados en instrumentos a bordo de animales y los obtenidos a partir de telemetría satelital de GPS en el Amazonas colombiano. Rev Biol Trop 64(4):1441–1451

Caceres-Martínez CH, Rincón AAA, González-Maya JF (2016) Terrestrial medium and large-sized mammal's diversity and activity patterns from Tamá National Natural Park and buffer zone, Colombia. Therya 7(2):285–398

Camargo-Sanabria AA, López-Arévalo HF, Sarmiento-Parra D (2019) Evaluación de la calidad del hábitat del venado cola blanca en un bosque seco tropical del municipio de Nilo, Cundinamarca, Colombia. In: López-Arévalo HF (ed) Ecología, uso, manejo y conservación del venado cola blanca en Colombia. Biblioteca José Jerónimo Triana No. 33. Instituto de Ciencias Naturales, Universidad Nacional de Colombia, Bogotá, p 199–216

Caparroz R, Mantellatto AM, Bertioli DJ, Figueiredo MG, Duarte JM (2015) Characterization of the complete mitochondrial genome and a set of polymorphic microsatellite markers through next-generation sequencing for the brown brocket deer *Mazama gouazoubira*. Genet Mol Biol 38:338–345

Corporación Autónoma Regional de Cundinamarca [CAR] (2016) Plan de manejo y conservación del venado soche (*Mazama rufina*) en la jurisdicción CAR. Corporación Autónoma Regional de Cundinamarca CAR Dirección de Modelamiento, Monitoreo y Laboratorio Ambiental, Bogotá

Carrillo JD, Amson E, Jaramillo C, Sánchez R, Quiroz L, Cuartas C, Rincón AF, Sánchez-Villagra MR (2018) The Neogene record of northern south American native ungulates. Smithsonian Institution Scholarly Press, Washington

Castiblanco J (2002) Uso y percepción de fauna de cacería por una comunidad negra en el golfo de Tribugá, Chocó, Colombia. Tesis pregrado en Biología, Universidad Nacional de Colombia, Bogotá

Cavelier J, Lizcano DJ, Yerena E, Downer C (2011) The mountain tapir (*Tapirus pinchaque*) and Andean bear (*Tremarctos ornatus*): two charismatic large mammals in south American tropical mountain cloud forest. In: Bruijnzeel LA, Scatena FN, Hamilton LS (eds) Tropical montane cloud forest. Cambridge University Press, New York, pp 172–181

[CDB] Convenio sobre la Diversidad Biológica (2010) Plan 2011–2020 Decenio de las Naciones Unidas sobre la Biodiversidad. https://www.cbd.int/undb/media/factsheets/undb-factsheets-es-web.pdf. Accessed 21 Feb 2019

Chávez Hoffmeister MF (2016) El origen de la fauna Sudamericana moderna: de Gondwana al Gran Intercambio Americano. In: Pino M (ed) El Sitio Pilauco Osorno, Patagonia noroccidental de Chile. Universidad Austral de Chile, Valdivia, pp 47–74

Colciencias (2019) Expedición BIO. https://www.colciencias.gov.co/portafolio/colombia-bio/expedicion-bio#. Accessed 15 Feb 2019

Constantino E (2005) Current distribution and conservation status of the Colombian lowland tapir (*Tapirus terrestris colombianus*) and the Baird's or central American tapir (*Tapirus bairdii*) in Colombia. Tapir Conservation, newsletter of the IUCN /SSC Tapir Specialist Group 14:15–18

Cornare, Corporación Autónoma Regional de las cuencas de los ríos Negro y Nare (2012) Resolución 112-53 del 6 de diciembre de 2012

Cornare, Corporación Autónoma Regional de las cuencas de los ríos Negro y Nare (2014) Exitosa cirugía de esterilización de dos hipopótamos. http://www.cornare.gov.co/sala-de-prensa/informativo/noticias-corporativas/301-exitosa-cirugia-de-esterilizacion-de-dos-hipopotamos. Accessed 15 Feb 2019

Cornare, Corporación Autónoma Regional de las cuencas de los ríos Negro y Nare (2018) Corantioquia y Cornare rechazan captura y supuesta venta de hipopótamos en el Magdalena medio. Available at http://www.cornare.gov.co/sala-de-prensa/informativo/noticias-corporativas/695-corantioquia-y-cornare-rechazan-captura-y-supuesta-venta-de-hipopotamos-en-el-magdalena-medio. Accessed 25 Jan 2019

Corpoguajira, Fundación Nativa (2007) Programa de implementación de la estrategia de conservación de la danta colombiana (*Tapirus terrestris colombianus*) en La Guajira. Corpoguajira and Fundación Nativa

Correal G, Van der Hammen T (1977) Investigaciones arqueológicas en los abrigos rocosos del Tequendama: 11000 años de prehistoria en la Sabana de Bogotá. Biblioteca Banco Popular, Bogotá

Correal Urrego G (1990) Aguazuque: evidencias de cazadores, recolectores y plantadores en la altiplanicie de la Cordillera Oriental. Fundación de Investigaciones Arqueológicas Nacionales, Bogotá

Cozzuol M, de Thoisy AB, Fernandes-Ferreira H, Rodrigues FHG, Santos FR (2014) How much evidence is enough evidence for a new species? J Mammal 95(4):899–905

Cozzuol MA, Clozato CL, Holanda EC, Rodrigues FHG, Nienow S, de Thoisy B, Redondo AF, Santos FR (2013) A new species of tapir from the Amazon. J Mammal 94(6):1331–1345

Cuesta-Ríos EY, Valencia-Mazo JD, Ortega AMJ (2007) Aprovechamiento de los vertebrados terrestres por una comunidad humana en bosques tropicales (Tutunendo, Chocó, Colombia). Rev Biodiv Neotrop 26(2):37–43

Cuvier M (1829) Memoire pour servir a l'histoire du tapir; et description d'úne espece nouvelle apartenent aux hautes regions de la Cordillere del Andes. Ann Sc Nat Paris (1re Serie) 17:107–112

David LDJ, Aguirre RNJ, Vélez MFJ (2017) Relación de las poblaciones humanas con los mamíferos silvestres del Sistema Cenagoso de Ayapel, Colombia. Biocenosis 31:1–2

de la Ossa J, de la Ossa LA (2011) Cacería de subsistencia en San Marcos, Sucre, Colombia. Rev Colombiana Cienc Anim 3(2):213–224

Delgadillo-Ordoñez N, Díaz-Rodríguez JV, Cadena-Morgante V, Carrillo-Villamizar JZ, López-Arévalo HF (2017) Catálogo de mamíferos de San José de Guaviare. Grupo estudiantil de conservación y manejo de la vida silvestre de la Universidad Nacional de Colombia. Universidad Nacional de Colombia, Bogotá

de Thoisy B, Da Silva AG, Ruiz-García M, Tapia A, Ramirez O, Arana M, Lavergne A (2010) Population history, phylogeography, and conservation genetics of the last Neotropical megaherbivore, the lowland tapir (*Tapirus terrestris*). BMC Evol Biol 10(1):278

Díaz A et al (2016) In: Instituto de Investigación de Recursos Biológicos Alexander von Humboldt (IAvH) (ed) Fototrampeo, Una herramienta para el muestreo de mamíferos medianos y grandes, ficha 104. Reporte de estado y tendencias de la Biodiversidad continental de Colombia, Bogotá

Downer CC (2001) Observations on the diet and habitat of the mountain tapir (*Tapirus pinchaque*). J Zool 254:279–291

Duarte JMB, Vogliotti A (2016) *Mazama americana*. The IUCN red list of threatened species 2016. https://doi.org/10.2305/IUCN.UK.2016-1.RLTS.T29619A22154827. Accessed 15 Jan 2019

Dudley J, Hang'ombe M, Leendertz F, Dorward L, de Castro J, Subalusky A, Clauss M (2016) Carnivory in the common *Hippopotamus amphibius*: implications for the ecology and epidemiology of anthrax in African landscapes. Mammal Rev 46:n/a–n/a. https://doi.org/10.1111/mam.12056

Eisenberg JF (1989) Mammals of the neotropics: Panama, Colombia, Venezuela, Guyana, Surinam, Frech Guiana Northen neotropics, vol 1. University of Chicago Press, Chicago

El Espectador (2016) Hipopótamos sueltos en Hacienda Nápoles, un peligro vigente. Available at https://www.elespectador.com/noticias/nacional/hipopotamos-sueltos-de-hacienda-napoles-un-peligro-vige-articulo-641502. Accessed 25 Jan 2019

Emmons L (1999) Mamíferos de los bosques húmedos de América tropical. Una guía de campo. Editorial F.A.N, Santa Cruz de la Sierra

Escobedo-Morales LA, Mandujano S, Eguiarte LE, Rodríguez-Rodríguez MA, Maldonado JE (2016) First phylogenetic analysis of Mesoamerican brocket deer *Mazama pandora* and *Mazama temama* (Cetartiodactyla: Cervidae) based on mitochondrial sequences: implications for neotropical deer evolution. Mamm Biol 81(3):303–313

Fernández-Llario P, Carranza J (2000) Reproductive performance of the wild boar in a Mediterraean ecosystem under drought conditions. Ethol Ecol Evol 12:335–343

Ferrer-Pérez A, Beltrán Gutiérrez M, Lasso C (2009) Mamíferos de la Estrella Fluvial de Inírida: ríos Inírida, Guaviare, Atabapo y Orinoco (Colombia). Biota Colomb 10:1–2

Focardi S, Gaillard J, Ronchi F, Rossi S (2008) Unexpected life-history variation in an unusual ungulate. J Mammal 89(5):1113–1123

García M, Jordan C, O'Farril G, Poot C, Meyer N, Estrada N, Leonardo R, Naranjo E, Simons Á, Herrera A, Urgilés C, Schank C, Boshoff L, Ruiz-Galeano M (2016) *Tapirus bairdii*. The IUCN red list of threatened species 2016. https://doi.org/10.2305/IUCN.UK.2016-1.RLTS.T21471A45173340. Accessed 25 Feb 2019

Gómez YH (2017) Efecto de la oferta de recurso en la distribución y densidad del venado cola blanca en tres zonas del ecotono del bosque altoandino y páramo. Maestría en manejo uso y conservación del bosque. Universidad Distrital Francisco José de Caldas, Bogotá

Gómez B, Montenegro OL (2012) Abundancia del pecarí de collar (*Pecari tajacu*) en dos áreas protegidas de la Guayana colombiana. Mastozool Neotrop 19:163–178

Gómez B, Montenegro OL, Sánchez-Palomino P (2016) Abundance variation of ungulates in two protected areas of the Colombian Guayana estimated with occupancy models. Therya 7(1):89–106

Gómez-Hoyos DA, Ríos-Franco CA, Marín-Gómez OH, Suarez-Joaqui T, González-Maya JF (2014) Representatividad de mamíferos amenazados en el Sistema Departamental de Áreas Protegidas (SIDAP). Mammal Notes 1(2):35–41

Góngora J, Morales S, Berl J, Moran C (2006) Phylogenetic divisions among collared peccaries (*Pecari tajacu*) detected using mitochondrial and nuclear sequences. Mol Phylogenet Evol 41:1–11

Góngora J, Reyna-Hurtado R, Beck H, Taber A, Altrichter M, Keuroghlian A (2011) Pecari tajacu. La lista roja de la UICN de especies amenazadas 2011. https://doi.org/10.2305/IUCN.UK.2011-2.RLTS.T41777A10562361.

González-B V (2016) Los palmares de pantano de *Mauritia flexuosa* en Suramérica: Una revisión. In: Lasso CA, Colonnello G, Moraes-R M (eds) Morichales, cananguchales y otros palmares inundables de Suramérica. Parte II: Colombia, Venezuela, Brasil, Perú, Bolivia, Paraguay, Uruguay y Argentina. Serie Editorial Recursos Hidrobiológicos y Pesqueros Continentales de Colombia. Instituto de Investigación de Recursos Biológicos Alexander von Humboldt (IAvH), Bogotá, pp 44–83

González-D TM, González-Trujillo JD, Palmer JRB, Pino J, Armenteras D (2017) Movement behavior of a tropical mammal: the case of *Tapirus terrestris*. Ecol Model 360:223–229

González-D TM, Kaston Flórez F, Armentera D (2014) Aportes al uso de coberturas de la danta de tierras bajas, *Tapirus terrestris colombianus* Hershkovitz 1954 (Perissodactyla: Tapiridae) en la sierra nevada de Santa Marta en las cuencas río Ancho y Palomino Norte de Colombia. Bol Cient Mus Hist Nat U Caldas 18(1):25–37

González-Maya JF, Schipper J (2009) Elevational distribution and abundance of Baird's tapir (*Tapirus bairdii*) at different protection areas in Talamanca region of Costa Rica. Tapir Conservation, newsletter of the IUCN /SSC Tapir Specialist Group 18:29–35

González-Zárate A, Aguirre-Santoro J, Balaguera NC, Rivera TG (2005) Evaluación del hábitat del Venado de Cola Blanca (*Odocoileus virginianus*) en diferentes comunidades vegetales del Parque Nacional Natural Chingaza (Colombia). Acta Biol Colomb 10(2):142

Groves CP, Grubb P (1993) The suborder Suiformes. In: WLR O (ed) Pigs, peccaries and hippos: status survey and conservation action plan. IUCN, The World Conservation Union, Gland, pp 1–4

Groves C, Grubb P (2011) Ungulate taxonomy. The John Hopkins University Press, Baltimore, MD

Guerrero EA, Montenegro OL (2012) Densidad poblacional y distribución de la especie invasora *Sus scrofa*, y su uso de hábitat en el municipio de Paz de Ariporo, Casanare. In: Universidad Cooperativa de Colombia, Sede Arauca, Universidad Nacional de Colombia, Sede Orinoquia (ed) Mem Segundo Congr Intern Producción, Desarrollo Sostenible y Conservación - Versión Sabanas Tropicales, Arauca, p 77

Gutiérrez EE, Maldonado JE, Radosavljevic A, Molinari J, Patterson BD, Martínez-C JM, Helgen KM (2015) The taxonomic status of *Mazama bricenii* and the significance of the Táchira depression for mammalian endemism in the cordillera de Mérida, Venezuela. PLoS One 10(6):e0129113

Guzmán JD (2005) Actividad de cacería y percepciones de la fauna en la comunidad de Punta Pava, reserva natural nacional Puinawai, Guainía, Colombia. Facultad de estudios ambientales y rurales. Tesis de pregrado en Ecología, PUJ. Bogotá

Guzmán-Lenis AR, López-Arévalo HF (2019) Historia natural del venado cola blanca (*Odocoileus virginianus*) en Colombia bajo condiciones de cautiverio. In: López-Arévalo HF (ed) Ecología, uso, manejo y conservación del venado cola blanca en Colombia. Biblioteca José Jerónimo Triana No. 33. Instituto de Ciencias Naturales, Universidad Nacional de Colombia, Bogotá, p 45–54

Hanson L (2006) Demography of feral pig populations at Fort Benning, Georgia. Doctoral thesis, Auburn University, Alabama

Heckeberg NS, Erpenbeck D, Wörheide G, Rössner GE (2016) Systematic relationships of five newly sequenced cervid species. PeerJ 4:e2307

Hernández C (1995) Ideas y prácticas ambientales del pueblo Emberá del Chocó. CEREC and ICAN-COLCULTURA, Bogotá

Hernández-Guzmán A, Payán E, Monroy-Vilchis O (2011) Hábitos alimentarios del *Puma concolor* (Carnivora: Felidae) en el Parque Nacional Natural Puracé, Colombia. Rev Biol Trop 59(3):1285–1294

Hershkovitz P (1954) Mammals of northern Colombia, preliminary report no. 7: tapirs (genus *Tapirus*), with a systematic review of American species. Proc US Natl Mus 103:465–496

Horovitz I (2012) Los ungulados autóctonos de América del Sur. In: Sánchez-Villagra MR (ed) Venezuela Paleontológica. Printwork Art, St. Gallen, pp 274–294

Humanez-López E, Racero-Casarrubia J, Arias-Alzate A (2016) Anotaciones sobre distribución y estado de conservación de los cerdos de monte *Pecari tajacu* y *Tayassu pecari* (Mammalia: Tayassuidae) para el departamento de Córdoba, Colombia. Mammal Notes 3:24–28

[IDEAM] Instituto de Hidrología, Meteorología y Estudios Ambientales (2015) Mapa de ecosistemas de Colombia 2015. http://www.ideam.gov.co/web/ecosistemas/mapa-ecosistemas-continentales-costeros-marinos. Accessed 11 Feb 2019

Kaston F, Fernández C, Peñalosa W, Rodríguez J, Torres J, Armenta MM (2008) Distribución histórica y actual de la población de danta de tierras bajas *Tapirus terrestris colombianus* (Hershkovitz 1954) más al norte de Sur América. Tapir Conservation, newsletter of the IUCN /SSC Tapir Specialist Group 24:22–25

Keuroghlian A, Desbiez A, Reyna-Hurtado R, Altrichter M, Beck H, Taber A, Fragoso JMV (2013) *Tayassu pecari*. The IUCN red list of threatened species 2013. https://doi.org/10.2305/IUCN.UK.2013-1.RLTS.T41778A44051115. Accessed 27 Dec 2018

Lewison R, Pluhaček J (2017) *Hippopotamus amphibius*. The IUCN red list of threatened species 2017. https://doi.org/10.2305/IUCN.UK.2017-2.RLTS.T10103A18567364. Accessed 21 Jan 2019

Lizcano D, Alvarez SJ (2016) *Mazama rufina*. The IUCN red list of threatened species 2016. https://doi.org/10.2305/IUCN.UK.2016-2.RLTS.T12914A22165586. Accessed 15 Jan 2019

Lizcano D, Álvarez S, Delgado VC (2010) Dwarf red brocket deer *Mazama rufina* (Pucheran 1951). In: Duarte JMB, Gonzalez S (eds) Neotropical Cervidology: biology and medicine of Latin American deer. FUNEP/IUCN, Jaboticabal, pp 177–180

Lizcano DJ, Cavelier J (2000) Daily and seasonal activity of the mountain tapir (*Tapirus pinchaque*) in the Central Andes of Colombia. J Zool 252:429–435

Lizcano DJ, Cavelier J (2004a) Using GPS collars to study mountain tapirs (*Tapirus pinchaque*) in the Central Andes of Colombia. Tapir Conservation, newsletter of the IUCN /SSC Tapir Specialist Group 13:18–23

Lizcano DJ, Cavelier J (2004b) Características químicas de salados y hábitos alimenticios de la danta de montaña (*Tapirus pinchaque* Roulin, 1829) en los Andes centrales de Colombia. Mastozool Neotrop 11:193–201

Lizcano DJ, Medici P, Montenegro OL, Carrillo L, Camacho A, Miller PA (eds) (2005) Taller de Conservación de Danta de Montaña. Reporte Final. IUCN/SSC Conservation Breeding Specialist Group. Apple Valley

Lizcano DJ, Pizarro V, Cavelier J, Carmona J (2002) Geographic distribution and population size of the mountain tapir (*Tapirus pinchaque*) in Colombia. J Biogeogr 29:7–15

Lizcano DJ, Prieto-Torres D, Ortega-Andrade HM (2015) Distribución de la danta de montaña (*Tapirus pinchaque*) en Colombia: importancia de las áreas no protegidas para la conservación en escenarios de cambio climático. In: Payan C, Lasso E, Castaño-Uribe CA (eds) I. Conservación de grandes vertebrados en áreas no protegidas de Colombia, Venezuela y Brasil. Instituto de Investigación de Recursos Biológicos Alexander von Humboldt (IAvH), Bogota, p 115–132

Lizcano DJ, Sissa A (2003) Notes on the distribution, and conservation status of mountain tapir (*Tapirus pinchaque*) in north Perú. Tapir Conservation, newsletter of the IUCN /SSC Tapir Specialist Group 12:21–24

López-Arévalo H, González-Hernández A (2006) Venado sabanero *Odocoileus virginianus*. In: Rodríguez JV, Alberico M (eds) Libro rojo de los mamíferos de Colombia. Serie libros rojos de especies amenazadas de Colombia, Bogotá, pp 114–120

López-Arévalo HF, Montenegro OL (2015) Implicaciones de la no erradicación del hipopótamo (*Hippopotamus amphibius*) en Colombia. In: Asociación Colombiana de Zoología (ed) La biodiversidad sensible: patrimonio natural irreemplazable. IV Congreso Colombiano de Zoología. Libro de resúmenes. Asociación Colombiana de Zoología, Cartagena, Colombia, p 315

López-Arévalo HF, Montenegro OL, Sánchez-Palomino P, Alba Mejía L, Cardona C, Jiménez JS, Mora-Beltrán C, Pérez HY, Serrano H, Tiboche A, Rojas D (2018) Caracterización de las poblaciones de cerdos asilvestrados (*Sus scrofa*) y su hábitat en la sabana inundable de Arauca, Casanare y Meta. Revista porkcolombia 243:24–31

López-Arévalo HF, Pardo Vargas LF, Pérez-Moreno H (2019) Generalidades de la especie. In: López-Arévalo HF (ed) Ecología, uso, manejo y conservación del venado cola blanca en Colombia. Biblioteca José Jerónimo Triana No. 33. Instituto de Ciencias Naturales, Universidad Nacional de Colombia, Bogotá, p 19–30

López-Arévalo HF, Velásquez-Carrillo KL, Mora-Beltrán C, Raz L, Checa ÁC, Páez-Torres AE, Agudelo-Zamora HD (2018) QUYN: Plataforma de registros digitales de fauna silvestre. Mammal Notes 5(2):36–38

Lozano CM (2004) Efectos de la acción humana sobre la frecuencia de uso de los salados por las dantas (*Tapirus terrestris*) en el sureste del trapecio Amazónico colombiano. Maestría en Estudios Amazónicos, Universidad Nacional de Colombia, Leticia

MacFadden BJ, Higgins P (2004) Ancient ecology of 15-million-year-old browsing mammals within C3 plant communities from Panama. Oecologia 140(1):169–182

Maldonado AM (2010) The impact of subsistence hunting by Tikunas on game species in Amacayacu National Park, Colombian Amazon. PhD thesis. Oxford Brookes University, Oxford

Mantilla-Meluk H, Mosquera-Guerra F, Lizcano D, Díaz V, Pérez-Amaya N, Trujillo F (2017) Plan de manejo para la conservación del venado soche (*Mazama rufina*) en el departamento del Quindío. Corporación Autónoma Regional del Quindío. Centro de Estudios de Alta Montaña CEAM Universidad del Quindío, Armenia

March IJ (1993) El pecarí labiado (*Tayassu pecari*). In: Oliver WI (Ed) Pigs, peccaries and hippos: status survey and conservation action plan. IUCN, Gland, Switzerland, p 15–28

Martínez-Polanco MF, Montenegro OL, Peña L, Germán A (2015) La sostenibilidad y el manejo de la caza del venado cola blanca (*Odocoileus virginianus*) por cazadores-recolectores del periodo precerámico de la sabana de Bogotá, en el yacimiento arqueológico de Aguazuque (Colombia). Caldasia 37(1):1–14

Martínez-Polanco MF, Montenegro OL, Peña-León GA, Rincón-Rodríguez LS (2019) La cacería del venado cola blanca en Aguazuque: un sitio de cazadores-recolectores tardíos en la sabana de Bogotá. In: López-Arévalo HF (ed) Ecología, uso, manejo y conservación del venado cola blanca en Colombia. Biblioteca José Jerónimo Triana No. 33. Instituto de Ciencias Naturales, Universidad Nacional de Colombia, Bogotá, p 119–142

Martínez-Salas M, López-Arévalo HF, Sánchez-Palomino P (2016) Cacería de subsistencia de mamíferos en el sector oriental de la Reserva de Biósfera El Tuparro, Vichada (Colombia). Acta Biol Colomb 21:151–166

Mateus Gutiérrez C (2014) Efecto de la estructura del hábitat sobre las características demográficas de dos poblaciones locales de Venado Cola Blanca, *Odocoileus virginianus goudotii*, en el Parque Nacional Natural Chingaza (Colombia). Tesis maestría, Universidad Nacional de Colombia, Bogotá

Mateus-Gutiérrez C, López-Arévalo HF (2019) Plantas consumidas por el venado cola blanca en colombia. In: López-Arévalo HF (ed) Ecología, uso, manejo y conservación del venado cola blanca en Colombia. Biblioteca José Jerónimo Triana No. 33. Instituto de Ciencias Naturales, Universidad Nacional de Colombia, Bogotá, p 55–66

Mateus-Gutiérrez C, López-Arévalo HF, Montenegro OL, Pérez-Moreno H (2019) Análisis de la información bibliográfica sobre el venado cola blanca en Colombia. In: López-Arévalo HF (ed) Ecología, uso, manejo y conservación del venado cola blanca en Colombia. Biblioteca José Jerónimo Triana No. 33. Instituto de Ciencias Naturales, Universidad Nacional de Colombia, Bogotá, p 31–44

Mayer J, Wetzel R (1987) Tayassu pecari. Mamm Species 293:1–7

Medici EP, Carrillo L, Montenegro OL, Miller PS, Carbonell F, Chassot O, Cruz-Aldán E, García M, Estrada-Andino N, Shoemaker AH, Mendoza A (2005) Taller de Conservación de la danta centroamericana (*Tapirus bairdii*): evaluación de viabilidad poblacional y del hábitat (PHVA). Zoológico de Belice y Centro de Educación Tropical Belice. 15 al 19 de agosto. Grupo de especialistas de Tapires del IUCN

Mejía-Correa S, Díaz-Martínez JA (2009) Primeros registros e inventario de mamíferos grandes y medianos en el Parque Nacional Munchique, Colombia. Mesoamericana 13(3):7–22

Mejia-Correa S, Diaz-Martinez JA (2014) Densidad y hábitos alimentarios de la danta *Tapirus bairdii* en el Parque Nacional Natural Los Katios, Colombia. Tapir Conservation, newsletter of the IUCN /SSC Tapir Specialist Group 23:16–23

Minambiente, Ministerio de Ambiente y Desarrollo Sostenible (2017a) Resolución 1912 del 15 de septiembre de 2017. Bogotá, D.C.

Minambiente, Ministerio de Ambiente y Desarrollo Sostenible (2017b) Plan de acción de biodiversidad para la implementación de la Política Nacional para la Gestión Integral de la Biodiversidad y sus Servicios Ecosistémicos 2016–2030. Comps. Ministerio de Ambiente y Desarrollo Sostenible. Dirección de Bosques, Biodiversidad y Servicios Ecosistémicos, Bogotá

Molina M, Molinari J (1999) Taxonomy of Venezuelan white-tailed deer (*Odocoileus*, Cervidae, Mammalia), based on cranial and mandibular traits. Can J Zool 77(4):632–645

Monsalve S, Ramírez A (2018) Estado actual de los hipopótamos (*Hippopotamus amphibius*) en Colombia: 2018. Rev CES Med Zootec 13(3):338–346

Montenegro OL (2005) Programa nacional para la conservación del género *Tapirus* en Colombia. Ministerio de Ambiente, Vivienda y Desarrollo Territorial. Dirección de Ecosistemas, Bogotá

Montenegro OL, Héctor Restrepo JL, Contreras WR, López JP (2017) Departamento de Guaviare, Colombia Mamíferos grandes y medianos de los alrededores de la Serranía de la Lindosa. Rapid Color Guide No. 880. Field Guides—Am Mus Nat Hist, New York

Montenegro OL, Roncancio N, Soler-Tovar D, Cortés-Duque J, Contreras-Herrera J, Sabogal S, Navas-Suárez PE (2018) Serologic survey for selected viral and bacterial swine pathogens in Colombian collared peccaries (*Pecari tajacu*) and feral pigs (*Sus scrofa*). J Wildl Dis 54(4):700–707

Montenegro OL, Sánchez-Palomino P, García L, Roncancio N, Soler D, Cortés J, Acevedo L (2009) Evaluación serológica y programa de vigilancia epidemiológica para las especies silvestres de sainos *Pecari tajacu* y *Tayassu pecari*, y cerdos ferales respecto a la peste porcina clásica (PPC), en áreas piloto de los departamentos de Arauca. Casanare. Informe final, Universidad Nacional de Colombia, Bogotá

Montenegro OL, Sánchez P, Mesa E, Sarmiento C, Gómez B (2008) Determinación de la distribución de las dos especies de saínos en Colombia como primer paso para caracterizar desde el punto de vista sanitario las poblaciones silvestres de saínos presentes en el territorio nacional, respecto a la presencia de peste porcina clásica, en el marco de la política nacional de sanidad pecuaria. Convenio de Cooperación Científica, Tecnológica y Financiera entre el Ministerio de Ambiente, Vivienda y Desarrollo Territorial y la Universidad Nacional de Colombia No. 225, Bogotá

Morales L (2017) La paz y la protección ambiental en Colombia: Propuestas para un desarrollo rural sostenible. Diálogo Interamericano, Washington

Morales-Jiménez AL, Sánchez F, Poveda K, Cadena A (2004) Mamíferos terrestres y voladores de Colombia. Guía de Campo, Ramos López Editorial, Bogotá

Morán EF (1997) La ecología humana en los pueblos de la Amazonia. Fondo de Cultura Económica, Madrid

Moscarella RA, Aguilera M, Escalante AA (2003) Phylogeography, population structure, and implications for conservation of white-tailed deer (*Odocoileus virginianus*) in Venezuela. J Mammal 84(4):1300–1315

Mosquera-Guerra F, Trujillo F, Diaz-Pulido AP, Mantilla-Meluk H (2018) Diversidad, abundancia relativa y patrones de actividad de los mamíferos medianos y grandes, asociados a los bosques riparios del río Bita, Vichada, Colombia. Biota Colomb 19(1):202–218

Naranjo EJ, Vaughan C (2000) Ampliación del ámbito altitudinal del tapir centroamericano (*Tapirus bairdii*). Rev Biol Trop 48(2-3):724–725

Naveda A, de Thoisy B, Richard-Hansen C, Torres DA, Salas L, Wallance R, Chalukian S, de Bustos S (2008) *Tapirus terrestris*. The IUCN red list of threatened species 2008. https://doi.org/10.2305/IUCN.UK.2008.RLTS.T21474A9285933. Accessed 24 Feb 2019

Noguera-Urbano EA, Montenegro-Muñoz SA, Lasso LL, Calderon-Leyton JJ (2014) Mamíferos medianos y grandes en el piedemonte Andes-Amazonía de Monopamba-Puerres, Colombia. Brenesia 81–82:111–114

Nores C, Fernández A, Corral N (2000) Estimación de la población de jabalí (*Sus scrofa*) por recuento de grupos familiares. Naturalia Cantabricae 1:53–59

Norman JE, Ashley MV (2000) Phylogenetics of Perissodactyla and test of the molecular clock. J Mol Evol 50:11–21

Ojasti J, Dallmeier F (eds) (2000) Manejo de fauna silvestre neotropical. SI/MAB Series No. 5. Smithsonian Institution/MAB Biodiversity Program, Washington, D.C

Orejuela JE (1992) Traditional productive systems of the Awa (Cuaiquer) Indians of southwestern Colombia and neighbouring Ecuador. In: Redford KH, Padoch C (eds) Conservation of Neotropical forests: working from traditional resource use. Columbia University Press, New York, pp 58–82

Ortega-Andrade HM, Prieto-Torres DA, Gómez-Lora I, Lizcano DJ (2015) Ecological and geographical analysis of the distribution of the mountain tapir (*Tapirus pinchaque*) in Ecuador: importance of protected areas in future scenarios of global warming. PLoS One 10:e0121137

Oslinger A, Muñoz J, Álvarez L, Ariza F, Moreno F, Posso A (2006) Caracterización de cerdos criollos colombianos mediante la técnica molecular RAMs. Acta Agron 55(4):45–50

Osorno M, Atuesta-Dimian N, Jaramillo LF, Sua S, Barona A, Roncancio N (2014) La despensa del Tiquié: Diagnóstico y manejo comunitario de la fauna de consumo en la Guayana Colombiana. Inst Amazón Inv Cient "SINCHI", Bogotá

Palacios-Mosquera L, Parra Ibarguen C, Mantilla-Meluk H (2010) Datos preliminares sobre la abundancia relativa y caracterización del hábitat de *Tayassu pecari* (Artiodactyla: Tayassuidae) en los municipios del Medio Baudó y Cértegui en el Departamento del Chocó, Colombia. Bioetnia 7(1):16–22

Palacios-Mosquera Y, Rodríguez-Bolaños A, Jiménez-Ortega AM (2008) Aprovechamiento de los recursos naturales por parte de la comunidad local en la cuenca media del Río Atrato, Chocó, Colombia. Revi Biodiv Neotrop 27(2):75–85

Parra-Colorado JW, Botero-Botero Á, Saavedra-Rodríguez CA (2014) Percepción y uso de mamíferos silvestres por comunidades campesinas andinas de Génova, Quindío, Colombia. Bol Cient Mus His Nat 18:78–93

Payán GE, Escudero-Páez S (2015) Densidad de jaguares (*Panthera onca*) y abundancia de grandes mamíferos terrestres en un área no protegida de Amazonas colombiano. In: Payán GE, Lasso CA, Castaño-Uribe C (eds) Conservación de grandes vertebrados en áreas no protegidas de Colombia, Venezuela y Brasil, 1st edn. Instituto de Investigación de Recursos Biológicos Alexander von Humboldt (IAvH), Bogotá, pp 225–242

Peña-León G, Pinto M (1996) Mamíferos más comunes en sitios precerámicos de la Sabana de Bogotá. Guía ilustrada para arqueólogos. Colección Julio Carrizosa Valenzuela N° 6. Acad Colomb Ciencias Exactas, Físicas y Naturales, Bogotá

Peña-León GA, Rincón-Rodríguez LS (2019) Revisión de los registros arqueológicos de venado cola blanca en la sabana de Bogotá. In: López-Arévalo HF (ed) Ecología, uso, manejo y conservación del venado cola blanca en Colombia. Biblioteca José Jerónimo Triana No. 33. Instituto de Ciencias Naturales, Universidad Nacional de Colombia, Bogotá, p 157–170

Pérez-Moreno HY, Montenegro OL, López-Arévalo HF, Sarmiento C (2019) Densidad poblacional y uso de hábitat del venado de cola blanca (*Odocoileus virginianus*), durante la época de lluvias en la Orinoquia colombiana. In: López-Arévalo HF (ed) Ecología, uso, manejo y conservación del venado cola blanca en Colombia. Biblioteca José Jerónimo Triana No. 33. Instituto de Ciencias Naturales, Universidad Nacional de Colombia, Bogotá, p 67–84

Plata Rangel AM (2005) Importancia de la fauna silvestre en la etnia Sikuani, comunidad de Cumarianae, selva de Matavén, Vichada, Colombia. Tesis Pontificia Universidad Javeriana, Bogotá

Racero-Casarrubia J, González-Maya JF (2014) Inventario preliminar y uso de mamíferos silvestres por comunidades campesinas del sector oriental del cerro Murrucucú, municipio de Tierralta, Córdoba, Colombia. Mammal Notes 1(2):25–28

Racero-Casarrubia JA, Vidal CC, Ruiz ÓD, Jesús BC (2008) Percepción y patrones de uso de la fauna silvestre por las comunidades indígenas Embera-Katíos en la cuenca del río San Jorge, zona amortiguadora del PNN-Paramillo. Rev Estud Soc 31:118–131

Ramírez RAC (2007) Patrones de uso de la fauna silvestre por parte de la población asentada en las veredas Alejandría, Cardozo y La Libertad (San Eduardo, Boyacá, Colombia). Tesis doctoral, Universidad Pedagógica y Tecnológica de Colombia, Bogotá

Ramírez-Chaves HE, Ortega-Rincón M, Pérez WA, Marín D (2011) Historia de las especies de mamíferos exóticos en Colombia. Bol Cient Mus Univ Caldas 15(2):139–156

Ramírez-Chaves HE, Suárez-Castro AF, González-Maya JF (2016) Cambios recientes a la lista de los mamíferos de Colombia. Mammal Notes 3(1):1–9

Ramírez-Mejía AF, Sánchez F (2016) Activity patterns and habitat use of mammals in an Andean forest and a *Eucalyptus* reforestation in Colombia. Hystrix J 27(2):1–7

Ramos D (1995) Estudio de la dieta y distribución del venado cola blanca (*Odocoileus virginianus*) en el Parque Nacional Natural Chingaza. Bogotá. Trabajo grado, Facultad de Ciencias Básicas, Pontificia Universidad Javeriana, Bogotá

Rayo Avendaño WA (2017) Densidad poblacional y abundancia relativa de venados cola blanca (*Odocoileus virginianus*) en el páramo de Siscunsí, Boyacá. Tesis Biología, Universidad El Bosque, Bogotá

Rincón-Rodríguez LS (2019) Arqueozoología del venado cola blanca en San Carlos, municipio de Funza, Sabana de Bogotá. In: López-Arévalo HF (ed) Ecología, uso, manejo y conservación del venado cola blanca en Colombia. Biblioteca José Jerónimo Triana No. 33. Instituto de Ciencias Naturales, Universidad Nacional de Colombia, Bogotá, p 143–156

Rivas-Pava MP, Ramírez-Chaves HE, Álvarez ZI, Niño V (2007) Catálogo de los mamíferos presentes en las colecciones de referencia y exhibición del Museo de Historia Natural de la Universidad del Cauca. Universidad del Cauca, Popayán

Rodríguez J (2004) Propuesta de áreas prioritarias para la conservación de la danta de tierras bajas *Tapirus terrestris* en la Amazonía y Orinoquía Colombiana. Tesis maestría, Universidad Nacional de Colombia, Bogotá

Rodríguez PAL (2014) Analisis filogeográfico y genético poblacional de la especie de pecarí (*Tayassu pecari* Tayassuidae, Artiodactyla) mediante una región mitocondrial y microsatelites Tesis Magister. Pontificia Universidad Javeriana. Facultad de Ciencias. Bogotá

Rodríguez-Castellanos O (2016) Modelación de la dinámica poblacional del venado cola blanca (*Odocoileus virginianus goudotii*) en el Parque Nacional Natural Chingaza (Tesis maestría). Universidad Distrital Francisco José de Caldas, Facultad de Medio Ambiente y Recursos Naturales, Bogotá

Rodríguez A, Gómez R, Moreno A et al (2014) Record of a mountain tapir attacked by an Andean bear on a camera trap. Tapir Conservation Newsletter of the IUCN/SSC Tapir Specialist Group 23:25–26

Rodríguez-Mahecha JV, Alberico M, Trujillo F, Jorgenson J (eds) (2006) Libro rojo de los Mamíferos de Colombia. Serie Libros Rojos de especies amenazadas de Colombia. Conservación Internacional Colombia, Ministerio del Medio Ambiente, Vivienda y Desarrollo Territorial, Bogotá

Rodríguez-Mahecha JV, Hernández-Camacho JI, Defler TR, Alberico M, Mast RB, Mitttermeier RA, Cadena A (1995) Mamíferos colombianos: sus nombres comunes e indígenas. Occ Pap Conserv Biol 3:1–56

Rojas L (2010) Evaluación del uso y calidad del hábitat en poblaciones del venado cola blanca *Odocoileus virginianus* en la reserva natural la Aurora, municipio de Hato Corozal, Casanare. Tesis grado, Pontificia Universidad Javeriana, Bogotá

Rosell C, Fernández-Llario P, Herrero J (2001) El Jabalí (*Sus scrofa* Linnaeus, 1758). Galemys 13(2):1–25

Rossi RV, Duarte JMB (2016) *Mazama nemorivaga*. The IUCN red list of threatened species 2016. https://doi.org/10.2305/IUCN.UK.2016-1.RLTS.T136708A22158407. Accessed 15 Jan 2019

Rubio-Torgler H (1997) Estrategia para el manejo de especies de caza en el área de influencia del Parque Nacional Natural Utría. In: Fang T, Bodmer RE, Aquino R, Valqui MH (eds) Manejo de fauna silvestre en la Amazonia. UNAP, University of Florida, UNDP/GEF and Instituto de Ecología, La Paz, pp 135–143

Ruiz-García M, Castellanos A, Bernal L, Navas D, Pinedo M, Shostell J (2015a) Mitochondrial gene diversity of the mega-herbivorous species of the genus *Tapirus* (Tapiridae, Perissodactyla) in South America and some insights on their genetic conservation, systematics and the Pleistocene influence on their genetic characteristics. Adv Genet Res 14:1–51

Ruiz-García M, Pinedo-Castro M, Luengas-Villamil K, Vergara C, Rodriguez JA, Shostell JM (2015b) Molecular phylogenetics of the white-lipped peccary (*Tayassu pecari*) did not confirm morphological subspecies in northwestern South America. Genet Mol Res 14(2):5355–5378

Ruiz-García M, Vásquez C, Pinedo-Castro MO, Sandoval S, Castellanos A, Kaston F, Thoisy B, Shostell J (2012) Phylogeography of the mountain tapir (*Tapirus pinchaque*) and the central American tapir (*Tapirus bairdii*) and the origins of the three Latin-American tapirs by means of mtCyt-B sequences. In: Anamthawat-Jónsson K (ed) Current topics in phylogenetics and phylogeography of terrestrial and aquatic systems. InTech, Rijeka, pp 83–116

Ruiz-García M, Vásquez-Carrillo C, Sandoval S, Kaston F, Luengas-Villamil K, Shostell J (2016) Phylogeography and spatial structure of the lowland tapir (*Tapirus terrestris*, Perissodactyla: Tapiridae) in South America. Mitochondrial DNA A DNA Mapp Seq Anal 27(4):2334–2342

Ruiz-García M, Randi E, Martínez-Agüero M, Alvarez D (2007) Relaciones filogenéticas entre géneros de ciervos neotropicales (Artiodactyla, Cervidae) mediante secuenciación de AND mitochondrial y marcadores microsatelitales. Rev Biol Trop 55:723–741

Sábato MAL, Melo LFB, Magni EMV, Young RJ (2008) Aportes a la historia natural de la danta colombiana (*Tapirus terrestris colombianus*) compilados en el norte de los Andes Centrales Colombianos. Mammalia 69:405–412

Sabogal S (2010) Filogeografía y conservación genética del pecarí de collar, *Pecari tajacu*, en cuatro departamentos de Colombia. Tesis maestría en Ciencias-Biología, Universidad Nacional de Colombia, Bogotá

Sandrin F, L'haridon L, Vanegas L, Ponta N, Gómez J, Cuellar JR, Van Vliet N (2016) Manejo comunitario de la cacería y de la fauna: Avances realizados por la asociación de cazadores airu-makiichi en Puerto Nariño, Amazonas Colombia. No. CIFOR Working Paper no. 213. Center for International Forestry Research (CIFOR), Bogor

Sarmiento AMD (2007) Patrones en la distribución de los lugares de captura del tapir (*Tapirus terrestris*) con base en el conocimiento tradicional de las comunidades indígenas Andoque y Nonuya, y el asentamiento de Santander-Araracuara, Amazonia Colombiana. Tesis maestría, Universidad Nacional de Colombia, Bogotá

Sarria Perea JA (2012) Taxonomia e filogenia de algumas espécies de *Mazama* (Mammalia; Cervidae) da Colômbia. Tesis maestría Universidade Estadual Paulista "Julio de Mesquita Filho", Rua Quirino de Andrade

Sarria Perea JA, Vargas Munar DSF (unpublished data) Health status survey of wild populations of the dwarf brocket (*Mazama rufina*) and the Andean tapir (*Tapirus pinchaque*) in the massif of Mamapacha (Boyacá, Colombia). Final Report, The Rufford Foundation. https://www.rufford.org/rsg/Projects/JavierAdolfoSarriaPerea. Accesed 21 Feb 2019

Saunders G (1995) Ecological comparison of two wild pig populations in semi-arid and sub-alpine Australia. Ibex J Moun Ecol 3:152–156

Schank C, Cove MV, O 'Farrill G, Estrada N, Poot C, Meyer N, Mendoza E, García Vettorazzi MJ, Leonardo R, Jordan CA, Lizcano DJ (2015) Integrating current range-wide occurrence data with species distribution models to map the potential distribution of Baird's tapir. Tapir Conservation Newsletter of the IUCN /SSC Tapir Specialist Group 24:15–25

Schank CJ, Cove KMJ, Mendoza E, O'Farrill G, Reyna-Hurtado R, Meyer N, Jordan CA, González-Maya JF, Lizcano DJ, Moreno R, Dobbins MT, Montalvo V, Sáenz-Bolaños C, Jimenez EC, Estrada, Cruz Díaz JC, Saenz J, Spínola M, Carver A, Fort J, Nielsen CK, Botello F, Pozo Montuy F, Rivero M, de la Torre JA, Brenes-Mora E, Godínez-Gómez O, Wood MA, Gilbert J, Miller JA (2017) Using a novel model approach to assess the distribution and conservation status of the endangered Baird's tapir. Divers Distrib 23:1459–1471

Semana (2009) Hay que reducir o anular los hipopótamos en Colombia. https://www.semana.com/nacion/medio-ambiente/articulo/hay-reducir-anular-hipopotamos-colombia/106952-3. Accessed 11 Feb 2019

[SMBYC] Sistema de Monitoreo de Bosques y Carbono, [IDEAM] Instituto de Hidrología, Meteorología y Estudios Ambientales (2017) Décimo Boletín de Alertas Tempranas de Deforestación (AT-D) Primer trimestre 2017. https://www.ambienteysociedad.org.co/es/decimo-boletin-de-alertas-tempranas-de-deforestacion-at-d-primer-trimestre-2017/. Accessed 11 Feb 2019

Solari S, Muñoz-Saba Y, Rodríguez-Mahecha JV, Defler T, Ramírez-Chaves HE, Trujillo F (2013) Riqueza, endemismo y conservación de los mamíferos de Colombia. Mastozool Neotrop 20:301–365

Taber A, Chalukian S, Altrichter M, Minkowski K, Lizarraga L, Sanderson E, Rumiz D, Venticinque E, Amorim E, de Angelo C, Antúnez M, Ayala G, Beck H, Bodmer R, Boher S, Cartes JL, de Bustos S, Eaton D, Emmons L, Estrada N, Zapata-Ríos G (2008) El destino de los arquitectos de los bosques neotropicales: evaluación de la distribución y el estado de conservación de los Pecaries Labiados y los Tapires de Tierras Bajas. Pigs, Peccaries and Hippos Specialist Group (IUCN/SSC); Tapir Specalist Group (IUCN/SSC); Wildlife Conservation Society; and Wildlife Trust

Tafur-Guarin MP (2010) Evaluación de la sostenibilidad de la cacería de mamíferos en la comunidad de Zancudo, Reserva Nacional Natural Puinawai, Colombia. Tesis maestría en Ciencias-Biología, Universidad Nacional de Colombia, Bogotá

Thomas O (1913) On certain of the smaller south American Cervidae. Ann Mag Nat Hist 8(11):585–589

Tinoco-Sotomayor AN (2018) Riqueza, uso y amenazas de mamíferos medianos y grandes en el Distrito de Cartagena de Indias, Colombia. Tesis de pregrado, Universidad de Cartagena, Cartagena de Indias

Tinoco-Sotomayor AN, Ramos-Guerra HD, Vides-Avilez HA, Rodríguez-Alarcón DC, González-Maya JF, Gómez-Estrada H (2016) Inventario preliminar y uso de mamíferos silvestres no voladores en la vereda Camarón, Montes de María (Bolívar-Colombia). Mammal Notes 3(1):32–36

Trujillo F, Mosquera-Guerra F (2016) Caracterización, uso y manejo de la mastofauna asociada a los morichales de los llanos orientales colombianos. In: Lasso CA, Colonnello G, Moraes-R M (eds) Morichales, cananguchales y otros palmares inundables de Suramérica. Parte II: Colombia, Venezuela, Brasil, Perú, Bolivia, Paraguay, Uruguay y Argentina. Serie Editorial Recursos Hidrobiológicos y Pesqueros Continentales de Colombia. Instituto de Investigación de Recursos Biológicos Alexander von Humboldt (IAvH), Bogotá, pp 191–219

Ulloa A, Rubio-Torgler H, Campos-Rozo C (2004) Conceptual basis for the selection of wildlife management strategies by the Embera people in Utria National Park, Choco, Colombia. In: Silvius KM, Bodmer RE, Fragoso MV (eds) People in nature: wildlife conservation in south and Central America. Columbia University Press, New York, pp 11–36

Universidad Nacional de Colombia, Corporinoquia (2015) Estado actual de las poblaciones de Chigüiro (*Hydrochoerus hydrochaeris*) y su hábitat bajo escenario de intervención antrópica y cambio climático, en el departamento de Arauca y Casanare en jurisdicción de Corporinoquia. Informe final, Bogotá

Valderrama C (2012) Wild hippos in Colombia. Aliens The Invasive Spec Bull 32:8–12

van Vliet N, Mesa MPQ, Cruz-Antia D, de Aquino LJN, Moreno J, Nasi R (2014) The uncovered volumes of bushmeat commercialized in the Amazonian trifrontier between Colombia, Peru & Brazil. Ethno Cons 3:1–11

Varela DM, Trovati RG, Guzmán KR, Rossi RV, Duarte JMB (2010) Red brocket deer Mazama americana (Erxleben 1777). In: Duarte JMB, Gonzalez S (eds) Neotropical Cervidology: biology and medicine of Latin American deer. FUNEP/IUCN, Jaboticabal, pp 151–159

Vélez J, Espelta J, Rivera O, Armenteras D (2017) Effects of seasonality and habitat on the browsing and frugivory preferences of *Tapirus terrestris* in North-Western Amazonia. J Trop Ecol 33(6):395–406

Voss RS, Helgen KM, Jansa SA (2014) Extraordinary claims require extraordinary evidence: a comment on Cozzuol et al. (2013). 2014. J Mammal 95(4):893–898

Wallace R, Ayala G, Viscarra M (2012) Lowland tapir (*Tapirus terrestris*) distribution, activity patterns and relative abundance in the greater Madidi–Tambopata landscape. Integr Zool 7:407–419

Weigl R (2005) Longevity of mammals in captivity; from the living collections of the world. Kleine Senckenberg-Reihe 48, Stuttgart

Wolf T, Conover M (2003) Feral pigs and the environment: an annotated bibliography. Jack H. Berryman Institute, Publication 21, Utah State University, Logan, and Mississippi State University, Starkville

Chapter 10
Tropical Ungulates of Venezuela

Alejandra Soto-Werschitz, Paolo Ramoni-Perazzi, and Guillermo Bianchi-Pérez

Abstract Venezuela is a megadiverse country which harbors seven species of wild ungulates, belonging to five genera, three families, and two orders. However, knowledge on this mammalian group in Venezuela is disparate in all aspects of their biology. Here, we present a comprehensive assessment of the literature regarding Venezuelan ungulates, with emphasis on the scientific production of the last 10 years. Feral ungulates (donkeys, horses, domestic pigs, goats, cows, etc.), with well-known stabilized populations in regions of the country such as Paraguaná Peninsula or Llanos, were not included in this review. We compiled the information from specialized bibliographic sources, including Scholar Google, SciELO and the Science Citation Index expanded. Most studies are mainly focused on their systematics and taxonomy, their importance as food source in rural areas, and their potential role in zoonoses. All Venezuelan species of ungulates are threated in some extent, nationally or locally, with poaching and habitat loss as the main affecting factors. This compilation reflects that Venezuelan ungulates are disproportionately poorly known, and that research in both the wild and in captivity is urgently required.

10.1 Introduction

Due to the geographical position of Venezuela, its flora and fauna have affinities to Caribbean, Andean, Amazonian, or Guyanese biogeographic regions (Rodríguez and Rojas-Suárez 2008). It includes some of the most threatened ecosystems and forest ecosystems considered to be the last frontiers of the tropical world (Rodríguez

A. Soto-Werschitz (✉) · P. Ramoni-Perazzi
Laboratorio de Zoología Aplicada, Departamento de Biología, Facultad de Ciencias, Universidad de Los Andes, Mérida, Venezuela

Laboratório de Ecologia e Conservação de Mamíferos, Setor Ecologia Aplicada, Universidade Federal de Lavras, Lavras, MG, Brazil
e-mail: alesoto@ula.ve; rpaolo@ula.ve

G. Bianchi-Pérez
Departamento de Biología, Facultad de Ciencias, Universidad de Los Andes, Mérida, Venezuela
e-mail: gbianchi@ula.ve

S. Gallina-Tessaro (ed.), *Ecology and Conservation of Tropical Ungulates in Latin America*, https://doi.org/10.1007/978-3-030-28868-6_10

et al. 2010). This megadiverse country is placed fifteenth in the world rank and the fifth in Latin America, along with Brazil, Colombia, Peru and Ecuador (Bisbal and Naveda-Rodríguez 2010). It has 27 climatic zones, 650 natural vegetation types and 38 geological units with more than 137,141 recognized species, grouped into 1775 families and more than 9200 genera (Aguilera et al. 2003). Moreover, the number of species of mammals described has increased in recent years from 351 species (Ochoa and Aguilera 2003) to 390 species, 30 of them endemic, representing 184 Genera, 47 Families, 14 Orders, placing Venezuela eighth in the world ranking (Sánchez and Lew 2012).

10.2 Species Account

In Venezuela, the Perissodactyla Order is represented by Tapiridae (one species), while the Artiodactyla are represented by the families Cervidae (with two genera: *Mazama* and *Odocoileus*) and Tayassuidae (also with two genera: *Tayassu* and *Pecari*).

10.2.1 Tapiridae Family

There are some discrepancies regarding the tapir's species richness in Venezuela (Naveda-Rodríguez and Torres 2002). The mountain tapir, *Tapirus pinchaque*, has been historically recorded in Venezuela, but lacks of recent records in this country (Lizcano et al. 2016), thus, is currently considered as Extinct at the Regional Level (Ojasti and Lacabana 2008). The controversial recently described little black tapir, *T. kabomani*, has been recorded from scattered localities in three countries in the rainforests of the Amazon river basin, in apparently sympatry with the lowland tapir (*T. terrestris*; Cozzuol et al. 2013), thus its occurrence is reasonably expected for the forests in Southern Venezuela. Hence, the lowland tapir (*T. terrestris*, Linnaeus 1758) is the unique species whose presence has been confirmed and counts on reliable recent records, represented by the subspecies *T. t. terrestris* and, perhaps, also *T. t. colombianus* (Naveda-Rodríguez and Torres 2002).

The danta or lowland tapir is considered the largest terrestrial mammal in Venezuela, weighing more than 200 kg. It has elongated face, is dark-colored except the whitish tips of the ears. It has a short mane that goes from the neck to the tail. The calves have a distinctive coloration of white bands alternated with white spots. It is a diurnal and terrestrial species usually associated to water bodies (Linares 1998; Linares and Rivas 2004) (Fig. 10.1).

The studies carried out in Venezuela have been focused on erosive ulcerative gastric syndrome associated with stress induced by captivity (Gómez et al. 2010) and other veterinary aspects (Cañizales and Guerrero 2010), genetics (Aguilera and Expósito 2010; Naveda-Rodríguez et al. 2012), population abundance at some localities (Naveda-Rodríguez et al. 2012; Marín-Wikander 2012), fossil record

Fig. 10.1 Adult *Tapirus terrestris*. Captive specimen kept at Chorros de Milla Zoo, Mérida State. 19 Mar 2009. Photo credit Paolo Ramoni Perazzi

(Holanda and Rincón 2012), models of habitat use (Isasi-Catalá 2013), and hunting pressure (Gondeles et al. 1981; Matallana et al. 2012).

Historically they were distributed in Venezuela in the humid forests of the Orinoco basin, in the foothills and gallery of the Sierra de Perijá and the Cordillera de los Andes and in the humid and cloudy forests of the Cordillera de la Costa (Torres et al. 2015). It is notorious for its absence in arid, insular areas and high mountains of the country with more than 1600 m (Gondeles et al. 1981; Naveda-Rodríguez et al. 2008). The most recent records are from the Guárico river basin, in the Northcenter region of the country (Bisbal and Naveda-Rodríguez et al. 2010; Naveda-Rodríguez et al. 2012; Isasi-Catalá 2013), whereas the most recent record of the species with an offspring was obtained in 2016 (Morán et al. 2018) (Fig. 10.2). Most records are associated to creeks, inundated forests, terrafirme forest, hill sides and hill tops (Salas 1996; Salas and Fuller 1996), as well as seasonal deciduous and semi-evergreen forests (Morán et al. 2018). Densities from 0.466 SD 0.464 individuals/km (Naveda-Rodríguez et al. 2012) to 1.26 individuals/km (Isasi-Catalá 2013) have been estimated in Northern Venezuela. In the Guatopo National Park, occupancy models applied using GIS, indirect traces and camera traps, found a high occupancy probability (0.90 SD 0.01) in extensive and contiguous areas of mature forest with little change in vegetation cover, and human activity does not seem to affect their occupation, except where roads have been established and fires are frequent (Isasi-Catalá 2013).

Fig. 10.2 Female with calf of *Tapirus terrestris*. Captive specimens kept at Chorros de Milla Zoo, Mérida State. 19 Mar 2009. Photo credit Paolo Ramoni Perazzi

At continental level, Naveda-Rodríguez et al. (2012) reported a range contraction for the species of approximately 14%, equivalent to 1,897,856 km^2, mainly on the coast of Venezuela, Argentina and the Northeast and South of Brazil. At national level, there are not recent publications focused in Venezuela contrasting information of the expected range against information where the species has been verified, thus, very little is known about the state of tapir populations in this country (Naveda-Rodríguez et al. 2008). Passos et al. (2016) estimated that the tropical moist broadleaf forests of southern Venezuela are widely suitable for tapirs, while other more open biomes, such as Llanos (savannas), are less suitable. However, Venezuelan populations can be currently isolated or dispersed, and the probabilities of local extinction concomitantly high due to the decrease of habitat quality (Taber et al. 2009) and poaching (Naveda-Rodríguez and Torres 2002; Naveda-Rodríguez et al. 2012), especially at North from the Orinoco river (Torres et al. 2015).

de Thoisy et al. (2010) found that nucleotide diversity estimates for lowland tapir varied among ecoregions and, although different populations show some degree of genetic structure, it is specially marked in populations such as that from Venezuelan Llanos indicating a significant isolation. From a conservation perspective, isolated populations harbor unique genetic diversities, which must to be conserved.

Holanda and Rincón (2012) reported fossil tapirs from two Pleistocene localities of Venezuela. One of these (skull, mandible, and postcrania), the Zumbador Cave, is located in the eastern portion of Falcón State, and can be assigned to the extant species *T. terrestris*, representing the northwesternmost fossil record of this species in South America. Interestingly, these authors highlight that palaeoenvironmental indicators of that area suggest a savannah–like landscape with patches of scattered trees, which in some extent contrasts with the environments where the species is mainly found nowadays.

Modern tapirs conserve an ancestral karyotype ($2n = 80$), that has remained constant with slight chromosomal arrangements. However, it has not been possible to compare cytogenetically the populations due to the lack of samples from wild animals (Aguilera and Expósito 2010). Naveda-Rodríguez and Torres (2002) registered 29 tapir individuals in the zoos of Venezuela, lacking of reliable information about their geographic origins, precluding the possibility of accurate phylogeographic studies on the species.

Globally, *T. terrestris* is listed on CITES Appendix II (Naveda-Rodríguez et al. 2008). In Venezuela, the lowland tapir is considered as Vulnerable (Torres et al. 2015), and hunting is indefinitely against law, and was declared as an Endangered Species (República de Venezuela 1996a, b). Some of the factors that have threaten the species are the proximity to large cities, oil exploitation, road construction, productive activities (Rodríguez et al. 2010), and hunting (Naveda-Rodríguez and López 2006; Matallana et al. 2012). In Venezuela, rural communities supply around 62% of their protein needs with the consumption of wild animals although their health may be affected (Gongora et al. 2011). The tapir is a reservoir of diseases such as foot-and-mouth disease, bluetongue, infectious bovine rhinotracheitis, equine encephalitis, nematodes and ectoparasites (Cañizales and Guerrero 2010) (Fig. 10.3).

Fig. 10.3 Adult male and female of *Tapirus terrestris*. Captive specimens kept at Chorros de Milla Zoo, Mérida State. 19 Mar 2009. Photo credit Paolo Ramoni Perazzi

In this way, conservation programs for the species could be established. As indicated by Naveda-Rodríguez and Torres (2002) it is impossible to establish an ex situ conservation program because the number of males in zoos exceeds the number of females combined with the fact that the origin of the animals is unknown. Moreover, the health of zoo specimens is also unexplored and their chances of survival are probably null. Finally, Torres et al. (2015) recommend to evaluate the system of protected areas and to promote corridors that favor the gene flow among isolated populations, along with environmental education programs that favor the conservation of the species.

10.2.2 Cervidae Family

Two genera of Cervidae have been reported to Venezuela: *Mazama* Rafinesque, 1817, and *Odocoileus* Rafinesque, 1832. The genus *Mazama* is represented by three species, while *Odocoileus* is represented by a single one.

10.2.2.1 *Mazama rufina* (Thomas, 1908)

A dwarf *Mazama* which according to Linares (1998) has 853–1078 mm of total length and weighs 8.2–13.6 kg, with no sexual dimorphism in size (Dietrich 1993; Lizcano et al. 2010a, b), even at craniometrical level (Gutiérrez et al. 2015). Coloration is variable, but usually reddish dorsally, with lighter underparts, throat is orange, and dark gray to black in legs and head (Linares 1998). The coat is long-haired, rather glossy, usually with two white spots in the upper lip and a white area at the tip of the mandible (Dietrich 1993). The antlers are short simple spikes (up to 50 mm), which tend to be kept a long time before being shed Czernay (1987). The species has chromosome complements of $2n = 60$ (FN = 70), differing in number and structure from other *Mazama* (Aguilera et al. 2010).

The *Mazama* genus has been object of taxonomic controversies. Allen (1915) recognized 18 species, which were further reduced to four (*M. chunyi*, *M. gouazoubira*, *M. nana*, and *M. rufina*) by Cabrera (1960). According to Czernay (1987) *M. americana* and *M. bricenii* are full valid species, proposition that was further echoed by other authors such as Bisbal (1991), Geist (1998), Weber and González (2003), Grubb (2005), Sánchez and Lew (2012), Molinari et al. (2015b), Lizcano and Alvarez (2016). In contrast, Eisenberg (1987, 1989) did not recognized *M. bricenii*, (Nowak 1999) only recognized four species for genus *Mazama* with *M. rufina* the one ranging the Venezuelan Andes, and Mattioli (2011) considered *M. bricenii* as subspecies of *M. rufina*.

More recently, Gutiérrez et al. (2015) found that the qualitative traits (depth of lacrimal fossa and shape of frontal bones) used to separate *M. bricenii* from *M. rufina* are extremely variable and, hence, of no taxonomic value. Moreover, these authors stated that *M. bricenii* is undifferentiable from *M. rufina* from morphologic and molecular point of view. On the other hand, their environmental niche models revealed the existence of suitable climatic conditions for *M. rufina* in the alleged biogeographical barrier, the Táchira depression, during both the last glacial maximum and at present, suggesting that gene flow between populations may continue nowadays. These authors conclude that *M. bricenii* should be regarded as synonym of *M. rufina*. Similarly, Heckeberg et al. (2016) found that *M. rufina* is a subclade of *M. bricenii*, and probably both taxa may represent the same species with *M. rufina* as a senior synonym, being more related to the *M. americana*-clade than to the *M. gouazoubira*-clade.

In Venezuela, *M. rufina* has been recorded in forested environments at El Tamá Macif, Sierra de Perijá, and Mérida Cordillera, between 1000 and 3500 m asl (Lizcano et al. 2010b). Apparently, is tolerant to habitat alteration since it can be observed in second growth thickets and shaded-coffee plantations far from primary forests (Soto-Werschitz, Bianchi-Pérez and Ramoni-Perazzi pers. obs.).

Venezuelan Andes are extensively protected by a network of protected areas (a list which includes the Sierra Nevada, Yacambú, Terepaima, El Tamá, Guaramacal, Dinira, General Juan Pablo Peñaloza, Chorro El Indio, Sierra La Culata, Tapo-Caparo, El Guache National Parks), covering a great portion of the range of this species. Moreover, brocket deer is actually considered an endangered species, and hunting is against the law, in Venezuela (República de Venezuela 1996a, b). However, the ongoing habitat modification, hunting pressure, and predation by dogs has depressed their populations (Bisbal 1991). Nevertheless, *M. rufina* is the most threatened species of the genus in Venezuela (Dietrich 1993), with their populations fragmented and diminished due to constant hunting and destruction of Andean forests (Linares 1998; Lizcano et al. 2010a, b). According to both IUCN and the Venezuelan Red Book of Threatened Species of Fauna, this deer is considered as Vulnerable (Lizcano and Alvarez 2016; Molinari et al. 2015b). For instance, some additional actions are required: (1) establishing corridors among the protected areas; and (2) consensus-building considering ecological, social, and political sustainability (Yerena et al. 2003).

10.2.2.2 *Mazama americana* (Erxleben, 1777)

According to Linares (1998) this is the largest species of the genus, attaining 1170–1590 mm of total length, and weighs 16–45 kg. Underparts white or cream and dorsum reddish brown, with the hind legs dark gray until black.

The taxonomy of *M. americana* is complex and still unsolved (Gutiérrez et al. 2017), given their extensive karyotype variation under a relatively unvarying morphotype (Abril et al. 2010), obscuring the definition about subspecies number and valid species delimitation. This deer is found in a wide variety of forested habitat (Eisenberg 1989; Linares 1998), perhaps reflecting its taxonomic complexity.

Eisenberg (1989) and Linares (1998) mention that *M. americana* is mainly crepuscular and nocturnal, contrasting with recent observations of Morán et al. (2018) who, based on a camera-trap sampling scheme in the Maracaibo's Lake basin, found that the species is both diurnal and nocturnal, but preferring full-light hours. Linares (1998) indicate that, in Venezuela, pregnant females have been reported in February and July, in agreement with the observations of Morán et al. (2018) who were able to photograph offspring in those same months, which also coincide with the drier periods of the Maracaibo's Lake basin, perhaps an adaptation to avoid risks due to heavy precipitations.

It is one of the species hunted for both subsistence and commercially (Fergusson-Laguna 2010). While in some regions are consumed only occasionally (García et al. 2012), in others *M. americana*, along with other vertebrates, is important integrating the estimated 62% of the protein needs in rural communities in Venezuela, and also a reservoir of some parasites (Cañizales and Guerrero 2010). Castellanos et al. (2010) reported intermediated abundances, lower than those of the peccaries, but greater than those of tapirs, which are some of the mammalian items preferred by the Ye'kwanas and Sanemas from Bolívar state, and also appreciated by the Baniva, Bare, Kurripaco, Warekena, and Yeral people of Amazonas State (Olivero et al. 2016), in Southern Venezuela. Major threats are hunting and habitat destruction.

Listed as Data Deficient by the IUCN due its high taxonomic uncertainty (Duarte and Vogliotti 2016), *M. americana* was included as Data Deficient in previous versions of the Venezuelan Red Book of Threatened Species of Fauna (Rodríguez and Rojas-Suárez 2008), but was not included in the newer editions of this book (Rodríguez and Rojas-Suárez 2015), omission which evidently must be corrected in this fundamental resource for education, research and conservation.

10.2.2.3 *Mazama nemorivaga* (G. Fischer, 1814)

The brown to grayish-brown color of this species, separates it from the other co-generics at a first glance in Venezuela. This is a medium sized representant of the genus that according to Linares (1998) has a total length of 970–1400 mm of total length, and weighs 11–18 kg, with females being slightly smaller than males.

Usually regarded as *M. gouazoubira* in the Venezuelan literature (i.e., Bisbal 1991; Linares 1998; Sánchez and Lew 2012). Gutiérrez et al. (2017) found that this is an isolated lineage which probably requires genus-level recognition.

Pairs of adults (coupling pairs?) were reported in August by Morán et al. (2018). Another of the species hunted both for subsistence and commercially (Fergusson-Laguna 2010), important as protein source for rural communities in Venezuela, and also a reservoir of some parasites (Cañizales and Guerrero 2010). Castellanos et al. (2010) reported intermediated abundances in South Venezuela, similar to those of *M. americana*, and interestingly Lew et al. (2009) pointed at this species as one of the characteristic elements of the savannas present at the Canaima National Park.

Included within the Least Concern category of the IUCN (Black-Decima and Vogliotti 2016; Rossi and Duarte 2016), in the Venezuelan Red Book of Threatened Species of Fauna has been considered in a similar way than *M. americana* (Rodríguez and Rojas-Suárez 2008, 2015).

10.2.2.4 *Odocoileus virginianus* (Zimmermann, 1780)

By far, the biggest deer in Venezuela, which exhibits a marked sexual dimorphism, with a total length ranging from 1260 to 2400 mm, and females weighing 30 kg and males 50 kg in average (Linares 1998), although size as well as coloration vary geo-

Fig. 10.4 Adult female and male of *Odocoileus virginianus lasiotis*. Laguna de Mucubají area, Mérida state. 11 Apr 2014. Photo credit Paolo Ramoni-Perazzi

graphically (Molinari 2007) (Fig. 10.4). Chromosome complements of $2n = 70$ (FN = 70; Aguilera et al. 2008, 2010). The species is found in open to semi-open environments throughout the country, from 0 to 3450 m asl, with two remarkably isolated populations, one in the Margarita Island and the other in the páramos of central portion of the Cordillera of Mérida (Eisenberg 1989; Linares 1998; Molinari 2007). However, a recent survey of the páramo population found this species ranging from 3243 to 4180 m asl Fernández (Unpublished) (Fig. 10.5). The term páramo refers to an Andean biogeographical province (Morrone 2014) of high elevation ecosystems located above the montane tree line and below the snowline in the mountains of northern Peru, Ecuador, Colombia and Venezuela (Luteyn 1999) (Fig. 10.6).

Molina and Molinari (1999) found that páramo specimens have distinctively shaped mandibles, while those from Margarita Island had smaller mandibles, smaller body sizes, and distinctive cranial and peculiar mandible traits. These authors considered that *O. lasiotis* and *O. margaritae*, from Andes and Margarita respectively, are separated full valid species, while the remnant Venezuelan populations must be regarded as *O. cariacou*, that is, separating them from the North American *O. virginianus* (Molinari 2007; Molinari et al. 2015a, c). Such classification was echoed by national researchers (i. e. Sánchez and Lew 2012). However, Moscarella et al. (2003), Aguilera et al. (2008) found no clear support for elevating any Venezuelan subspecies to the species rank, and the use of *O. virginianus* as the unique valid species prevails in the international specialized literature (i.e., Grubb 2005; Gallina et al. 2010; Ortega et al. 2011; Gallina and Lopez Arevalo 2016).

The bibliographic revision made by Ortega et al. (2011) demonstrates the scarcity of studies on the biology of this species in Venezuela, especially during the

Fig. 10.5 Panoramic view of a páramo in the Laguna de Mucubají area, Mérida State, Venezuela, at approximately 3600 m asl. Pines are allochthonous vegetation. 04 Jun 2015. Photo credit Paolo Ramoni-Perazzi

Fig. 10.6 Pellets of *Odocoileus virginianus lasiotis* in the Laguna de Mucubají area, Mérida State, Venezuela. 09 Aug 2016. Photo credit Paolo Ramoni-Perazzi

present century. In Southern Venezuelan Llanos, *O. virginianus* can be natural reservoirs of *Anaplasma marginale* and *Trypanosoma evansi*, which had been also detected in the capibaras (*Hydrochoerus hydrochaeris*) as well as cows and horses extensively produced in that region (Silva-Iturriza et al. 2013). In an ongoing pilot study, which evaluates the occupancy of the *O. v. lasiotis* in some páramos of the Venezuelan Andes, Fernández (Unpublished) found that the null model suggests an occupancy probability of 0.82 and a detection probability of 0.61, being positively correlated to the distance to park guard stations, negatively correlated to cow presence, and with deer being absent where the habitat was substituted by crops (Fig. 10.7). These values are similar to those reported for other American Cervidae in areas affected by poaching (Ferreguetti et al. 2015; Waldstein Parsons et al. 2017).

Fig. 10.7 Páramo vegetation and forest of *Polylepis sericea* Wedd., 1861 in the Pico Mucuñuque area. Mérida State, Venezuela. 04 Jun 2015. Photo credit Paolo Ramoni-Perazzi

Divergency among IUCN and national red lists can exist (Bender et al. 2012; Hidasi-Neto et al. 2013), and *O. virginianus* is an excellent example of such situation. Considered as of Least Concern at global level by the IUCN (Gallina and Lopez Arevalo 2016), the situation is radically different in Venezuela. There, Moscarella et al. (2003) found moderated levels of genetic diversity and significative genetic structuration of the Venezuelan subspecies. Such situation, in addition to the evident degree of geographic isolation, hunting pressures, predation by feral dogs and habitat loss of the populations from páramos and the Margarita Island, with the concomitant risk of genetic diversity loose indicate, undoubtedly, that these groups deserve special consideration and must be considered as Endangered, as indicated by Molinari et al. (2015a, c). This highlights the urgency of the implementation of conservation actions at local levels.

10.2.3 Tayassuidae Family

10.2.3.1 *Pecari tajacu* (Linnaeus, 1758)

The collared peccary, known in Venezuela as báquiro, ranges from southeastern USA in Arizona, New Mexico, and Texas, through Central America, until Northern Argentina in South America (Gongora et al. 2011). In Venezuela it is widely distributed, found in the South of the Orinoco River, the Deltaic System, Eastern mountain ranges, Central mountain ranges, east of the Coro System, and Eastern Region of the Andes, as well as in the Llanos (Linares 1998; Aguilera et al. 2010; Gongora et al. 2011).

Research on its biology and ecology in Venezuela is scarce, in the seventies, basically consisted of collections of specimens by the Smithsonian Institution and the Wildlife Service MARNR (Castellanos 1983). Little information has been generated in Venezuela during the last 10 years and it was focused on: use of habitat at some localities (García et al. 2012), short chromosome studies (Aguilera and Expósito 2010; Aguilera et al. 2010), zoonosis potentially associated to poaching (Cañizales and Guerrero 2010), consumption of wild fauna (Ferrer et al. 2012; Matallana et al. 2012).

In the Yurubí National Park, located at Sierra de Aroa, Yaracuy State, the peccaries use different types of vegetation such as evergreen forests (197 m), semi-deciduous forests (100–230 m), and cloud forests (1446 m; García et al. 2012). Aguilera and Expósito (2010) reported chromosome complement of $2n = 30$ (FN = 44), similar to those reported by other authors. They consist of seven pairs of metacentric chromosomes, a submetacentric pair and six acrocentric pairs. The sex chromosomes are acrocentric contrasting with other studies.

A social species, in Venezuela forms groups of 25–54 individuals, depending on the season or even hour of the day. A peak of births has been observed between August and September and between January and February, coinciding with the beginning of the rainy seasons (Castellanos 1983; Barreto et al. 1997).

P. tajacu is one of the most poached species in Venezuela, for subsistence by both rural and indigenous communities (Pérez and Ojasti 1996; Aguilera et al. 2010; Cañizales and Guerrero 2010; García et al. 2012; Matallana et al. 2012). For example, at El Caura Forest Reserve, Bolivar State in Southern Venezuela, the exorbitant annual hunting of 2399 individuals has been recorded (average weight 15 kg, 35,985 kg of biomass harvested; Ferrer et al. 2012). Pradera et al. (2007) studied some reproductive aspects of the species, in order to contribute to the establishment of commercial breeding sites, the recovery of their populations and the conservation of the species.

Globally, the species is considered at Least Concern, and being included in Appendix II of CITES out of Mexico and the USA (Gongora et al. 2011). This species has been excluded from Venezuelan Red Book of Threatened Species of Fauna (Rodríguez and Rojas-Suárez 2008, 2015).

10.2.3.2 *Tayassu pecari* (Link, 1795)

White-lipped peccary (*Tayassu pecari*) occurs from Mexico to Argentina (Altrichter et al. 2011; Keuroghlian et al. 2013), dwelling mainly in humid and deciduous undisturbed forests, savannas and dry forest (Mayer and Wetzel 1987). Very little is known about the biology and ecology of the species, especially in Venezuela, where the publications involving this species are very scarce: diet (Barreto et al. 1997), bio-ecological aspects (Barreto and Hernández 1988), mammal inventories (García et al. 2012), hunt and consume their meat (Matallana et al. 2012; García et al. 2012), transmissible diseases (Cañizales and Guerrero 2010; Ferrer et al. 2012), besides studies regarding the distribution of the species and mentioning Venezuela (Altrichter et al. 2011).

In the Yurubí National Park, Sierra de Aroa in Yaracuy State, the peccaries use different types of vegetation such as an evergreen forest (197 m), semi-deciduous forest ranging between (100–230 m), and cloud forest at (1446 m) (García et al. 2012). A social species which, during the dry season, seems to disappear from some areas (Barreto and Hernández 1988), probably because it performs migratory movements (Fragoso 2004). Some authors found major range declines in Argentina, Paraguay, southern Brazil, Colombia, Venezuela, north-east Brazil, Mexico, and Costa Rica (Altrichter et al. 2011). Taber et al. (2009) provide some estimates of the extent of the original and extirpated ranges of the species for the whole country and in several Venezuelan ecosystems. According these authors, the species has been extinct from 37% of its original range in Venezuela, being populations from the Llanos savannas the most threatened (extinct in 77% of the original range), and the Guayana's mountain forest the most preserved (0.03%). More actualized information could depict a worst panorama, considering the anthropic affection of the ecosystems in Venezuela already reported by Rodríguez et al. (2010), which could result in more population fragmentation.

T. pecari it is one of the most threatened mammals in the Neotropics, due essentially to the loss of its natural habitat and hunting pressure (Smythe 1987). Some authors recommend conservation efforts on landscape conservation of large, continuous and ecologically intact areas containing a mosaic of different habitat types (Altrichter et al. 2011). In Delta Amacuro, 28.9% of the specimen hunted are for self-consumption and the remaining 71% is for illegal trade, both nationally and internationally (Matallana et al. 2012), being the white-lipped peccary one of these. Fragoso (2004) mentions a population of white-lipped peccaries in the mountains of Parima, on the range that divides Venezuela (Orinoco basin) from Brazil, but is unknown how poaching has impacted this population.

Globally, the species is listed in the Appendix II of CITES and in the category of Least Concern category of the IUCN (Keuroghlian et al. 2013). This species is not included in the Venezuelan Red Book of Threatened Species of Fauna (Rodríguez and Rojas-Suárez 2008, 2015).

Acknowledgments We thank Luis Angel Niño Barreat and Alexis Araujo Quintero, for providing us with valuable bibliographic information. Marcelo Passamani critically read the manuscript and provided valuable comments.

References

Abril VV, Carnelossi EAG, González S, Duarte JMB (2010) Elucidating the evolution of the red brocket deer *Mazama americana* complex (Artiodactyla; Cervidae). Cytogenet Genome Res 128:177–187. https://doi.org/10.1159/000298819

Aguilera MM, Expósito Á (2010) Cariotipos de *Tapirus terrestris* (Perissodactyla, Tapiridae) y *Pecari tajacu* (Artiodactyla, Tayassuidae) presentes en Venezuela. Mem Fund La Salle Cien Nat 171:7–18

Aguilera MM, Azócar A, González EJ (2003) Venezuela: un país megadiverso. In: Aguilera MM, Azócar A, González EJ (eds) Biodiversidad en Venezuela, vol 2. Fundación Polar, Ministerio de Ciencia y Tecnología, Fondo Nacional de Ciencia, Tecnología e Innovación (FONACYT), Caracas, pp 1–20

Aguilera MM, Expósito A, La Rocca O (2008) Cytogenetics of two subspecies of white-tailed deer (*Odocoileus*) from Venezuela. Caryologia 61(1):19–25. https://doi.org/10.1080/00087114.2008.10589606

Aguilera MM, Expósito A, Caldera T (2010) Citogenética de mamíferos cinegéticos de Venezuela. In: Machado-Allison A, Hernández O, Aguilera M, Seijas AE, Rojas F (eds) Simposio Investigación y Manejo de Fauna Silvestre en Venezuela en homenaje al Dr. Juhani Ojasti, Finland Embassy, Academia de Ciencias Físicas, Matemáticas y Naturales, FUDECI, Instituto de Zoología y Ecología Tropical (IZET), UNELLEZ, USB, PROVITA, Fundación La Salle de Ciencias Naturales, PDVSA. Fundación Jardín Botánico de Caracas "Dr. Tobías Lasser", Caracas, pp 55–68

Allen GM (1915) Notes on American deer of the genus *Mazama*. Bull Am Mus Nat Hist 34:521–553

Altrichter M, Taber A, Beck H et al (2011) Range-wide declines for a key Neotropical ecosystem architect, the near-threatened white-lipped peccary *Tayassu pecari*. Oryx 46:87–98. https://doi.org/10.1017/S0030605311000421

Barreto GR, Hernández OE (1988) Aspectos bioecológicos de los báquiros (*Tayassu tajacu* and *T. pecari*) en el estado Cojedes: estudio comparativo. Dissertation, Universidad Central de Venezuela

Barreto GR, Hernández OE, Ojasti J (1997) Diet of peccaries (*Tayassu tajacu* and *T. pecari*) in a dry forest of Venezuela. J Zool 241:279–284. https://doi.org/10.1111/j.1469-7998.1997.tb01958.x

Bender MG, Floeter SR, Ferreira CEL et al (2012) Mismatches between global, national and local red lists and their consequences for Brazilian reef fish conservation. Endanger Species Res 18:247–254. https://doi.org/10.3354/esr00449

Bisbal FJ (1991) Distribución y taxonomía del venado Matacán (*Mazama* sp) en Venezuela. Acta Biol Venez 13:89–104

Bisbal FJ, Naveda-Rodríguez A (2010) Mamíferos de la cuenca del río Guárico, estados Aragua, Carabobo y Guárico, Venezuela. Mem Fund La Salle Cien Nat 172:69–89

Black-Decima PA, Vogliotti A (2016) *Mazama gouazoubira*. In: The IUCN red list of threatened species 2016: e.T29620A22154584. https://doi.org/10.2305/IUCN.UK.2016-2.RLTS.T29620A22154584.en. Accessed 12 Dec 2018

Cabrera A (1960) Catálogo de los mamíferos de América del Sur, vol 2. Rev Mus Argent Cienc Nat 4:309–732

Cañizales I, Guerrero R (2010) Parásitos y otras enfermedades transmisibles de la fauna cinegética en Venezuela. In: Machado-Allison A, Hernández O, Aguilera M, Seijas AE, Rojas F (eds) Simposio Investigación y manejo de fauna silvestre en Venezuela en homenaje al Dr. Juhani Ojasti, Finland embassy, academia de Ciencias Físicas, Matemáticas y Naturales, FUDECI, Instituto de Zoología y Ecología Tropical (IZET), UNELLEZ, USB, PROVITA, Fundación La Salle de Ciencias Naturales, PDVSA. Fundación Jardín Botánico de Caracas "Dr. Tobías Lasser, Caracas, pp 97–108

Castellanos HG (1983) Aspectos de la organización social del báquiro de collar, *Tayassu tajacu* L., en el Estado Guárico Venezuela. Acta Biol Venez 11(4):127–143

Castellanos H, Bertsch C, Veit A et al (2010) Cosecha de fauna silvestre y acuática por comunidades Ye'kwana y Sanema del alto río Caura. In: Machado-Allison A, Hernández O, Aguilera M, Seijas AE, Rojas F (eds) Simposio investigación y manejo de fauna silvestre en Venezuela en homenaje al Dr. Juhani Ojasti, Finland Embassy, Academia de Ciencias Físicas, Matemáticas y Naturales, FUDECI, Instituto de Zoología y Ecología tropical (IZET), UNELLEZ, USB, PROVITA, Fundación La Salle de Ciencias Naturales, PDVSA. Fundación Jardín Botánico de Caracas "Dr. Tobías Lasser", Caracas, pp 133–148

Cozzuol MA, Clozato CL, Holanda EC et al (2013) A new species of tapir from the Amazon. J Mammal 94(6):1331–1345. https://doi.org/10.1644/12-MAMM-A-169.1

Czernay S (1987) Spiesshirsche und Pudus: die Gattungen *Mazama* und *Pudu*. Ziemsen Verlag, Wittenberg

de Thoisy B, da Silva AG, Ruiz-García M et al (2010) Population history, phylogeography, and conservation genetics of the last Neotropical mega-herbivore, the lowland tapir (*Tapirus terrestris*). BMC Evol Biol 10:278. https://doi.org/10.1186/1471-2148-10-278

Dietrich JR (1993) Biology of the brocket deer (genus *Mazama*) in Northern Venezuela. Dissertation, University of Basel

Duarte JMB, Vogliotti A (2016) *Mazama americana*. In: The IUCN red list of threatened species 2016: e.T29619A22154827. https://doi.org/10.2305/IUCN.UK.2016-1.RLTS.T29619A22154827.en. Accessed 12 Dec 2018

Eisenberg JF (1987) The evolutionary history of the Cervidae with special reference to the South American radiation. In: Wemmer C (ed) Biology and management of the Cervidae. Smithsonian Institution Press, Washington, pp 60–64

Eisenberg JF (1989) Mammals of the Neotropics: the northern Neotropics, vol 1. University of Chicago Press, Chicago

Fergusson-Laguna A (2010) El aprovechamiento sustentable de la diversidad biológica en Venezuela. In: Machado-Allison A, Hernández O, Aguilera M, Seijas AE, Rojas F (eds) Simposio Investigación y manejo de fauna silvestre en Venezuela en homenaje al Dr. Juhani Ojasti, Finland Embassy, Academia de Ciencias Físicas, Matemáticas y Naturales, FUDECI, Instituto de Zoología y Ecología Tropical (IZET), UNELLEZ, USB, PROVITA, Fundación La Salle de Ciencias Naturales, PDVSA. Fundación Jardín Botánico de Caracas "Dr. Tobías Lasser", Caracas, pp 185–204

Fernández DC (Unpublished) Lineamientos estratégicos para la conservación del venado del páramo (*Odocoileus virginianus lasiotis*) Dissertation, Universidad de Los Andes
Ferreguetti ÁC, Tomás WM et al (2015) Density, occupancy, and activity pattern of two sympatric deer (*Mazama*) in the Atlantic Forest, Brazil. J Mammal 96(6):1245–1254. https://doi.org/10.1093/jmammal/gyv132
Ferrer A, Romero V, Lew D (2012) Consumo de fauna silvestre en el eje agrícola Guarataro, Reserva Forestal El Caura, Estado Bolívar, Venezuela. Mem Fund La Salle Cien Nat 173–174:239–251
Fragoso JMV (2004) A long-term study of white-lipped peccary (*Tayassu pecari*) population fluctuations in northern Amazonia anthropogenic versus "natural" causes. In: Silvius KM, Bodmer RE, Fragoso JMV (eds) People in nature: wildlife conservation in South and Central America. Columbia University Press, New York, pp 286–296
Gallina S, Lopez Arevalo H (2016) *Odocoileus virginianus*. In: The IUCN red list of threatened species 2016: e.T42394A22162580. https://doi.org/10.2305/IUCN.UK.2016-2.RLTS.T42394A22162580.en. Accessed 12 Dec 2018
Gallina S, Mandujano S, Bello J et al (2010) White-tailed deer *Odocoileus virginianus* (Zimmermann 1780). In: Duarte JMB, González S (eds) Neotropical cervidology: biology and medicine of Latin American deer. IUCN, Jaboticabal, Brazil: Funep and Gland, Switzerland, pp 101–118
García JF, Jaramillo MD, Machado M, Aular L (2012) Preliminary inventory of mammals from Yurubí National Park, Yaracuy, Venezuela with some comments on their natural history. Rev Biol Trop 60(1):459–472
Geist V (1998) Deer of the world: their evolution, behaviour, and ecology. Stackpole Books, Mechanicsburg
Gómez RMS, Morales BAA, Garcia F et al (2010) Síndrome erosivo ulceroso gástrico asociado a estrés en una danta (*Tapirus terrestres*). Analecta Vet 30(2):41–44
Gondeles AR, Medina GP, Méndez JLA et al (1981) Nuestros animales de caza, guía para su conservación. Fundación de Educación Ambiental, Caracas
Gongora J, Reyna-Hurtado R, Beck H et al (2011) *Pecari tajacu*. In: The IUCN red list of threatened species 2011: e.T41777A10562361. https://doi.org/10.2305/IUCN.UK.2011-2.RLTS.T41777A10562361.en. Accessed 12 Dec 2018
Grubb P (2005) Order artiodactyla. In: Wilson DE, Reeder DM (eds) Mammal species of the world: a taxonomic and geographic reference, 3rd edn. Johns Hopkins University Press, Baltimore, pp 655–657
Gutiérrez EE, Maldonado JE, Radosavljevic A et al (2015) The taxonomic status of *Mazama bricenii* and the significance of the Táchira depression for mammalian endemism in the Cordillera de Mérida, Venezuela. PLoS One 10(6):e0129113. https://doi.org/10.1371/journal.pone.0129113
Gutiérrez EE, Helgen KM, McDonough MM et al (2017) A gene-tree test of the traditional taxonomy of American deer: the importance of voucher specimens, geographic data, and dense sampling. ZooKeys 697:87–131. https://doi.org/10.3897/zookeys.697.15124
Heckeberg NS, Erpenbeck D, Wörheide G, Rössner GE (2016) Systematic relationships of five newly sequenced cervid species. PeerJ 2016(4):e2307. https://doi.org/10.7717/peerj.2307
Hidasi-Neto J, Loyola RD, Cianciaruso MV (2013) Conservation actions based on red lists do not capture the functional and phylogenetic diversity of birds in Brazil. PLoS One 8(9):e73431. https://doi.org/10.1371/journal.pone.0073431
Holanda EC, Rincón AD (2012) Tapirs from the Pleistocene of Venezuela. Acta Palaeontol Pol 57(3):463–473
Isasi-Catalá E (2013) Modelos de uso de hábitat de *Tapirus terrestris* en un área montañosa de la Cordillera de La Costa, Venezuela. In: Tirira DG (ed) 1er Congreso Latinoamericano de Tapires and 2do Congreso Ecuatoriano de Mastozoología, Universidad Estatal Amazónica, Asociación Ecuatoriana de Mastozoología. Grupo de Especialistas de Tapires del Ecuador, Puyo, pp 8–11

Keuroghlian A, Desbiez A, Reyna-Hurtado R et al (2013) *Tayassu pecari*. In: The IUCN red list of threatened species 2013: e.T41778A44051115. https://doi.org/10.2305/IUCN.UK.2013-1.RLTS.T41778A44051115.en. Accessed 15 Dec 2018

Lew D, Rivas BA, Rojas H et al (2009) Mamíferos del Parque Nacional Canaima. In: Senaris JC, Lew D, Lasso C (eds) Biodiversidad del Parque Nacional Canaima: bases técnicas para la conservación de la Guayana Venezolana. Fundación La Salle de Ciencias Naturales and The Nature Conservancy, Caracas, pp 153–179

Linares OJ (1998) Mamíferos de Venezuela. Sociedad Conservacionista Audubon, Caracas

Linares OJ, Rivas B (2004) Mamíferos del sistema Deltaico (Delta del Orinoco–Golfo de Paria), Venezuela. Mem Fund La Salle Cien Nat 159–160:27–104

Lizcano DJ, Alvarez SJ (2016) *Mazama bricenii*. In: The IUCN red list of threatened species 2016: e.T136301A22165039. https://doi.org/10.2305/IUCN.UK.2016-2.RLTS.T136301A22165039.en. Accessed 12 Dec 2018

Lizcano DJ, Álvarez SJ, Delgado VC (2010a) Dwarf red brocket *Mazama rufina* (Pucheran 1951). In: Duarte JMB, González S (eds) Neotropical cervidology: biology and medicine of Latin American deer. IUCN, Jaboticabal, Brazil: Funep and Gland, Switzerland, pp 177–180

Lizcano DJ, Edgard YDJ, Silva J et al (2010b) Mérida bocket *Mazama bricenii* (Thomas 1908). In: Duarte JMB, González S (eds) Neotropical cervidology: biology and medicine of Latin American deer. IUCN, Jaboticabal, Brazil: Funep and Gland, Switzerland, pp 181–184

Lizcano DJ, Amanzo J, Castellanos A et al (2016) *Tapirus pinchaque*. In: The IUCN red list of threatened species 2016: e.T21473A45173922. https://doi.org/10.2305/IUCN.UK.2016-1.RLTS.T21473A45173922.en. Accessed 12 Dec 2018

Luteyn J (1999) Páramos: a checklist of plant diversity, geographic distribution, and botanical literature. New York Botanical Garden Press, New York

Marín-Wikander S (2012) Densidad poblacional y uso de hábitat de la danta de tierras bajas (*Tapirus terrestris*) en el Parque Nacional Guatopo. Dissertation, Universidad Simón Bolívar

Matallana C, Lasso CS, Baptiste MP (eds) (2012) Carne de monte y consumo de fauna silvestre en la Orinoquia y Amazonia (Colombia y Venezuela). Instituto de Investigaciones de Recursos Biológicos Alexander von Humboldt. Universidad Nacional de Colombia, Bogotá

Mattioli S (2011) Family Cervidae (deer). In: Wilson DE, Mittermeier RA (eds) Handbook of the mammals of the world, vol 2. Lynx Edicions, Barcelona

Mayer JJ, Wetzel RM (1987) Tayassu pecari. Mamm Species 293:1–7

Molina M, Molinari J (1999) Taxonomy of Venezuelan white-tailed deer (*Odocoileus*, Cervidae, Mammalia), based on cranial and mandibular traits. Can J Zool 77:632–645

Molinari J (2007) Variación geográfica en los venados de cola blanca (Cervidae: *Odocoileus*) de Venezuela, con énfasis en *O. margaritae*, la especie enana de la Isla de Margarita. Mem Fund La Salle Cien Nat 167:29–72

Molinari J et al (2015a) Venado margariteño, *Odocoileus margaritae*. In: Rodríguez JP, García-Rawlins A, Rojas-Suárez F (eds) Libro Rojo de la Fauna Venezolana, 4th edn. Provita, Fundación Empresas Polar, Caracas. http://animalesamenazados.provita.org.ve/content/venado-margariteno. Accessed 12 Dec 2018

Molinari J, Lew D, Sánchez-Hernández J (2015b) Venado matacán candelillo, *Mazama bricenii*. In: Rodríguez JP, García-Rawlins A, Rojas-Suárez F (eds) Libro Rojo de la Fauna Venezolana, 4th edn. Provita, Fundación Empresas Polar, Caracas. http://animalesamenazados.provita.org.ve/content/venado-matacan-candelillo. Accessed 13 Dec 2018

Molinari J, Lew D, Sánchez-Hernández J (2015c) Venado paramero, *Odocoileus lasiotis*. In: Rodríguez JP, García-Rawlins A, Rojas-Suárez F (eds) Libro Rojo de la Fauna Venezolana, 4th edn. Provita, Fundación Empresas Polar, Caracas. http://animalesamenazados.provita.org.ve/content/venado-paramero. Accessed 11 Dec 2018

Morán L, García L, Ferrebuz JD et al (2018) Interannual and daily activity patterns of mid-sized mammals in Maracaibo Lake Basin, Venezuela. Therya 9(3):227–236

Morrone JJ (2014) Biogeographical regionalisation of the Neotropical region. Zootaxa 3782:1–110. https://doi.org/10.11646/zootaxa.3782.1.1

Moscarella RA, Aguilera M, Ananias A et al (2003) Phylogeography, population structure, and implications for conservation of white-tailed deer (*Odocoileus virginianus*) in Venezuela. J Mammal 84:1300–1315

Naveda-Rodríguez A, López A (2006) Etnozoologia de la danta (*Tapirus terrestris*) en Venezuela. Tapir conservation, newsletter of the IUCN/SSC tapir specialist. Group 15(1):36–38

Naveda-Rodríguez A, Torres DA (2002) Situación actual y registro genealógico de las dantas o tapires en los zoológicos de Venezuela. Fundación Andígen A, EarthMatters.org. IUCN/SSC Tapir Specialist Group, Houston Zoo

Naveda-Rodríguez A, de Thoisy B, Richard-Hansen C, et al (2008) *Tapirus terrestris*. In: The IUCN red list of threatened species 2016: e.T21474A9285933. https://doi.org/10.2305/IUCN.UK.2008.RLTS.T21474A9285933.en. Accessed 12 Dec 2018

Naveda-Rodríguez A, Bermúdez PA, Bisbal F (2012) Abundancia de *Tapirus terrestris* (Perissodactyla, Tapiridae) en la Cordillera de la Costa Central. Venezuela Anartia 24:74–82

Nowak RM (ed) (1999) Walker's mammals of the world, 6th edn. Johns Hopkins University Press, Baltimore

Ochoa GJ, Aguilera M (2003) Mamíferos. In: Aguilera M, Azocar A, González Jiménez EE (eds) Biodiversidad en Venezuela. Fundación Polar-FONACIT, Caracas, pp 651–672

Ojasti J, Lacabana P (2008) Danta de montaña, *Tapirus pinchaque*. In: Rodríguez JP, Rojas-Suárez F (eds) Libro Rojo de la Fauna Venezolana, 3rd edn. Provita, Shell Venezuela SA, Caracas, p 104

Olivero J, Ferri F, Acevedo P et al (2016) Using indigenous knowledge to link hyper-temporal land cover mapping with land use in the Venezuelan Amazon: "The Forest Pulse". Rev Biol Trop 64(4):1661–1682. https://doi.org/10.15517/rbt.v64i4.21886

Ortega SJÁ, Mandujano S, Villarreal J et al (2011) Managing white-tailed deer: Latin America. In: Hewitt DG (ed) Biology and management of white-tailed deer. Taylor Francis Press, Boca Raton, pp 565–597

Passos CJL, Fragoso JMV, Crawshaw D et al (2016) Lowland tapir distribution and habitat loss in South America. PeerJ 4:e2456. https://doi.org/10.7717/peerj.2456

Pérez EM, Ojasti J (1996) Utilización de la fauna silvestre en América Latina: situación y recomendaciones para su manejo sustentable en las sabanas. Ecotropicos 9(2):71–82

Pradera JD, Ruiz AZ, González M et al (2007) Vascularization of the uterus and ovaries of the collared peccary. Rev Cientif FCV-LUZ 48(1):1–10

República de Venezuela (1996a) Decreto 1485: Animales vedados para la caza. Gaceta Oficial No. 36.059-7 de octubre de 1996. Caracas

República de Venezuela (1996b) Decreto 1486: Lista de especies en peligro de extinción en Venezuela. Gaceta Oficial No. 36.062, Caracas

Rodríguez JP, Rojas-Suárez F (eds) (2008) Libro Rojo de la fauna venezolana, 3rd edn. Provita, Shell Venezuela SA, Caracas

Rodríguez JP, Rojas-Suárez F (eds) (2015) Libro Rojo de la Fauna Venezolana, 4th edn. Provita, Shell Venezuela SA, Caracas. http://animalesamenazados.provita.org.ve/search/node/Mazama. Accessed 13 Dec 2018

Rodríguez JP, Rojas-Suárez F, Hernández GD (eds) (2010) Libro Rojo de los ecosistemas terrestres de Venezuela. Provita, Shell Venezuela SA, Lenovo Venezuela, Caracas

Rossi RV, Duarte JMB (2016) *Mazama nemorivaga*. In: The IUCN red list of threatened species 2016: e.T136708A22158407. https://doi.org/10.2305/IUCN.UK.2016-1.RLTS.T136708A22158407.en. Accessed 12 Dec 2018

Salas L (1996) Habitat use by lowland tapirs (*Tapirus terrestris* L.) in the Tabaro River Valley, Southern Venezuela. Can J Zool 74:1452–1458

Salas L, Fuller T (1996) Diet of the lowland tapir (*Tapirus terrestris* L.) in the Tabaro river valley, Southern Venezuela. Can J Zool 74:1444–1451

Sánchez J, Lew D (2012) Lista actualizada y comentada de los mamíferos de Venezuela. Mem Fund La Salle Cien Nat 173–174:173–238

Silva-Iturriza A, Panier E, Reyna-Bello A et al (2013) Evaluación parasitológica de infecciones hemotrópicas en venados cola blanca (*Odocoileus virginianus*) en Venezuela. Rev Cientif FCV-LUZ 23(1):37–41

Smythe N (1987) The importance of mammals in neotropical forest management. In: Figueroa JC (ed) Management of the forests of tropical America: prospects and technologies. USDA Forest Service. Universidad de Puerto Rico, Rio Piedras, pp 79–98

Taber A, Chalakian SC, Altrichter M et al (2009) El destino de los arquitectos de los bosques neotropicales: evaluación de la distribución y el estado de conservación de los pecaríes labiados y los tapires de tierras bajas. Wildlife Conservation Society, Wildlife Trust

Torres D, Isasi-Catalá E, Marín Wikander S (2015) Danta, *Tapirus terrestris*. In: Rodríguez JP, García-Rawlins A, Rojas-Suárez F (eds) Libro Rojo de la fauna venezolana, 4th edn. Provita, Fundación Empresas Polar, Caracas. http://animalesamenazados.provita.org.ve/content/danta. Accessed 05 Dec 2018

Waldstein Parsons A, Forrester T, McShea WJ et al (2017) Do occupancy or detection rates from camera traps reflect deer density? J Mammal 98(6):1547–1557. https://doi.org/10.1093/jmammal/gyx128

Weber M, González S (2003) Latin American deer diversity and conservation: a review of status and distribution. Ecoscience 10(4):443–454. https://doi.org/10.1080/11956860.2003.11682792

Yerena E, Padron J, Vera R et al (2003) Building consensus on biological corridors in the Venezuela Andes. Mt Res Dev 23:215–218. https://doi.org/10.1659/0276-4741(2003)023[0215:BCOBCI]2.0.CO;2

Chapter 11
Tropical Ungulates of Ecuador: An Update of the State of Knowledge

Diego G. Tirira, Carlos A. Urgilés-Verdugo, Andrés Tapia, Carlos A. Cajas-Bermeo, Xiomara Izurieta V., and Galo Zapata-Ríos

Abstract In Ecuador, there are 12 recognized species and 16 taxa of neotropical ungulates. The tapirs (Tapiridae) include two species: *Tapirus pinchaque* and *T. terrestris*, present in highlands and humid tropical and subtropical forest in the Amazon, respectively; traditionally, a third species of tapir was added to the Ecuadorian fauna, *Tapirus bairdii*; however, there are no concrete records and its presence is not supported by any form of evidence. Within the peccaries (Tayassuidae), two species are recognized: *Pecari tajacu* and *Tayassu pecari*; both are sympatric in most of their range, and niche overlap is high. The camelids (Camelidae) have three species, two domesticated and with historical presence in the country for more than 500 years: *Lama glama* and *Vicugna pacos*; and *Vicugna vicugna*, a species that was considered extinct in Ecuador and initially treated as reintroduced; however, there is no evidence to confirm the historical presence of this species in Ecuador. The situation of deer (Cervidae) is more complex. This family includes three genera and five species, but with several unresolved taxonomic problems. The genus *Mazama* is widely distributed in Ecuador, and three species have been known to occur in Ecuador: *M. americana*, *M. nemorivaga*, and *M. rufina*. The taxonomic knowledge of the *Mazama americana* sensu lato in Ecuador is still incomplete, and a detailed revision is still pending, especially for the specimens

D. G. Tirira (✉)
Fundación Mamíferos y Conservación, Quito, Ecuador
e-mail: info@murcielagoblanco.com

C. A. Urgilés-Verdugo
Instituto para la Conservación y Capacitación Ambiental, Quito, Ecuador

A. Tapia
Centro Ecológico Zanja Arajuno, Puyo, Pastaza, Ecuador

C. A. Cajas-Bermeo
Facultad de Recursos Naturales, Escuela Superior Politécnica de Chimborazo, Riobamba, Ecuador

X. Izurieta V.
Corporación Ecopar, Quito, Ecuador

G. Zapata-Ríos
Wildlife Conservation Society–Ecuador, Quito, Ecuador

S. Gallina-Tessaro (ed.), *Ecology and Conservation of Tropical Ungulates in Latin America*, https://doi.org/10.1007/978-3-030-28868-6_11

west of the Andes. The taxonomic status of *Odocoileus virginianus* and its subspecies is also incomplete and needs a detailed review. In Ecuador, the two recognized subspecies (*O. v. ustus* and *O. v. peruvianus*) have been treated as valid species by some authors. Finally, *Hippocamelus antisensis* was a species reported in Ecuador since 1851, and with a series of records in the country in the following decades. It was excluded from Ecuadorian fauna in the absence of supporting evidence.

11.1 Introduction

In Ecuador, there are 12 recognized species of neotropical ungulates, 9 native species, 2 domesticated, and a species considered "reintroduced." Ungulates in Ecuador belong to two orders and four families (Table 11.1). For several taxa, there are still gaps in knowledge about the number of species, and a series of taxonomic problems that have not been solved.

Currently, two species of odd-toed ungulates (Perissodactyla) occur in Ecuador: *Tapirus pinchaque* and *T. terrestris* (family Tapiridae, tapirs), present in Andean forests and páramo of the Andean mountain range, and humid tropical and subtropical forest in the Amazon, respectively (Tirira 2017). Traditionally, and for more than 50 years, a third species of tapir was included as part of the Ecuadorian fauna, the Central American tapir (*Tapirus bairdii*) (Albuja 1983, 1991; Tirira 1999, 2007, 2008). However, there is no current concrete evidence of its presence in Ecuador (Tirira et al. 2011a; Tirira 2017). The first record of this species appeared in Hershkovitz (1954); and since then, it was considered that any additional tapir records from coastal Ecuador (some of them anecdotic) corresponded to this species (Patzelt 1978; Albuja 1983; Suárez and García 1986; Tirira and Castellanos 2001). However, the presence of the Central American tapir in Ecuador is not supported by archaeological or paleontological evidence (Hoffstétter 1952; D. G. Tirira, unpublished data), nor by the modeling of its potential distribution (Schank et al. 2017). Confirmed records of tapirs in Guayas and Azuay provinces are now being attributed to the Andean tapir (*Tapirus pinchaque*) (TSG 2011; D. G. Tirira, unpublished data).

The situation of even-toed ungulates (Artiodactyla) is somewhat more complex. Within the peccaries (Tayassuidae), two species are recognized in Ecuador: the collared peccary (*Pecari tajacu*) and the white-lipped peccary (*Tayassu pecari*). Both species are sympatric in most of their range, and niche overlap is high; however, each species respond to different levels of spatial organization in their habitats due to differences in body size, morphology (e.g., bite force), and behavior (Kiltie 1982; Fragoso 1999; Desbiez et al. 2009). Groves and Grubb (2011) recognize morphological differences among collared peccary populations across their distribution, which at their discretion justifies the recognition of three species, two of which would be present in Ecuador: *Pecari crassus* (western population) and *Pecari tajacu* (eastern population).

Table 11.1 Species of Ungulates in Ecuador (updated from Wilson and Mittermeier 2011; Tirira 2017)

Species and subspecies	Inhabits	Altitudinal range	Status[a]
Order Perissodactyla			
Family Tapiridae			
Tapirus pinchaque (Roulin, 1829)	Subtropical, temperate, and high-Andean forests; páramo	1200–4700 m	Native. Threatened (critically endangered)
Tapirus terrestris (Linnaeus, 1758) *T. t. aenigmaticus* (Gray, 1872)	Eastern tropical and subtropical forests	200–1500 m	Native. Threatened (endangered)
Order Artiodactyla			
Family Tayassuidae			
Pecari tajacu (Linnaeus, 1758)			Native. Near threatened
P. t. niger (J. A. Allen, 1913)	Western tropical, subtropical, and temperate forests	0–2255 m	
P. t. tajacu (Linnaeus, 1758)	Eastern tropical and subtropical forests	200–1600 m	
Tayassu pecari (Link, 1795)			
T. p. aequatore (Lönnberg, 1921)	Western tropical and subtropical forests	0–1600 m	Native. Threatened (critically endangered)
T. p. pecari (Link, 1795)	Eastern tropical and subtropical forests	200–1800 m	Native. Threatened (endangered)
Family Camelidae			
Lama glama (Linnaeus, 1758)	All the country, preferably in the Andes	0–4700 m	Native. Domesticated
Vicugna pacos (Linnaeus, 1758)	Andes	2000–4700 m	Native. Domesticated
Vicugna vicugna (Molida, 1782) *V. v. mensalis* (Thomas, 1917)	Central Andes	3300–4700 m	Introduced. Protected (Appendix II of CITES)
Family Cervidae			
Mazama americana (Erxleben, 1777)			Native. Near threatened
M. a. gualea J. A. Allen, 1915	Western tropical, subtropical, and temperate forests	0–2196 m	
M. a. zamora J. A. Allen, 1915	Eastern tropical and subtropical forests	200–1900 m	
Mazama nemorivaga (F. Cuvier, 1817)	Eastern tropical forests	200–900 m	Native. Near threatened

(continued)

Table 11.1 (continued)

Species and subspecies	Inhabits	Altitudinal range	Status[a]
Mazama rufina (Pucheran, 1851)	Temperate and high-Andean forests; páramo	1900–3600 m	Native. Threatened (vulnerable)
Odocoileus virginianus Zimmermann, 1780			
O. v. peruvianus (Gray, 1875)	Southwestern dry tropical, subtropical, temperate forests	0–2400 m	Native. Threatened (endangered)
O. v. ustus (Trouessart, 1910)	Páramo	3000–4600 m	Native. Near threatened
Pudu mephistophiles (de Winton, 1896)	Temperate and high-Andean forests, páramo	2196–4000 m	Native. Threatened (vulnerable)

[a]Status according to Tirira (2011a, 2017)

The camelids (Camelidae) have three species in Ecuador, two domesticated and with historical presence in the country for more than 500 years (Cieza de León 1553; de Velasco 1841; Estrella 1990; Ortiz 1999; White 2001a): the llama (*Lama glama*) and the alpaca (*Vicugna pacos*). The populations of these two species of camelids decreased considerably during the sixteenth and twentieth centuries due initially to indiscriminate killing, and later to poor management and little interest in their breeding (Almeida 2014). The government of Ecuador undertook in the 1980s the project "Promotion of South American camelids in Ecuador" (MAE 2013) that has allowed to recover the populations of these two species. The third additional species is the vicuña (*Vicugna vicugna*), a species that was considered extinct in Ecuador and that was initially treated as reintroduced (Gallo 1981; Tirira 1999, 2007; MAE 2013); however, there is no paleontological or archaeological evidence to confirm the historical presence of the vicuña in Ecuador (Tirira 2017), while the written evidence (Cieza de León 1553; de Velasco 1841) is considered unreliable (Ortiz 1999; White 2001a).

Cervidae (deer) includes three genera and five species in Ecuador, but with several unresolved taxonomic problems. The brocket deer (genus *Mazama*) are widely distributed in Ecuador, on both sides of the Andes, occurring from low elevations to 3600 m in altitude (Tirira 2017; Palacios et al. 2017). Traditionally, three species of brocket deer have been known to occur in Ecuador (Grubb 2005; Tirira 2007; Duarte and González 2010): red brocket (*Mazama americana*), Amazonia brown brocket (*Mazama nemorivaga*), and little red brocket (*Mazama rufina*). Tirira (2017), following Groves and Grubb (2011), indicates that four species of brocket deer inhabit Ecuador: *Mazama gualea* and *Mazama zamora*, both threated as subspecies or junior synonyms of *M. americana* by Grubb (2005) and Varela et al. (2010); *Mazama murelia*, treated as a synonym of *M. nemorivaga*; and *M. rufina*.

The taxonomic knowledge of the red brocket deer (*Mazama americana* sensu lato) in Ecuador is still incomplete, and a detailed revision is pending, especially for the specimens west of the Andes, a population that has even been omitted from the distribution maps of the species in several review publications (e.g., Varela et al. 2010; Mattioli 2011). Photographic records and comments from local inhabitants indicate that two forms of *Mazama* live in the tropical and subtropical forests of western Ecuador, one is dark chestnut brown, and found mostly in humid forests; and another, yellowish red brown found in dry forests (Tirira 2017). In addition, unconfirmed records from the upper part of the Chogón-Colonche mountain range, in the provinces of Santa Elena and Manabí, mention the two forms live in sympatry (Albuja and Muñoz 2000; D. G. Tirira, unpublished data).

The taxonomic status of white-tailed deer (*Odocoileus virginianus*) and its subspecies is also incomplete and needs a detailed review (Mattioli 2011). In Ecuador, the two recognized subspecies (*O. v. ustus* and *O. v. peruvianus*) (Gallina et al. 2010; Tirira 2011a) have been treated as valid species by Tirira (2017), based on Molina and Molinari (1999) and Molinari (2007) criteria.

Finally, the North Andean huemul (*Hippocamelus antisensis*), a species of deer reported in Ecuador since 1851, and with a series of records in the country in the following decades, was excluded of Ecuadorian fauna in the absence of supporting evidence in the form of specimens or convincing literature reports (current, archaeological, and paleontological), and ecological niche models of current and past distribution that conclude that the species never inhabited Ecuador (Pinto et al. 2016). Huemul records in Ecuador surely represent a case of misidentification of white tailed-deer (Tirira 2017).

11.2 Andean Tapir (*Tapirus pinchaque*)

Carlos A. Urgilés-Verdugo
Diego G. Tirira

11.2.1 Distribution

At present, the distribution of the Andean tapir (*Tapirus pinchaque*) (Figs. 11.1 and 11.2) is restricted mainly to hard-to-access areas of the Eastern Cordillera of the Andes, with recent records in the provinces of Carchi, Sucumbíos, Imbabura, Pichincha, Napo, Tungurahua, Pastaza, Chimborazo, Morona Santiago, Loja, and Zamora Chinchipe (Arcos 2009; Bermúdez-Loor and Reyes-Puig 2011; Castellanos 2013; Narváez 2013; Zapata-Ríos and Branch 2016). In addition, there are few historical records that confirm that the Andean tapir inhabited some places in the Western Cordillera, in the provinces of Cañar and Azuay (Schauenberg 1969; Tapia et al. 2011a), and still surviving in the foothills of the Azuay province (F. Sánchez-Karste, pers. comm., 2017). Recently, an unexpected record in the highlands of El Oro province (Brito et al. 2018) extends its distribution to the southwestern Ecuador;

Fig. 11.1 *Tapirus pinchaque*. Credit: Luis Gómez. Location: Cuyuja, Cayambe-Coca National Park, Napo province, Ecuador. Note: Infant

Fig. 11.2 *Tapirus pinchaque*. Credit: Patricio Pillajo. Location: Papallacta, Cayambe-Coca National Park, Napo province, Ecuador. Note: Adult

however, all populations in western Ecuador are considered relict and probably will not survive in the future. Historical records in the Western Cordillera in the provinces of Carchi, Imbabura and Chimborazo have been not confirmed (Downer 1997). The altitudinal range of the Andean tapir in Ecuador is 1200–4700 m, although it is recorded more frequently between 2300 and 4300 m (Downer 1996; Tirira 2017). The model of environmental suitability estimated an extent of occurrence for species of 21,729 km^2 in all of Ecuador, mainly occurring along the corridor of the eastern Ecuadorian Andes (Ortega-Andrade et al. 2015). Records in Cordillera del Cóndor (Mittermeier et al. 1975; Downer 1996, 1997; Tirira 2007; Padilla et al. 2010; Tapia et al. 2011a) are considered to be erroneous (Tirira 2017).

11.2.2 Natural History

Diet. The Andean tapir feeds on a wide variety of herbs, branches, and leaves of shrubs and trees, as well as fruits and berries (Downer 1996, 2001; Bermúdez-Loor and Reyes-Puig 2011; Urgilés-Verdugo and Gallo 2012; Yépez and Urgiles-Verdugo 2013). It prefers shoots and young leaves (Bermúdez-Loor and Reyes-Puig 2011; Gallo 2012). Studies carried out in Cayambe-Coca, Llanganates and Sangay national parks, and the Antisana Ecological Reserve, determined that the Andean tapir feeds on at least 322 botanical species, belonging to 71 families, among them the most representative were Solanaceae (26 species), Poaceae (23), Rosaceae (15), Melastomataceae (15), and Urticaceae (12) (Castellanos 1994; Downer 1996, 2001; Bermúdez-Loor and Reyes-Puig 2011; Urgilés-Verdugo and Gallo 2012). It was also observed that it can eat certain fungi, mosses and hepatic plants (Downer 1996).

Sexual Composition. In the Sangay National Park a proportion of 2.5:1 females have been estimated with respect to males (Downer 1996); in the Cayambe-Coca National Park the proportion was 0.75:1 females with respect to males (Castellanos 2013).

Reproduction. The only reproductive information on Andean tapir in Ecuador was provided by Schauenberg (1969), who documented that a gravid female captured in Llanganates in April gave a birth to a single baby in August, with a weight, at that moment, of 6.2 kg.

Activity Patterns. This species is diurnal and nocturnal. In a study in páramo ecosystem, in the Sangay National Park, indicates that Andean tapir forage from 15:00 to 21:00 h, with activity diminishing by midnight. They rest between midnight and dawn; then there are some activity till 09:00 h, when the animals graze along rivers or o soak up the sun on open slopes (Downer 1995, 1996). In another study, in mountain forest, in the northeast of Sangay National Park, it was observed that the highest activity of the Andean tapir occurs between 12:00 and 18:00 h and between 18:00 and 24:00 h; then it has a low activity between 24:00 and 06:00 h; while between 06:00 and 12:00 it practically has no activity (Santacruz 2012). In this same locality was determined that the Andean tapir is more active on full moon nights (42%), followed by the waning moon nights (24%); nights of new (18%) and crescent moons (15%) have the lowest activity (Santacruz 2012).

Behavior. Although the Andean tapir is a shy species that avoids human presence (Downer 1997), there are reports that it has bitten people when they invaded its territory or have tried to capture or manipulate it (Schauenberg 1969; Downer 1996; Castellanos 2015; C. A. Urgilés-Verdugo, pers. comm.).

11.2.3 Population Dynamics

Home Range. In the Sangay National Park it has been estimated that the Andean tapir has a home range of 8.8 km^2, estimated average value based on three individuals (from 7.75 to 10.2 km^2), two males and one female (Downer 1996). In the Cayambe-Coca National Park, the home range has an important variation among individuals, from 0.62 to 9.16 km^2, with an average area of 3.97 km^2, according to data from five individuals, three males and two females (Castellanos 2013). It has been seen that the home range of adult females is somewhat larger and is often associated with the territory of adult males (Downer 1996, 2001). Four of five Andean tapirs that were monitored in the Cayambe-Coca National Park showed overlaps between their home ranges area, which shows intraspecific tolerance (Castellanos 2013). In the Sangay National Park was determined that adult males maintain territories adjacent to each other, overlapping by as much as a third, particularly along riverine meadows; while adult females are loosely tied to the male's territory (Downer 1995).

Density Estimates. During the 1960s and 1970s, the Andean tapir was considered a common species in many areas of its historical distribution in Ecuador (Stummer 1971). Studies that have determined its population density in Ecuador have used different sampling methodologies and form of data analysis. The first approach was presented for Sangay National Park, with an estimate of 0.19 individuals/km^2 (Stummer 1971); another estimate for this same protected area was 0.17 individuals/km^2 (Downer 1996). There are three studies for the Cayambe-Coca National Park, with values of 0.04 individuals/km^2 (Zapata-Ríos and Branch 2010), 0.29 individuals/km^2 (Castellanos 2013), and 0.12 individuals/km^2 (Urgilés-Verdugo et al. 2018). There are two estimates for the Antisana Ecological Reserve: 0.12 individuals/km^2 (Gallo 2012) and 0.24 individuals/km^2 (Urgilés-Verdugo et al. 2018). On average for Ecuador, a density of 0.17 individuals/km^2 is estimated.

Population Structure. The Andean tapir is a solitary species (Downer 1995); however, pairs of two females or two males, or family groups of an adult male and female and their offspring have been observed (Stummer 1971; Downer 1996; C. A. Urgilés-Verdugo, pers. obs.).

There are intraspecific interactions between males of the Andean tapir, related to competition for territory or reproduction, when males can fight with each other (Schauenberg 1969; Downer 1995, 1996). Interactions between females and males, and between females and offspring were also observed (Downer 1996). It was observed that the infant accompanies the mother until the first year of life. The mother can abandon the young for short periods of time (1–3 days) while searching for food; the reencounter between mother and baby is given by means of a system of acute whistles between both (Downer 1996; C. A. Urgilés-Verdugo, pers. obs.).

Mortality Factors. Natural predators of the Andean tapir are Andean bear (*Tremactos ornatus*), and puma (*Puma concolor*), and probably jaguar (*Panthera onca*) in subtropical forests (Schauenberg 1969; Stummer 1971; Downer 1996). In the páramo it has been determined that the presence of puma, and to a lesser degree of Andean bear, decreases the population density of the Andean tapir (Gallo 2012; Urgilés-Verdugo et al. 2018). Another negative interaction exists with certain introduced animals, such as cattle and feral dogs (Downer 2003; Zapata-Ríos and Branch 2016). The presence of cattle contributes to the increase of bears and pumas, which has also been related to an increase in the mortality of young tapirs (Downer 1996, 2003). On the other hand, the presence of feral dogs negatively affects the abundance of the Andean tapir and alters its activity patterns; however, it does not affect the use of habitat (Zapata-Ríos and Branch 2016).

11.2.4 Habitat Requirements

The Andean tapir occurs in páramo, high-Andean, temperate, and subtropical forests. It occupies different habitats, in order of preference: Andean forest, 29%; riverside prairies (vegetation along rivers and other bodies of water), 23%; scrubland (between forest and páramo), 22%; páramo, 20%; and pampas (modified prairies for fires and crops), 6% (Downer 1996, 1997). A recent study has determined that the Andean tapir occupies ten different Andean ecosystems, with herbaceous páramo, montane evergreen forest, and upper montane evergreen forest being the most representative (Ortega-Andrade et al. 2015). In addition, Andean tapir is a frequent visitor of mineral licks, with several studies in Llanganates and Sangay National Park (Downer 1999; Reyes-Puig et al. 2007; Tapia A. et al. 2008a; Santacruz 2012).

In studies of habitat use in relation to the availability of habitats, it has been determined that Andean tapir prefers high montane evergreen forest and herbaceous páramo to montane cloud forest, shrub páramo, and cushion páramo (Gallo 2012; Zapata-Ríos 2013). This differentiated use of habitat preference is due to several factors, such as seasonal immigration, availability of resources, introduction of exotic species, and anthropogenic disturbance (Stummer 1971; Downer 1996; Gallo 2012; Narváez 2013; Zapata-Ríos 2013; Urgilés-Verdugo et al. 2012, 2018).

There is evidence that in certain areas, such as the Cayambe-Coca National Park, the Andean tapir prefers to use the páramo to other nearby habitats (Castellanos 2013); however, this preferential use of the habitat is suspected to be due to the availability of food: during the dry season, it often frequents páramo; but during the rainy season more often forest (Downer 1996).

11.2.5 Ecological Importance

The Andean tapir acts as a key species in ecological terms for its extensive herbivorous and frugivorous diet, which makes it an effective seed disperser (Downer 2001; Urgilés-Verdugo et al. 2018), it can disperse seeds of more than 50 plant species (Downer 1996). In projects of germination of seeds present in tapir feces collected in Llanganates and Sangay national parks, it was recorded that between 41% and 42% of the ingested species germinated, corresponding to 131 and 86 species, respectively (Downer 1996, 1997). This condition makes the Andean tapir play an important role in the dynamics and restoration of cloud and montane forests and in the páramos of Ecuador (Downer 1996). The selection in their feeding of nitrogen-fixing plants, such as *Lupinus* spp. and *Gunnera brephogea*, suggests that there is a mutualism between the Andean tapir and the flora of the Ecuadorian Andes. The highly significant correlations between the preference quotient and the frequency of germination of seeds in fecal samples of species corroborate this (Downer 2001).

11.2.6 Threats and Conservation

The Andean tapir is one of the most critically endangered large mammals in the world (Medici 2011). In Ecuador it is classified as Critically Endangered, according to the *Red Book Mammals of Ecuador* (Tirira 2011a). It considers that the species has disappeared from many areas in which it lived until a few decades ago (Stummer 1971; Downer 1996). Most of its current distribution is within protected areas or in places of difficult access, with the best populations possibly in the central Andes of the country (within the national parks Llanganates and Sangay) (Tapia et al. 2011a); while the most threatened populations would be those of the provinces of the center-north (Chimborazo, Tungurahua, Cotopaxi, Pichincha, Carchi) and south of the country (Azuay and Loja) (Ortega-Andrade et al. 2015); here the relict population found in the province of El Oro should be included (Brito et al. 2018).

The loss and fragmentation of the habitat are considered the main threats that the species faces in Ecuador (Tapia et al. 2011a). These impacts are linked to the intolerance of the species to disturbed environments and to its shyness in the face of human presence (Downer 1997). Hunting is a threat that has diminished in the last decades, a frequent activity in the past, whether for the consumption of its meat, the use of its fat or of some part of its body (like the hooves) (Downer 1997; Tirira 2007; Tapia et al. 2011a). Indirect threats are the extractive activities that have appeared within its distribution area, such as mining, or productive activities, such as the burning of páramo and the overgrazing of sheep and cattle (Tapia et al. 2011a). It is estimated that in Ecuador the effects of climate change and habitat loss, under current conditions of land use, could reduce the distribution of Andean tapir by an additional 20–45% (Ortega-Andrade et al. 2015).

Currently, there is the “National Strategy for the Conservation of Tapirs in Ecuador” (TSG 2011). The objectives of this strategy are the conservation of core areas, the restoration of ecosystems, providing information on hunting, the development of training and environmental education plans, increasing research on species, updating ex situ conservation programs, and strengthening interinstitutional cooperation.

11.3 Lowland Tapir (*Tapirus terrestris*)

Diego G. Tirira
Andrés Tapia

11.3.1 Distribution

The lowland tapir (*Tapirus terrestris*) (Figs. 11.3 and 11.4) inhabits Ecuador to the east of the Andes, between 200 and 1500 m of altitude, although it is usually recorded at less than 700 m (Tirira 2017); it is also present in the Cordillera del Cóndor, in Morona Santiago and Zamora Chinchipe provinces (Tirira 2017). It is estimated that 51% of the current lowland tapir distribution in Ecuador is within

Fig. 11.3 *Tapirus terrestris*. Credit: Aldo Fernando Sornoza. Location: Yasuní National Park, Orellana province, Ecuador. Note: Adult

Fig. 11.4 *Tapirus terrestris*. Credit: Diego G. Tirira. Location: Yasuní National Park, Orellana province, Ecuador. Note: Infant

protected areas by the State, mainly the Yasuní National Park and the Cuyabeno Wildlife Reserve; while the remaining 49% corresponds to indigenous territories and non-protected areas within the provinces of Pastaza, Morona Santiago and Zamora Chinchipe (Correa and Torres 2005; TSG 2011). The subspecies presents in Ecuador is *T. t. aenigmaticus* (Medici 2011; Tirira 2017).

11.3.2 Natural History

Diet. Studies in Ecuador indicate that the lowland tapir consumes a wide variety of plant species, including herbs, leaves, branches, buds, seeds, and fallen fruits (Tapia 1998; Tapia and Arias 2011). Records in semi-captivity, within secondary forest, document the consumption of more than 100 plant species belonging to more than 20 families, mostly species from Araceae, Asteraceae, and Melastomataceae (Tapia 1998, 1999; Tapia and Machoa 2006). The food is taken mainly from the understory, but it can also consume aquatic vegetation on the banks of rivers and lakes (Tapia 1998). It feeds by browsing leaves and branches with the help of the proboscis of the muzzle and without causing greater damage to the plant species (Tapia 1998). Lowland tapir takes fruits directly from the soil that mainly correspond to the families Annonaceae, Arecaceae, Malvaceae, Moraceae, Myrtaceae and Sapotaceae (Tapia and Arias 2011).

This species ingests certain plants, such as *Miconia* spp. (the leaves) (Melastomataceae), *Piptocoma discolor* (stem) (Asteraceae), and *Xanthosoma violaceum* (leaves) (Araceae), whose ingestion allows it to counteract bacteria or parasites and prevent diseases (Tapia 1998, 1999).

Mineral salts are a fundamental component of the diet of the lowland tapir, which is why it often visits natural salt licks (Sandoval-Cañas 2004; Tapia and Arias 2011); salts contribute to the biochemical compensation of this mammal which, due to its monogastric character, has a permanent energy deficit that needs to be compensated with food and mineral salts (Naranjo et al. 2014).

Sexual Composition. No information is available.

Reproduction. In conditions of semi-captivity, the female reaches sexual maturity between 2 and 3 years of age, and the male at 3 (Tapia 1998, 1999). Reproduction occurs at 3 years and copulation can occur in any month of the year (Tapia 1998). The estrus cycle of the female lasts approximately 15 days (Tapia 1998) and can be repeated every 2 months (López and Merino 1994); during this period, the female shows changes in behavior, such as restlessness, emission of squeals louder than normal, nibbles the ears and feet of her fellows and runs in circles with increasing rapidity; the estrous cycle is accompanied by vaginal secretions of yellowish-white color (Tapia 1998).

The copulation is carried out on solid ground, from one to two times (Tapia 1998) and the gestation time is 390–400 days and a single calf is born (López and Merino 1994; Tapia 1998). The birth usually occurs in water sources or nearby (Medici 2011; A. Tapia, pers. obs.); but in semi-captivity a birth occurred on ground, despite the existence of nearby water sources (Tapia 1998).

Breastfeeding occurs exclusively during the night and takes from 5 to 6 months, after which the offspring is weaned and separated from the mother (López and Merino 1994; Tapia 1998). The female can give birth every 18 months during her adult stage (López and Merino 1994).

In the wild, little information is available. It has been seen that the birth of the young coincide with the rainy season and the copulation is synchronized with the abundance of fruits fallen in the forest (Tapia 1998).

Activity Patterns. A study in the Yasuní National Park determined that the lowland tapir is mainly a nocturnal animal, with two peaks of higher activity, one between 20:00 and 22:00 h, and the other between 03:00 and 04:00 h (Durango-Cordero 2011; Espinosa and Salvador 2017).

Behavior. In semi-captive conditions, the lowland tapir has shown to have a territorial behavior, mainly during the breeding season, when two or more males have been observed fighting to mate with a fertile female (Tapia 1999). During the gestation and lactation periods, the females can be uncouth and become aggressive in the presence of their congeners or other mammals (including humans). This aggressive behavior has also been reported in wild animals (Sirén 2004; Tapia and Machoa 2006). In the Pastaza province has shown lowland tapir females to kill other females when competing for territory in semi-captive conditions (A. Tapia, pers. obs.).

Communication between members of the species is through high-pitched whistles, like singing a bird. The calf also communicates with the mother through mild whistling (Tapia 1998).

The lowland tapir always defecates in the water or on the banks, for which it sometimes must travel long stretches in search of streams where to defecate (Bianchi 1988; López and Merino 1994; Tapia 1998).

11.3.3 Population Dynamics

Home Range. It has been estimated for the Yasuní National Park a home range of 1.02–13.6 km^2, with an average of 3–4 km^2 (Sandoval-Cañas 2004). Another estimate to Sarayaku (Pastaza province) reports a home range of 25 km^2 (Sirén 2004).

Density Estimates. There are several estimates of the population density of the lowland tapir in Ecuador. In Sucumbíos province, in Cuyabeno Wildlife Reserve, it was estimated 1.6 individual/km^2 (Prieto 2007). In Pastaza province a density of 0.4 individuals/km^2 was estimated in Sarayaku (Tapia 2010a, 2013), and 0.5 individuals/km^2 in Victoria and Curaray, on the Curaray River basin (Durango-Cordero 2011; Tapia 2011). In Morona Santiago province, in the eastern foothills of the Kutukú mountain range, a population density of 0.87 individuals/km^2 was estimated (Zapata-Ríos et al. 2009).

Population Structure. The lowland tapir is a solitary animal, but couples can be observed (mainly mothers with offspring, or male and female adults during the intercourse period); in addition, up to four tapirs have been observed in salt licks (Sandoval-Cañas 2004).

Mortality Factors. The main natural predator of the lowland tapir in Ecuador is the jaguar (*Panthera onca*), as well as boas, anacondas, and alligators; however, the highest mortality occurs in infants and juveniles; whereas, in adults there are not greater natural threats (Tapia 1999; TSG 2011).

The main cause of lowland tapir mortality in Ecuador is fragmentation of habitat and hunting. Several studies indicate that this species represents between 4% and 35% of the total biomass captured by indigenous communities (Freire 1997; Mena-Valenzuela et al. 1997, 2000; Zapata-Ríos 2001; Sirén 2004; Zapata-Ríos et al. 2009; Tirira 2018).

In semi-captivity the death of individuals due to foot-and-mouth disease or fighting between males has been reported (Tapia 1999).

11.3.4 Habitat Requirements

In Ecuador, the lowland tapir inhabits in tropical humid and subtropical low forests (Tirira 2017), mainly primary, in terra firme, seasonal or permanent flood (marshes) forests, in flat or hill areas, in gallery forests, and river borders (Tirira 2007, 2017; TSG 2011). In studies conducted in the central Amazon it has been determined that it prefers the forests of the terra firme (34%) and hillsides (34%), followed by flooded forests (19%) and hill forests (8%) (Tapia and Machoa 2006; Tapia 2010a, 2011, 2013). It is tolerant to disturbed habitats in areas with little human pressure, where it can be found even near abandoned orchards or pastures (TSG 2011).

11.3.5 Ecological Importance

The lowland tapir fulfills important ecological functions by acting as a seed disperser, in enriching the soil, and recycling nutrients (Fragoso et al. 2003). Studies in Ecuador indicate that the lowland tapir consumes a wide variety and quantity of fruits, among them *Mauritia flexuosa* and *Oenocarpus bataua*, which implies its importance as seed disperser at short distances (Borgtoft and Balslev 1993; Tapia and Arias 2011; Ojeda 2016). Other ecological functions are the creation of trails that lead to sources of mineral salts and that are used by other animals (Tapia 1998; TSG 2011), and the contribution of feces to the coprofauna and the disintegrating organisms where it participates in the enrichment of the soil and contributes important amounts of food substrate for 13 species of coprophagous beetles (Tapia 2005).

11.3.6 Genetics

There are some genetic studies conducted in Ecuador or that have mainly used Ecuadorian material, with the following results:

Based on molecular markers, it was determined that the separation between the Central American tapir (*T. bairdii*) and the two South American species would have occurred 19–20 million years ago; while the Andean tapir (*T. pinchaque*) may have diverged from the lowland tapir about 3 million years ago (Tapia 2010b; de Thoisy et al. 2010).

Cytogenetic studies have determined that the lowland tapir has a karyotype with 80 chromosomes, composed of a pair of metacentric chromosomes, seven pairs of acrocentric/subtelocentric chromosomes and 32 pairs of acrocentric chromosomes; sex chromosomes are metacentric in females and acrocentric in males (Tapia 2007).

Studies have also been carried out on the cryopreservation of semen with a view to the conservation of germplasm (Almeida and Guevara 2018).

11.3.7 Threats and Conservation

The lowland tapir is a protected species in Ecuador, currently categorized as Endangered according to the *Red Book of Mammals of Ecuador* (Tirira 2011a) because it is estimated that its populations have experienced a reduction of more than 50% (Tapia et al. 2011b); previous evaluations in Ecuador with IUCN criteria indicate that the species was treated as Least Concern (Tirira 1999) and Near Threatened (Tirira 2001a).The main direct threats identified in Ecuador are the loss and fragmentation of habitat and hunting.

The species has disappeared from areas in which it was previously common to observe (Tirira 2001b; Tapia et al. 2011b). At present, its presence is discontinuous in its range of distribution due to the colonization and expansion of the agricultural frontier, especially near towns and cities located between 500 and 1200 m altitude (Tapia M et al. 2008b; TSG 2011). The best populations below 500 m are towards the border with Peru; in the north, within the Cuyabeno Wildlife Reserve; in the center, south of the Napo River, within the Yasuní National Park and around the hydrographic basins of the Pastaza and Tigre rivers; and in the south, in the Morona and Santiago basins (Tapia et al. 2011b; TSG 2011). Above 1000 m altitude, forests with continuity of habitat to maintain viable populations persist; among them, the buffer zone of the eastern part of the Llanganates National Park (Tapia M et al. 2008b; Arias-Gutiérrez et al. 2016).

In some areas, especially in the northern Amazon, the destruction of native forests to make way for development projects, particularly oil extraction, and the consequent colonization, is critical and could cause the loss of genetic flow in tapir populations, which increases their vulnerability to extinction (Tirira 2011a; Tapia et al. 2011b).

There are also analyzes that indicate that the hunting of large mammals, including the lowland tapir, is not sustainable within indigenous territories (Zapata-Ríos 2001, 2002; Zapata-Ríos and Jorgenson 2003; Zapata-Ríos et al. 2009). It has been determined that the area of influence of hunting in the area of the Yasuní National Park covers an approximate area of 486 km^2, mainly along the Pompeya Sur-Iro-Ginta highway (Jorgenson and Coppolillo 2001), but also includes the road to Auca, the banks of the Napo, Tiputini, and Yasuní rivers, and in the Nuevo Rocafuerte sector (TSG 2011).

Analysis of population viability and habitat availability in several localities of the Ecuadorian Amazon (Desbiez et al. 2007), show that in areas with hunting pressure and without control of this activity, lowland tapir populations decline and disappear after several decades; unlike areas that exercise strategies to control hunting and habitat care, where populations remain stable and manage to recover in the medium and long term, with a population viability that can reach even hundreds of years (Machoa et al. 2007). Among these areas is the community of Sarayaku, in the province of Pastaza, which has implemented restriction and hunting models to cause a minor impact and ensure the survival of the lowland tapir (Sirén 2004; Sirén et al. 2004). In fact, during the last assembly of the Sarayaku people (December 2018) in Pastaza province, hunting prohibition for lowland tapir was ratified once more in a territory of approximately 150.000 ha. Something similar has been implemented in Shuar communities in the province of Morona Santiago, where despite having implemented a series of management strategies, such as restriction and zoning of the territory, tapir populations have not been able to recover (Zapata-Ríos et al. 2009).

It has been documented that there is a trade in bushmeat, which includes the lowland tapir, in several Amazonian locations (Zapata-Ríos 2001, 2002; Zapata-Ríos and Jorgenson 2003; WCS 2007; Zapata-Ríos et al. 2009; TSG 2011). Commercial hunting has benefited from the fuel subsidy that indigenous communities have within the oil areas (Suárez et al. 2009). This scenario of hunting for commercial purposes has decreased in the last decade thanks to controls implemented by the Ministry of the Environment (Tirira 2018), although it is not ruled out that it still exists.

Given the biological characteristics of the species (low reproduction rate, long gestation period, and high mortality during the first year of life) (Brooks et al. 1997) and the main threats indicated, the survival of the lowland tapir in Ecuador is not assured while no primary forests are conserved that allow the genetic flow, and the hunting is controlled (Zapata-Ríos 2001; Tapia et al. 2011b).

Climate change is also considered to be a major threat to this species in Ecuador under current conditions of land use, since it is estimated that up to 81% of its natural habitat can be lost by 2080 (Iturralde-Pólit 2010).

Currently, there is the "National Strategy for the Conservation of Tapirs in Ecuador" (TSG 2011). The objectives of this strategy are the conservation of core areas, the restoration of ecosystems, providing information on hunting, the development of training and environmental education plans, increasing research on species, updating ex situ conservation programs, and strengthening interinstitutional cooperation. Since 2009, May 9 has been declared the National Tapir Day in Ecuador.

11.4 Peccaries (*Pecari tajacu* and *Tayassu pecari*)

Galo Zapata-Ríos
Diego G. Tirira

11.4.1 Distribution

Both peccary species are widely distributed in Ecuador, showing a disjunct distribution on both sides of the Andes (Tirira 2017). The collared peccary (*Pecari tajacu*) (Fig. 11.5) inhabits humid and dry forest, tropical and subtropical, from sea level to 1600 m in altitude (Tirira 2017); and is rarely found at higher altitude. The highest records in Ecuador include Otonga Protected Forest, northwestern Cotopaxi province, at 2200 m (Jarrín 2001); and Yunguilla, northwestern Pichincha province, at 2255 m (Urgilés-Verdugo et al. 2019), both localities in the western slopes of the

Fig. 11.5 *Pecari tajacu*. Credit: Diego G. Tirira. Location: Parque Historico de Guayaquil, Guayas province, Ecuador. Note: An adult female and its infant in captivity

Andes. Two subspecies are recognized in Ecuador: *P. t. niger* (western populations) and *P. t. tajacu* (eastern populations) (Taber et al. 2011; Tirira 2017).
Ref. "Urgilés-Verdugo et al. 2019" is cited in text but not provided in the reference list. Please provide details in the list or delete the citation from the text. This reference is the same as 2018 (2019). The publication year was 2019. I corrected in the references

The white-lipped peccary (*Tayassu pecari*) (Fig. 11.6) inhabits lowlands and adjacent slopes on both sides of the Andes, in humid and dry forest, tropical and subtropical, from sea level to 1600 m (Tirira 2017). There is an exceptional record at a higher altitude, in Sangay National Park, in Morona Santiago province, reaching 1800 m (Brito and Ojala-Barbour 2016). Two subspecies are recognized in Ecuador: *T. p. aequatore* (western) and *T. p. pecari* (eastern) (Taber et al. 2011; Tirira 2017); however, Ruiz-García et al. (2015), based on genetic data, did not recognized subspecies for white-lipped peccaries.

11.4.2 Natural History

Diet. There is limited information about the diet of collared peccary in Ecuador, while there is no information for while-lipped peccary. Peccaries are frugivores feeding on fallen fruit, both soft and hard; and palm nuts form an important part of

Fig. 11.6 *Tayassu pecari*. Credit: Archivo Ecuambiente Consulting Group. Location: Yasuní National Park, Orellana province, Ecuador. Note: group in a mineral lick

their diets in the Amazon, mainly *Astrocaryum chambira* and *Mauritia flexuosa* (Tapia 1998; Tirira 2017). However, peccaries also dig in the ground for fungi, roots and bulbs (Robinson and Eisenberg 1985). Especially in the case of the white-lipped peccary, it complements its diet with leaves, insects, earthworms, and small vertebrates (Mayer and Wetzel 1987; Barreto et al. 1997; Tapia 1998). Information for collared peccary comes mainly from captivity and semi-captivity experiences, with the report of 23 botanical species eaten that belong to the families Araceae, Arecaceae, Convolvulaceae, Euphorbiaceae, Fabaceae, Malvaceae, Moraceae, Musaceae, Piperaceae, Poaceae, and Urticaceae (Tapia 1998). Also, there is a report of a collared peccary feeding on a newborn capybara (*Hydrochoerus hydrochaeris*) (Tapia 1998).

Sexual Composition. Our understanding of sex ratios in wild populations of Neotropical mammals is still very limited, and it is usually assumed to be 1:1 (Bodmer and Robinson 2004). From hunting data in the Ecuadorian and Peruvian Amazon, it is known white-lipped and collared peccary sex ratios are very close to 1:1, and have an average of 0.9 and 1.1 gestations per year respectively (Zapata-Ríos et al. 2009; Mayor et al. 2009, 2010).

Reproduction. Reproduction can occur year-round on both sides of the Andes, but it is more frequent in response to local changes in food availability, February–March in the Amazon, and September–October in the Chocó in northwestern Ecuador (Gottdenker and Bodmer 1998; Zapata-Ríos and Toasa 2004; Mayor et al. 2007; Zapata-Ríos et al. 2009).

In semi-captivity conditions, females reach reproductive age between 8 and 9 months, while males reach the reproductive age at 12 months (Da Silva et al. 2016). Mating usually occur with the beginning of the rainy season, when a female can mate with several males (Tapia 1998). The dominant male or males will prevent other subordinate males from copulating (Tapia 1998). A short period of courtship has been observed before copulation (Tapia 1998). Gestation takes about 5 months and they give birth to one or two offspring per birth, with an average weight of 0.60 kg (Tapia 1998). The period of lactation is 8 weeks, although the young eat solid food at 3 or 4 weeks after birth. After 2–3 months a new period of estrus occurs (Tapia 1998).

Activity Patterns. Peccary activity is mainly diurnal with activity peaks at dawn and dusk, and a significant decrease around noon (Durango-Cordero 2011), but may vary significantly due to environmental factors, like precipitation and temperature, and in areas where hunting is pervasive. In addition, no consistent relationship between activity and moon phases has been reported (Blake et al. 2012; Espinosa and Salvador 2017).

Behavior. There is documented information on behavior only for collared peccary in captivity and semi-captivity conditions (Tapia 1998). During the food ingestion events, it is aggressive with other members of the species and in front of other species, sounding its fangs and teeth (Tapia 1998). When the animals are frightened, they curl the bristles on their back, and to defend themselves against a dominant animal the offended individuals kneel and lie down, thus avoiding being bitten (Tapia 1998). The females jealously guard their territory (dominant female) of the intromission of females of other groups, not observing the same behavior in case the intruder is a male; in such a circumstance it will be the dominant male who defends his territory (Tapia 1998).

11.4.3 Population Dynamics

Home Range. No information is available.

Density Estimates. Information about peccary abundance and carrying capacity for different habitats is very limited. There are a few abundance estimates for the Amazon and Chocó lowlands. In the Amazon rainforests, population density estimates for white-lipped peccaries vary between 9.25 and 41.98 individuals/km^2 in areas without hunting, or where hunting levels are very low (Durango-Cordero 2011; Suárez et al. 2013; WCS unpublished information); while 1.2 individuals/km^2 has been estimated in areas with hunting (Durango-Cordero 2011). Meanwhile, in the Chocó lowlands (western Ecuador), white-lipped peccary population density has been estimated at 0.23 individuals/100 km^2 (Zapata-Ríos and Araguillin 2013).

There are no population density estimates for collared peccaries in western Ecuador; however, for the Amazon lowlands, estimates of population density vary significantly between 0.75 and 36.25 individuals/km^2, depending on hunting levels (Zapata-Ríos et al. 2006, 2009; Durango-Cordero 2011; Suárez et al. 2013; WCS unpublished information). In a year-long mammal survey in northern Esmeraldas (Awá territory), collared peccaries were rarely encountered (0.04 records/km), in an area where overhunting had been severe during several decades (Zapata-Ríos and Toasa 2004). However, in a survey in 2016 the south of Esmeraldas province, the collared peccary was common in a protected area (Mache Chindul Ecological Reserve), while it was rare in unprotected areas (Urgilés-Verdugo et al. 2019).

Population Structure. In Ecuador, collared peccary forms small groups up to 20 individuals, and usually no more than 9 (Tirira 2017). In Yasuní National Park, average group size is 2.9 individuals in areas without hunting, and 1.7 individuals per group in hunting areas (Durango-Cordero 2011). White-lipped peccary forms large groups in Ecuador, up to 500 individuals (Tirira 2017). In Yasuní National Park, average groups size has been estimated at 55.9 individuals per group in areas without hunting; and 1.5 individuals per group in areas with hunting (Durango-Cordero 2011).

Mortality Factors. There is no information about natural predators. However, across the country, peccary species (especially white-lipped peccary) have suffered significant mortality due to subsistence and commercial hunting (Zapata-Ríos 2001; Suárez et al. 2009; Tirira 2011a; Suárez et al. 2013; Zapata-Ríos and Araguillin 2013).

11.4.4 Habitat Requirements

The white-lipped and collared peccaries inhabit a wide range of habitats (Tirira 2017). The collared peccary, in the Pacific Coast, occurs in a range of habitats from pluvial to dry forests (Zapata-Ríos and Araguillin 2013). In the Ecuadorian Amazon, occurs in a variety of habitats, but prefers *terra firme* forests, and it can be found in both primary and secondary forest (Tirira 2017).

In the Pacific Coast, white-lipped peccary occurs only in northwestern Ecuador, mostly in uplands in pluvial forest in Esmeraldas province (Zapata-Ríos and Araguillin 2013); however, historical evidence suggests that this species also inhabited dry forests, in Cerro Blanco Protected Forest, in Guayas province (Parker III and Carr 1992); and Portovelo, in El Oro province (Catalogue of the American Museum of Natural History). In the Ecuadorian Amazon, white-lipped peccary occurs in a variety of habitats but prefers palm tree (*Mauritia flexuosa*) swamps (Fragoso 1999). It can be found in both primary and secondary forest (Tirira 2017).

Both species are frequently found in salt licks that provide them with much needed minerals (Tapia 1998; Tirira 2017), that are important for their reproduction and complement their diet (Gittleman and Thompson 1988). In addition, is common to find both species in clayey soils or "bathing" in swampy areas (Tapia 1998; Tirira 2017).

11.4.5 Threats and Conservation

Currently, the white-lipped peccary is considered threatened on the *Red Book of Mammals of Ecuador* (Tirira 2011a); while the collared peccary is classified as Near Threatened (Zapata-Ríos and Tirira 2011a). Although Ruiz-García et al. (2015) did not recognize subspecies for white-lipped peccaries, for conservation planning and threat categorization purposes, in Ecuador the species has been divided into two subspecies with different assessments: *T. pecari aequatoris* for the populations that occur in western Ecuador (Critically Endangered), and *T. pecari pecari* for the populations that occur in the Amazon (Endangered) (Zapata-Ríos and Tirira 2011b, c; MAE, WCS 2018).

Across the country, and their whole distribution range, peccary species (especially white-lipped peccaries) have suffered significant population declines due to habitat loss and fragmentation, subsistence and commercial hunting, and transmission of infectious diseases (Zapata-Ríos 2001; Suárez et al. 2009; Tirira 2011a; Suárez et al. 2013; Zapata-Ríos and Araguillin 2013). Habitat loss and fragmentation represent severe threats to peccary's long-term persistence. Habitat loss and fragmentation not only restrict habitat availability but weakens the viability of isolated populations (Taber et al. 2008; Altrichter et al. 2011). White-lipped peccaries use of large-scale landscape-level mosaics of different vegetation types, while collared peccaries use single large-scale vegetation types. This difference explains in part the higher susceptibility of white-lipped peccaries to habitat loss and fragmentation (Fragoso 1999). Forest cover maps for the 1990–2008 period provide deforestation rates for western Ecuador and the Ecuadorian Amazon. The derived annual deforestation rates (expressed as the percentage of remaining forest that is cleared per year), in the Ecuadorian Amazon, were −0.30% for the interval 1990–2000, and −0.26 for 2001–2008. The same analysis for western Ecuador shows more dramatic figures. The deforestation rates in western Ecuador, for the same periods, were −2.49% and −2.19% respectively (MAE 2012). In this context, effective protection of natural reserves and increased connectivity between remaining tracts of forests are essential for peccary long-term survival.

The white-lipped peccary and collared peccary are two of the most preferred game species in Ecuador, on both sides of the Andes, often heading the list of wild meat extracted by indigenous peoples (Zapata-Ríos et al. 2011). Collared peccary constitutes 4–55% of the wild meat consumed by indigenous peoples in the Ecuadorian Amazon, and the white-lipped peccary 4–49% (Yost and Kelley 1983; Zapata-Ríos 2001; Sirén et al. 2004; Franzen 2006; Zapata-Ríos et al. 2009). This represents an average extraction rate for the Ecuadorian Amazon of 11.8 kg/km^2/year (range = 0.16–29.53) and 13.4 kg/km^2/year (range = 3.79–22.98) for the collared peccary and white-lipped peccary, respectively. During a 2-year survey (2005–2007) of an illegal wild meat market in northern Yasuní National Park, collared peccary and white-lipped peccary meat accounted for 7% (586 kg) and

48% (3855 kg), respectively of all the transactions in terms of biomass (Suárez et al. 2009). Similar information about hunting extraction rates, for western Ecuador, is nonexistent. Reproduction rates of white-lipped peccary are lower than collared peccary (Gottdenker and Bodmer 1998). This difference has implications for management strategies, making the white-lipped peccary more susceptible to the effects of hunting than collared peccaries (Gottdenker and Bodmer 1998). Currently, hunting (both for subsistence and commercial purposes) is not sustainable, and community-based wildlife management programs are urgently needed throughout their distribution range in Ecuador (Zapata-Ríos et al. 2011).

Periodical disappearances of white-lipped peccaries, from areas where they were once common, have been reported occasionally in the literature (Kiltie 1981; Emmons 1987; Vickers 1991; Fragoso 2004). Factors causing these disappearances are not well understood, but disease transmission has been proposed as a potential explanation (Fragoso 1997, 2004). Peccaries, especially white-lipped peccaries, appear to be very sensitive to disease outbreaks transmitted by domestic animals (Corn et al. 1987; Gruver and Guthrle 1996; Noon et al. 2003; Gerber et al. 2012). However, more research is needed to understand the role of disease in the long-term persistence of peccaries in the landscape.

Climate change is also considered an important threat for the collared and the white-lipped peccaries in Ecuador under current conditions of land use, since it is estimated that up to 58% and 74% of their natural habitats can be lost by 2080, respectively (Iturralde-Pólit 2010).

11.5 Llama (*Lama glama*) and Alpaca (*Vicugna pacos*)

Diego G. Tirira

11.5.1 Distribution

Llama (*Lama glama*) (Fig 11.7) and alpaca (*Vicugna pacos*) (Fig. 11.8) are found in cold temperate and high-Andean climates in the Andes of Ecuador. Their naturally inhabit the páramo, an Alpine grass and shrub terrain; however, they have adapted to a wide range of environments. Domesticated animals have been introduced to other life zones, such as the inter-Andean valleys and, not that uncommonly, warm and tropical climates as a tourist attraction (Tirira 2017). Both species are particularly abundant in the indigenous communities of Central Andes of Ecuador, within the provinces of Cotopaxi, Tungurahua, Bolívar, and Chimborazo (White 2001a; FAO 2005; Tirira 2017).

Fig. 11.7 *Lama glama*. Credit: Diego G. Tirira. Location: El Angel, Ecological Reserve, Carchi province, Ecuador

Fig. 11.8 *Vicugna pacos*. Credit: Diego G. Tirira. Location: El Angel, Ecological Reserve, Carchi province, Ecuador

11.5.2 Reproduction

In llama and alpaca, males can breed after 3 years, while females can reproduce at 2 years of age. In the llama, when the female reaches 60% of the adult weight, it can reproduce with the best selected males and it is possible to obtain 70–85% of success in the birth; in both species, the female calves one young per year (Bravo 2008).

For the alpaca, it has been estimated in the province of Cañar (between 1985 and 1995) an annual fertility rate higher than 80% and a mortality rate of 4.6% in adults, and 8.5% in offspring (White 2001b).

11.5.3 Population

Both species are protected since the 1970s (Registro Oficial 193 of 1974), when their populations reached numbers far lower than the historical references (Ortiz 1999; White 2001a). For the llama, it was estimated that less than 2000 individuals lived in Ecuador until 1980 (Gallo 1981); number that in 25 years increased to 10,356 animals, with the largest populations in the provinces of Bolívar (2750 individuals), Chimborazo (2606), Cotopaxi (2141) and Tungurahua (1150) (FAO 2005).

For the alpaca, it is estimated that in Ecuador inhabits 4500 animals in 2000 (White 2001b), and 6685 animals in 2010 (MAGAP 2013), with the largest populations in the provinces of Cotopaxi (3402 individuals) and Pichincha (1816); other provinces with smaller populations are Azuay, Carchi, Cañar, Chimborazo, Bolívar and Tungurahua (MAGAP 2013).

The llamas present in Ecuador are ancestral native populations (Estrella 1990; Baptista-Vargas 2009); on the contrary, the majority of alpacas that currently live in Ecuador are descendants of a group of individuals imported in 1985 from Peru and Chile (White 2001a); in 2013, the most recent import of 200 alpacas was made to be delivered in indigenous communities of the provinces of Azuay, Cañar, Chimborazo, Bolívar, Cotopaxi, Tungurahua, and Imbabura in order to motivate their breeding and improve the genetic variety of existing individuals (MAGAP 2013).

11.5.4 Use

The llamas and alpacas are managed mainly by indigenous communities with scarce economic resources (Baptista-Vargas 2009; Quispe et al. 2009). Five forms of use have been identified: as pack animals (only llama) and as producers of fiber, meat, leather or fertilizer (White 2001a, b; Baptista-Vargas 2009; Córdova-Ruiz 2015).

The use as pack animals has lost importance in Ecuador, since the llama has been replaced by stronger animals (horses, mules or donkeys) or by vehicles (Baptista-Vargas 2009).

The raising of llamas and alpacas and the production of fiber for textile use is managed under a traditional system, which implies low productivity due to various factors, such as the discrepancy in land possession, lack of training and technical assistance in adequate management, little knowledge for adequate marketing, among others (Coeli 2012; Córdova-Ruiz 2015).

Despite this, the production of alpaca fiber has increased in the country, but it is still insufficient, since only 22% of the products are dedicated to the commercialization of processed fiber or finished products, compared to 17% that only commercializes it in its raw state, and 60% of the producers that have no interest or do not know how to take advantage of the fiber (FAO 2005).

Llama fiber, being of lower quality than that of alpaca (White 2010; Córdova-Ruiz 2015) has a lower price in the domestic market, similar to the price of sheep wool and, therefore, there is little interest in its production and marketing. The indigenous communities of the provinces of Tungurahua and Chimborazo that make clothes with this fiber use it only for family use, since they have no interest in its commercialization (Baptista-Vargas 2009).

The use of both species as producers of meat, leather or fertilizer is limited and little exploited in Ecuador (Baptista-Vargas 2009).

11.5.5 Threats

Feral dogs are possibly the main threat to these species, particularly for individuals that are kept in the páramo, in conditions of "semi-captivity."

11.6 Vicuña (*Vicugna vicugna*)

Diego G. Tirira
Carlos A. Cajas-Bermeo

11.6.1 Distribution

The vicuña (*Vicugna vicugna*) (Fig. 11.9) is an introduced species to the Andes of central Ecuador, between 3300 and 4700 m in altitude (Tirira 2017). In Ecuador there are two populations, one in the Chimborazo Fauna Production Reserve, in the provinces of Bolívar, Chimborazo, and Tungurahua, introduced in 1988 and 1993; and another in the community of San José de Tipín, Palmira parish, Guamote can-

Fig. 11.9 *Vicugna vicugna*. Credit: David Torres Costales. Location: Chimborazo Wildlife Reserve, Chimborazo province, Ecuador

ton, central-south of the Chimborazo province, introduced in 1999 (MAE 2013). The subspecies present in Ecuador is *V. v. mensalis* (MAE 2013).

11.6.2 Natural History

Diet. Studies conducted at the Chimborazo Wildlife Reserve determined that the vicuña feeds on about 20 plant species (Albán 2009; Caranqui and Pino 2013), but has preference for two species of Poaceae (*Elymus cordilleranus* and *Calamagrostis intermedia*); other species that are consumed regularly, although with less preference, are *Hypochaeris sessiliflora* (Asteraceae), *Werneria nubigena* (Asteraceae), and *Geranium ecuadoriense* (Geraniaceae) (Caranqui and Pino 2013). They are dedicated to grazing about 8 h a day, approximately between 06:00 and 14:00 h, but the time may be longer when food is scarce in the dry season (Rodríguez-González and Morales 2017); vicuñas consume water frequently; therefore, they do not go further than 2 km from the sources (Rodríguez-González and Morales 2017).

Sexual composition. In the Chimborazo Wildlife Reserve a ratio of 1:1 for males and females is estimated (MAE 2013).

Reproduction. In Ecuador the gestation of the vicuña takes about 340 days and the birth occurs in the morning; births can occur throughout the year, but two peaks are recorded, between April and May, and between August and September (MAE

2013; Rodríguez-González and Morales 2017). The babies at birth have a weight of 2.07–2.23 kg (Albán 2009). The females return to breed within a few days of the birth (up to a week later) (MAE 2013).

Activity patterns. This is a diurnal species, with the highest activity in the early hours of the morning (Rodríguez-González and Morales 2017).

Behavior. In the Chimborazo Wildlife Reserve, it has been observed that the adult male of a family group stays alert in the periphery of the group, a few meters from the females; this distance increases when there is the presence of a predator, the human, or the intrusion of a single male. When a single male wants to enter to the group there is aggression on the part of the dominant male to expel him. During the fight, the males in dispute are spit, kick, hit the chest, and can bite different parts of their opponent's body (such as the neck, head, legs, and even the testicles). The meeting ends with the estrangement of the losing male (Rodríguez-González and Morales 2017). The young are expelled from the family group before they reach their first birthday (Rodríguez-González and Morales 2017).

Vicuñas usually defecate in specific places, which appear as small mounds of feces of a few meters in diameter; they also use places where they wallow to clean their fur of impurities (Rodríguez-González and Morales 2017).

11.6.3 Population Dynamics

Home Range. In the Chimborazo Wildlife Reserve, it is estimated that the groups occupy an area of no more than 40 ha, but the size may be smaller depending on the size of the family group, the availability and quality of the food, and water sources (Rodríguez-González and Morales 2017).

Density Estimates. No information is available.

Population. In Ecuador, vicuñas have been introduced at three different times. In 1988, 200 individuals from Chile and Peru were brought; and in 1993, 77 individuals from Bolivia, all were located within the Chimborazo Wildlife Reserve (MAE 2013). A fourth shipment in 1999 included the entry of 100 vicuñas that were located in San José de Tipín (MAE 2013). At the beginning there was a high mortality of vicuñas (between 40% and 50%) (MAE 2013; Garrido 2016).

The population of vicuñas in Ecuador is estimated at 7185 individuals, distributed 38% in the province of Chimborazo, 42% in Tungurahua, and 20% in Bolívar (MAE 2016). The annual growth rate of the vicuña is 5.5%, that is approximately 176 new vicuñas every year (MAE 2016).

Population Structure. The vicuña is a social animal (Tirira 2017). In the Chimborazo Wildlife Reserve three population structures have been observed (Rodríguez-González and Morales 2017): family groups, groups of single males, and solitary individuals.

Family groups are composed of an adult male, a harem of several adult females (up to five) and their offspring younger than 1 year; the family group is stable and remains united not only during the breeding season; adult males usually maintain the family group for at least 6 years (Rodríguez-González and Morales 2017).

The groups of single males are composed of young males over 1 year old, expelled from their groups, and old adult males; these groups can be formed by a few (less than 5) up to a 100 individuals, but on average around 20; they are unstable groups that vary in their composition and number permanently (Albán 2009; Rodríguez-González and Morales 2017).

Solitary individuals are adult males who have decided not to join a group; this condition is the least frequent in the Chimborazo Wildlife Reserve (Albán 2009).

Mortality Factors. Puma (*Puma concolor*) is considered the main predator of the vicuña (Yensen and Tarifa 1993; Pacheco et al. 2004); however, this species has not been registered in the Chimborazo Wildlife Reserve for at least 50 years (Tirira 2019). Actually, main predator of the vicuña is considered to be the culpeo (*Pseudalopex culpaeus*), especially attacking young individuals (Albán 2009; MAE 2013). Feral dogs are another cause of mortality of young and solitary vicuñas; it has been reported that attacks occur mostly at night (MAE 2013). There is also a report of the attack of a group of domestic dogs on a newborn vicuña in the province of Bolívar (Albán 2009).

Another cause of mortality are run-overs on the main roads surrounding the Chimborazo Wildlife Reserve, although there are no data of this threat (Rodríguez-González and Morales 2017).

In 2000 the death of 29 vicuñas was recorded due to the consumption of grass contaminated with the ash that expelled the Tungurahua Volcano and covered much of the eastern part of the Chimborazo Wildlife Reserve (MAE 2013).

11.6.4 Habitat Requirements

In the Chimborazo Wildlife Reserve, the vicuña inhabits dry páramo, herbaceous páramo, and bofedales (high Andean wetlands) (Siavichay-Mendoza 2016; Rodríguez-González and Morales 2017). The highest density of vicuñas has been recorded in bofedales, where they attend to hydrate and look for food (Siavichay-Mendoza 2016; Rodríguez-González and Morales 2017). Vicuña prefers flat areas at daytime hours to look for food; but at night it prefers slopes to rest (Rodríguez-González and Morales 2017).

It is estimated that the potential habitat for the vicuña in the Chimborazo Wildlife Reserve is 41,846 ha (MAE 2013).

11.6.5 Use

Although the vicuña is a protected species in Ecuador, shearing and use of its fiber is allowed, but it cannot be sacrificed (MAE 2013). Since the change of the Ecuadorian vicuña populations from Appendix I to II of CITES, international trade of vicuña fibers and textiles is possible (MAE 2013). However, the use of vicuña fiber is limited due to several factors, mainly there are gaps in the ownership of the land where the vicuñas live (MAE 2013; Garrido 2016), an important issue to be resolved and which determines the indigenous communities that will benefit from this management (Ulloa 2015).

Valdivieso-Morocho (2018), identifies three possibilities of sustainable use of the vicuña in Ecuador: the sale of fiber as raw material, the elaboration of textiles, and the development of tourism focused on the traditional capture and shearing of animals and the processing of the fiber.

11.6.6 Threats and Conservation

One of the objectives of the creation of the Chimborazo Wildlife Reserve was the "reintroduction" of the vicuña in Ecuador, as a mechanism for the recovery of the country's natural and cultural heritage (Baptista-Vargas 2009). For this reason, in 2004 the "Regulation for the management and conservation of the vicuña" was issued, where it is indicated that the vicuñas are state heritage, that their conservation is of public interest, and that their management must be supervised by the environmental authority of the country (Registro Oficial 430 of 2004). To ensure this management, the government of Ecuador requested CITES in 2013 that the population of vicuñas in the country be changed from Appendix I to Appendix II (MAE 2013).

Among the threats that affect the vicuña in Ecuador, death by feral dogs and run-over on the roads are considered the main ones (MAE 2013); other secondary or potential threats are forest fires and páramo burning, the presence of feral cats, and the transmission of diseases due to the proximity with domestic livestock (mainly cattle and sheep). Poaching is sporadic and does not represent a threat to the vicuña population in Ecuador (MAE 2013).

11.7 Brocket Deer (Genus *Mazama*)

Galo Zapata-Ríos
Diego G. Tirira

11.7.1 Distribution

The brocket deer (genus *Mazama*) is widely distributed in Ecuador, with four taxa: Gualea red brocket (*Mazama americana gualea*) (Fig. 11.10), Zamora red brocket (*Mazama americana zamora*) (Fig. 11.11), Amazonian brown brocket (*Mazama nemorivaga*) Figs. 11.12 and 11.13), and little red brocket (*Mazama*

Fig. 11.10 *Mazama americana gualea*. Credit: Francisco Sornoza. Location: Canandé Biological Reserve, Esmeraldas province, Ecuador. Note: Female captured by camera trap

Fig. 11.11 *Mazama americana zamora*. Credit: Aldo Fernando Sornoza. Location: Yasuní National Park, Orellana province, Ecuador. Note: Male

Fig. 11.12 *Mazama nemorivaga*. Credit: Archivo Ecuambiente Consulting Group. Location: Yasuní National Park, Orellana province, Ecuador. Note: Male captured by camera trap

Fig. 11.13 *Mazama nemorivaga*. Credit: Archivo Ecuambiente Consulting Group. Location: Cuyabeno Wildlife Reserve, Sucumbíos province, Ecuador. Note: Female captured by camera trap

Fig. 11.14 *Mazama rufina*. Credit: Aldo Fernando Sornoza. Location: El Boliche National Recreation Area, Cotopaxi province, Ecuador. Note: Male

rufina) (Fig. 11.14). Gualea Red Brocket is distributed in the lowlands and foothills west of the Andes, from sea level to 2196 m; inhabits humid and dry forest, tropical and subtropical (Tirira 2017; Urgilés-Verdugo et al. 2019); this subspecies is considered endemic to Ecuador, but its presence in western Colombia is expected (Groves and Grubb 2011; Tirira 2017). Zamora red brocket occurs east of the Andes, from 200 to 1900 m in altitude, in humid tropical and subtropical forest (Tirira 2017). Amazonian brown brocket inhabits areas east of the Andes, from 200 to 900 m in altitude, in humid tropical; in most of their tropical range, both Amazonian species are sympatric (Tirira 2017). In Ecuador, there are no resource partitioning studies for these two species, but a differential use of habitats would be expected (Rossi et al. 2010). Finally, the fourth taxa in Ecuador, the little red brocket is the only species of the genus restricted to the Andean highlands, from 1900 to 3600 m in altitude; inhabits high-Andean and temperate forest, but also reaches the lower páramo (Groves and Grubb 2011; Zapata-Ríos and Branch 2016; Tirira 2017).

11.7.2 Natural History

Diet. Brocket deer species are mainly frugivore and seed eaters, relying heavily on fruits and seeds, and complement their diet by eating leaves and fibers (Rossi et al. 2010; Varela et al. 2010). No specific information for Ecuadorian species is available.

Sexual Composition. Our understanding of sex ratios in wild populations of Neotropical mammals is still very limited, and it is usually assumed to be 1:1 (Bodmer and Robinson 2004). From hunting data in the Ecuadorian Amazon, it is known *M. americana zamora* sex ratio is very close to 1:1 and has an average of 1.5 gestations per year (Zapata-Ríos et al. 2009).

Reproduction. Reproduction can occur year-round, but it is more frequent in response to local changes in food availability, February–March in the Amazon, and September–October in the Chocó in northwestern Ecuador (Zapata-Ríos and Toasa 2004; Zapata-Ríos et al. 2009; Mayor et al. 2011).

Activity Patterns. Brocket deer activity patterns typically are crepuscular but vary according to a series of environmental factors (cloud cover, wind, precipitation), including human disturbance (Di Bitetti et al. 2008; Zapata-Ríos and Branch 2016; Espinosa and Salvador 2017). No consistent relationship between activity and moon phases has been reported (Norris et al. 2010; Zapata-Ríos and Branch 2016).

In a study in Yasuní National Park with camera traps (Durango-Cordero 2011), the red brocket (*M. americana*) showed diurnal and nocturnal activity, with the highest peak after 06:00 h, and reduced activity during 07:00 to 18:00 h; for the Amazonian brown brocket (*M. nemorivaga*), the busiest activity period varies between 14:00 and 17:00 h.

Behavior. No information is available.

11.7.3 Population Dynamics

Home Range. No information is available.

Density Estimates. Comparative data of *Mazama* abundance, and carrying capacity, for different habitats is currently nonexistent in Ecuador. There are few abundance estimates for *M. americana zamora* in lowland Amazon rainforests. These estimates vary widely from 0.22 to 10.4 individuals/km^2, and the differences are caused by different levels of unsustainable hunting (Zapata-Ríos et al. 2006; Prieto 2007; Durango-Cordero 2011; Suárez et al. 2013). For *Mazama nemorivaga* estimates vary from 0.8 to 10.5 individuals/km^2 in Curaray, in areas without hunting (Prieto 2007; Durango-Cordero 2011). In a year-long mammal survey in northern Esmeraldas (Awá territory), *M. americana gualea* was relatively one of the most abundant species (0.3 records/km) (Zapata-Ríos and Toasa 2004). In the Llanganates National Park an abundance of 0.82 photographic events/100 trap-nights was estimated for *Mazama rufina* (Palacios et al. 2017).

Population Structure. *Mazama* are solitary species, except during reproductive periods, territorial, and show scent-marking behavior (Rossi et al. 2010; Tirira 2017). Dispersion, social interactions, and behavior may be a result of intense selective pressure to reduce predation and hunting (Varela et al. 2010; Zapata-Ríos and Branch 2016).

Mortality Factors. Puma (*Puma concolor*) is the primary predator of *Mazama*; Jaguar (*Panthera onca*), ocelot (*Leopardus pardalis*), and tayra (*Eira barbara*) in the lowlands; and Andean bear (*Tremarctos ornatus*) in the highlands are probably incidental predators, especially of fawns (Galef et al. 1976; Peyton 1980; Garla et al. 2001; Lizcano et al. 2010). Culpeo (*Pseudalopex culpaeus*) is an important predator of *Mazama rufina* in Podocarpus National Park (Guntiñas et al. 2017).

11.7.4 Habitat Requirements

Brocket deer inhabit a wide range of habitats in Ecuador, from pluvial to dry forests in the lowlands on both sides of the Andes, to Andean forest and páramos in the Andean highlands (Zapata-Ríos and Branch 2016; Palacios et al. 2017; Tirira 2017). In general, *Mazama* species inhabit forested areas, and nearly all the activities of *M. a. gualea*, *M. a. zamora*, and *M. nemorivaga* occur in forests (Tirira 2011a, 2017). *Mazama americana* is found in primary, secondary, and gallery forests, and at forest edges. It prefers dry areas along rivers and streams, and it is possible to find them in open areas near forest. The species avoids swamps and flooded forests (Tirira 2017). In southwestern Ecuador, there is evidence that prefers humid or semi-humid forests in the low-elevation mountain ranges along the Coast, and does not use adjacent dry forests, where the Peruvian white-tailed deer (*Odocoileus virginianus peruvianus*) occurs (D. G. Tirira, unpublished data).

Amazonian brown brocket (*Mazama nemorivaga*) prefers primary forest, but also occurs in old secondary growth forest. It is found in terra firme forests and in areas with dense vegetation, including the vicinity of rivers; it is unlikely that it will go out into open areas (Tirira 2017; D. G. Tirira, pers. obs.).

Habitat use in the little red brocket (*Mazama rufina*) is different to other *Mazama* species. It mainly occurs in the high Andes, in Andean and cloud forest, in shrub páramo and lower herbaceous páramo. It is frequent in gullies and steep slopes and prefers dense bushy places to open areas (Zapata-Ríos and Branch 2016; Palacios et al. 2017; Tirira 2017).

11.7.5 Threats and Conservation

According to the *Red Book of Mammals of Ecuador*, *Mazama americana* and *M. nemorivaga* are considered Near Threatened, and *M. rufina* is classified as Vulnerable (Tirira 2011a). Across the country, brocket deer species have suffered population declines due to habitat loss and fragmentation, subsistence and commercial hunting, introduction of exotic species, and probably transmission of infectious diseases from cattle (Lizcano et al. 2010; Varela et al. 2010; Rossi et al. 2010; Tirira 2011b, c; Tirira and Zapata-Ríos 2011).

Habitat loss and fragmentation represent a severe threat to brocket deer's long-term survival in Ecuador. Habitat loss and fragmentation not only restrict habitat availability but weakens the viability of isolated populations (Yackulic et al. 2011;

Kosydar et al. 2014; Benchimol and Peres 2015). Forest cover maps for the 1990–2008 period provide deforestation rates for western Ecuador and the Ecuadorian Amazon (MAE 2012). The derived annual deforestation rates (expressed as the percentage of remaining forest that is cleared per year), in the Ecuadorian Amazon, were −0.3% for the interval 1990–2000 and −0.26% for 2001–2008. The same analysis for western Ecuador show more dramatic figures. The deforestation rates in western Ecuador, for the same periods, were −2.5% and −2.2%, respectively (MAE 2012). In the context of deforestation, the classification of *M. americana* in two different subspecies requires a reassessment of both, having *M. a. gualea* a more urgent need because it occurs in the Ecuadorian Chocó (western Ecuador), a highly endangered ecosystem (Myers 1988; Dodson and Gentry 1991; Brooks et al. 2002).

In the Ecuadorian lowlands (less than 1500 m), brocket deer species are hunted extensively, and make a significant contribution to rural household diets. The meat is highly appreciated throughout the range of the species and is prized by local people on both sides of the Andes, ranking behind only that of paca (*Cuniculus paca*) and white-lipped peccary (*Tayassu pecari*). Zamora Red brocket (*Mazama a. zamora*) constitutes 3–14% (878 kg extracted by three Waorani communities to 3717 kg extracted by four Shuar communities) of the wild meat consumed by indigenous peoples, in a 1-year period, in the Ecuadorian Amazon (Mena-Valenzuela et al. 2000; Zapata-Ríos 2001; Sirén et al. 2004; Franzen 2006; Zapata-Ríos et al. 2009). This represents an average extraction rate for the Ecuadorian Amazon of 7.9 kg/km^2/year (IC 95% = ±5.2; range = 1.22–15.3). During a 2-year survey (2005–2007) of an illegal wild meat market in northern Yasuní National Park, *Mazama americana zamora* and *M. nemorivaga* meat accounted for 5% (402 kg) and 0.2% (15 kg), respectively, of all the transactions in terms of biomass (Suárez et al. 2009). Similar information about hunting extraction rates, for western Ecuador, is nonexistent. Currently, hunting (both for subsistence and commercial purposes) is not sustainable, and community-based wildlife management programs are urgently needed (Zapata-Ríos et al. 2011).

Currently, domestic dogs are the most widespread predator of little red brocket (*Mazama rufina*) (Lizcano et al. 2010; Zapata-Ríos and Branch 2016). In the high Andes of Ecuador, relative abundances of *M. rufina* in areas where feral dogs were present were lower compared to areas without dogs. Also, *M. rufina* significantly altered activity patterns where feral dogs were present (Zapata-Ríos and Branch 2016). Given the existing threats to *Mazama* species, particularly *M. rufina*, effective strategies for reducing problems with dogs are needed urgently in other areas where *Mazama* species occur, as well as elsewhere feral dogs and wildlife interact. In addition, there is no comprehensive disease information for *Mazama* species in Ecuador. However, evidence of *Mazama* exposure to domestic species' infectious diseases (e.g., bovine respiratory syncytial virus, epizootic hemorrhagic disease virus, malignant catarrhal fever) has been reported in other countries (Driemeier et al. 2002; Deem et al. 2004; Favero et al. 2013). In addition, climate change is also considered an important threat for these species in Ecuador under current conditions of land use, since it is estimated that up to 63% (*Mazama americana*) and 18% (*Mazama rufina*) of their natural habitat can be lost by 2080 (Iturralde-Pólit 2010).

In this context, there is an urgent need for conservation and management interventions, based on information gathered in the field, to reduce and mitigate negative impacts of human activities, and to ensure effective long-term conservation of ecologically functional populations of *Mazama* species.

11.8 White-Tailed Deer (*Odocoileus virginianus*)

X. Izurieta V.
Diego G. Tirira

11.8.1 Distribution

There are two subspecies in Ecuador: Andean white-tailed deer (*Odocoileus virginianus ustus*) (Fig. 11.15) lives in the páramo regions of the country along the Andean mountain range, between 3000 and 4600 m, but its preferred habitat is between 3300 and 4100 m (Izurieta 1992; Albuja 2007; Tirira 2017). Peruvian white-tailed deer (*Odocoileus virginianus peruvianus*) (Figs. 11.16 and 11.17) inhabits the lowlands of southwestern Ecuador, in the provinces of Manabí, Santa Elena,

Fig. 11.15 *Odocoileus virginianus ustus*. Credit: Juan Manuel Carrión. Location: Antisana Ecological Reserve, Napo province, Ecuador. Note: Male

Fig. 11.16 *Odocoileus peruvianus peruvianus*. Credit: Diego G. Tirira. Location: Parque Histórico de Guayaquil, Guayas province, Ecuador. Note: Male in captivity

Fig. 11.17 *Odocoileus virginianus peruvianus*. Credit: Diego G. Tirira. Location: Zoo El Pantanal, Guayaquil, Guayas province, Ecuador. Note: Juvenile in captivity

Guayas, Loja, El Oro, and in western Los Ríos. The preferred habitat of *O. virginianus peruvianus* is between sea level and 650 m but is most likely to be found in areas lower than 200 m. In addition to the lower elevations, in Loja province it also occasionally reaches subtropical and temperate zones to an altitude of 2400 m (Jiggins et al. 1999; Morocho and Romero 2003; Tirira 2017). Unverified records mentioned by Morocho and Romero (2003) suggest that the two subspecies could be sympatric in the slopes of Loja province).

11.8.2 Natural History

Diet. Information on diet is only available for the Andean white-tailed deer. This information comes through direct observations of deer in semi-captivity in Guaytaloma, between Oyacachi and Papallacta, at altitudes of 3700–4000 m; and, in the páramo of Antisana, between 3600 and 4100 m (Albuja 2007). Both localities are found in the province of Napo. It was determined that this subspecies includes in its diet 72 species of plants of different habits, as well as fungi and lichens. During the study it was evident that the deer preferred some plants over others and that the selection varied according to the site, the time of year, the availability and the state of the preferred part of each plant. In Guaytaloma 45 species of plants belonging to 20 families were registered as deer forage. The preferred plants were (indicated by the consumed portion of the plant): *Niphogeton ternata* (leaves), *Senecio tephrosioides* (flowers) *Ranunculus praemorsus* (leaves and stems), *Miconia latifolia* (flowers and fruit), *Gaultheria foliolosa* (leaves), *Panicum* sp. (grass) and *Armillaria* sp. (a fungus). In Antisana 33 species of plants belonging to 17 families were identified as deer forage, as well as three fungi and one lichen. The preferred forage species in this locality were *Senecio tephrosioides* and the three fungi, among them the *Armillaria* sp.

In another study in Cotopaxi National Park and the Boliche National Recreation Area (Izurieta 1992) it showed that the white-tailed deer had a preference for the native grasses, particularly *Lachemilla orbiculata* as well as young shoots of Poaceae. Also noted was the frequent consumption of the fungus *Suillus luteus*, a species associated with the introduced pine forests.

Sexual Composition. Information is only available for the Andean white-tailed deer. In a study carried out in the Chimborazo Wildlife Reserve, it was determined that females are more abundant than males in a 4:1 ratio (Guano 2016).

Reproduction. Information is only available for the Andean white-tailed deer. Reproduction data was obtained from observing two separate individuals kept in semi-captivity. One individual was observed in Guaytaloma and a second individual in Antisana. The two ecosystems are comparable, and both are in the province of Napo (Albuja 2007). In Guaytaloma, copulation occurred in June, and the doe gave birth in January. The birth of the fawns in Guaytaloma coincided with a period of lower rainfall and medium-high temperatures. On the contrary, in Antisana copulation occurred at the end of November and the doe gave birth in the first weeks of June. The birth of the Antisana fawns coincided with the rainy season and lower temperatures. The gestation period in both cases took around 200 days.

The spotted fur of newborns disappeared at 4 months of age; at this time the males began the growth of small antlers (about 5 mm and a single point). These antlers were shed about 4.5–5 months later (or at approximately 9 months of age). At 11 months bumps began to regrow on the forehead of the males. The bucks acquired a pair of four-pointed antlers at 1 year of age (Albuja 2007).

Activity Patterns. Information is only available for the Andean white-tailed deer. In a study carried out in Cayambe-Coca National Park, at elevations of between 3400 and 3900 m, it was observed that this subspecies shows diurnal and nocturnal behavior. Two well-defined peaks of activity were observed: one at dawn, between 05:00 and 09:00 h, with most activity between 07:00 and 08:00 h; and another in the afternoon/evening, between 14:00 and 20:00 h, with the highest activity recorded between 16:00 and 18:00 h (Zapata-Ríos and Branch 2016). In another study carried out in the Chimborazo Wildlife Reserve it was determined that the highest activity of this subspecies occurs between 06:00 and 12:00 h; while between 12:00 and 24:00 h the activity is lower and between 24:00 and 06:00 h there is practically no activity (Guano 2016).

Behavior. No information is available.

11.8.3 Population Dynamics

Home Range. No information is available.

Density Estimates. Albuja (2007) estimated the size of the Andean white-tailed deer population from 2.0 to 3.8 individuals/km^2 in the Antisana páramo, and 1.6 individuals/km^2 in Guaytaloma. In Guaytaloma was estimated the presence of 19 individuals in an area of 1200 ha and of 528 individuals for the entire Oyacachi-Papallacta area (approximately 33,000 ha).

Population Structure. The white-tailed deer is a solitary species, although it is common to observe them in pairs or small groups (notably an adult female and her offspring) (Albuja 2007; Tirira 2017). In a study carried out in the Chimborazo Wildlife Reserve, it was determined that adults Andean white-tailed deer are more abundant than juveniles and infants in a 5:1 ratio (Guano 2016).

Mortality Factors. Andean white-tailed deer remains have been found in the feces of the Andean bear (*Tremactos ornatus*) in Oyacachi, Cayambe-Coca National Park (Troya et al. 2004). It is suspected that the puma (*Puma concolor*) and the jaguar (*Panthera onca*) are also predators of this species in Ecuador. Zapata-Ríos and Branch (2016) determined in a study carried out in the same national park, that there are no significant differences between the population of Andean white-tailed deer in areas with, or without, the presence of feral dogs. It can be concluded at this time that the presence of feral dogs is not a threat to this species. The greatest cause of mortality for the two subspecies of white-tailed deer in Ecuador is considered to be human hunting (Tirira 2011d, e).

11.8.4 Habitat Requirements

Andean white-tailed deer in Ecuador is found exclusively in páramo (Tirira 2017). This subspecies has preference for grassland paramo; also occupy open areas of humid páramo and low shrub vegetation and riverbanks (Izurieta 1992; Albuja 2007; Zapata-Ríos and Branch 2016). There is no evidence that this subspecies enters the Andean forests near the páramo, but it does visit the surrounding areas in search of food (Albuja 2007).

The study carried out on the western side of the Cotopaxi National Park and in the El Boliche National Recreation Area, in the province of Cotopaxi, the researchers identified three factors that define the habitat preference of this subspecies living in the páramo ecosystem (Izurieta 1992). These factors were: sufficient vegetative cover, human presence, and topography. The variables that were considered less significant were altitude, slope, proximity to water, and vegetation type. The evidence used to come to this conclusion included direct observations, presence of excrement, fallen antlers, and trees scraped with antlers.

These results also demonstrated that Andean white-tailed deer occupied habitats with a variety of physical and biological characteristics (Izurieta 1992). Data was collected principally at lower altitudes (less than 4000 m) and on regular, irregular or only slightly sloped terrain. The animals preferred medium-high environmental protection to help them hide but allow them to escape quickly if necessary. The deer prefers little human disturbance and distances greater than 300 m from the water. It was also determined that in the El Boliche National Recreation Area the Andean white-tailed deer is a frequent visitor to the introduced pine forest: 70% of the pine forest sites sampled contained signs that the deer had visited the area (Izurieta 1992).

Peruvian white-tailed deer inhabits tropical, subtropical, and temperate dry forests. It is found in both primary and secondary growth forests, disturbed areas, and areas near human presence (Tirira 2017). The evidence indicates that this subspecies does not enter the humid or semi-humid forests of the small mountain ranges of costal Ecuador, where it is replaced by the Red Brocket (*Mazama americana*) (D. G. Tirira, unpublished data).

11.8.5 Threats and Conservation

Historical evidence indicates that white-tailed deer was a much more common species than it is today (Albuja 1981, 2007). In a study carried out between 1977 and 1980 in the provinces of Carchi, Imbabura, Pichincha, Tungurahua, Chimborazo, Bolivar, Azuay, and Loja, it was estimated that the populations of Andean white-tailed deer were in good condition in all localities visited, except in the páramos of the volcano Pichincha (homonymous province) and El Ángel (province of Carchi), places where the hunting had been intense and, therefore, the populations of deer had been exterminated (Albuja 2007). It is known that in places where its habitat is adequate and where hunting has not been intense, as in areas far from human presence and inside certain protected areas, it is a common and easy to observe this species, mainly in the Eastern Cordillera of the Andes; while populations of the Western Cordillera and southwest of the country are more affected (Tirira 2011d, e).

The Andean white-tailed Andean deer is considered a Near Threatened species according to the *Red Book of Mammals of Ecuador* (Tirira 2011d). The conservation status of the Peruvian white-tailed deer populations is more complicated; this subspecies has been categorized as Endangered (Tirira 2011e).

The greatest threat that affects both subspecies is the destruction of their natural habitat due to the increase of agricultural activity in the area. Human encroachment into the deer's habitat has also increased the effects of fragmentation and isolation of the populations. Both subspecies are also hunted heavily but hunting pressure has somewhat decreased during the last decade thanks to their inclusion in the *Red Book of Mammals of Ecuador* (Tirira 2011a). Trophy hunting affects mainly the Andean subspecies, but both subspecies are hunted for their meat and pelts (Tirira 2011d, e). Other suspected threats to affect these animals are wild fires in the vegetation forage (Tirira 2011d, e). Climate change is not considered an important threat for the white-tailed deer in Ecuador, since it is estimated that only 2% of its natural habitat can be lost by 2080 (Iturralde-Pólit 2010).

11.9 Northern Pudu (*Pudu mephistophiles*)

Diego G. Tirira

11.9.1 Distribution

In Ecuador the distribution of Northern pudu (*Pudu mephistophiles*) (Fig. 11.18) is discontinuous (Tirira et al. 2011b); inhabits cold climates in Andes mountains, from 2196 up to 4000 m above sea level (Tirira 2017; Urgilés-Verdugo et al. 2019); records in Ecuador at higher altitudes, up 4500 m (Tirira 2007) are erroneous. There is evidence of its presence in most of the provinces influences of the Cordillera de los Andes (Carchi, Imbabura, Pichincha, Napo, Cotopaxi, Tungurahua, Chimborazo, Azuay, Loja, and Zamora Chinchipe), especially in the Cordillera Oriental (Escamilo et al. 2010; Tirira et al. 2011b; Tirira 2017).

11.9.2 Natural History

Diet. The Northern pudu is assumed to feed on leaves and fruits (Tirira 2017).

Fig. 11.18 *Pudu mephistophiles*. Credit: Patricio Pillajo. Location: Guamaní, Cayambe-Coca National Park, Napo province, Ecuador

Sexual Composition. No information is available.

Reproduction. No information is available.

Activity Patterns. The Northern pudu has crepuscular and nocturnal habits (Escamilo et al. 2010). An individual was recorded in northwestern of Pichincha province at 18:17 h (C. A. Urgilés-Verdugo, obs. pers.); another was found at 18:44 h in Papallacta, Napo province (P. Pillajo, com. pers.). However, there is a record by day, at 11:28 h, in Llanganates National Park (Palacios et al. 2017).

Behavior. No information is available.

11.9.3 Population Dynamics

Home Range. No information is available.

Density Estimates. In Ecuador it is a rare species and surely it is the rarest ungulate in the country (Tirira 2017). Only few records of solitary individuals are known (Palacios et al. 2017; Tirira 2017; Urgilés-Verdugo et al. 2019). In a study in three mountain forest locations in the northwestern of Pichincha province, of 470 records of mammals in trap camera, only two (0.4%) belonged to this species (Urgilés-Verdugo et al. 2019). In the Llanganates National Park an abundance of 0.69 photographic events/100 trap-nights was estimated (Palacios et al. 2017); while near the Yacuri National Park, an abundance of 0.1 events/km was estimated (Arcos 2009).

Population Structure. The Northern pudu is solitary (Tirira 2017; Palacios et al. 2017; Urgilés-Verdugo et al. 2019).

Mortality Factors. Culpeo (*Pseudalopex culpaeus*) was determined as an important predator of Northern pudu in Podocarpus National Park (Guntiñas et al. 2017).

11.9.4 Habitat Requirements

The Northern pudu inhabits cold areas and apparently prefers open herbaceous páramo (Tirira 2017), but there are some records in mountain and cloud forests (Urgilés-Verdugo et al. 2019). This species prefers primary and well-preserved areas (Tirira et al. 2011b). Inside forest it moves through the undergrowth (C. A. Urgilés-Verdugo, pers. obs.).

11.9.5 Threats and Conservation

Northern pudu is Vulnerable according to the *Red Book of Mammals of Ecuador* (Tirira 2011a). The conservation status of this species remains unknown and apparently it disappeared from many areas. Most of its current distribution is within protected areas and in places of difficult access. Protected areas with records in last two decades are: Cajas National Park, Cayambe-Coca National Park, Llanganates National Park, Podocarpus National Park, Yacuri National Park, and Antisana Ecological Reserve (Arcos et al. 2007; Guntiñas et al. 2017; Palacios et al. 2017; P. Pillajo, F. Sánchez-Karste, and C. A. Urgilés-Verdugo, pers. comm.).

The loss and fragmentation of the habitat are considered the main threats that the species faces in Ecuador (Tirira et al. 2011b) due to changes in land use to develop and increase agricultural and livestock activities (Escamilo et al. 2010). Another major threat is the burning of páramo (Tirira et al. 2011b). Climate change is also considered to be an important threat to this species in Ecuador under current conditions of land use, since it is estimated that up to 20% of its natural habitat can be lost by 2080 (Iturralde-Pólit 2010). Currently, feral dogs are important predators of Little Red Brocket (*Mazama rufina*) in some locations (Lizcano et al. 2010; Zapata-Ríos and Branch 2016), so it is considered that their presence could also affect the Northern pudu. Hunting is considered not to be a major threat at present (Tirira 2017; Tirira et al. 2011b).

References

Albán GMF (2009) Estudio del hábitat y costumbres de las vicuñas en la Reserva de Producción de Fauna Chimborazo. BSc thesis, Escuela Superior Politécnica de Chimborazo, Riobamba

Albuja VL (1981) Estado actual de los ciervos en el Ecuador. In: Memorias, V Jornadas Nacionales de Biología. Pontificia Universidad Católica del Ecuador, Quito, pp 36–37

Albuja L (1983) Mamíferos ecuatorianos considerados raros o en peligro de extinción. In: Programa Nacional Forestal. Ministerio de Agricultura y Ganadería, Quito, pp 35–67

Albuja VL (1991) Lista de vertebrados del Ecuador: mamíferos. Rev Politécnica 16(3):163–203

Albuja VL (2007) Biología y ecología del venado de cola blanca (*Odocoileus virginianus ustus* Gray, 1874) en los páramos de Oyacachi-Papallacta y Antisana, Ecuador. Rev Politécnica (Biología 7) 27(4):34–57

Albuja VL, Muñoz BR (2000) Fauna del Parque Nacional Machalilla. In: Iturralde M, Josse C (eds) Compendio de investigaciones en el Parque Nacional Machalilla. Centro de Datos para la Conservación (CDC-Ecuador) y Fundación Natura, Quito, pp 32–41

Almeida RE (2014) Los camélidos sudamericanos en la historia del Ecuador. Propuesta Universitaria, Quito. http://docenteconvoz.blogspot.com/2014/01/los-camelidos-sudamericanos-en-la.html

Almeida R, Guevara S (2018) Evaluación de la criopreservación de semen fresco para la conservación de germoplasma de tapir amazónico (*Tapirus terrestris*). VMD thesis, Universidad Central del Ecuador, Quito

Altrichter M, Taber A, Beck H, Reyna-Hurtado R, Lizárraga L, Keuroghlian A, Sanderson EW (2011) Range-wide declines of a key Neotropical ecosystem architect, the near threatened white-lipped peccary *Tayassu pecari*. Oryx 46:87–98

Arcos DR (2009) Abundancia relativa de medianos y grandes mamíferos en el Bosque Protector Colambo-Yacuri, suroriente del Ecuador. Rev Ecuatoriana Med Cienc Biol 30(1–2):78–93

Arcos R, Albuja VL, Moreno P (2007) Nuevos registros y ampliación del rango de distribución de algunos mamíferos del Ecuador. Rev Politécnica (Biología 7) 27(4):126–132

Arias-Gutiérrez RI, Reyes-Puig JP, Tapia A, Terán A, López-de-Vargas K, Bermúdez D, Rodríguez X (2016) Desarrollo local y conservación en la vertiente oriental andina: corredor ecológico Llanganates-Sangay-valle del Anzu. Rev Amazónica Cienc y Tecnol 5(1):52–68

Baptista-Vargas VC (2009) Los camélidos en la Reserva de Producción de Fauna Chimborazo: ¿Una alternativa para la sustentabilidad del páramo? Estudio de caso en torno a la organización campesina, la economía y la gobernanza ambiental. MSc thesis, Facultad Latinoamericana de Ciencias Sociales, Quito

Barreto GR, Hernandez OE, Ojasti J (1997) Diet of peccaries (*Tayassu tajacu* and *T. pecari*) in a dry forest of Venezuela. J Zool 241:279–284

Benchimol M, Peres CA (2015) Predicting local extinctions of Amazonian vertebrates in forest islands created by a mega dam. Biol Conserv 187:61–72

Bermúdez-Loor D, Reyes-Puig JP (2011) Dieta del tapir de montaña (*Tapirus pinchaque*) en tres localidades del corredor ecológico Llangantes-Sangay. Bol Téc 10 Zool 7:1–13

Bianchi C (1988) El Shuar y el ambiente. Conocimiento del medio y cacería no destructiva, 2nd edn. Ediciones Abya-Yala, Quito

Blake JG, Mosquera D, Loiselle BA, Swing K, Guerra J, Romo D (2012) Temporal activity patterns of terrestrial mammals in lowland rainforest of eastern Ecuador. Ecotropica 18:137–146

Bodmer RE, Robinson JG (2004) Evaluating the sustainability of hunting in the nootropics. In: Silvius KM, Bodmer RE, Fragoso JMV (eds) People in nature, wildlife conservation in South and Central America. Columbia University Press, New York, pp 299–323

Borgtoft H, Balslev H (1993) Palmas útiles, especies ecuatorianas para agroforestería y extractivismo. Editorial Abya-Yala, Quito

Bravo E (2008) Estudio sobre las propiedades organolépticas de la carne de llama y su aplicación en la gastronomía como sustituto de carnes rojas. MSc thesis, Universidad Tecnológica Equinoccial, Quito

Brito MJ, Ojala-Barbour R (2016) Mamíferos no voladores del Parque Nacional Sangay, Ecuador. Pap Avulsos Zool 56(5):45–61

Brito MJ, Garzón-Santomaro C, Mena-Valenzuela P, González-Romero D, Mena-Jaén J (eds) (2018) Mamíferos de la provincia de El Oro: una guía de identificación de especies de mamíferos del páramo al mar. Publ Misc 8. GADPEO, INABIO, Quito

Brooks DM, Bodmer RM, Matola S (eds) (1997) Tapirs-status survey and conservation action plan. IUCN/SSC Tapir specialist group, Gland, Switzerland

Brooks TM, Mittermeier RA, Mittermeier CG, da Fonseca GAB, Rylands AB, Konstant WR, Flick P, Pilgrim J, Oldfield S, Magin G, Hilton-Taylor C (2002) Habitat loss and extinction in the hotspots of biodiversity. Conserv Biol 16:909–923

Caranqui J, Pino M (2013) Especies alimenticias de la vicuña en la Reserva de Producción Faunística Chimborazo, Ecuador. Technical report, Dirección Provincial de Chimborazo del Ministerio de Ambiente del Ecuador, Escuela Superior Politécnica de Chimborazo, Riobamba

Castellanos A (1994) El tapir andino (*Tapirus pinchaque* Roulin): crianza de un ejemplar en el Bosque Protector Pasochoa y notas ecológicas en el Parque Nacional Sangay, Ecuador. BSc thesis, Universidad Central del Ecuador, Quito

Castellanos A (2013) Iridium/GPS Telemetry to study home range and population density of mountain tapirs in the Río Papallacta watershed, Ecuador. Tapir Conserv 22(31):20–25

Castellanos A (2015) Reintroduced Andean tapir attacks a person in the Antisana Ecological Reserve, Ecuador. Tapir Conserv 24(33):11–12

Cieza de León P (1553 [2005]) Crónica del Perú: el señorío de los Incas. Biblioteca Ayacucho, Caracas

Coeli E (2012) Difusión y sistematización de buenas prácticas con énfasis en todos los eslabones de la cadena de valor de la alpaca en Ecuador. Pastores andinos: tejedores de espacio económico e inclusión alimentaria alto-andina. http://www.pastoresandinos.org/ Pastoresandinos.org

Córdova-Ruiz LM (2015) Comparación de la calidad de las fibras de *Vicugna pacos* (alpaca) y *Lama glama* (llama). BSc thesis, Escuela Superior Politécnica de Chimborazo, Riobamba

Corn JL, Lee RM, Erickson GA, Murphy CD (1987) Serological survey for evidence of exposure to vesicular stomatitis virus, pseudorabies virus, brucellosis and leptospirosis in collared peccaries from Arizona. J Wildl Dis 23:551–557

Correa A, Torres N (2005) Estrategia Nacional Preliminar para la Conservación de los Tapires (género *Tapirus*) en el Ecuador. BSc thesis, Universidad Técnica Particular de Loja, Loja

Da Silva SSB, Le Pendu Y, Ohashi OM, Oba E, de Albuquerque NI, García AR, Mayor P, Guimarães DAA (2016) Sexual behavior of *Pecari tajacu* (Cetartiodactyla: Tayassuidae) during periovulatory and early gestation periods. Behav Process 131:68–73

de Thoisy B, Gonçalves da Silva A, Ruiz-García M, Tapia A, Ramírez O, Arana M, Quse V, Paz-y-Miño C, Tobler M, Pedraza C, Lavergne A (2010) Population history, phylogeography, and conservation genetics of the last Neotropical mega-herbivore, the lowland tapir (*Tapirus terrestris*). BMC Evol Biol 10(278):1–16

De Velasco J (1841). Historia del Reino de Quito en la América Meridional. Tomo 1, Parte 1: Historia natural. Ed Casa de la Cultura Ecuatoriana, Quito (1977)

Deem SL, Noss AJ, Villaroel R, Uhart MM, Karesh WB (2004) Disease survey of free-ranging grey brocket deer (*Mazama gouazoubira*) in the Gran Chaco, Bolivia. J Wildl Dis 40:92–98

Desbiez A, Gonçalves da Silva A, Lacy B (2007) Population dynamics and modeling report. In: Medici EP, Desbiez A, Gonçalves da Silva A, Jerusalinsky L, Chassot O, Montenegro OL, Rodríguez JO, Mendoza A, Quse VB, Pedraza C, Gatti A, Oliveira-Santos LGR, Tortato MA, Ramos V, Reis ML, Landau-Remy G, Tapia A, Morais AA (eds) Workshop para a Conservação da Anta Brasileira: Relatório Final. IUCN/SSC Tapir Specialist Group and IUCN/SSC Conservation Breeding Specialist Group, Sao Paulo

Desbiez A, Santos SA, Keuroghlian A, Bodmer RE (2009) Niche partitioning among white-lipped peccaries (*Tayassu pecari*), collared peccaries (*Pecari tajacu*), and feral pigs (*Sus scrofa*). J Mammal 90:119–128

Di Bitetti MS, Paviolo A, Ferrari CA, de Angelo C, Di Bianco Y (2008) Differential responses to hunting in two sympatric species of brocket deer (*Mazama americana* and *M. nana*). Biotropica 40:636–645

Dodson CH, Gentry AH (1991) Biological extinction in western Ecuador. Ann Mo Bot Gard 78:273–295

Downer CC (1995) The gentle botanist. Wildl Conserv 98(4):30–35

Downer CC (1996) The mountain tapir, endangered 'flagship' species of the high Andes. Oryx 30(1):45–58

Downner CC (1997) Status and Action Plan of the mountain tapir (*Tapirus pinchaque*). In: Brooks DM, Bodmer RE, Matola S (eds) Tapirs: Status survey and conservation action plan. IUCN/SSC Tapirs Specialist Group, Gland, Switzerland, pp 10–22

Downer CC (1999) Un caso de mutualismo en los Andes: observaciones sobre dieta-hábitat del tapir de montaña. In: Fana TG, Montenegro OL, Bodmer RE (eds) Manejo y conservación de fauna silvestre en América Latina. Wildlife Conservation Society and New York Zoological Society, Santa Cruz de la Sierra, pp 415–427

Downer CC (2001) Observations on the diet and habitat of the mountain tapir (*Tapirus pinchaque*). J Zool 254(3):279–291

Downer CC (2003) Ambito hogareño y utilización de hábitat del tapir andino e ingreso de ganado en el Parque Nacional Sangay, Ecuador. Lyonia 4(1):31–34

Driemeier D, Brito MF, Traverso SD, Cattani C, Cruz CEF (2002) Outbreak of malignant catarrhal fever in brown brocket deer (*Mazama gouazoubira*) in Brazil. Vet Rec 151:271–272

Duarte JMB, González S (eds) (2010) Neotropical Cervidology: biology and medicine of Latin American deer. Funep and IUCN, Jaboticabal, Brazil, and Gland, Switzerland

Durango-Cordero MF (2011) Abundancia relativa, densidad poblacional y patrones de actividad de cinco especies de ungulados en dos sitios dentro de la Reserva de la Biosfera Yasuní, Amazonia-Ecuador. BSc thesis, Pontificia Universidad Católica del Ecuador, Quito

Emmons LH (1987) Comparative feeding ecology of felids in a Neotropical rainforest. Behav Ecol Sociobiol 20:271–283

Escamilo BLL, Barrio J, Benavides FJ, Tirira DG (2010) Chapter 13: Northern Pudu, *Pudu mephistophiles* (De Winton, 1896). In: Duarte JMB, González S (eds) Neotropical cervidology: biology and medicine of Latin American deer. FUNEP/IUCN, Jaboticabal, Brazil, pp 133–139

Espinosa S, Salvador J (2017) Hunters' landscape accessibility and daily activity of ungulates in Yasuní Biosphere Reserve, Ecuador. Therya 8:45–52

Estrella E (1990) El pan de américa. Ediciones Abya-Yala, Quito

FAO (2005) Situación actual de los camélidos sudamericanos en el Ecuador. Organización de las Naciones Unidas para la Agricultura y la Alimentación, Roma

Favero CM, Matos ACD, Campos FS, Cândido MV, Costa EA, Heinemann MB, Barbosa-Stancioli EF, Lobato ZIP (2013) Epizootic hemorrhagic disease in brocket deer, Brazil. Emerg Infect Dis 19:346–348

Fragoso JMV (1997) Desapariciones locales del báquiro labiado (*Tayassu pecari*) en la Amazonía: migración, sobre-cosecha, o epidemia? In: Fang TG, Bodmer RE, Aquino R, Valqui MH (eds) Manejo de fauna silvestre en la Amazonía. Instituto de Ecología, La Paz, pp 309–312

Fragoso JMV (1999) Perception of scale and resource partitioning by peccaries: behavioral causes and ecological implications. J Mammal 80:993–1003

Fragoso JMV (2004) A long-term study of white-lipped peccary (*Tayassu pecari*) population fluctuations in northern Amazonia: anthropogenic vs "natural" causes. In: Silvius KM, Bodmer RE, Fragoso JMV (eds) People in nature: wildlife conservation in South and Central America. Columbia University Press, New York, pp 286–298

Fragoso JMV, Silvius KM, Correa JA (2003) Long-distance seed dispersal by tapirs increases seed survival and aggregates tropical trees. Ecology 84:1998–2006

Franzen M (2006) Evaluating the sustainability of hunting: a comparison of harvest profiles across three Huaorani communities. Environ Conserv 33:36–45

Freire M (1997) La cacería de mamíferos, aves y reptiles en una comunidad quichua y en destacamentos militares, Lorocachi-Pastaza 1995–1996. BSc thesis, Universidad del Azuay, Cuenca

Galef BG, Mittermeier RA, Bailey RC (1976) Predation by the tayra (*Eira barbara*). J Mammal 57:760–761

Gallina S, Mandujano S, Bello J, López AHF, Weber M (2010) Chapter 11: White-tailed deer *Odocoileus virginianus* (Zimmermann 1780). In: Duarte JMB, González S (eds) Neotropical cervidology: biology and medicine of Latin American deer. FUNEP/IUCN, Jaboticabal, Brazil, pp 101–118

Gallo N (1981) Fomento de camélidos sudamericanos en el Ecuador. In: Memorias, V Jornadas Nacionales de Biología. Pontificia Universidad Católica del Ecuador, Quito, pp 46–47
Gallo F (2012). Estado poblacional, selección de hábitat y actividad estacional del tapir de montaña (*Tapirus pinchaque* Roulin, 1829), Cuyuja, Andes Tropicales del Norte del Ecuador. BSc thesis, Universidad Central del Ecuador, Quito
Garla RC, Setz EZF, Gobbi N (2001) Jaguar (*Panthera onca*) food habits in Atlantic rainforest of southeastern Brazil. Biotropica 33:691–696
Garrido PAM (2016) Propuesta de líneas de acción de la vicuña (*Vicugna vicugna*) en el Ecuador para el aprovechamiento turístico del patrimonio cultural. MSc thesis, Escuela Superior Politécnica de Chimborazo, Riobamba
Gerber PF, Galinari GCF, Cortez A, Paula CD, Lobato ZIP, Heinemann MB (2012) Orbivirus infections in collared peccaries (*Tayassu tajacu*) in southeastern Brazil. J Wildl Dis 48:230–232
Gittleman JL, Thompson SD (1988) Energy allocation in mammalian reproduction. Am Zool 28(3):863–875
Gottdenker N, Bodmer RE (1998) Reproduction and productivity of white-lipped and collared peccaries in the Peruvian Amazon. J Zool 245:423–430
Groves C, Grubb P (2011) Ungulate taxonomy. Johns Hopkins University Press, Baltimore, MD
Grubb P (2005) Order Artiodactyla. In: Wilson DE, Reeder DM (eds) Mammal species of the world, a taxonomic and geographic reference, 3rd edn. The John Hopkins University Press, Baltimore, MD, pp 637–722
Gruver KS, Guthrle JW (1996) Parasites and selected diseases of collared peccaries (*Tayassu tajacu*) in the Trans-Pecos region of Texas. J Wildl Dis 32:560–562
Guano VMA (2016) Programa de manejo sostenible para el venado de cola blanca *Odocoileus virginianus* (Zimmermann, 1780) para la Reserva de Producción de Fauna Chimborazo. BSc thesis, Escuela Superior Politécnica de Chimborazo, Riobamba
Guntiñas M, Lozano J, Cisneros R, Narváez C, Armijos J (2017) Feeding ecology of the culpeo in southern Ecuador: wild ungulates being the main prey. Contr Zool 86:169–180
Hershkovitz P (1954) Mammals of northern Colombia. Preliminary report No. 7: Tapirs (genus *Tapirus*), with a systematic revision of American species. Proc U S Natl Mus 103:465–496
Hoffstétter R (1952) Les mammifères pléistocènes de la République de l'Equateur. Mem Soc Geol Fr 31(66):1–391
Iturralde-Pólit P (2010) Evaluación del posible impacto del cambio climático en el área de distribución de especies de mamíferos del Ecuador. BSc thesis, Pontificia Universidad Católica del Ecuador, Quito
Izurieta X (1992) Preferencias de hábitat del venado de cola blanca (*Odocoileus virginianus ustus*) en el sector occidental del Parque Nacional Cotopaxi y en el Área Nacional de Recreación El Boliche. BSc thesis, Pontificia Universidad Católica del Ecuador, Quito
Jarrín VP (2001) Mamíferos en la niebla. Otonga, un bosque nublado del Ecuador. Publ Espec 5. Museo de Zoología, Pontificia Universidad Católica del Ecuador, Quito
Jiggins C, Andrade PA, Cueva E, Dixon S, Isherwood I, Willis J (1999) The conservation of three forests in south-west Ecuador. Biosphere Publ Res Rep 2. Otley, UK
Jorgenson JP, Coppolillo P (2001) Análisis de amenazas. In: Jorgenson JP, Coello-Rodríguez M (eds) Conservación y desarrollo sostenible en el Parque Nacional Yasuní y su área de influencia. Memorias del seminario-taller. Ministerio del Ambiente del Ecuador. UNESCO, Wildlife Conservation Society, Editorial Simbioe, Quito
Kiltie RA (1981) Distribution of palm fruits on a rain forest floor: why white-lipped peccaries forage near objects. Biotropica 13:141–145
Kiltie RA (1982) Bite force as a basis for niche differentiation between rainforest peccaries (*Tayassu tajacu* and *T. pecari*). Biotropica 14(3):188–195
Kosydar AJ, Rumiz DI, Conquest LL, Tewksbury JJ (2014) Effects of hunting and fragmentation on terrestrial mammals in the Chiquitano forests of Bolivia. Trop Conserv Sci 7:288–307
Lizcano DJ, Álvarez SJ, Delgado CA (2010) Chapter 19: Dwarf red brocket deer *Mazama rufina* (Pucheran 1951). In: Duarte JMB, González S (eds) Neotropical cervidology: biology and medicine of Latin American deer. FUNEP/IUCN, Jaboticabal, Brazil, pp 177–180

López J, Merino H (1994) Determinación de constantes fisiológicas, hematológicas, aspectos productivos y reproductivos de la danta (*Tapirus terrestris*), pecarí (*Tayassu tajacu*), capibara (*Hydrochaeris hydrochaeris*), guanta (*Agouti paca*), y guatuza (*Dasyprocta punctata*). BSc thesis, Escuela Superior Politécnica de Chimborazo, Riobamba

Machoa JD, Tapia A, Desbiez ALJ (2007) El modelo de Vortex como una herramienta para la conservación del territorio y los recursos naturales en las comunidades indígenas de la Amazonía ecuatoriana. In: Memorias, XXXI Jornadas Nacionales de Biología. Escuela Superior Politécnica del Litoral, Guayaquil

MAE (2012) Línea base de deforestación del Ecuador continental. Ministerio del Ambiente del Ecuador, Programa Socio Bosque, Quito

MAE (2013) Plan de acción nacional para el manejo y conservación de la vicuña en el Ecuador. Technical report to CITES. Ministerio de Ambiente del Ecuador, Quito

MAE (2016) XXXII Reunión del Convenio para la Conservación y Manejo de la vicuña en las provincias de Chimborazo, Tungurahua y Bolívar. Technical report. Ministerio de Ambiente del Ecuador, Riobamba

MAE, WCS (2018) Plan de Acción para la conservación del pecarí de labio blanco (*Tayassu pecari*) en el Ecuador. Ministerio del Ambiente del Ecuador, Wildlife Conservation Society, Quito

MAGAP (2013) Actividades turísticas en Ecuador con camélidos sudamericanos. Ministerio de Agricultura, Ganaderia, Acuacultura y Pesca, Quito

Mattioli S (2011) Family Cervidae. In: Wilson DE, Mittermeier RA (eds) Handbook of the mammals of the world, Hoofed mammals, vol 2. Lynx Edicions, Barcelona, pp 350–443

Mayer JJ, Wetzel RM (1987) Tayassu pecari. Mamm Spec 293:1–7

Mayor PG, Guimarães DA, Le Pendu Y, Da Silva JV, Jori F, López-Béjar M (2007) Reproductive performance of captive collared peccaries (*Tayassu tajacu*) in the eastern Amazon. Anim Reprod Sci 102:88–97

Mayor PG, Bodmer RE, López-Béjar M (2009) Reproductive performance of the wild white-lipped peccary (*Tayassu pecari*) female in the Peruvian Amazon. Eur J Wildl Res 55:631–634

Mayor PG, Bodmer RE, López-Béjar M (2010) Reproductive performance of the wild collared peccary (*Tayassu tajacu*) female in the Peruvian Amazon. Eur J Wildl Res 56:681–684

Mayor P, Bodmer RE, López-Béjar M, López-Plana C (2011) Reproductive biology of the wild red brocket deer (*Mazama americana*) female in the Peruvian Amazon. Anim Reprod Sci 128:123–128

Medici P (2011) Family Tapiridae. In: Wilson DE, Mittermeier RA (eds) Handbook of the mammals of the world, Hoofed mammals, vol 2. Lynx Edicions, Barcelona, pp 182–204

Mena-Valenzuela P, Regalado J, Cueva R (1997) Oferta de animales en el bosque y cacería en la comunidad huaorani de Quehueire'ono, zona de amortiguamiento del Parque Nacional Yasuní, Napo, Ecuador. In: Mena PA, Soldi A, Alarcón R, Chiriboga C, Suárez L (eds) Estudios biológicos para la conservación: diversidad, ecología y etnobiología. EcoCiencia, Quito, pp 395–426

Mena-Valenzuela P, Stallings J, Regalado J, Cueva R (2000) The sustainability of current Hunting practices by the Huaorani. In: Robinson J, Bennett E (eds) Hunting for sustainability in tropical forests. Columbia University Press, New York, pp 57–78

Mittermeier RA, Macedo Ruiz H, Luscombe A (1975) A wooly monkey rediscovered in Peru. Oryx 13:41–46

Molina M, Molinari J (1999) Taxonomy of Venezuelan white-tailed deer (*Odocoileus*, Cervidae, Mammalia), based on cranial and mandibular traits. Can J Zool 77:632–645

Molinari J (2007) Variación geográfica en los venados de cola blanca (Cervidae, *Odocoileus*) de Venezuela, con énfasis en *O. margaritae*, la especie enana de la Isla de Margarita. Mem Fund La Salle Cienc Nat 167:29–72

Morocho D, Romero JC (2003) Bosques del Sur. El estado de 12 remanentes de bosques andinos de la provincia de Loja. Fundación Ecológica Arcoíris, PROBONA, DICA, Loja

Myers N (1988) Threatened biotas: "hotspots" in tropical forests. Environmentalist 8:187–208

Naranjo B, Suárez A, Arias R, Tapia A (2014) Caracterización químico-nutritiva de especies vegetales consumidas por el tapir amazónico (*Tapirus terrestris*). Rev Amazónica Cienc Tecnol 3(1):21–32

Narváez C (2013) Análisis de estacionalidad y abundancia relativa de oso de anteojos (*Tremarctos ornatus*), lobo de páramo (*Lycalopex culpaeus*) y tapir andino (*Tapirus pinchaque*) en los páramos del Parque Nacional Podocarpus. BSc thesis, Universidad Técnica Particular de Loja, Loja

Noon TH, Heffelfinger JR, Olding RJ, Wesche SL, Reggiardo C (2003) Serological survey for antibodies to canine distemper virus in collared peccary (*Tayassu tajacu*) populations in Arizona. J Wildl Dis 39:221–223

Norris D, Michalski F, Peres CA (2010) Habitat patch size modulates terrestrial mammal activity patterns in Amazonian forest fragments. J Mammal 91:551–560

Ojeda MI (2016) Dispersores primarios y secundarios de *Oenocarpus bataua* y *Mauritia flexuosa* en el bosque tropical Yasuní, Amazonía Ecuatoriana. MSc thesis, Pontificia Universidad Católica del Ecuador, Quito

Ortega-Andrade HM, Prieto-Torres DA, Gómez-Lora I, Lizcano DJ (2015) Ecological and geographical analysis of the distribution of the mountain tapir (*Tapirus pinchaque*) in Ecuador: importance of protected areas in future scenarios of global warming. PLoS One 10(3):1–20

Ortiz F (1999) Los camélidos sudamericanos en los Andes del norte. In: Memorias, III Simposio Internacional de Desarrollo Sustentable de Montañas: entendiendo las interfaces ecológicas para la gestión de los paisajes culturales en los Andes, Quito, pp 257–263

Pacheco LF, Lucero A, Villca M (2004) Dieta del puma (*Puma concolor*) en el Parque Nacional Sajama, Bolivia y su conflicto con la ganadería. Ecol Bolivia 39(1):75–83

Padilla M, Dowler RC, Downer CC (2010) *Tapirus pinchaque* (Perissodactyla: Tapiridae). Mamm Spec 42(863):166–182

Palacios J, Naveda-Rodríguez A, Zapata-Ríos G (2017) Large mammal richness in Llanganates National Park, Ecuador. Mammalia 82:309–314

Parker TA III, Carr JL (eds) (1992) Status of forest remnants in the Cordillera de la Costa and adjacent areas of southwestern Ecuador. Conservation International. Rapid Assessment Program (RAP), Working Papers 2, Washington, DC

Patzelt E (1978) Fauna del Ecuador, 1st edn. Editorial Las Casas, Quito

Peyton B (1980) Ecology, distribution, and food habits of spectacled bears, *Tremarctos ornatus*, in Peru. J Mammal 61:639–652

Pinto CM, Soto-Centeno JA, Núñez Quiroz AM, Ferreyra N, Delgado-Espinoza F, Stahl PW, Tirira DG (2016) Archaeology, biogeography, and mammalogy do not provide evidence for tarukas (Cervidae: *Hippocamelus antisensis*) in Ecuador. J Mammal 97(1):41–53

Prieto AF (2007) Diagnóstico de la Cacería de Subsistencia en Comunidades Indígenas de la Reserva de Producción Faunística Cuyabeno, Amazonía Ecuatoriana. BSc thesis, Pontificia Universidad Católica del Ecuador, Quito

Quispe EC, Rodríguez TC, Iñiguez LR, Mueller JP (2009) Producción de fibra de alpaca, llama, vicuña y guanaco en Sudamérica. Anim Genet Res 45:1–14

Registro Oficial (1974) Prohíbase la caza, comercialización y sacrificio de llamas. Presidencia de la República del Ecuador, Registro Oficial 506 del 6 de marzo de 1974 (No. 193)

Registro Oficial (2004) Reglamento para el Manejo y Conservación de la Vicuña en el Ecuador, Registro Oficial 430 del 28 de septiembre de 2004 (No. 2093)

Reyes-Puig JP, Tapia A, Palacios N (2007) Tungurahua volcano: a strategic refuge for mountain tapir in Ecuador. Tapir Conserv 16/1(21):16–17

Robinson JG, Eisenberg JF (1985) Group size and foraging habits of the collared peccary *Tayassu tajacu*. J Mammal 66(1):153–155

Rodríguez-González NF, Morales A (2017) La vicuña ecuatoriana y su entorno. Ministerio del Ambiente de Ecuador, PNUD, ESPOCH, Quito

Rossi RV, Bodmer R, Duarte JMB, Trovati RG (2010) Chapter 23: Amazonian brown brocket deer *Mazama nemorivaga* (Cuvier, 1817). In: Duarte JMB, González S (eds) Neotropical cervidology: biology and medicine of Latin American deer. FUNEP/IUCN, Jaboticabal, Brazil, pp 202–210

Ruiz-García M, Pinedo-Castro M, Luengas-Villamil K, Vergara C, Rodríguez JA, Shostell JM (2015) Molecular phylogenetics of the white-lipped peccary (*Tayassu pecari*) did not confirm morphological subspecies in northwestern South America. Genet Mol Res 14:5355–5378

Sandoval-Cañas LF (2004) Abundancia relativa del tapir (*Tapirus terrestris*) en un gradiente de intervención humana en el Parque Nacional Yasuní, Amazonía ecuatoriana. BSc thesis, Universidad Central del Ecuador, Quito

Santacruz SLJ (2012) Patrón de actividad de *Tapirus pinchaque* en distintos hábitats y fases lunares, en la hacienda San Antonio, flanco oriental del volcán Tungurahua, noviembre 2010–mayo 2011. BSc thesis, Universidad Central del Ecuador, Quito

Schank CJ, Cove MV, Kelly MJ, Mendoza E, O'Farrill G, Reyna-Hurtado R, Meyer N, Jordan CA, González-Maya JF, Lizcano DJ, Moreno R, Dobbins MT, Montalvo V, Sáenz-Bolaños C, Carillo Jimenez E, Estrada N, Cruz Díaz JC, Saenz J, Spínola M, Carver A, Fort J, Nielsen CK, Botello F, Pozo Montuy G, Rivero M, de la Torre JA, Brenes-Mora E, Godínez-Gómez O, Wood MA, Gilbert J, Miller JA (2017) Using a novel model approach to assess the distribution and conservation status of the endangered Baird's tapir. Div Distrib 23:1459–1471

Schauenberg P (1969) Contribution à l'étude du Tapir pinchaque, *Tapirus pinchaque* Roulin 1829. Rev Suisse Zool 76(8):211–255

Siavichay-Mendoza CA (2016) Comportamiento de la vicuña alrededor de los bofedales y frente al ganado: un fenómeno pertinente para el manejo de la Reserva de Producción de Fauna Chimborazo. BSc thesis, Escuela Superior Politécnica de Chimborazo, Riobamba

Sirén AH (2004) Changing interactions between humans and nature in Sarayaku, Ecuadorian Amazon. Acta Univ Agriculturae Sueciae (Agraria) 447:191–217

Sirén AH, Hambäck P, Machoa JD (2004) Including spatial heterogeneity and animal dispersal when evaluating hunting: a model analysis and an empirical assessment in an Amazonian community. Conserv Biol 18(5):1315–1329

Stummer M (1971) Wolltapire, *Tapirus pinchaque* (Roulin), in Ecuador. Der Zoologische Garten 40(3):148–159

Suárez L, García M (1986) Extinción de animales en el Ecuador, descripción de 60 especies amenazadas. Fundación Natura, Quito

Suárez E, Morales M, Cueva R, Utreras BV, Zapata-Ríos G, Toral EC, Torres J, Prado W, Vargas J (2009) Oil industry, wild meat trade and roads: indirect effects of oil extraction activities in a protected area in north-eastern Ecuador. Anim Conserv 12:364–373

Suárez E, Zapata-Ríos G, Utreras BV, Strindberg S, Vargas J (2013) Controlling access to oil roads protects forest cover, but not wildlife communities: a case study from the rainforest of Yasuní Biosphere Reserve (Ecuador). Anim Conserv 16:265–274

Taber AB, Chalukian SC, Altrichter M, Minkowski K, Lizárraga L, Sanderson E, Rumiz D, Venticinque E, Moraes EA, de Angelo C, Antúnez M, Ayala G, Beck H, Bodmer RE, Boher S, Cartes JL, de Bustos S, Eaton D, Emmons LH, Estrada N, de Oliveira LF, Fragoso JMV, García R, Gómez C, Gómez H, Keuroghlian A, Ledesma K, Lizcano DJ, Lozano C, Montenegro O, Neris N, Noss A, Palacio VJA, Paviolo A, Perovic P, Portillo H, Radachowsky J, Reyna-Hurtado RA, Rodríguez OJ, Salas L, Sarmiento DA, Sarría Perea JA, Schiaffino K, de Thoisy B, Tobler M, Utreras BV, Varela D, Wallace RB, Zapata-Ríos G (2008) El destino de los arquitectos de los bosques neotropicales: Evaluación de la distribución y el estado de conservación de los pecaríes labiados y los tapires de tierras bajas. Wildlife Conservation Society, Tapir Specialist Group, Wildlife Trust, New York

Taber AB, Altrichter M, Beck H, Gongora J (2011) Family Tayassuidae. In: Wilson DE, Mittermeier RA (eds) Handbook of the mammals of the world, Hoofed mammals, vol 2. Lynx Edicions, Barcelona, pp 292–307

Tapia M (1998) Manejo de mamíferos amazónicos en cautiverio y semicautiverio en el Centro Experimental Fátima. In: Tirira DG (ed) Biología, sistemática y conservación de los mamíferos del Ecuador, Publicación Especial sobre los mamíferos del Ecuador 1. Pontificia Universidad Católica del Ecuador, Quito, pp 155–198

Tapia M (1999) Guía para crianza, manejo y conservación de tapir amazónico *Tapirus terrestris*. Centro Tecnológico Recursos Amazónicos, Puyo, Ecuador

Tapia A (2005) Preferencia por fecas de tapir amazónico (*Tapirus terrestris*) de escarabajos estercoleros (Coleoptera: Scarabaeidae: Scarabaeinae) en bosque secundario amazónico. Tapir Conserv 14(17):24–28

Tapia A (2007) Caracterización cromosómica y molecular del tapir amazónico *Tapirus terrestris aenigmaticus* en la Amazonía ecuatoriana. BSc thesis, Universidad Central del Ecuador, Quito
Tapia A (2010a) Estudio de la diversidad, abundancia, distribución, etnozoología y conservación de los mamíferos, aves y reptiles de caza en el territorio Kichwa de la cuenca del Bobonaza. Technical report. Pueblo Kichwa de Sarayaku, Sarayaku, Pastaza, Ecuador
Tapia A (2010b) Relaciones filo geográficas del tapir amazónico *Tapirus terrestris*. MSc thesis, Universidad Internacional Menéndez-Pelayo, Madrid
Tapia A (2011) Estudio de la diversidad, abundancia, distribución, etnozoología y conservación de los mamíferos, aves y reptiles de caza en el territorio kichwa de la comunidad Victoria, cuenca del Curaray-Pastaza. Instituto Quichua de Biotecnología Sacha Supai, Quito
Tapia A (2013) Monitoreo de la biodiversidad del territorio Kichwa de la cuenca del Bobonaza. Technical report. Pueblo Kichwa de Sarayaku, Sarayaku, Pastaza, Ecuador
Tapia M, Arias R (2011) Guía para el manejo y cría de tapir *Tapirus terrestris* (Linnaeus, 1758). Centro Fátima-Zanja Arajuno, Puyo, Ecuador
Tapia A, Machoa D (2006) Ethnozoology of the Amazonian Tapir (*Tapirus terrestris* Linnaeus 1758) in the Sarayaku Community, Ecuador. Tapir Conserv 15(19):28–31
Tapia A, Reyes-Puig JP, Sandoval-Cañas L, Palacios N, Bermúdez D (2008a) Proyecto de conservación del tapir andino (*Tapirus pinchaque*) en la vertiente oriental de los Andes centrales del Ecuador. Rev Parque Nac Sangay:63–66
Tapia M, Arias R, Tapia A (2008b) Conservación y manejo en semicautiverio de fauna silvestre en la región amazónica ecuatoriana. In: Resúmenes, VIII Congreso Internacional de Manejo de Fauna Silvestre. Rio Branco, Brazil, pp 172–188
Tapia A, Nogales F, Castellanos A, Tapia M, Tirira DG (2011a) Tapir andino (*Tapirus pinchaque*). In: Tirira DG (ed) Libro Rojo de los mamíferos del Ecuador, Publicación Especial sobre los mamíferos del Ecuador 8, 2nd edn. Fundación mamíferos y conservación, Pontificia Universidad Católica del Ecuador, Ministerio del Ambiente del Ecuador, Quito, pp 98–100
Tapia A, Nogales F, Tapia M, Zapata-Ríos G, Tirira DG (2011b) Tapir amazónico (*Tapirus terrestris*). In: Tirira DG (ed) Libro Rojo de los mamíferos del Ecuador, Publicación Especial sobre los mamíferos del Ecuador 8, 2nd edn. Fundación Mamíferos y Conservación, Pontificia Universidad Católica del Ecuador, Ministerio del Ambiente del Ecuador, Quito, pp 140–141
Tirira DG (1999) Mamíferos del Ecuador, Publicación Especial sobre los mamíferos del Ecuador 2. Pontificia Universidad Católica del Ecuador, SIMBIOE, Quito
Tirira DG (2001a) Libro Rojo de los mamíferos del Ecuador. Publicación Especial sobre los mamíferos del Ecuador 4, 1st edn. SIMBIOE, EcoCiencia, Ministerio del Ambiente del Ecuador, IUCN, Quito
Tirira DG (2001b) Tapir amazónico (*Tapirus terrestris*). In: Tirira DG (ed) Libro rojo de los mamíferos del Ecuador, Publicación Especial sobre los mamíferos del Ecuador 4, 1st edn. SIMBIOE, EcoCiencia, Ministerio del Ambiente del Ecuador, IUCN, Quito
Tirira DG (2007) Guía de campo de los mamíferos del Ecuador, Publicación Especial sobre los mamíferos del Ecuador 6, 1st edn. Ediciones Murciélago Blanco, Quito
Tirira DG (2008) Mamíferos de los bosques húmedos del noroccidente de Ecuador, Publicación Especial sobre los mamíferos del Ecuador 7. Ediciones Murciélago Blanco, Proyecto PRIMENET, Quito
Tirira DG (2011a) Libro Rojo de los mamíferos del Ecuador, Publicación Especial sobre los mamíferos del Ecuador 8, 2nd edn. Fundación Mamíferos y Conservación, Pontificia Universidad Católica del Ecuador y Ministerio del Ambiente del Ecuador, Quito
Tirira DG (2011b) Venado marrón amazónico (*Mazama nemorivaga*). In: Tirira DG (ed) Libro Rojo de los mamíferos del Ecuador, Publicación Especial sobre los mamíferos del Ecuador 8, 2nd edn. Fundación Mamíferos y Conservación, Pontificia Universidad Católica del Ecuador, Ministerio del Ambiente del Ecuador, Quito, p 272
Tirira DG (2011c) Venado colorado enano (*Mazama rufina*). In: Tirira DG (ed) Libro Rojo de los mamíferos del Ecuador, Publicación Especial sobre los mamíferos del Ecuador 8, 2nd edn. Fundación Mamíferos y Conservación, Pontificia Universidad Católica del Ecuador, Ministerio del Ambiente del Ecuador, Quito, pp 226–227

Tirira DG (2011d) Venado de cola blanca de páramo (Odocoileus virginianus ustus). In: Tirira DG (ed) Libro Rojo de los mamíferos del Ecuador, Publicación Especial sobre los mamíferos del Ecuador 8, 2nd edn. Fundación Mamíferos y Conservación, Pontificia Universidad Católica del Ecuador, Ministerio del Ambiente del Ecuador, Quito, p 273

Tirira DG (2011e) Venado de cola blanca de la Costa (Odocoileus virginianus peruvianus). In: Tirira DG (ed) Libro Rojo de los mamíferos del Ecuador, Publicación Especial sobre los mamíferos del Ecuador 8, 2nd edn. Fundación Mamíferos y Conservación, Pontificia Universidad Católica del Ecuador, Ministerio del Ambiente del Ecuador, Quito, pp 143–144

Tirira DG (2017) A field guide to the mammals of Ecuador, Publicación Especial sobre los mamíferos del Ecuador 10, 2nd edn. Asociación Ecuatoriana de Mastozoología, Editorial Murciélago Blanco, Quito

Tirira DG (2018) Uso de la fauna por el pueblo Waorani, Amazonía del Ecuador. In: Tirira DG, Rios M (eds) Monitoreo Biológico Yasuní, Uso de flora y fauna por el pueblo Waorani, Amazonía del Ecuador, vol 8. Ecuambiente Consulting Group, Quito

Tirira DG (2019) Red Noctilio: unpublished data base on mammals of Ecuador. Grupo Murciélago Blanco, Quito

Tirira DG, Castellanos A (2001) Tapir de la Costa (*Tapirus bairdii*). In: Tirira DG (ed) Libro Rojo de los mamíferos del Ecuador, Publicación Especial sobre los mamíferos del Ecuador 4, 1st edn. SIMBIOE, EcoCiencia, Ministerio del Ambiente del Ecuador, IUCN, Quito, p 96

Tirira DG, Zapata-Ríos G (2011) Venado colorado (*Mazama americana*). In: Tirira DG (ed) Libro Rojo de los mamíferos del Ecuador, Publicación Especial sobre los mamíferos del Ecuador 8, 2nd edn. Fundación Mamíferos y Conservación, Pontificia Universidad Católica del Ecuador, Ministerio del Ambiente del Ecuador, Quito, p 271

Tirira DG, Tapia A, Nogales F, Castellanos A, Tapia M (2011a) Tapir del Chocó (Tapirus bairdii). In: Tirira DG (ed) Libro Rojo de los mamíferos del Ecuador, Publicación Especial sobre los mamíferos del Ecuador 8, 2nd edn. Fundación Mamíferos y Conservación, Pontificia Universidad Católica del Ecuador, Ministerio del Ambiente del Ecuador, Quito, p 343

Tirira DG, Arcos DR, Zapata-Ríos G (2011b) Ciervo enano (*Pudu mephistophiles*). In: Tirira DG (ed) Libro Rojo de los mamíferos del Ecuador, Publicación Especial sobre los mamíferos del Ecuador 8, 2nd edn. Fundación Mamíferos y Conservación, Pontificia Universidad Católica del Ecuador, Ministerio del Ambiente del Ecuador, Quito, pp 228–229

Troya V, Cuesta F, Peralvo M (2004) Food habits of Andean bears in the Oyacachi River Basin, Ecuador. Ursus 15(1):57–60

TSG (2011) Estrategia Nacional para la Conservación de los Tapires en el Ecuador. Tapir Specialist Group-Ecuador, Quito

Ulloa SMJ (2015) Estado de conservación de la vicuña (*Vicugna vicugna*) en Ecuador: el cambio de categoría en la CITES ¿Éxito o fracaso? BSc thesis, Pontificia Universidad Católica del Ecuador, Quito

Urgilés-Verdugo CA, Gallo VF (2012) In: Instituto para la Conservación y Capacitación Ambiental, Fundación EcoFondo (ed) Análisis multivariado de agrupamiento de las localidades donde se han realizado estudios florísticos para la dieta del tapir de montaña (*Tapirus pinchaque* Roulin, 1829). Technical report. Quito

Urgilés-Verdugo C, Gallo F, Borman R (2012) Abundancia relativa y selección de hábitat del tapir de montaña (*Tapirus pinchaque* Roulin, 1829), en una localidad de la Reserva Ecológica Antisana, Andes tropicales del norte del Ecuador. In: III Congreso de la Sociedad Peruana de Mastozoología. Universidad Nacional de Piura, Piura, p 84

Urgilés-Verdugo CA, Gallo VF, Trávez BH (2019 [2018]) Composición y estado de conservación de los mamíferos medianos y grandes del Corredor Biológico Tropi-Andino, Ecuador. Instituto para la Conservación y Capacitación Ambiental, Quito

Urgilés-Verdugo CA, Gallo VF, Borman R, Cisneros R, Lizcano D (2018) What factors of habitat or anthropic disturbance determine the density of the Andean tapir (*Tapirus pinchaque*) in the tropical Andes of northern Ecuador? Technical report. Instituto para la Conservación y Capacitación Ambiental, Quito

Valdivieso-Morocho J (2018) Estrategias de desarrollo local bajo el enfoque de biocomercio para el aprovechamiento sostenible de la vicuña (*Vicugna vicugna*), en la Reserva de Producción de Fauna Chimborazo. MSc thesis, Escuela Superior Politécnica de Chimborazo, Riobamba

Varela DM, Trovati RG, Guzmán KR, Rossi RV, Duarte JMB (2010) Chapter 15: Red brocket deer *Mazama americana* (Erxleben, 1777). In: Duarte JMB, González S (eds) Neotropical cervidology: biology and medicine of Latin American deer. FUNEP/IUCN, Jaboticabal, Brazil, pp 151–159

Vickers WT (1991) Hunting yields and game composition over ten years in an Amazon Indian territory. In: Robinson JG, Redford KH (eds) Neotropical wildlife use and conservation. The University of Chicago Press, Chicago, pp 53–81

WCS (2007) El tráfico de carne silvestre en el Parque Nacional Yasuní: caracterización de un mercado creciente en la Amazonía norte del Ecuador. Boletín de Wildlife Conservation Society, programa Ecuador 2:1–8

White S (2001a) Los camélidos sudamericanos en los páramos ecuatorianos: presente, historia y futuro. In: Mena PA, Medina G, Hofstede R (eds) Los páramos del Ecuador: particularidades, prioridades y perspectivas. Proyecto Páramo, Quito, pp 139–140

White S (2001b) Perspectivas para la producción de alpacas en el páramo ecuatoriano. In: Medina G, Mena PA (eds) La agricultura y la ganadería de los páramos. Serie Páramo 8. GTP. Abya-Yala, Quito, pp 33–58

White S (2010) Alpacas y llamas como herramientas de conservación del páramo. J Field Archaeol 17:49–68

Wilson DE, Mittermeier RA (2011) Handbook of the mammals of the world, Hoofed mammals, vol 2. Lynx Edicions, Barcelona

Yackulic CB, Sanderson EW, Uriarte M (2011) Anthropogenic and environmental drivers of modern range loss in large mammals. Proc Natl Acad Sci U S A 108:4024–4029

Yensen E, Tarifa T (1993) Reconocimiento de los mamíferos del Parque Nacional Sajama. Ecol Bolivia 21(1):45–66

Yépez J, Urgiles-Verdugo C (2013) Diversidad florística de los recursos vegetales usados por el tapir de montaña (*Tapirus pinchaque* Roulin, 1829) en la localidad de Cuyuja, zona de amortiguamiento de la Reserva Ecológica Antisana. Technical report. Instituto para la Conservación y Capacitación Ambiental, Quito

Yost JA, Kelley PM (1983) Shotguns, blowguns, and spears: the analysis of technological efficiency. In: Hames RB, Vickers WT (eds) Adaptive responses of native Amazonians. Academic, New York, pp 189–224

Zapata-Ríos G (2001) Sustentabilidad de la cacería de subsistencia: el caso de cuatro comunidades quichuas de la Amazonía ecuatoriana. Mastozool Neotrop 8:59–66

Zapata-Ríos G (2002) Evaluación del impacto de la cacería de subsistencia en la mastofauna de la Amazonía nororiental ecuatoriana. In: Bussmann RW, Lange S (eds) Conservación de la biodiversidad en los Andes y la Amazonía. Universidad de Bayreuth, Inka e.V., Munich, pp 595–600

Zapata-Ríos G (2013) Uso y diponibilidad de hábitat del tapir de montaña en el Parque Nacional Cayambe-Coca, Ecuador. In: Resúmenes, I Congreso Latinoamericano de Tapires y II Congreso Ecuatoriano de Mastozoología. Universidad Estatal Amazónica, Puyo, Ecuador, pp 115–116

Zapata-Ríos G, Araguillin E (2013) Estado de conservación del jaguar y el pecarí de labio blanco en el Ecuador occidental. Biodivers Netrop 3:21–29

Zapata-Ríos G, Branch LC (2010) Densidad poblacional de dos especies andinas amenazadas: tapir de montaña (*Tapirus pinchaque*) y guanta de altura (*Cuniculus taczanowskii*) en la Reserva Ecológica Cayambe-Coca. In: Resúmenes, XXXIV Jornadas Nacionales de Biología. Sociedad Ecuatoriana de Biología, Quito

Zapata-Ríos G, Branch LC (2016) Altered activity patterns and reduced abundance of native mammals in sites with feral dogs in the high Andes. Biol Conserv 193:9–16

Zapata-Ríos G, Jorgenson JP (2003) La utilización del autorregistro en los estudios de cacería de subsistencia: el ejemplo de los Shuar del suroriente de Ecuador. In: Campos C, Ulloa A (eds) Fauna socializada: tendencias en el manejo participativo de la fauna en América Latina. Fundación Natura, MacArthur Foundation. Instituto Colombiano de Antropología e Historia, Bogotá, pp 131–144

Zapata-Ríos G, Tirira DG (2011a) Pecarí de collar (*Pecari tajacu*). In: Tirira DG (ed) Libro Rojo de los mamíferos del Ecuador, Publicación Especial sobre los mamíferos del Ecuador 8, 2nd edn. Fundación Mamíferos y Conservación, Pontificia Universidad Católica del Ecuador, Ministerio del Ambiente del Ecuador, Quito, p 270

Zapata-Ríos G, Tirira DG (2011b) Pecarí de labio blanco de occidente (*Tayassu pecari aequatoris*). In: Tirira DG (ed) Libro Rojo de los mamíferos del Ecuador, Publicación Especial sobre los mamíferos del Ecuador 8, 2nd edn. Fundación Mamíferos y Conservación, Pontificia Universidad Católica del Ecuador, Ministerio del Ambiente del Ecuador, Quito, pp 101–102

Zapata-Ríos G, Tirira DG (2011c) Pecarí de labio blanco de oriente (*Tayassu pecari pecari*). In: Tirira DG (ed) Libro Rojo de los mamíferos del Ecuador, Publicación Especial sobre los mamíferos del Ecuador 8, 2nd edn. Fundación Mamíferos y Conservación, Pontificia Universidad Católica del Ecuador, Ministerio del Ambiente del Ecuador, Quito, p 142

Zapata-Ríos G, Toasa G (2004) Plan de Manejo Comunitario del Centro Awá Río Bogotá, Esmeraldas, Ecuador. Federación de Centros Awá del Ecuador, Missouri Botanical Garden, Wildlife Conservation Society, Ibarra, Ecuador

Zapata-Ríos G, Araguillin E, Jorgenson JP (2006) Caracterización de la comunidad de mamíferos no voladores en las estribaciones orientales de la cordillera del Kutukú, Amazonía ecuatoriana. Mastozool Neotrop 13:227–238

Zapata-Ríos G, Urgilés-Verdugo CA, Suárez E (2009) Mammal hunting by the Shuar of the Ecuadorian Amazon: is it sustainable? Oryx 43:375–385

Zapata-Ríos G, Suárez E, Utreras BV, Cueva R (2011) Uso y conservación de fauna silvestre en el Ecuador. In: Krainer A, Mora MF (eds) Retos y amenazas en Yasuní. Facultad Latinoamericana de Ciencias Sociales, Quito, pp 97–116

Chapter 12
Neotropical Ungulates of Uruguay

Natalia Mannise, Federica Moreno, and Susana González

Abstract Uruguay is located in the Neotropical region in a transitional zone with assemblage of diverse ecotypes being the austral limit of the distribution for several species of tropical plants and animals. The aim of this chapter is to provide a review of the Uruguayan ungulate species knowledge: the current distribution patterns, ecology, and evolutionary and demographic history. In Uruguay inhabit three species of native ungulates: collared peccary (*Pecari tajacu*), pampas deer (*Ozotoceros bezoarticus*) and gray brocket deer (*Mazama gouazoubira*). The introduction of livestock produced dramatically land uses changes, being a main factor in the population decline of native ungulate species; particularly of the in situ collared peccary and marsh deer (*Blastocerus dichotomus*) population extinction. Conservation measures needs to be implemented for the pampas deer and gray brocket deer, which are known that had a high population decline directly linked to the anthropogenic activities that involved the land use changes, disease transmissions, and poaching.

12.1 Introduction

Uruguay is the South American country located between 30°–35° S latitude and 53°–58° W longitude with an extension of 176,215 km^2. At the west, limited with Argentina and at north east with Brazil; being on the southeast the Atlantic Ocean, while the Rio de la Plata is the south frontier. The climate is temperate, having an average annual temperature of 17.5 °C, and an annual mean rainfall of 1300 mm (Bidegain and Caffera 1997).

This country is located between two geological Gondwanian formations, along irregular or fragmented escarpments (Escudero et al. 2004). It is originated by three main formations: the Chaqueño (through the spinal), Paraná (by the Uruguay River) and the central depression of Santa Catarina and Rio Grande do Sul of Brazil

N. Mannise · F. Moreno · S. González (✉)
Departamento de Biodiversidad y Genética, Instituto de Investigaciones Biológicas Clemente Estable- Ministerio de Educación y Cultura, Montevideo, Uruguay
e-mail: nmannise@iibce.edu.uy; fmoreno@iibce.edu.uy; sgonzalez@iibce.edu.uy

S. Gallina-Tessaro (ed.), *Ecology and Conservation of Tropical Ungulates in Latin America*, https://doi.org/10.1007/978-3-030-28868-6_12

(Brussa and Grela 2007). The main vegetation areas are open grassland and pampas. Currently, only the 50% of the native forests remains, which comprise 3.7% of the national territory (Escudero et al. 2004). The types of forest are riparian, mountain, park, coastal, and scrubland (Brussa and Grela 2007). The ravines and mountains of eastern and northeaster of Uruguay, originated from the Santa Catarina and Rio Grande do Sul depression, are composed of great diversity and richness of shrubs and woody species of significant ecological importance. They exhibit a lower flow dynamic than the Paraná formation corresponding to the forest of the Uruguay River, due to their great fragmentation and recent adaptation to climatic conditions (Escudero et al. 2004).

The location in the Neotropical region is interesting for being in a transitional zone including grassland, pampas, wetlands, forests, and coastal lagoons (Gonzaléz and Lessa 2014). The assembly of diverse ecotypes result in a very peculiar biodiversity component, being the austral limit of distribution for several species of tropical plants and animals (Gonzaléz and Lessa 2014). Moreover, the middle of the country was identified as a possible Neotropical species dispersion node (Cracco et al. 2005).

The Neotropical ungulates species that inhabit in Uruguay have a wide geographic range. Over the last 400 years, two of the four ungulate wild species have become locally extinct: the Tayassuidae collared peccary (*Pecari tajacu*) and the marsh deer (*Blastocerus dichotomus*). In this scenario wild ungulates coinhabit with introduced domestic ones as sheep, horses, pigs, and livestock.

Nowadays, there are three species of native ungulates: one Tayassuidae and two Cervidae. Deer species are pampas deer (*Ozotoceros bezoarticus*) and gray brocket deer (*Mazama gouazoubira*), which are also represented in the archaeological record (Moreno 2016; Moreno et al. 2016). The introduction of livestock produced dramatically land uses changes, being one main factor for the decline and the extinction process of the two native ungulates species previously mentioned.

In addition, during the eighteenth century was documented the introduction of several wild exotic ungulate species with sport hunting purposes: wild boar (*Sus scrofa*), fallow deer (*Dama dama*), and spotted deer (*Axis axis*) being currently widely distributed in Uruguay and producing several damages to the Uruguayan ecosystems and native species (González and Seal 1997).

The aim of this chapter is to provide a review of the Uruguayan ungulate species state of knowledge. We will describe distribution patterns, ecology, and evolutionary and demographic history.

12.2 Collared Peccary (*Pecari tajacu*)

The collared peccary (*Pecari tajacu*) is a pig like ungulate medium sized, between 84 and 106 cm total body length and 60 cm height to shoulders. Despite there are no systematic analysis if there are significant morphological differences between sexes, the males could be bigger than females (Nowak and Walker 1999; González and

Martínez-Lanfranco 2010). Their weight is quite variable and ranges from 16 to 25 kg (Nowak and Walker 1999; González and Martínez-Lanfranco 2010). Fur is harsh and dark grey with a yellowish white collar around the neck. The species have small round ears with a noticeable snout; legs are short, and the tail is vestigial. A gland area is located on the back and it secretes a musk oily substance; meanwhile preorbital glands are also present. The dorsal scent gland is scrubbed against trees trunks, rocks, and other individuals for territorial marks (Nowak and Walker 1999).

12.2.1 Habitat and Distribution

Collared peccary (or javelina) is widely distributed in Neotropics (Fig. 12.1). The southern limit of its current geographic distribution is located in Argentina, where it is already extinct in the eastern and southern areas of this country. In the recent past, the southern limit of its geographical distribution was in Uruguay (Fig. 12.1, González et al. 2013).

12.2.2 In Situ Population

Collared peccary populations were earlier reported in the south and east of Uruguay through exploration journeys in the seventeenth and eighteenth centuries (Mones 2001). Moreover, archaeological remains were found in the east and it was considered frequent around 600 BP (González and Martínez-Lanfranco 2010; Prigioni et al. 2018). In Uruguay, the last records of collared peccary were in the north and east during 1894 (Fig. 12.1, González et al. 2013; Cravino et al. 2018).

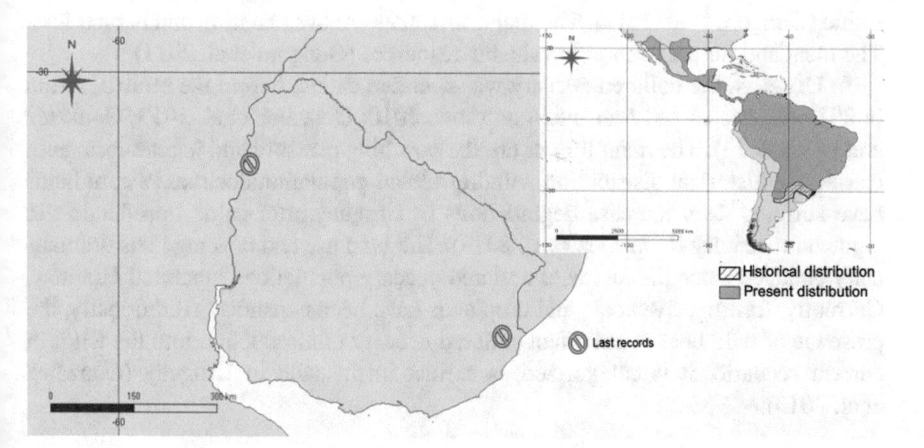

Fig. 12.1 Collared peccary. In situ last records reported in Uruguay; historic and present global distribution range (upper inset)

During 2017, a reintroduction experience took place at the northwest region. A total of 140 animals from an ex situ population were reintroduced after health and habitat assessment (Cravino et al. 2018). One year after the reintroduction experiment, the presence of animals was documented through camera trap records and activity traces. In addition, puppies were found near the release area (Cravino et al. 2018). Despite it was reported a high poaching pressure after reintroduction, the animals remain inside of the release protected area. Future monitoring efforts of collared peccary in that locality will be conducted in order to evaluate the species survival and adaptation (Cravino et al. 2018).

12.2.3 Ex Situ Population

The ex situ population was established 17 years ago with four founders exported from Argentina to Bioparque M'Bopicuá-Rio Negro, Uruguay (Cravino et al. 2018). The animals were maintained and reproduced successfully in a private captive center at the *Bioparque M'Bopicuá*, a native fauna reserve which belongs to *Montes del Plata* forestry company in the northwest of Uruguay. The other collared peccary captive population is in a public captive center at the *Estación de Cría de Fauna Autóctona* (Pan de Azúcar, Maldonado Uruguay) at the south-eastern Uruguay. This center is also planning a future reintroduction of the animal's surplus. Currently it is estimated a number of 400 individuals living in captivity in Uruguay.

12.2.4 Conservation Status

The IUCN Red List™- categorized as Least Concern, being the population trend stable (Gongora et al. 2011). The major threatens are over-hunting and habitat loss. The meat and fur are economic valuable resources (Gongora et al. 2011).

In Uruguay, the collared peccary was declared extinct before the reintroduction in 2017 (González and Martínez-Lanfranco 2010; González et al. 2013; Gonzaléz and Lessa 2014). The main threats are the poaching pressure and habitat loss; both caused the historical distribution withdrawal and population decline. Pasture lands have suffered deep increase degradations in Uruguay after cattle introduction in eighteenth century (González et al. 2013). The land use and coverage has dramatically changed since the last wild collared peccary populations inhabited Uruguay. Currently, forestry, livestock, and croplands have been expanded. Additionally, the presence of wild boar could affect collared peccary reintroduction into the wild. In current scenario, it is categorized as extinct in the wild in Uruguay (González et al. 2013).

12.3 Gray Brocket Deer (*Mazama gouazoubira*)

The gray brocket deer (*Mazama gouazoubira* Fischer, 1814) is a small-to-medium–sized deer, with an average shoulder height of 65 cm and 18 kg of weight (Black-Decima and Vogliotti 2016). The skin coloration varies according to the different habitat. The animals that inhabited on open areas have light brown color, meanwhile those from close forest regions are darker and greyish (Black-Decima and Vogliotti 2016). Only the adult males have antlers, that are spiky and simple with a single straight tine of 8 cm length, directed backward (Black-Décima et al. 2010). As other Cervidae it has several cutaneous glands, some well-developed, especially the tarsals and interdigital (Black-Decima and Vogliotti 2016). Gray brocket deer is mainly a solitary and selective feeder species (Black-Décima et al. 2010; Black-Decima and Vogliotti 2016). It has not been detected a well define reproductive seasonality. Gestation term is about 7 months long and the fawns are spotted until 3 months of life (Black-Décima et al. 2010).

12.3.1 *Habitat and Distribution*

The current distribution ranges from the pre-Andean regions (Argentina and Bolivia) to the Atlantic coast in the south and the southern region of the Amazon in the north (Black-Décima et al. 2010). Uruguay is the southern limit of the species geographical distribution (Fig. 12.2). The gray brocket deer ranges from open to leafy areas,

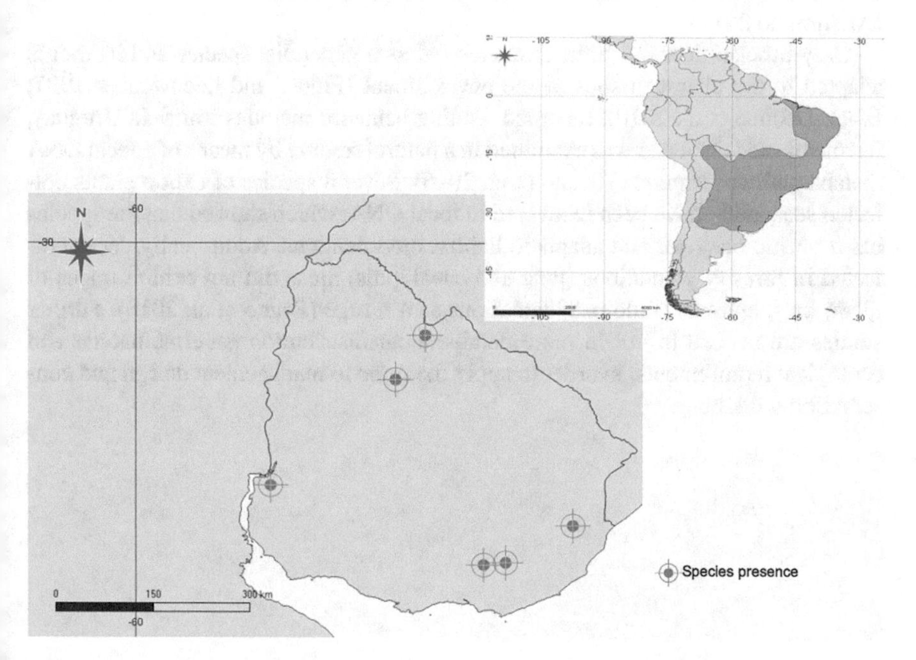

Fig. 12.2 Grey brocket deer geographic distribution range in Uruguay and global distribution (upper inset)

from moderately humid to dry regions, and avoids very open areas, swampy meadows and dense forests (Black-Décima et al. 2010, Black-Decima and Vogliotti 2016).

In Uruguay was reported in small forestry areas, riparian, gallery and range forest, as well in exotic plantations (Fig. 12.2, González 2001).

12.3.2 In Situ Population

This species lives in riverside and mountain range forests, as well as in small wooded areas. It is an elusive mammal and direct observation studies such as census or behavior are difficult to perform. In Uruguay, the species is cited in ten departments based on reports from scientific collections, studies with remote cameras and feces (González and Elizondo 2010).

Genetic analyses were conducted through non-invasive sampling methods: fecal DNA, tissue of road killed animals and hair of individuals captured from the wild (living afterwards in captivity). Mitochondrial DNA approaches were carried out using DNA fragments of cytochrome b (*Cyt b*) gene and *D-loop* control region (Bidegaray-Batista 2003; Elizondo 2010; Aristimuño 2013, 2017; Aristimuño et al. 2015). These studies described high polymorphism levels with discrete genetic substructure. Results were supported by the analysis of ancient DNA from archaeological samples, dated 2000 years ago in which two new haplotypes were described (Moreno et al. 2016). This supports the hypothesis that the gray brocket had high levels of gene flow and high population sizes in the recent past; additionally an event of population expansion in the middle Pleistocene was estimated (Aristimuño 2017).

Gray brocket deer has been characterized as a generalist species and its diet is adapted to the characteristics of the environment (Pinder and Leeuwenber 1997; Black-Décima et al. 2010). Its broad feeding behavior includes fruits. In Uruguay, its role of seed disperser was examined in a natural reserve by means of a fecal DNA metabarcoding technique (Bruno et al. 2016). Several species of exotic plants collected seasonally have been recovered in fecal DNA, which showed that the species has a plastic behavior and adapts to habitat modifications. Additionally, feces collected in forestry plantations (pine and eucalyptus) areas did not exhibit traces of them; thus, animals could use plantations as a refuge (Bruno et al. 2016). Further studies are needed to obtain more details about distribution patterns, habitat and ecological requirements, in order to apply the same to management design and conservation guidelines.

12.3.3 *Ex Situ Populations*

The species is in captivity, alone or as couples, in Uruguay in few zoos (Rodríguez and Rohrer 2014). Their presence is mainly due to sporadic and random episodes of capture. Currently there is no management program or information available about the captive brown brocket deer. In this scenario, it is necessary to carry out an ex situ animal census in order to generate a first database.

12.3.4 *Conservation Status*

The IUCN Red List™ classified the species as Least Concern with a decreasing trend of the population (Black-Decima and Vogliotti 2016). In Uruguay the available data is not enough to make an adequate categorization of the species (González 2001; Aristimuño 2017). As gray brocket deer exhibited high levels of genetic diversity, it probably has the potential of being widely distributed in the native and exotic forests of Uruguay. The main threats are the poaching pressure, the land use change and habitat fragmentation, as well the expansion of the urbanized areas (Black-Decima and Vogliotti 2016; Aristimuño 2017).

12.4 Pampas Deer (*Ozotoceros bezoarticus*)

The Pampas deer (*Ozotoceros bezoarticus*, Linnaeus, 1758), weights from 20 to 40 kg and the shoulder height varies from 62 to 70 cm (González et al. 2010). The coat color is bay and differs geographically according to subspecies. Fawns are spotted until 3 months of age (González et al. 2010). Males have typically antlers with three tines, being one of the tines closer to the base; meanwhile the other two generates a branch (González et al. 2010; Ungerfeld 2015).

The pampas deer has a preorbital gland with a characteristic scent of onion or garlic (González et al. 2010; Ungerfeld 2015).

Group size ranges from 5 to 17 individuals of both sexes and all ages (González and Cosse 2000; Moore 2001; Sturm 2002). Moreover, larger feeding groups were reported in agricultural areas such as Los Ajos (González 1997; Cosse and González 2013).

In recent past, the jaguar (*Panthera onca*) and the puma (*Puma concolor*) were their natural predators in Uruguay. Actually, the pampas fox (*Pseudalopex gymnocercus*), the ocelot (*Leopardus pardalis*), and the introduced wild boar (*Sus scrofa*) would be potential predators for fawns (Jackson and Langguth 1987).

The evaluation of reproductive condition on captured females in Los Ajos population showed high reproductive success levels with 87.5% pregnant or lactating females between 18 months and 5-years old. This information indicates that the reproductive maturity would be reached at 11 months in this population (González and Duarte 2003).

In Uruguay, fawns are generally born in spring (between October and December). This seasonality is clearer in El Tapado population (Jackson and Langguth 1987; González 1997; Sturm 2002) than in Los Ajos population, where new-born fawns have been seen all year-round including midwinter (González 1997; Cosse and González 2013).

12.4.1 Habitat and Distribution

Formerly, pampas deer had a wide distribution range, occupying open habitats of Brazil, Paraguay, Argentina, Bolivia, and Uruguay (Cabrera 1943; Jackson 1987; Merino et al. 1997; González et al. 2002; Weber and González 2003; Cosse and González 2013). The species has suffered a sharp population decline from the beginning of the twentieth century.

Habitat loss due to land modification for agriculture, combined with the introduction of wild and domestic ungulates, and urbanization have been identified as the main factors affecting the pampas deer populations. Likewise, overhunting and animal diseases contributed to population decline (Giménez Dixon 1987). Currently, the species inhabits small and isolated populations (Fig. 12.3, González et al. 1998). However, pampas deer retain high levels of genetic diversity, showing a potential for recovery and expansion (González et al. 1998). In Uruguay, were described two subspecies *O. b. arerunguaensis* and *O. b. uruguayensis* (Fig. 12.3, González et al. 2002).

12.4.2 In Situ Population

Two isolated populations remain in Uruguay: "El Tapado" in the northwest of Uruguay (Arerunguá, Salto Department) and "Los Ajos" in the southeast (Rocha Department, Fig. 12.3). Both are located in several private cattle ranches (González et al. 2002).

González et al. (1998) recognized phylogenetic patterns using genetic mitochondrial markers. Latter González et al. (2002) based on significant craniometric differences, described two subspecies in Uruguay: *O. b. arerunguaensis* and *O. b. uruguayensis*. Each subspecies occupies a different isolated population, at El Tapado and Los Ajos respectively (Fig. 12.3).

Pampas deer at El Tapado: The area is characterized by an open pasture with scattered trees (mostly eucalyptus), mainly used for the extensive breeding of sheep and cattle. The population is about 1000–1500 individuals that cohabit with cattle (Cosse and González 2013). The population size of pampas deer is related with

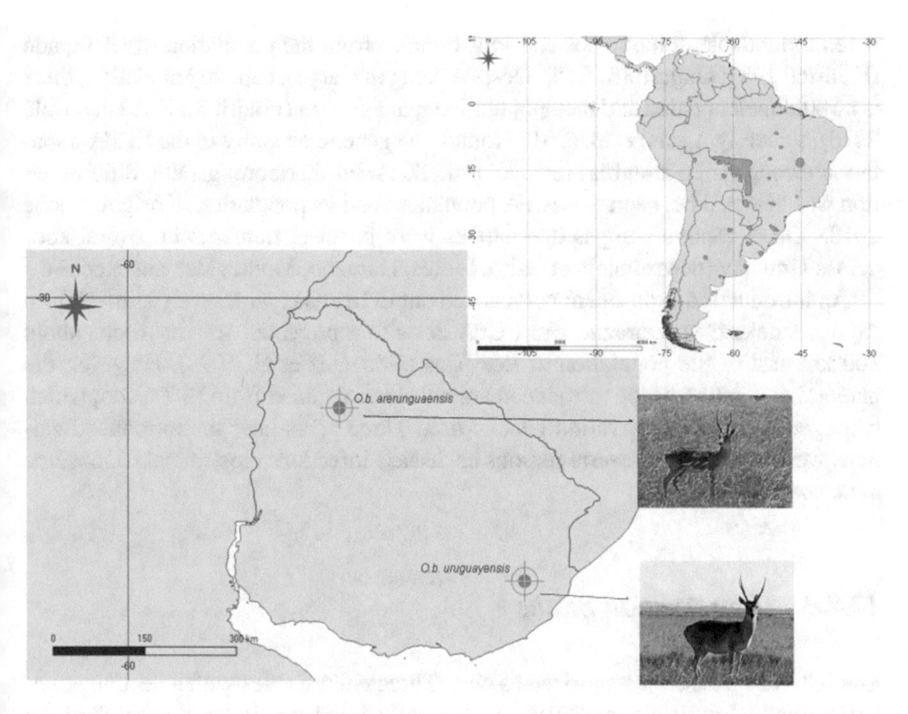

Fig. 12.3 Pampas deer geographic range and main populations (upper inset). For Uruguay is shown male pictures of subspecies *Ozotoceros bezoarticus arerunguaensis* and subspecies *Ozotoceros bezoarticus uruguayensis*

sheep stock size; the larger the sheep stock the smaller the pampas deer population size (Sturm 2002).

Pampas deer at Los Ajos: It is estimated that 350–400 animals inhabit the area, and almost the entire population live in a ranch on the "Bañados del Este" Biosphere Reserve MAB/UNESCO (Cosse and González 2013). However, some animals are expanding their home range to adjacent ranches (González *pers. comm*). The land use in the area is mostly based on livestock and crops. It has been reported higher densities of animals on natural grasslands or on rye grass pastures rather than on crops or sheep presence (Cosse et al. 2009). A low-level competition for food with sheep was documented, therefore agriculture activities are compatible with pampas deer presence but caution on cattle density is needed (Cosse et al. 2009).

12.4.3 Ex Situ Populations

The largest captive stock of *O. b. arerunguaensis*, with an average in the last years of 100 individuals, is located at the "Estación de Cría y Fauna Autóctona, Tabaré González" (ECFA) in Maldonado department, Uruguay (González Sierra 1985; Ungerfeld 2015; Leone 2018). This ex situ population was founded in 1981 with seven individuals, three males and four females from the population of El Tapado (Frädrich 1987; Ungerfeld 2015). Despite being the largest captive population, there is no studbook updated and metapopulation management (Frädrich 1987; Ungerfeld 2015). Recently, Leone et al. (2018) found low genetic diversity in the ECFA population through mitochondrial molecular markers. Furthermore, genetic differentiation was detected between the ECFA population and its population of origin (Leone 2018). Other captive pampas deer stocks were in lower numbers in several zoos across Uruguay (departments of Salto, Flores, Durazno, Montevideo and Rocha).

An introduction experiment was carried out in Uruguay with seven animals from captive stocks from Durazno and ECFA Zoos to a protected area in Rocha about 100 km east of the population in Los Ajos (González et al. 1998). However, the parental generation of the introduced animals was obtained from El Tapado, which belonged to other conservation genetic unit. Three years later the introduced animals were extinct, due several reasons as disease infections clostridiosis (González *pers. comm*).

12.4.4 Conservation Status

The IUCN Red List™ categorized as Near Threatened species with a decline population trend (González et al. 2016). Additionally, it is listed in "Appendix I" of the CITES Convention (CITES 2017). In Uruguay, pampas deer was declared a Natural Monument in 1985, which implies the legal protection of the species (Decree 12/9/85).

The Pampas deer is endangered in Uruguay, with fewer than 1800 individuals (González et al. 2016). The main threats to survival are: habitat loss, the introduction of domestic and wild ungulates and their associated diseases and excessive hunting by farmers who believe that deer compete with livestock (González and Sans 2009; González et al. 2016). Additionally, human–wildlife conflicts emerged in livestock areas for being considered a pasture competitor (González et al. 2010).

The landowners have an important role for conservation policies in both subspecies because in Uruguay the species inhabit in private ranches. A successful program that involved the landowners and offered them an incentive to perform habitat restoration that safeguard the species was implemented during 2008 (Proyecto "Producción Responsable-Ministerio Ganaderia, Agricultura y Pesca," González and Sans 2009; Bruno and Parrilla 2011). This project has concluded and there are no other promotion actions for landowners to develop sustainable activities friendly with wildlife. Conservation actions in private lands should be promoted for pampas deer conservation in Uruguay.

The current levels of genetic diversity suggest that historic population sizes were larger in several orders of magnitude, thus recently populations have decreased dramatically providing a strong mandate for restoration and growth (González et al. 1998). Population size might increase if they were protected from poaching in areas where natural habitats remain and if some grazing land, as a buffer, could be designated for dual use by deer and livestock. The genetic data suggest that pampas deer have the potential to exist over a much greater area and historical data demonstrate a much wider distribution for the species. Therefore, if the goal of conservation is to maintain long-term population stability and preserve genetic variation, conservation efforts should focus on the restoration of deer habitats and their reintroduction over a wide geographic area.

12.5 Marsh Deer (*Blastocerus dichotomus*)

The marsh deer (*Blastocerus dichotomus* Illiger, 1815) is the largest South American deer, with a body length of 91–153 cm, shoulder height of 110–127 cm and weight of 80–150 kg (Cabrera and Yepes 1960; Pinder and Grosse 1991). Adult males have branched antlers that normally have five points per side, up to 60 cm in length along the beam, and a spread of 50 cm (de Miranda Ribeiro 1919: Miller 1930; Cabrera 1945; Hoffmann et al. 1976; Whitehead 1993). Old animals have been reported with up to 28 points forming asymmetrical shelves (Cabrera 1945). The marsh deer's coat is brownish red, with paler chest, neck and flanks. The muzzle and legs up to the radius and tibia are black. As an adaptation to swamp habitats, they have the toes connected by an interdigital membrane that can spread up to 10 cm (Hoffmann et al. 1976).

12.5.1 Habitat and Distribution

Inhabit swamp habitats with water depth up to 70 cm (Schaller and Vasconcelos 1978; Tomas 1986; Beccaceci 1994; Tomas et al. 1997). Tomas and Salis (2000) suggested that primary marsh deer habitat can be best defined as the zone between upland-wetland and wetland-open water ecotones of large floodplains in South America.

The species is well represented in the archaeological record, showing its human exploitation during the late Holocene (Moreno 2016). Historically it occupied a vast region in central South America, covering most of the lowland areas located east from the Andes to the southern headwaters of the Amazon basin to the delta of Río de la Plata (Fig. 12.4). In Uruguay was cited in the departments of Rocha and Treinta y Tres in the wetlands of the "*Bañados del Este*" and in the west along the Uruguay River (González 1994).

The last scientific record in collections corresponds to the "*Bañado de Los Indios*," in the department of Rocha, where a specimen was hunted in 1959 (Fig. 12.4, González 1994). Another specimen poached in 1991 was recorded near the town of "Villa Soriano" in the department of Soriano (Fig. 12.4). A skull was recovered in 2010 on the border of the coast of the Solís Grande stream (border of

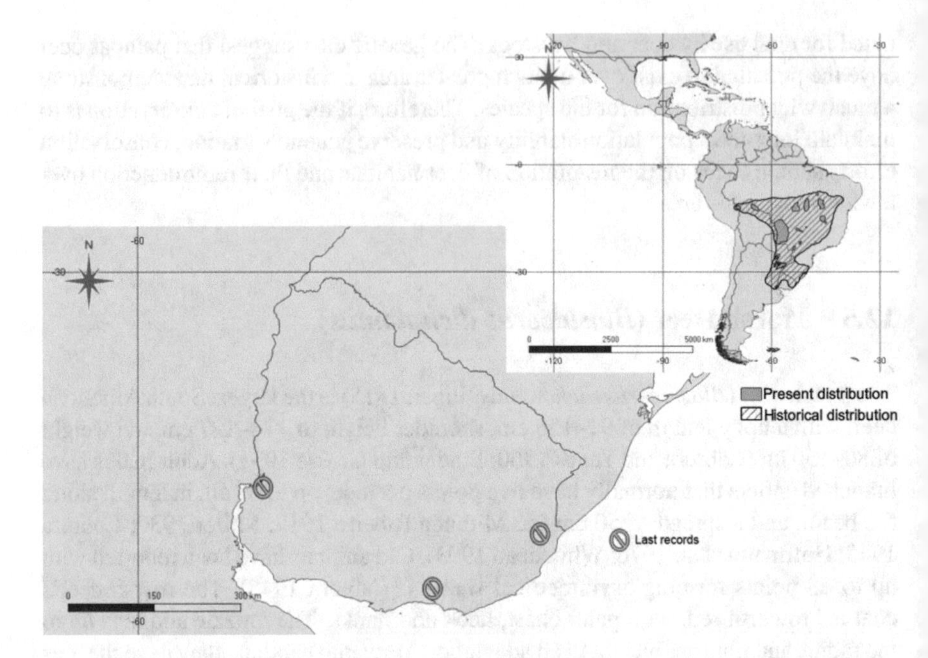

Fig. 12.4 Marsh deer geographic distribution range. Historic and present global distribution (upper inset), including the last records reported in Uruguay

the departments of Canelones and Maldonado, Fig. 12.4, González and Aristimuño 2014). Currently there are nor ex situ or in situ populations in Uruguay.

12.5.2 Conservation Status

The IUCN Red List™ classified as "Vulnerable" (Duarte et al. 2016), and also is included in the Appendix I of CITES (CITES 2017). In Uruguay was considered extinct (González 1994).

Reservoirs for hydroelectric power plants have caused a considerable loss of habitat for marsh deer (Tomas et al. 1997; Weber and González 2003; Piovezan 2004; Márquez et al. 2006; Piovezan et al. 2010) but also intensive agricultural use of flood plains and poaching are important threats (Tiepolo et al. 2004).

12.6 Uruguayan Ungulate Conservation Implications and Future Perspectives

Many variables affect the long-term viability and survival of the ungulate species. Extrinsic factors such as the habitat transformation ratio by human activities, the transmission of diseases and the intrinsic factors that must be evaluated for each particular case, such as genetic variability and demography. Habitat transformation is generally associated with the decline of wild populations because it generates a reduction in the connectivity between populations (Rey Benayas et al. 2009). It is important to evaluate if there are for each species demographic partitions that could affect gene flow patterns and determine the hierarchical distribution of the genetic diversity among populations (Moritz 1995; Crandall et al. 2000). Additionally it is necessary to analyze the geographic range to accurately assess the conservation status and the main threats.

In situ conservation guidelines are based on the successful implementation of natural protected areas and biological corridors as the primary tool to conserve biological diversity. In the Uruguayan current scenario, in which the advance of urbanization and human activities have been accelerating ecosystem fragmentation, it is crucial to integrate conservation actions in public and private lands devoted to agriculture, livestock, and forestry.

In particular, conservation measures needs to be implemented for pampas deer and gray brocket deer, which have populations decline that are directly linked to anthropogenic activities (Black-Décima et al. 2010; González et al. 2010, 2017).

In general, ecosystem conservation in private lands is a complement to the national protected public areas system. In Uruguay, in which protected areas represent less than 1% of the territory, it is crucial to design strategies to integrate private lands in conservation programs. We recommend to base the strategies on the following issues: (1) identification of indicator species, such as large mammals, to assess conservation status and ecosystem health, (2) involve local communities through dissemination campaigns and environmental education to achieve the effective integration with the National Protected Areas System, (3) establish new friendly production systems and alternatives business using certification schemes, and (4) inter and trans-disciplinary cooperation to establish private protected areas and design biological corridors.

Acknowledgments We acknowledge Dr. Sonia Gallina for inviting us to contribute to this book. We are grateful to the following agencies that funded our research: Agencia Nacional de Investigación e Innovación (ANII), Comisión Sectorial de Investigación Científica de Universidad de la República (CSIC-UDELAR), Programa de Desarrollo de Ciencias Básicas (PEDECIBA), Whitley Fund for Nature, and Lóreal-Unesco- ANII-MEC from Uruguay.

References

Aristimuño MP (2013) Variabilidad genética en poblaciones de *Mazama gouazoubira* (Mammalia: Cervidae Fischer, 1814) de Uruguay y la región. Bachelor dissertation. Facultad de Ciencias, Universidad de la República Oriental del Uruguay

Aristimuño MP (2017) Análisis genético de poblaciones de *Mazama gouazoubira* (Fischer, 1814). Implicancias para su conservación. Master degree dissertation. PEDECIBA-Facultad de Ciencias, Universidad de la República Oriental del Uruguay

Aristimuño MP, González S, Duarte JMB (2015) Population structure and genetic variability of the gray-brocket deer (*Mazama gouazoubira*; Mammalia: Cervidae) in Uruguayan populations. Deer Specialist Group Newsl 27:58–66

Beccaceci MD (1994) A census of marsh deer in Iberá Natural Reserve, its Argentine stronghold. Oryx 28:131–134

Bidegain M, Caffera R (1997) Clima del Uruguay. https://www.rau.edu.uy/uruguay/geografia/Uy_c-info.htm

Bidegaray-Batista L (2003) Variación genética de *Mazama gouazoubira*. Bachelor dissertation. Facultad de Ciencias, Universidad de la República del Uruguay

Black-Decima PA, Vogliotti A (2016) *Mazama gouazoubira*. The IUCN red list™ of threatened species 2016: T29620A22154584. https://doi.org/10.2305/IUCN.UK.2016-2.RLTS.T29620A22154584.en. Accessed 06 Feb 2019

Black-Décima P, Rossi RV, Vogliotti A, Cartes JL, Maffei L, Duarte J, González S, Juliá JP (2010) Brown brocket deer *Mazama gouazoubira* (Fischer 1814). In: Duarte J, González S (eds) Neotropical Cervidology. Biology and medicine of Latin American deer. FUNEP/IUCN, Jaboticabal, pp 190–201

Bruno A, Parrilla MA (2011) Proyecto Producción Responsable: seis años de ejecución. Informe Técnico. Ministerio Ganadería Agricultura y Pesca

Bruno A, Mannise N, Iriarte A, Rodríguez A, De Mello A, Méndez E, Cosse M (2016) Utilización de una herramienta molecular Para el análisis de la dieta del guazubirá (*Mazama gouazoubira*) en Parque Salus. Lavalleja. Libro de Resúmenes IV Congreso Uruguayo de Zoología, Maldonado, Uruguay

Brussa C, Grela I (2007) Flora arbórea del Uruguay. Con énfasis en las especies de Rivera y Tacuarembó. COFUSA, Montevideo

Cabrera A (1943) Sobre la sistemática del venado y su variación individual y geográfica. Rev Mus de La Plata (NS) Tomo III Zool 18:5–41

Cabrera A (1945) Sobre algunas cornamentas de *Blastocerus dichotomus*. Notas Mus de La Plata 10:221–231

Cabrera A, Yepes J (1960) Mamíferos Sudamericanos: Vida, costumbres y descripción. Ediar Buenos Aires

CITES (2017) Appendices I, II and III, valid from 4 October 2017. https://www.cites.org/esp/app/appendices.php. Accessed 19 Mar 2019

Cosse M, González S (2013) Demographic characterization and social patterns of the Neotropical pampas deer. Springer Plus 2:259

Cosse M, González S, Gimenez-Dixon M (2009) Feeding ecology of *Ozotoceros bezoarticus*: conservation implications in Uruguay. Iheringia Sér Zoologia 99:158–164

Crandall KA, Bininda-Emonds ORP, Mace GM, Wayne RK (2000) Considering evolutionary processes in conservation biology. TREE 15:290–295

Cravino A, Giordano H, Villalba-Macías J (2018) Liberación experimental de pecarí de collar: conocimientos, información obtenida y dificultades tras un año de monitoreo. Libro de resúmenes del V Congreso de Zoología de Uruguay, Montevideo, Uruguay, p 169

Cracco M, García L, González E, Rodríguez L, Quintillán A (2005) Importancia global de la biodiversidad de Uruguay. Proyecto Fortalecimiento de las Capacidades para la Implementación del Sistema Nacional de Áreas Protegidas (URU/05/001). DINAMA/MVOTMA, PNUD/GEF.

De Miranda Ribeiro A (1919) Os veados do Brasil. Rev Mus Paul 11:209–307

Duarte JMB, Varela D, Piovezan U, Beccaceci MD, Garcia JE (2016) Blastocerus dichotomus. The IUCN Red List of Threatened Species 2016: e.T2828A22160916. https://doi.org/10.2305/IUCN.UK.2016-1.RLTS.T2828A22160916.en. Downloaded on 03 September 2019.

Elizondo MC (2010) Estudio de la variabilidad genética del *Mazama gouazoubira* (Mammalia: Cervidae) Fischer 1814. Bachelor dissertation. Facultad de Ciencias, Universidad de la República Oriental del Uruguay

Escudero R, Brussa C, Grela I (2004) Compilación, sistematización y análisis de la información disponible publicada o en proceso, descripción de la situación actual y propuestas de intervención. Informe de Proyecto Combinado GEF/IBRD “Manejo Integrado de Ecosistemas y Recursos Naturales en Uruguay”

Frädrich H (1987) Internationales Zuchtbuch für den Pampashirsch. Berlin

Giménez Dixon M (1987) La conservación del venado de las pampas. Min As Agr, Dir Rec Nat y Ecología, Pcia Buenos Aires

Gongora J, Reyna-Hurtado R, Beck H, Taber A, Altrichter M, Keuroghlian A (2011) *Pecari tajacu*. The IUCN red list™ of threatened species 2011: e.T41777A10562361. https://doi.org/10.2305/IUCN.UK.2011-2.RLTS.T41777A10562361.en. Accessed 17 Jan 2019

González S (1994) Marsh deer in Uruguay: population and habitat viability assessment. Workshop for the marsh deer (*Blastocerus dichotomus*). CBSG Publ Sect 4:1–6

González S (1997) Estudio de la variabilidad morfológica, genética y molecular de poblaciones relictuales de venado de campo (*Ozotoceros bezoarticus* L. 1758) y sus consecuencias para la conservación. PhD dissertation, Universidad de la República Oriental del Uruguay

González E (2001) Guía de campo de los mamíferos de Uruguay: introducción al estudio de los mamíferos. Vida Silvestre Ed, Montevideo

González S, Aristimuño MP (2014) Registro de ciervo de los pantanos (*Blastocerus dichotomus*) en el departamento de Canelones-Uruguay. Paper presented at the III Congreso de Zoología del Uruguay

González S, Cosse M (2000) In: Cabrera E, Mercoli C, Resquín R (eds) Alternativas para la conservación del venado de campo en el Uruguay. Manejo de fauna silvestre en Amazonia y Latino América, Paraguay, pp 205–218

González S, Duarte JMB (2003) Emergency pampas deer capture in Uruguay. Deer Specialist Group News Uruguay 18:16–17

Gonzaléz EM, Lessa EP (2014) Historia de la mastozoología en Uruguay. In: Ortega J, Martínez JL, Tirira D (eds) Historia de la mastozoología en Latinoamérica, las Guayanas y el Caribe. Editorial Murciélago Blanco y Asociación Ecuatoriana de Mastozoología, Quito y México DF, pp 381–404

González E, Martínez-Lanfranco A (2010) Mamíferos de Uruguay: guía de campo e introducción a su estudio y conservación. Ediciones de la Banda Oriental, Montevideo

González S, Sans C (2009) Diagnóstico del Área Prioritaria Arerunguá. Informe presentado al Proyecto de Manejo Integral de los Recursos Naturales y la Biodiversidad–Producción Responsable-Ministerio Ganadería Agricultura y Pesca

González S, Seal US (1997) El manejo del ciervo axis (*Cervus axis*) en la Residencia Presidencial de Colonia-Uruguay. IUCN/SSC Conservation Breeding Specialist Group, Apple Valley

González Sierra UT (1985) Venado de campo- *Ozotoceros bezoarticus*- en semi cautividad. Comunicaciones de estudios de comportamiento en la Estación de Cría de Fauna Autóctona de Piriápolis 1(1):1–21

González S, Maldonado JE, Leonard JA, Vilá C, Duarte JMB, Merino M, Brum-Zorrilla N, Wayne RK (1998) Conservation genetics of the endangered pampas deer (*Ozotoceros bezoarticus*). Mol Ecol 7:47–56

González S, Álvarez-Valin F, Maldonado JE (2002) Morphometric differentiation of endangered pampas deer (*Ozotoceros bezoarticus*), with description of new subspecies from Uruguay. J Mammal 83:1127–1140

González S, Cosse M, Goss Braga F, Vila A, Merino ML, Dellafiore C, Cartes JL, Maffei L, Gimenez-Dixon M (2010) Pampas deer *Ozotoceros bezoarticus* (Linnaeus 1758). In: Duarte JMB, González S (eds) Neotropical Cervidology. Biology and medicine of Latin American deer. FUNEP/IUCN, Jaboticabal, pp 119–132

González EM, Martínez-Lanfranco JA, Juri E, Rodales AL, Botto G, Soutullo A (2013) Mamíferos. In: Soutullo A, Clavijo C, Martínez-Lanfranco JA (eds) Especies prioritarias para la conservación en Uruguay. Vertebrados, moluscos continentales y plantas vasculares. SNAP/DINAMA/MVOTMA/DICYT/MEC, Montevideo, pp 175–207

González S, Jackson JE, Merino ML (2016) *Ozotoceros bezoarticus*. The IUCN red list™ of threatened species 2016: e.T15803A22160030. https://doi.org/10.2305/IUCN.UK.2016-1.RLTS.T15803A22160030.en. Accessed 06 Feb 2019

González S, Duarte JMB, Cosse M, Repetto L (2017) Conservation genetics, taxonomy and management applications in neotropical deer. In: Alonso Aguirre A, Sukumar R (eds) Tropical conservation. Perspectives on local and global priorities. Oxford University Press, New York, pp 238–254

Hoffmann RK, Ponde del Prado CF, Otte KC (1976) Registro de dos nuevas especies de mamíferos para el Perú, *Odocoileus dichotomus* (Illiger, 1811) y *Chrysocyon brachyurus* (Illiger, 1811) con notas sobre su hábitat. Rev For Peru 5:61–81

Jackson JE (1987) Ozotoceros bezoarticus. Mamm Species 295:1–5

Jackson JE, Langguth A (1987) Ecology and status of the pampas deer in the Argentinian pampas and Uruguay. In: Wemmer C (ed) Biology and management of the Cervidae. Smithsonian Institute, Washington, DC, pp 402–409

Leone Y (2018) Variabilidad genética en poblaciones uruguayas de venado de campo (*Ozotoceros bezoarticus*). Master degree dissertation PEDECIBA-Facultad de Ciencias, Universidad de la República Oriental del Uruguay

Leone Y, Cosse M, Elizondo C, González S (2018) Patrones de estructuración genética de poblaciones de venado de campo (*Ozotoceros bezoarticus* L. 1758). Libro de resúmenes V Congreso Uruguayo de Zoología, Montevideo, Uruguay

Márquez A, Maldonado JE, González S, Beccaceci MD, Garcia JE, Duarte JMB (2006) Phylogeography and Pleistocene demographic history of the endangered marsh deer (*Blastocerus dichotomus*) from the Río de la Plata Basin. Conserv Genet 7:563–575

Merino ML, González S, Leeuwenberg F, Rodrigues FHG, Pinder L, Tomas WM (1997) Veado-campeiro (*Ozotoceros bezoarticus*). In: Duarte JMB (ed) Biologia e conservação de cervídeos Sul-americanos: *Blastocerus*, *Ozotoceros* e *Mazama*. FUNEP, Jaboticabal

Miller FW (1930) Notes on some mammals of southern Matto Grosso, Brazil. J Mammal 11(1):10–22

Mones A (2001) La mastozoología en el Uruguay: pasado y presente. Com Zool Mus de Hist Nat Montevideo XIII:197

Moore DE (2001) Aspects of the behavior, ecology and conservation of the Pampas deer. PhD dissertation, University of New York

Moreno F (2016) La gestión animal en la prehistoria del Este de Uruguay: de la economía de amplio espectro al control de animales salvajes. Tessituras 4(1):161–187

Moreno F, Figueiro G, Mannise N, Iriarte A, González S, Duarte JMB, Cosse M (2016) Use of next-generation molecular tools in archaeological neotropical deer sample analysis. J Archaeol Sci Rep 10:403–410. https://doi.org/10.1016/j.jasrep.2016.11.006

Moritz C (1995) Uses of molecular phylogenies for conservation. Philos Trans R Soc Lond B 349:113–118

Nowak RM, Walker EP (1999) Walker's mammals of the world, Vol. 2. JHU Press, Baltimore

Pinder L, Grosse AP (1991) Blastocerus dichotomus. Mamm Species 380:1–4

Pinder L, Leeuwenberg F (1997) Veado catingueiro (Mazama gouazoubira, Fisher, 1814). In J. M. Barbanti Duarte (ed.) Biologia e Conservação de Cervídeos Sul-Americanos: Blastocerus, Ozotoceros e Mazama. FUNEP. Jaboticabal, p. 60–68.

Piovezan U (2004) História natural, área de vida, abundância de *Blastocerus dichotomus* (Illiger, 1815) (Mammalia, Cervidae) e monitoramento de uma população à montante da hidrelétrica Sérgio Motta, rio Paraná, Brasil. PhD dissertation, Universidade de Brasília

Piovezan U, Tiepolo LM, Tomas WM, Duarte JMB, Varela D, Marinho Filho JS (2010) Marsh deer (*Blastocerus dichotomus* Illiger, 1815). In: Duarte JMB, González S (eds) Neotropical Cervidology. Biology and medicine of Latin American deer, Jaboticabal, pp 12–17

Prigioni CM, Sappa A, Villalba Macías JS (2018) Algunas consideraciones sobre la posible presencia del tapir (*Tapirus terrestris*) Mammalia, Perissodactyla, en el Uruguay, en tiempos históricos. Acta Zool Platense 2(12)

Rey Benayas JM, Newton AC, Diaz A, Bullock JM (2009) Enhancement of biodiversity and ecosystem services by ecological restoration: a meta-analysis. Science 235:1123–1124

Rodríguez E, Rohrer MV (2014) Relevamiento de mamíferos presentes en zoológicos del Uruguay. Graduate dissertation. Facultad de Veterinaria, Universidad de la República

Schaller GB, Vasconcelos JMC (1978) A marsh deer census in Brazil. Oryx 14:345–351

Sturm M (2002) Pampas deer (*Ozotoceros bezoarticus*) habitat vegetation analysis and deer habitat utilization, Salto, Uruguay. PhD dissertation, University of New University of New York College of Environmental Science and Forestry, Syracuse, New York, p 109
Tiepolo LM, Fernandez FA, Tomas WM (2004) A conservação da população de cervo-do-pantanal *Blastocerus dichotomus* (Illiger 1815) (Mammalia, Cervidae) no Parque Nacional de Ilha Grande e entorno (PR/MS). Natureza e Conservação 2:56–66
Tomas WM (1986) Observações preliminares sobre a biologia do cervo-do-pantanal (*Blastocerus dichotomus* Illiger 1811) (Mammalia Cervidae) no Pantanal de Poconé, MT. Bachelor dissertation. Universidade Federal d Mato Grosso
Tomas WM, Salis SM (2000) Diet of the marsh deer (*Blastocerus dichotomus*) in the Pantanal wetland, Brazil. Stud Neotropical Fauna Environ 35:165–172
Tomas WM, Beccaceci MD, Pinder L (1997) Cervo-do-pantanal (*Blastocerus dichotomus*). In: Duarte JMB (ed) Biologia e conservação de cervídeos sul-americanos. FUNEP, Jaboticabal, pp 24–40
Ungerfeld R (2015) Reproducción en los cérvidos: una revisión con énfasis en el venado de campo (*Ozotoceros bezoarticus*). Rev Bras Reprod Anim 39:66–76
Weber M, González S (2003) Latin American deer diversity and conservation: a review of status and distribution. Ecoscience 10(4):443–454
Whitehead GK (1993) The Whitehead Encyclopedia of deer. Swan Hill Press, Shrewsbury, UK

Chapter 13
Tropical Ungulates of Argentina

Patricia Black-Decima, Micaela Camino, Sebastian Cirignoli, Soledad de Bustos, Silvia D. Matteucci, Lorena Perez Carusi, and Diego Varela

Abstract Argentina has an extensive and diverse terrain classified into 11 ecoregions. Seven of these ecoregions, occupying the north and north-central parts of the country, house the 11 tropical ungulate species found here. The ecoregions are lowland and subtropical, some beginning in the tropics, some extending to temperate climates. The principal topographical characteristics, hydrology, climate, vegetation and fauna are described for these seven ecoregions. Each of the 11 species is then treated in detail with respect to its ecology and conservation. Emphasis is placed on distribution, habitat and density, feeding ecology, threats and conservation in Argentina, based on the most recent studies. Data on reproductive biology and behaviour are included where information is relatively recent and unlikely to be covered elsewhere. The species include the following: the Brazilian tapir (*Tapirus*

P. Black-Decima (✉)
Universidad Nacional de Tucuman, San Miguel de Tucuman, Tucuman, Argentina

M. Camino
Centro de Ecología Aplicada del Litoral - Consejo Nacional de Investigaciones Científicas y Técnicas, Corrientes, Argentina

S. Cirignoli
Asociación Civil Centro de Investigaciones del Bosque Atlántico (CeIBA), Bertoni 85, Puerto Iguazú, Misiones, Argentina

S. de Bustos
Secretaría de Ambiente y Desarrollo Sustentable de Salta y Fundación Biodiversidad Argentina, Salta, Argentina

S. D. Matteucci
CONICET-GEPAMA, Buenos Aires, Argentina

L. Perez Carusi
Administración de Parques Nacionales, Ciudad Autónoma de Buenos Aires, Buenos Aires, Argentina

D. Varela
Asociación Civil Centro de Investigaciones del Bosque Atlántico (CeIBA), Bertoni 85, Puerto Iguazú, Misiones, Argentina

Instituto de Biología Subtropical (IBS), CONICET - Universidad Nacional de Misiones, Puerto Iguazu, Misiones, Argentina

S. Gallina-Tessaro (ed.), *Ecology and Conservation of Tropical Ungulates in Latin America*, https://doi.org/10.1007/978-3-030-28868-6_13

terrestris), found in northern subtropical ecoregions, three species of peccary (*Tayassu pecari*, *Pecari tajacu* and *Parachoerus wagneri*) from northern subtropical and drier regions, of which the Chacoan peccary (*P. wagneri*) is endemic while the other two species have more extensive distributions. The guanaco (*Lama guanicoe*) occurs only in relict populations in the ecoregions considered. The taruca (*Hippocamelus antisensis*) occupies the eastern boundary between the Yungas and drier, high altitude ecoregions. Three species of brocket deer (*Mazama americana*, *M. gouazoubira* and *M. nana*) occupy the northern tropical, subtropical and Chacoan areas. The marsh deer (*Blastocerus dichotomus*), the largest South American deer, has small populations occupying wetlands from the northern border to the Parana delta, while the pampas deer (*Ozotocerus bezoaticus*) is found in four isolated populations from Ibera to Buenos Aires province. Argentina represents the southern limit to the distribution of all these species and thus threats are often magnified. Ongoing conservation activities include the maintenance of protected areas, promotion (difusion, education, sensitization), investigation and the reintroduction of some species of formerly extinct ungulates into the Ibera wetlands area.

13.1 Introduction

Argentina occupies the southern cone of South America, extending somewhat less than half the latitudinal extent of the continent. Only a small part of the provinces of Formosa, Salta and Jujuy are technically tropical, that is, located between the boundaries of the Tropic of Cancer and the Tropic of Capricorn. However, many of the ecoregions are continuations of more tropical regions to the north. Different species of tropical ungulates with centres of distribution in more tropical areas have extensive distributions in Argentina. These include one species of tapir, all three species of peccary, six species of deer and, marginally, one species of camelid with relict populations.

The South American or lowland tapir, *Tapirus terrestris*, is found throughout the Amazon basin in the eastern part of South America and up to northern Argentinat in the west. The southern end of its distribution is in Argentina. The Chacoan peccary, *Parachoerus wagneri*, has the most limited distribution of the peccaries; it is endemic to the Gran Chaco in Bolivia, Paraguay and Argentina. The collared peccary, *Pecari tajacu*, is found from the southwestern USA through Central and South America, to northern Argentina. The white lipped peccary, *Tayassu pecari*, is found through Central and South America to northern Argentina, mainly in rain forests, but also in other habitats, e.g. swamps or dry ecosystems. The deer species use a variety of habitats. The taruca, *Hippocamelus antisensis*, is found in Andean habitats in Peru and Bolivia, but also in the northwest of Argentina. The marsh deer, *Blastocerus dichotomus*, was originally found over a wide area of river basins and floodplains in the southern half of Brazil, parts of Peru, Bolivia, Paraguay, Uruguay, and in a wide area of northern and central Argentina, down to the Parana Delta. It currently is found in the same geographical area, with the exception of Uruguay, but in isolated populations. Three species of brocket deer: *Mazama americana*, the red brocket, *M. gouazoubira*, the brown or gray brocket, and *M. nana*, the Brazilian dwarf brocket deer, inhabit mainly forested regions of northern Argentina. The red brocket is found from Venezuela and Colombia south to northern Argentina, east of the Andes. It probably consists of several cryptic species. The gray brocket is found south of Amazonia in Brazil, Paraguay and Bolivia, down to central Argentina, and occurs in drier areas than the other brockets. The Brazilian dwarf brocket deer is found in the Atlantic rain forest in Brazil, Paraguay and the northern part of the province in Misiones, Argentina. The pampas deer, *Ozotocerus bezoarticus*, was

Fig. 13.1 Subtropical ungulates from Argentine Ecoregions: (**a**) *Tapirus terrestris* (Darío Podestá); (**b**) *Lama guanicoe* (Darío Podestá); (**c**) *Tayassu pecari* (Sebastian Cirignoli); (**d**) *Parachoerus wagneri* (Micaela Camino); (**e**) *Pecari tajacu* (CeIBA/Diego Varela); (**f**) *Ozotoceros bezoarticus* (Sebastian Cirignoli); (**g**) *Blastocerus dichotomus* (Darío Podestá); (**h**) *Hippocamelus antisensis* (Marcelo Ortíz); (**i**) *Mazama nana* (CeIBA/Diego Varela); (**j**) *Mazama gouazoubira* (Darío Podestá); (**k**) *Mazama americana* (CeIBA/Diego Varela)

extremely abundant in grassland areas of Brazil, Bolivia, Paraguay, Uruguay and Argentina. There are still reasonably large populations in Brazil, but only small and isolated populations exist in the rest of the distribution (Duarte and González 2010). Argentine tropical ungulates are shown in Fig. 13.1.

Argentina also contains other ungulates that live in cooler climates or are mainly distributed in the temperate south. There are two camelid species, the guanaco (*Lama guanicoe*) and the vicuña (*Vicugna vicugna*). The guanaco is mainly found in Andean areas in western Argentina and throughout Patagonia. It was also found in the Gran Chaco in Bolivia and Argentina, but now only a relict population remains in Bolivia. The vicuña is found in the puna, the high Andean plains, from Peru to northern Argentina and Chile. In the south two more native deer species are found: the huemul (*Hippocamelus bisulcus*) and the southern pudu (*Pudu puda*). The huemul was probably originally found throughout Patagonia, but it is now confined to mountainous regions around the Andes in Argentina and Chile (Huemul Task Force 2012). The southern pudu is found in southern temperate rainforests in Argentina and Chile.

This chapter considers the characteristics of the ecoregions in which the tropical ungulates of Argentina are found and then provides current information on these ungulates, which include 11 of the 14 species mentioned for Argentina (Teta et al. 2018). It is intended to update the information available on Neotropical deer species provided by the book Neotropical Cervidology (Duarte and González 2010). It also provides a review of the status and the most recent information on all tropical ungulates found in Argentina.

13.2 Argentine Ecoregions Inhabited by Subtropical Ungulates

Tropical ungulates inhabit seven of the 16 Argentine ecoregions; specifically those in lowlands and subtropical climates. An ecoregion is a territory dominated by uniform characteristics of relief, geology, large groups of soil, geomorphogenetic processes, vegetation types and animal species complexes. From the evolutionary point of view, the ecoregion is characterized by homogeneous ecological responses to tectonics and climate. Landscape units within an ecoregion exchange organisms, matter and energy to a greater or lesser degree; exchanges are much restricted among ecoregions. Each of them extends widely from North to South or from West to East, or both; therefore, they house varied ecosystems, according to local natural and social conditions. This implies that the ungulates of a certain species are generally not found all over the ecoregion, but only in those landscape units in which the conditions for their survival are met. Argentine ecoregions containing tropical ungulates are shown in Fig. 13.2.

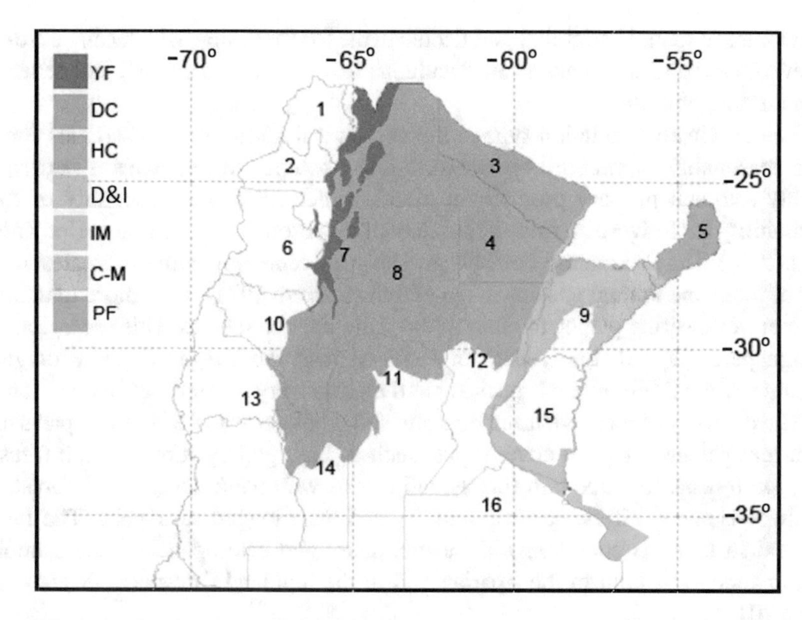

Fig. 13.2 Argentine Ecoregions inhabited by subtropical ungulates. Ecoregions: YF: Yungas Forest; DC: Dry Chaco; HC: Humid Chaco; PF: Paranaense Forest; IM: Ibera Marshes; C-M: Campos and Malezales Ecoregion; D&I: Delta and Islands of the Paraná and Uruguay rivers. Numbers identify the provinces: (1) Jujuy; (2) Salta; (3) Formosa; (4) Chaco; (5) Misiones; (6) Catamarca; (7) Tucumán; (8) Santiago del Estero; (9) Corrientes; (10) La Rioja; (11) Córdoba; (12) Santa Fe; (13) San Juan; (14) San Luis; (15) Entre Ríos; (16) Buenos Aires

13.2.1 Paranaense Forest Ecoregion

The Paranaense Forest Ecoregion is the southern extension of the Great Biogeographic Unit of the Atlantic Forest. In Argentina it occupies the basins of the Paraná, Iguazú and Uruguay rivers. It is the second smallest ecoregion, with 27,127 km^2. It is one of the two Argentine ecoregions with humid tropical and subtropical forests, together with the Yungas Forest.

The relief comprises relatively flat areas with deep soils, between 150 and 250 m above sea level, up to a flat plateau between 550 and 800 m above sea level. The spaces between the main rivers and the plateau have steep slopes and are exposed to soil erosion when they are cleared. It has an extensive hydrographic system, with numerous permanent water courses with springs in the mountains (López and Cámara 2007).

The climate is hot and humid, with great thermal and rainfall amplitudes, whose values range depend on the location in the N-S direction. The spatial variation of temperature depends on factors that operate at very different scales, such as latitude, elevation, and forest cover. In the winter months (June to August) frosts occur. Rainfall varies between 1000 and 2000 mm and decreases from N to S. In some areas there may be up to 5 dry months, usually in winter. The thermal and pluviometric spatial variability is reflected in the forest canopy, which can be closed, of evergreen, deciduous and semi-deciduous leaves, and late or early fall of leaves or a mixture of both.

The dominant vegetation type is the subtropical humid semi-deciduous forest. The seasonality of rainfall, temperature and photoperiod generates a pattern of highly seasonal primary productivity that is reflected in the seasonality of food availability for folivorous, frugivorous and insectivorous species (Manso Hernández et al. 2010). The Paranaense Forest is probably the ecoregion with the greatest number of plant and animal species in Argentina (Cabrera 1971), with more than 2000 species of flowering plants, of which about 10% are tree species. This ecoregion has a high percentage of tree genera in common with the Yungas Forest ecoregion, together with closely related species, but they have very few species in common.

The diversity of local environments and soil types generates a complex pattern of recurrent patches of plant cover types, such as high gallery forests, flood forests, bamboo forests, low tree fern forests, tall forests with palm trees, mixed forests of hardwoods and conifers, pastures and dry and waterlogged savannahs. The forest vertical structure is complex, with several dense and heterogeneous tree canopies and understories, both in the riparian and in the highland forests (Rodríguez and Silva 2012a).

13.2.2 Yungas Forest Ecoregion

The Yungas Forest Ecoregion, with almost 45,000 km^2, is covered by subtropical moist forests. It is located in the mountain system of NW Argentina, between 400 and 3000 m above sea level. It stretches from N to S about 600 km, of just under 100 km width. The landscape consists of patches of jungles in a matrix of forests and xerophilous savannahs. The west facing slopes are covered by arid ecosystems. These characteristics, caused by the heterogeneity of topography, give rise to a heterogeneous vegetation cover and richness of endemisms.

The climate is warm and humid to subhumid. Rainfall varies from east to west, partly due to the rise of winds caused by the relief; moisture is discharged as the temperature decreases with height. In addition, the temperature varies seasonally with average maxima between 20 and 30 °C and average minima between 10 and 15 °C, according to altitude. Average annual rainfall also varies with altitude, from 500 to 1000 mm in the piedmont to 1550 mm at 1350 m above sea level, and 2500 mm in the cloud forest zone.

The main vegetation types in the east–west rainfall gradient are warm and humid pedemontane or transitional forests; temperate warm and humid montane or evergreen forests; temperate with frequent humid winter frosts montane forest or mist forest; temperate cold and subhumid montane grasslands, alternating with patches of montane forests and shrubbery, to pure herbaceous communities at the peaks.

The foothill jungles are located at an altitude between 400 and 700 m, and constitute an interface between the Humid Forest and the Dry Chaco Forest. Foothill forests represent the floor of the Yungas Forests, with the highest percentage of exclusive tree species, with more than 70% of deciduous individuals, making it one of the most seasonal forest systems in South America, and with the largest number of timber species, of which 12 are exploited.

The montana jungle is located on the slopes between 700 and 1500 m above sea level. It comprises two forest types: Laurel Forest at the foot of the mountain ranges, and the Myrtaceae Forest between 800 and 1500 m above sea level, in both of which evergreen species of tropical origin predominate.

The montane forest is located at the cloud forest altitude, with almost continuous presence of clouds, between 1500 and 3000 m above sea level. The characteristic formations are the Pine Forests; the Alder Forests; and the Queñoa forests, which are low (4–6 m).

From 1800 to 3000 m above sea level, misty pastures with montane forests are found. In the foothill jungle there are lentic wetlands, near the angle formed by the mountain slope and the Chaco plain. They include dry river courses (locally called madrejones), lagoons, marshes and reservoirs of an average surface of 78 ha (Rodríguez and Silva 2012b).

13.2.3 Dry Chaco Ecoregion

The Dry Chaco Ecoregion, with an area of 498,953 km^2, is located in the north of Argentina. It is a vast sedimentary plain with a very slight slope to the East, which has generated gaps in horseshoe formed by meanders of the main rivers that come down from the Andes. Due to its great extension and amplitude from N to S, as well as from E to W, it includes a great variety of environments and climatic conditions, with extensive plains, alternating with occasional mountain ranges, large rivers with old channels and oxbow lakes, dry and flooded savannahs, estuaries, marshes and saltpeter deposits, all of which house different types of ecosystems. All these characteristics determine the presence of high diversity of animal and plant species, and a great variety of plant formations.

The climate is subtropical warm continental. It houses the South American heat pole where absolute maximum temperatures exceed 47 °C. The annual average temperature at the N end is 23 °C and in the South, 10 °C. The daily thermal amplitude is high, associated with the great seasonal variation. In winter, the entrance of cold fronts causes frosts; the absolute minimum temperatures oscillate around 6 °C in the plains.

The climax community is the forest dominated by two species of great size that are found throughout the region: Quebracho Santiagueño (*Schinopsis quebracho colorado*) and the White Quebracho (*Aspidosperma quebracho blanco*). The White Quebracho has persistent foliage and the Quebracho Santiagueño is deciduous, with the leaves falling at the beginning of spring; it maintains its foliage throughout the winter, therefore, it protects the ecosystem from the harsh weather during this period of scarce rains. The second stratum is formed mainly by trees of the genus *Prosopis*;

in the third stratum are *Cercidium praecox*, *Zizyphus mistol* and *Geofroea decorticans*. A high density bushy layer with shrubs of various sizes completely covers the low level. The soil is covered with grasses of various sizes, with several species of grasses and herbs. In the upper layer, evergreen and deciduous species are found, and among the herbaceous layer there are annual and evergreen species. There are also some vines and epiphytes (Ledesma 1992). There is a great variety of edaphic communities, formed by different species and life forms, depending on local soil conditions, topography, water availability, microclimate and intensity of use by local communities, which can alter the specific composition, forest height or their extension, or the three qualities simultaneously.

The native forest cover is significantly reduced due to long-standing deforestation. The exploitation of the Chaco forests began in the 1870s and lasted until 1950, with the development of the tannery industry in the hands of La Forestal and other English companies (Zarrilli 2004). "The entire national railway system settled its tracks on rot-proof sleepers of a single species, the red quebracho santiagueño (*Schinopsis lorentzii*). The entire national tanned leather industry depended on tannin, processed material from another red quebracho, the chaqueño (*Schinopsis balansae*), and the entire wiring system of an agro-exporter country depended on two or three native trees: ñandubay (*Prosopis affinis*), quebracho and carob trees" (Zarrilli 2004). Today, forest relicts alternate in a matrix of soybean plantations in large areas of the Dry Chaco.

13.2.4 Humid Chaco Ecoregion

Towards the East of the Dry Chaco is the Humid Chaco ecoregion, with an area of 123,153 km^2. It is a sunken block filled with the sediments of the Pilcomayo, Bermejo and Juramento rivers. It includes a variety of environments, such as large alluvial plains and large gullies; side hills along river coasts; inter-river areas; estuaries; hills interrupted by low riparian forests; concave plains with deficient drainage; flooding depressions; river floodplains and elevated blocks that function as water divides.

The presence and dynamics of rivers and streams, and the richness of wetlands justifies the ecoregion's name. The rainfall is monsoon, with 90% and 75% of the rainfall between October and April, respectively. The annual rainfall shows a gradient from W to E, from 750 to 1300 mm. The average annual temperatures occur in a N–S gradient, from 23 to 24 °C to 18–19 °C. Winters tend to be dry and the summers humid. Sometimes the winter drought is accentuated and prolonged, favouring savannah and grassland fires.

The Humid Chaco concentrates a great wealth of vegetation types, both wetlands (totorales, pirizales, pajonales, savannahs with palms, vinalares, "espinillares"), as well as forests, including riparian forests, high coastal forests and tall forests locally called "Monte Fuerte". The raleras (symmetric disposition of trees next to water courses) are located in depressions, other forests occupy the highlands, or an intermediate position, each one dominated by a different tree species. The greatest specific richness is found in the great alluvial plains, in the highlands (hills and hillocks) and in the prolongation of the alluvial plain of the Iberá estuary, in the eastern end of the ecoregion, where flooding grasslands, riparian jungles, albardón jungles, savannahs with forest islands, semideciduous forests, ñandubay and palmar savannahs and hard wood forests are found. The northern zone houses the greatest diversity of species, habitats, and land units originated by fluvial modelling, types of interfluvial water bodies and floating vegetation types (Morello 2012).

The adaptive strategies of the biota respond to: (a) normal or extraordinary winter droughts associated with fires; (b) irregular rainfall; (c) floods that may last a week to a semester; (d) deposition of sedimentary load and silting of estuaries; (e) redesign of the drainage network; (f) changes in the salinity-alkalinity of soils in some areas.

13.2.5 Campos and Malezales Ecoregion

This Ecoregion of Northeastern Argentina is an extension of the Paranaense Forest, which occupies the Southern Brazil and Eastern Paraguay plains. In Argentina it comprises 26,204 km^2. It differs from the neighbouring Ecoregions by its vegetation cover, of vast grasslands barely interrupted by small forest patches (Matteucci 2018a).

The climate is humid subtropical. The Ecoregion is enclosed by the 1800 mm and 1300 mm isopluvial lines to the Northeast and Southwest, respectively, with uniform rains all year round. Average temperatures fluctuate between 20 and 22 °C.

To the North, the relief is an over-raised platform, which is the extension of the basaltic Brazilian Planalto. The erosive modelling produces hillocks in the form of domes, bordered by low ridges crossed by short rivers; the elements of the landscape are plains, hills and waterlogged depressions. The Southern portion is formed by a sedimentary plain with slow runoff, without defined channels and presence of marshes; the predominant elements are the floodplains, marshes, longitudinal swamps and scrublands.

The differences of topographic pattern and hydrological regime between the North and South portions are reflected in the vegetation. To the North savannahs with patches of thickets, denseflower cordgrass scrublands, grasslands and red straw scrublands predominate, composed of herbaceous communities of a metre to a metre and a half in height, whose physiognomy and specific composition depend on their location on hills and slopes. The grasslands are interrupted by marshes and pastures dominated by *Paspalum* spp., flooded during much of the year. There are also gallery forests and patches of hydrophilic forests. They constitute the formation locally called "Campos", which, due to its subtropical and humid condition, lodges a great wealth of herbaceous species. Grasslands and pastures are interrupted by thin strips of river gallery forest and by small islands of forests locally called "capones" or "mogotes". Towards the South the "Malezales" are located, in which almost pure and very uniform grasslands predominate due to persistent soil flooding, with limitations caused by poor drainage. It is characterized by hosting a fauna adapted to flooded conditions.

The Campos and Malezales Ecoregion is very rich in species in relative terms, since its extension represents 0.2% of the Argentine territory, and it contains 31% of the genera and 51% of the families of the Argentinean flora. It is also very rich in animal species, partly due to the interaction of this Ecoregion with the neighbouring ones, since it includes Chaco and Parana species. Among the birds there are several species in danger of extinction. There is a great wealth of ophidians, with 45 species. Among the mammals there are worldwide vulnerable species, such as the pampas deer (*Ozotoceros bezoarticus leucogaster*) and the maned wolf (*Chrysocyon brachyurus*).

The Campos and Malezales Ecoregion is heavily disturbed by productive activities with forest, mate, tea and rice plantations in the North where acid soils prevail. Towards the South, ranching activities include fire and the drainage and channelling of wetlands. The capture and illegal trade of wildlife species is also practiced.

13.2.6 Ibera Marshes Ecoregion

The Ibera Marshes ecoregion is an extensive and diverse wetland system, among which there is a macro-wetland unique in South America, formed by a group of estuaries, marshes, shallow lakes and interconnected courses of different kinds (Neiff 2004). The ecoregion covers an area of 40,334 km^2 (Matteucci 2018b). It includes several functionally related ecosystems, among which the marshy environments (estuaries and marshes) that interconnect vast shallow lakes, joined by water courses of different order, predominate. It is one of the main superficial sources of clean water in Argentina (Neiff 2004).

The climate is humid subtropical and presents differences between localities at the North and South extremes. The average monthly minimum temperatures, recorded in June and July, are 16 °C and 17 °C respectively. The absolute minimum is −2 °C, and there is a low frequency of annual frosts. The maximum average temperature of 27 and 28 °C is recorded in January and February. Absolute maximums reach 44 °C.

The landscape elements include lagoons, reservoirs of floating vegetation, estuaries and marshes. The reservoirs are true floating islands that border the lagoons and are formed by the accumulation of dead or decaying organic matter. The estuaries are separated from each other by extensive sandy borders, which are the main elements of positive relief of the area (Neiff 1999). Both the Iberá and the other estuaries represent abandoned channels of the Paraná River, carved and remodelled by it in the past; the sandy borders were formed with alluvial trawl materials from the same river. At the Northwest edge of the Ecoregion, sandy slopes appear oriented in the Northeast–Southwest direction, as relics of fluvial modelling.

From the biological point of view, it is one of the least known and most conserved systems because of its great extension and difficult access. The most important contribution to the knowledge of the structure and functioning of the Iberá macrosystem was that of the IBERAQUA project, which gathered the existing information and carried out research with new technologies and the participation of more than 25 researchers from five institutions.

The typical plant associations of the various environments of the Ecoregion are the communities of *Cyperus giganteus* and other types of associations with deep-rooted marshy vegetation, the dams with water hyacinth and other plants that form drifting floating islands, the scrublands in gullies and marshes, the waterlogged meadows in sandbars, forest islets locally called mogotes, grasslands, savannahs with isolated trees, reed-baths covered with a scrubland (Neiff 2004), totorales, peguajosales and yellow straw grasslands. Open savannahs in lowlands that are seasonally flooded for several months, with waters up to 1.5–2 m deep, have a low stratum usually dominated by large grasses and/or several species of *Cyperus*, with little or no presence of palms. Riparian forests are dense, 10–20 m high, from semi-deciduous to evergreen.

On soils moderately drained to poorly drained in the lower horizons, dense to semi-dense forests are developed, with canopies of 12–18 m in height, and a tendency to form patches distributed in a matrix of savannahs and flooded palm groves.

In 2004, there were 624 species of vertebrates, among which there were 57 mammals.

The ecoregion is in a potentially critical situation due to the construction of the Yaciretá dam on the Paraná River, at a short distance from the Esteros del Iberá, which causes water levels to rise, increases in water flow and sediment drag, alters the set of processes and their interactions, to the detriment of the persistence and structure of plant and animal populations. The population of deer is also impacted by the loss of habitat quality, by decreasing the area of pastures and dry areas used for rest.

13.2.7 Delta and Islands of the Paraná and Uruguay river Ecoregion

This ecoregion, with 55,344 km^2, includes the flood valleys of the middle and lower reaches of the Paraná River and its tributary, the Paraguay River, the old marine estuary occupied by the Paraná River delta, the Rio de la Plata channel, the Bay of Samborombón southern portion and the Uruguay River from its confluence with the Pepirí Guazú River to its mouth in the Paraná Guazú. These rivers drain a continental basin formed by diverse regions of contrasting characteristics of hundreds of thousands of square kilometres (Matteucci 2018c).

Although this Ecoregion is not the most extensive, it has large latitudinal amplitude, extending from 25.42 to 36.33 °S Lat., and therefore has wide gradients of temperatures and rainfall. The climate is temperate and humid, with little daily and seasonal thermal amplitude, due to the presence of permanent water bodies. January temperatures (warmest month) vary from 25 to 27.5 °C between the extreme South and North of the Ecoregion; those of July (coldest month) vary between approximately 12 and 18 °C between both extremes. Thus, the climate varies between humid subtropical in the North to humid temperate in the South.

The Paraguay and Paraná rivers are an excellent biogeographical corridor, as shown by the presence of Amazon lineage species in all the gallery forests of the ecoregion, including the Monte Blanco in the Paraná Delta Complex, which is located more than 1200 km south of the Tropic of Capricorn (Oakley et al. 2005). Chaco species are also found in the temperate latitudes of the province of Buenos Aires.

The native vegetation presents a recurrent pattern determined by geomorphology and hydrological conditions, especially the frequency, depth and duration of the floods. Forests and shrubs alternate in the narrow riparian hillocks (locally called albardones), grasslands and pastures in depressions, and hygrophilous and aquatic communities on the banks and in the interior lagoons. The characteristic vegetation type is the fluvial forest, in which the distribution and abundance of the tree species is modelled by water runoff in the courses and by the length and alternation of periods of flooded soil and dry soil. Since the variables that modulate the pulses (frequency, intensity, seasonality, etc.), vary spatially, there are diverse types of fluvial forests (Neiff 2005).

Other characteristic vegetation types, ordered from those less frequently flooded to those permanently flooded, are albardon forests, tall pastures, herbaceous communities such as cataysales, canutillares or camalotales, depending on the dominant species. The albardón forest, locally called Monte Blanco, has a jungle appearance, with abundance of vines and epiphytes, but it is not a fluvial forest because it does not depend on hydrometric fluctuations although it is located on the edge of the river (Neiff 2005); however, if the albardones are low they can share species from the riparian forests.

In other sectors of the Ecoregion there are scrublands, while in periodically flooded island depressions, legume or white and red sarandi thickets are developed. At slightly higher elevations there are open forests with scrublands as undergrowth. The biodiversity of the communities subjected to hydrological pulses is very variable and depends fundamentally on the characteristics of the pulses; however, this potential wealth is not verified in the current diversity, which is much lower, probably due to the high degree of fluctuation of the system and the functional diversity of plant and animal organisms, which differ in their ability to adapt to the phases of hydrological pulses (Casco et al. 2005). The spatial and temporal variability of species richness is the clearest manifestation of the great complexity of the Ecoregion. The climatic and biogeographic influence is also evident in the fauna, which is very rich in comparison with that of the neighbouring Ecoregions.

The riparian system of this ecoregion, which reaches 3400 km in length, has great importance as a natural corridor through contrasting landscapes, such as tropical rain forest, savannahs, steppes and shrublands. All the ecosystems of the corridor are interconnected by horizontal flows of nutrients, sediments, eggs and seeds, on which depends the organization of the landscape and the assemblages of populations in each moment and place. The stability of the system and the great biological diversity in it depend entirely on the maintenance of these flows (Neiff et al. 2005).

In this Ecoregion, perhaps more than in the others, climate change becomes very important, due to the rapid fluctuations of the hydrological regime and the biological, ecological and social responses to these changes. The climatic events not only have to do with the floods but also with the great downspouts.

It should be noted that the islands of the Paraná and Uruguay rivers constitute the subsystem with the greatest diversity of geoforms and, therefore, of ecosystems and habitats. The combination of abandoned water courses (madrejones), albardones, temporary and permanent lagoons, estuaries, streams, sandy beaches and turns of meanders, and their respective gradients, give rise to numerous environments that differ in type and characteristics of the vegetal cover (Matteucci et al. 2004).

13.3 Species Account

13.3.1 Tapirus terrestris

13.3.1.1 Common Names

Anta, Tapir, Mboreví, Gran Bestia, Lowland Tapir, Brazilian Tapir, South American Tapir.

13.3.1.2 Distribution in Argentina

The tapir has the southern limit of its distribution in the northwest of Argentina. The extent of its distribution in Argentina is less than 2% of its total distribution in South America; it is also where the greatest reduction of its historic distribution occurred (Taber et al. 2009). The species occupies Salta, Jujuy, Formosa, Chaco, Misiones and small portions of the northwest of Santiago del Estero and Santa Fé provinces (Chalukian et al. 2009; de Bustos et al. 2019a). It is estimated that the distribution area reaches 189.954 km^2, which represents about 41.06% of its historical distribution (de Bustos et al. 2019a). This implied the species' extinction in the provinces of Tucumán and Corrientes.

13.3.1.3 Ecoregions

Yungas Forest, Paranaense Forest, Dry and Humid Chaco, a small portion of the Delta and Islands of the Paraná and Uruguay rivers.

13.3.1.4 Habitat and Density

Tapirus terrestris is the largest native terrestrial mammal in Argentina. The tapir occupies a great diversity of environments within an altitudinal range from 100 to 1800 m (Soler 2005; Chalukian and Merino 2006). The population density is quite variable throughout its geographic distribution. Up until now, densities have been estimated in Argentina only in the Paranense Forest (Misiones), in Iguazu National Park: 0.2–0.3 individuals/km^2 and in Uruguai Provincial Park: 0.05–0.08 individuals/km^2 (Cruz 2012). The relationship between the ways the different biomas influence density in this species is unclear, since it can adapt to different types of vegetation, and variations in the availability of water and food, with similar densities in dry and humid forests. However, human activities, such as hunting, and habitat loss and fragmentation have an enormous negative effect on its abundance. The tapir's low abundance in areas with high human presence is common to different environments (e.g. Paranense Forest) (Cruz 2012). In this ecoregion, the tapir uses areas with low hunting pressures in continuous primary and secondary forests, but also uses pine plantation stands near native forest blocks (Iezzi et al. 2018a). In the Yungas, in El Rey National Park (Salta, Argentina) and in a neighbouring cattle ranch, the species uses areas with feral cattle less frequently than areas without these animals (Chalukian et al. 2004). Although habitat preference changes according to resource availability, in the Yungas, de Bustos (2018) observed selection towards regenerating forests. This preference is probably associated with a higher food availability in regenerating forests, especially the fruit of *Vachellia aroma*.

13.3.1.5 Diet and Feeding Ecology

The tapir is herbivorous, a browser and fruit-eater. Its diet includes different types of plants, such as herbs, grasses, bushes, palms, trees, cacti and aquatic plants, and includes different parts of the plants such as leaves, flowers, shoots, stems, bark, and dry and fleshy fruits. It is a selective browser with respect to the available species, such that its diet can vary with the region and the season. It is considered the most important disperser of seeds of the genus *Tapirus* in South America, since it consumes the greatest number of plant species (194); the Moraceae and Rubiaceae are the most represented (Hibert et al. 2011; O'Farrill et al. 2013). Some studies have been done with respect to the diet and the usefulness of the tapir as a seed dispersor in Argentina. Sallenave (2009) found that the composition of various items in the faeces did not vary seasonally, that tapirs eat 22 species of fruits and damage 0.9% of the seeds through mastication, in the Paranaense Forest of Iguazu National Park (Misiones). Furthermore, Giombini et al. (2009) and Giombini et al. (2016), in the same area, found that the large seeds of the palm tree *Syagrus romanzoffiana* are the mainstay of the tapir's diet throughout the year and that it is the principal disseminator of this species, recruiting groups of seedlings in the faeces and favouring the mixture of genotypes in the landscape. Chalukian et al. (2013) mentioned that in El Rey National Parks, the tapir consumed different plant parts from 57 species. Also in El Rey, de Bustos (2018) showed that the composition of items in the faeces varied seasonally and that tusca fruits (*Vachellia aroma*) are a key dietary resource, mainly during the dry season and in secondary forests. Moreover, it disperses 26 seed species, nine from trees, and damage of seeds by mastication is no more than 22%, which is highly correlated with seed size (0.5–1 cm). A particular component of browsing behaviour of this species involves the breaking of seedlings, saplings and bushes to reach the more tender leaves at the top. This occurs more often when the availability of forage in the understory is scarce and mainly affects the main trunk of the plant; consequently, this affects the growth and structure of these plants (de Bustos 2018); these effects of the tapir on forest structure have also been observed in Paranaense Forest (D. Varela, pers. obs.).

13.3.1.6 Reproductive Biology

Under wild conditions, reproductive behaviour is observed in winter-spring in Paranaense and Yungas Forest (Cruz 2012, S. de Bustos pers. obs.).

13.3.1.7 Behaviour

The tapir is a solitary species, but male-female pairs are seen during the reproductive period and mother-young pairs during the rearing period. It is mainly nocturnal and crepuscular with most activity occurring between 18:00 and 7:00 h (de Bustos et al. 2009; Cruz et al. 2014; Albanesi et al. 2016) and apparently this range of hours is longer when the nights are longer (Cruz 2012; Albanesi et al. 2016). Tapirs can be less active during warmer days (Cruz 2012) or on very cold and cloudy days (Cuellar and Noss 2003). Given its size and mobility, the tapir needs a large home range to maintain its life cycle. The use of space within the home range can vary spatially and temporally according to the availability of food and water. In Argentina, estimates of home range and possible size differences between males and females do not exist. From the ecological point of view, the tapir is considered to be a "landscape species" (Sanderson et al. 2002), since it occupies a large territory with a great variety of interconnected and functional habitats. Consequently, by conserving populations of this species, one also preserves other species with fewer requirements (Coppolillo et al. 2003; Medici 2010). Moreover, because of its great mobility, biomass and influence on forest structure by means of its interactions with the vegetation and because it can modify the structure of the habitat, it has been called an "architect species" (Taber et al. 2009; de Bustos 2018), "the forest gardener" (Painter 1998) and even an "ecosystem engineer" (de Bustos 2018).

13.3.1.8 Threats and Conservation

The threats to this species in Argentina are no different from those that have been identified in other areas of its distribution. The loss and degradation of habitat are the main causes of its population decline, followed by hunting, cattle ranching carried out over extensive areas, and road kills (Chalukian et al. 2009; de Bustos et al. 2019a). The transformation of forests for lumbering, increasing agricultural areas and cattle ranching has ravaged the regions of the Chaco, the Pedemontane Yungas forests and the Paranaense Forest, directly affecting tapir environments. As a consequence, its distribution is greatly reduced, with local extinctions. The tapir is a species that is particularly sensitive to hunting since it takes time to reach sexual maturity, has a low reproductive rate and its populations usually occur at low densities (Novaro et al. 2000). In Misiones, hunting has an enormous negative effect on its abundance, reducing its populations to critical levels (Cruz et al. 2014); it probably was also the principal cause of its disappearance from the province of Corrientes (Di Martino et al. 2015). Cattle species are exotic in the forest environments of Yungas and Chaco, where traditionally ranching is done on an extensive scale. This activity is directly associated with loss and degradation of the landscape and it is usually associated with opportunistic hunting (Chalukian et al. 2004; Camino et al. 2018). Furthermore, cattle interfere with the presence of the tapir and, also transmit diseases (Chalukian et al. 2004). Roadkills have direct consequences for tapir population sizes. In Misiones, recurring vehicle hits occur on paved highways, mainly highways RN 12, RN 101 and RP 19, which cross protected areas like Iguazu National Park and Uruguai Provincial Park (Nigro and Lodeiro Ocampo 2009, Varela in litt.). Wildlife underpasses in Misiones, are an effective mitigation measure to reduce roadkill and enhance connectivity among the population (Varela 2015). In the Yungas, irrigation channels associated with agricultural fields (mainly sugar cane) have become a serious danger (Nicolossi 2004; Chalukian et al. 2009; Nicolossi and Baldo 2011). In spite of certain mitigation measures instituted to reduce these dangers, such as escape ladders, with positive effects (Albanesi et al. 2016), they are insufficient since every year, especially in the dry season, individuals of various species including tapirs are rescued (S. de Bustos obs. pers., Nicolossi and Baldo 2011). Again in Yungas, using potential habitat distribution to evaluate the role of protected areas and other covariates, it was concluded that tapir habitat use is positively associated with the proximity of national protected areas. This indicates that National Parks act as refuges and sources of dispersion for tapirs (Rivera et al. 2019). Taber et al. (2009) identified 11 conservation units for this species in Argentina, in nine of which, the tapir would have a high survival probability. In Ibera National Park, Corrientes, the only reintroduction experiment in the country is taking place, directed by "The Conservation Land Trust Argentina" (Di Martino et al. 2015; Zamboni et al. 2017). This project began in 2016, and up to now 11 individuals have been liberated, with two wild born young. The individuals released came from zoos and faunal rescue centres in the provinces of Salta, Tucuman, Mendoza, and the City of Buenos Aires (Di Martino S. com. pers.). The project had a second phase, evaluating the possibility of translocating wild tapirs from the area of The Impenetrable Chaco, but this project was abandoned. At a national scale, the species is categorized as Vulnerable (de Bustos et al. 2019a). It has been declared a Natural Monument in the provinces of Salta (Decreto No. 4625/11), Chaco (Ley No. 5887/07), Formosa (Ley No. 1582/12) and Misiones (Ley No. 2589/88).

13.3.2 Parachoerus wagneri

13.3.2.1 Common Names

Quimilero, Taguá, Chancho moro, collarejo, Chacoan peccary.

13.3.2.2 Distribution in Argentina

The Chacoan peccary is the largest mammal endemic to the Chaco (Nori et al. 2016). In Argentina, it occupies the east of Salta, western Formosa, northwest Chaco, forested portions of Santiago del Estero and northern Cordoba. It may also be present in Catamarca and La Rioja. It was recently found in Cordoba, more than 650 km south of its previously known distribution limit (Torres et al. 2017). Then, Torres et al. (2019) detected the species in Santiago del Estero province, which suggests that populations from Formosa to Cordoba may be connected.

13.3.2.3 Ecoregions

The species is present in the Dry Chaco. There are records in the Humid Chaco but its presence in that region has not been documented recently.

13.3.2.4 Habitat and Density

It is well adapted to inhabiting arid and semiarid areas. Torres et al. (2019) found that its main habitat is old-growth, well-preserved, forests. Yet other authors found that areas covered by secondary forests have suitable habitat and high probabilities of being occupied (Camino 2016; Ferraz et al. 2016). Camino (2016) studied habitat selection in the core area of its distribution and found that territories with primary and secondary native forest-cover have higher probabilities of being occupied by the species. The positive association between the presence of Chacoan peccary and secondary forests may be due to competitive exclusion by the other peccary species. A second hypothesis is that Chacoan peccaries select secondary forests because they have higher food availability (Camino 2016), that is, a diversity of plant species, a high renewal rate and a greater number of fruits per plant (López de Casenave et al. 1995; Tálamo and Caziani 2003). Besides, secondary forests may be a consequence of past selective tree extraction, but they share characteristics with natural forests that conform the edge of Chaco's forests and natural grasslands, now in extinction (López de Casenave et al. 1995; Grau et al. 2015). Natural edge forests may be selected by the Chacoan peccary. Forest cover decreases the probability of detecting Chacoan peccaries, which suggests that this habitat type is an important refuge (Camino et al. in litt.). The Chacoan peccary is absent in areas that lack forest cover (Altrichter and Boaglio 2004). They are also absent in areas dominated by intensive soy-production, even when forest curtains are left (Núñez-Regueiro et al. 2015). Altrichter and Boaglio (2004) found that the Chacoan peccary's presence in a site is inversely related to human presence. This result is expected, as near populated areas, the habitat is more degraded and hunting pressure is higher. Yet two recent studies found a positive relation between human presences and the occupancy probability (Saldivar 2014; Camino 2016). This unexpected association supports the hypothesis of competitive exclusion of the Chacoan peccary by the two other peccary species. Fossil records of the Chacoan peccary suggest that it may have been adapted to a wider range of habitats in the Pleistocene and Holocene eras (Gasparini et al. 2013).

13.3.2.5 Behaviour

The Chacoan peccary is territorial although females overlap their territories; territory size does not change with the season (Taber et al. 1993; Camino 2016). In Argentina, estimates of home range and differential territory use by the sexes do not exist.

13.3.2.6 Threats and Conservation

Chacoan peccaries are categorized as Endangered by IUCN (EN, Altrichter et al. 2015) and its conservation is important given that it represents a unique evolutionary path (Edge of Existence 2018). It was estimated that its distribution range decreased 40% (Altrichter et al. 2015). Yet recent discoveries of the species in new territories showed that we have little information about its current or past distribution (Torres et al. 2017, 2019). Despite new records, it remains categorized as Endangered (Torres and Camino 2019), because it is seriously threatened by habitat loss and fragmentation, as the Chaco has one of the highest deforestation rates in the world (Hansen et al. 2013). Chacoan peccaries cannot inhabit productive matrices (Altrichter and Boaglio 2004; Núñez-Regueiro et al. 2015). It is also threatened by high hunting pressures (Altrichter 2006; Camino et al. 2018). Local inhabitants of the Dry Chaco consume Chacoan peccary meat and hunting has an enormous negative impact on populations (Altrichter and Boaglio 2004; Altrichter 2005, 2006; Camino et al. 2018). Commercial hunting also exists (Quiroga, pers. com.). Thus, its preference for territories near roads or cities may be an ecological trap and threaten its conservation (Camino 2016; Saldivar 2014). These forces may lead to Argentinian populations' being genetically disconnected (Torres et al. 2019). Although conservation and management measures have been suggested by researchers, these have not been applied (Altrichter et al. 2016; Camino 2016).

13.3.3 Pecari tajacu

13.3.3.1 Common Names

Pecarí de collar, Rocillo, Morito, Tateto, Chancho moro, Collared Peccary.

13.3.3.2 Distribution in Argentina

North of Santiago del Estero, portions of Santa Fe, Chaco, Formosa, Salta, Tucumán, Misiones, Jujuy and San Juan, southeast Catamarca, south of La Rioja, west and north of Córdoba, San Luis and northeast Mendoza (Camino et al. 2019). The species is considered extinct in areas of Santa Fe, South of Córdoba, Southeast of Santiago del Estero, Entre Ríos and Corrientes. In Corrientes it is being reintroduced in the Iberá Wetlands (Zamboni et al. 2017).

13.3.3.3 Ecoregions

Paranaense Forest, Yungas, Dry and Humid Chaco, Campos y Malezales, Monte de Llanuras y Bolsones, Puna (P. Perovic, pers. comm, Camino et al. 2019, de Bustos y Alderete 2019).

13.3.3.4 Habitat and Density

Studies on habitat selection found that collared peccaries may be opportunistic but in the Paranaense Forest and in the Dry Chaco, they are associated with woody forests (Camino 2016; Puechagut et al. 2018). In the Dry Chaco, besides being positively associated with forests, the probability of a site's being occupied by collared peccaries increases with the availability of fruit, cacti and Palo-santo (*Bulnesia sarmientoi*) and decreases in areas with bare soil (Camino 2016). In the Dry Forests of Bañado del Quirquincho (Salta), this species had the highest camera trapping rate in relation to other medium and large-sized mammals (2.84 ± 0.57 records per camera trap) and more frequently in the Bañadero Forest with respect to Palo-santal, Quebrachal and Palm forests (Puechagut et al. 2018).

13.3.3.5 Diet and Feeding Ecology

The diet of the collared peccary changes depending on the region and, despite its wide distribution in Argentina, very few studies have focused on its diet. Periago (2017) found that the preferred species in northwest Córdoba included Acacias, *Bromelia urbaniana*, *Celtis ehrenbergiana* and that most consumed plants were in the Fabaceae family. When considering the whole species's distribution range, Fabaceae species are the second favourite plants, after Arecaceae (Beck 2005). There are also anecdotal mentions of the species's diet; for example, in the dry Chaco in Salta province, the species consumes *Ipomoea* sp., *Opuntia quimilo*, *Bromelia serra*, *B. hyeronymi*, *Gnaphalium* sp. and *Pennisetum frutescens* (Barbarán 2017). Collared peccaries have an important role as a seed dispersor and predator, regulating beetles' populations from the top and contributing to the persistence of water ponds and their associated biodiversity (Beck et al. 2010).

13.3.3.6 Behaviour

In the Dry Chaco the size of the groups seems to depend on hunting pressure and habitat degradation (M. Camino pers. obs.). In this region, they are mainly nocturnal and crepuscular, while in Yungas they seem to be cathemeral (Albanesi et al. 2016; Reppucci et al. 2017), although with greater activity between dawn and midday, and then near sunset (Reppucci et al. 2017).

13.3.3.7 Threats and Conservation

The species is heavily hunted throughout its entire distribution range (Altrichter and Boaglio 2004; Altrichter 2005; Paviolo et al. 2009; Camino et al. 2018, 2019). In Argentina, international commerce reached 425,305 peccary hides between 1988 and 2000, in spite of the fact that exportations had been prohibited since 1987. At the present, commerce has been reduced to local sales of hides from subsistence hunters (Barbarán 2017). The species is also negatively affected by habitat loss and fragmentation. In the South American Chaco deforestation rates are among the highest in the world (Baumann et al. 2017; Camino et al. 2019). Thus, although collared peccaries tolerate large habitat changes, habitat loss is threatening its persistence in the Gran Chaco (Periago et al. 2014). Additionally, deforestation is usually associated with intentional fires that directly kill individuals (M. Camino pers. obs.). In Corrientes province, *P. tajacu* has been locally extinct for over 50 years. As part of a multi-species rewilding project, captive-born collared peccaries were released into Ibera National Park in 2015. Up to the present, four populational nuclei have formed and now more than 100 animals live in the wild (Hurtado Martinez 2017; Hurtado et al. 2018; Zamboni et al. 2018). In Argentina the collared peccary was categorized as Near Threatened (Camino et al. 2019).

13.3.4 Tayassu pecari

13.3.4.1 Common Names

Pecarí labiado, Maján, Majano, Queixada, White-lipped peccary.

13.3.4.2 Distribution in Argentina

The area ocupied by the species in Argentina is estimated at 172,258 km^2, around 1.54% of the South American distribution (Taber et al. 2009; de Bustos et al. 2019b). It occupies Salta, Jujuy, Chaco, Formosa, north of Santiago del Estero and Misiones provinces.

13.3.4.3 Ecoregions

Atlantic Forest, Humid and Dry Chaco, Yungas. In the Dry Chaco, Camino (2016) generated distribution models using data gathered between 2011 and 2013. She worked on a 54,000 km^2 study area and found that there are about 43,000 km^2 with >0.8 probability of being used by white-lipped peccaries.

13.3.4.4 Habitat and Density

The presence of white-lipped peccaries is positively associated with forested areas (Altrichter and Boaglio 2004; Saldivar 2014; Camino 2016). In Yungas, El Rey National Park, they use mostly mature forests in relation to secondary forests (de Bustos et al. 2009). However in the Dry Chaco, secondary forests have a higher probability of being used than primary forests (Camino 2016). This may be because seedlings and fruits are more abundant in secondary forests (Coley 1983; López de Casenave et al. 1995; Tálamo and Caziani 2003) but alternative explanations are (1) competitive exclusion by cattle from primary forests and (2) that secondary forests are a better refuge against hunters (Camino 2016). There is controversy about the threshold of forest-cover tolerated by the species. In the late 1990s Altrichter and Boaglio found that it was absent when forest cover was under 80%. Between 2011 and 2013 Camino detected white-lipped peccaries in areas with a minimum of 10% forest-cover. Camino concluded that differences with Altrichter and Boaglio's findings could be due to differences in landscape and land-cover spatial configuration. Camino's results are explained if detections correspond to animals that, to access required resources or conditions, need to travel great distances and cross unsuitable areas. Despite differences, both studies coincide with its absence in totally converted areas. Research in intensive-productive landscapes also found that white-lipped peccaries are absent (Núñez-Regueiro et al. 2015). Forest curtains are insufficient to maintain white-lipped peccaries in productive matrices (Núñez-Regueiro et al. 2015).

The density of the species in the Dry Chaco is low. In the Argentinean Chaco, Camino (2016) found that although widely distributed, its density is probably low in a 54,000 km^2 area. These results support the findings of Núñez-Regueiro et al. (2015), Periago et al. (2014), and Perovic et al. (2017), who did not detect white-lipped peccaries in their study areas (Chaco) and concluded that the species is absent or exists at very low densities in these areas. By early 2000 Altrichter and Boaglio (2004) had already found low densities in the Dry Chaco. In the Paranaense Forest and the Yungas, the abundances in general are also low. In the Paranaense Forest, its situation was critical until 2010, when it almost disappeared (Paviolo 2010). However, now the population is recovering (D. Varela obs. pers.). De Bustos et al. (2018) evaluated its situation in Yungas between 2011 and 2017 in an area of approximately 3200 km^2 of the Upper Bermejo River watershed using camara traps, transects and indirect indices. They found a significantly low detection rate (0.00003, records/camara per trap night and 0.0179 records/km), which indicates a notable decline in its geographic range with respect to reports from previous years (Taber et al. 2009). Bardavid et al. (2019) obtained similar results in this area using camara traps, and mentioned that the situation is less critical in Yungas in the Lower River Bermejo watershed, where it is recorded more frequently, especially in more inaccessible and better conserved areas such as El Rey National Park (Salta). Yungas's subpopulations are only weakly connected by Yungas-Chaco transition environments and by Chaco forests.

13.3.4.5 Diet and Feeding Ecology

In Argentina, no studies have focused on the diet of the species. Research suggests that fruits are an important dietary element as areas with higher periods of availability of fruits across the year have a higher probability of being used. However fruit and water availability were correlated in this study and, thus, results are not conclusive (Camino 2016).

13.3.4.6 Behaviour

Herds of white-lipped peccaries of the Dry Chaco are smaller than herds described in other regions and consisted of 15–30 individuals (Altrichter and Boaglio 2004; Altrichter 2005). Nevertheless, in well conserved areas, the herds can be larger, as indicated by observations in El Rey National Park of 67 individuals (de Bustos obs. pers.). This species is more susceptible to human activity than its relatives, as it disappears from inhabited or heavily used territories, or uses them to a lesser degree (Altrichter and Boaglio 2004).

13.3.4.7 Threats and Conservation

The species has been categorized as Endangered at the national level (de Bustos et al. 2019b). Taber et al. (2009) identified three conservation units of this species, with intermediate levels of survival probability for long periods. Two of these are located in Yungas and the third, more extensive area, in the Chaco (shared with Paraguay and Bolivia).

In Argentina, the species' historical distribution suffered one of its greatest reductions in the last 100 years (Taber et al. 2009). Habitat loss, degradation and fragmentation combined with high hunting pressure threaten white-lipped peccary's long term persistence in the country (de Bustos et al. 2019b). Other threats that impact this peccary are disease transmission, the influence of exotic species and kills by domestic dogs (de Bustos et al. 2019b). The Chaco is a deforestation hotspot, and natural ecosystems are rapidly and completely converting to intensive agricultural productive systems (Baumann et al. 2016). This probably combines with unsustainable hunting on the species (Altrichter and Boaglio 2004) but we cannot be certain as Argentina has no estimates of present hunting pressure. Local inhabitants of the Dry Chaco hunt peccaries mainly for subsistence (Altrichter and Boaglio 2004; Camino et al. 2018). There are portions of the Dry Chaco where white-lipped peccary disappeared and local inhabitants blame overhunting and other anthropogenic disturbances (e.g. habitat loss) (Camino et al. 2016). This is highly possible as in the late 1990s subsistence hunting was already unsustainable and its presence was inversely correlated with *puestos*, that is, small local inhabitants' houses isolated in natural ecosystems (Altrichter and Boaglio 2004, Altrichter 2005). Camino (2016) found no relation between the species's presence and this variable and showed that the probability of use of an area was inversely correlated with the distance to large towns, cities and paved roads. White-lipped peccaries seem to be negatively affected by habitat degradation near these populated centres and by their inhabitants' intensive hunting. People from large towns and cities enter natural ecosystems with cars and trucks with freezers and hunt white-lipped peccaries and other wild species. These results and observations suggest commercial hunting of the species. Commercial hunting was reported in Salta Province (Barbarán 2017).

Argentina is the southern extreme of the white-lipped distribution and low densities and small herd size may be explained by this factor. Yet local inhabitants of the Chaco remember large herds, of 100 individuals, and higher densities in the past. In well preserved areas, these herds may still be present (Quiroga pers. com.). Thus, small herd-sizes and low densities are probably the result of high hunting pressure and poor conservation status. Although the white-lipped peccary is still present over a large surface area of the Dry Chaco, its density is probably low and it is locally extinct in some areas (Camino et al. 2016). Its wide distribution in the Dry Chaco does not provide information about its fitness; there may be areas where mortality is higher than reproduction rate.

13.3.5 *Lama guanicoe*

13.3.5.1 Common Names

Guanaco.

13.3.5.2 Distribution in Argentina

Wide distribution in Argentina but with only relict populations in Dry Chaco in Cordoba and San Luis provinces.

13.3.5.3 Ecoregions

Altoandina, Puna, Prepuna, Monte de Sierras y Bolsones y de Llanuras y Mesetas, Espinal, Estepa y Bosques Patagónicos, Islas del Atlántico Sur, for the purposes of this publication, in Dry Chaco (relict) (Carmanchahi et al. 2019).

13.3.5.4 Habitat and Density

It is believed that the species occupied open areas of the Dry Chaco, dominated by grasses. These areas are prehistoric river beds and are in extinction at the present. The disappearance of open habitats is explained by many factors: (1) these grasslands were maintained by indigenous nations through periodic fires and these practices were abandoned (Grau et al. 2015), (2) extensive ranching occurs in these grasslands, and cattle contribute to their conversion into bush lands, (3) intensive agriculture also occurs in these areas. In Argentina, especially in the province of Cordoba, guanaco occupy marginal bush lands near salt flats. The individuals reintroduced into Quebrada del Condorito National Park selected terrains with low growing vegetation with a high percentage of grasses and herbs, avoiding grasslands and bushy forests (Flores et al. 2013; Costa and Barri 2018), and, in the northeast of San Luis province, in the ecotone of the ecoregions Chaco and Monte (Gomez Vinassa and Beatríz Nuñez 2016).

13.3.5.5 Diet and Feeding Ecology

The diet of the guanacos reintroduced into Cordoba province consisted mainly of short grasses and cyperaceae species, with five species representing 71–93% of the total, according to the season (Barri et al. 2014). Plants of *Sorghastrum pellitum*, *Chascolytrum subaristatum*, *Carex fuscula*, *Eleocharis pseudoalbibracteata* and

Lachemilla pinnata were the most common; for this reason it is assumed that they behaved like selective grazers (Barri et al. 2014). In the population in the northwest of the province, the diet was more varied (57 species from 35 genera and 19 families), with a predominance of grasses, followed by *Geoffroea decorticans*, *Atriplex cordobensis* and several species of the genus *Prosopis* (Geisa et al. 2018). Both studies showed a small seasonal variation in the diet. Finally, in the arid lands of San Luis province, the diet had a more marked seasonal variation related to the availability of species during the wet season. The proportion of epiphytes and cacti increased proportionally during the dry season (Gomez Vinassa and Beatríz Nuñez 2016).

13.3.5.6 Threats and Conservation

The species is almost extinct in the Dry Chaco. There are relict populations in the Bolivian and Paraguayan Chaco (Cuéllar et al. 2017). In the Dry Chaco of Argentina there are relict populations in the province of Cordoba and in the northeast of San Luis. Illegal hunting threatens the species and limits its presence in areas of suitable habitat. Habitat conversion for intensive agriculture and extensive ranching is also a problem (Cuéllar 2011).

In Argentina in 2007, 113 individuals were reintroduced, with 25 more added in 2011, to the Quebrada del Condorito National Park, in the Dry Chaco Ecoregion (Tavarone et al. 2007; Barri and Cufré 2014; Aprile 2016). However, in 2015, only 25 individuals remained, divided into three groups (Barri 2016); only continuous supplementation of the animals could assure that they survive for the next 100 years (Barri 2016). The recent creation of Traslasierra National Park (105,000 ha), in Cordoba province, is an important step towards the conservation of the last individuals in the northwest of the province, which are estimated to be around 100 (Barri 2016).

13.3.6 Hippocamelus antisensis

13.3.6.1 Common Names

Taruca, Taruka, Chacu, Venado andino, Huemul del norte, North Andean Deer, North Andean Huemul.

13.3.6.2 Distribution in Argentina

Jujuy, Salta, Tucumán, La Rioja and Catamarca.

13.3.6.3 Ecoregions

Altos Andes, Puna, Yungas Forest and Montes de Sierras y Bolsones.

13.3.6.4 Habitat and Density

The taruca population in Argentina is found between 1800 and 5500 masl, mainly in high altitude pastures and mountainous bushy areas, rocky ravines and steep slopes of the subandean mountains (Guerra et al. 2019). It has a fragmented distribution (Chebez 1994; Bertonatti and Pastore 1998) determined by the terrain that it uses; data indicates that there are 8–10 loosely connected subpopulations. Although no recent information exists on population density in Argentina, between 1997 and 2003 the population was estimated to be between 1900 and 3000 individuals (Regidor et al. 1997; Regidor and Costilla 2003); from these data one can suppose that some subpopulations could be below 250 adults (Guerra et al. 2019). Additionally, in the valleys of the Capillas and Toro Rivers (Salta), a density of 0.7–1 individuals/km^2 was estimated (Regidor et al. 1997).

13.3.6.5 Diet and Feeding Ecology

D'Angelo (2012) supplied the only information available in the country with respect to the taruca's diet, which he studied at Cerro Santa Ana, Jujuy, during one dry season (March-August) and two wet seasons (September-February). He found 93 plant genera of different types, including trees, bushes, dicotyledonous herbs, cacti, grasses, graminoids (Juncaceae and Cyperaceae), bromeliads, and other monocotyledons, plants in mats, pteridophytes, bryophytes and lichens. The most common genera were *Phacelia* sp., *Trifolium* sp., *Coniza* sp., *Lupinus* sp., *Austrocylindropuntia* sp., *Salvia* sp., *Senecio* sp., *Calceolaria* sp., *Tillandsia* sp., *Microchloa* sp., *Plantago* sp., *Carex* sp., *Verbesina* sp., *Festuca* sp., *Deyeuxia* sp. and *Ephedra* sp. Although dicotyledonous herbs were the principal item in both seasons, the diet varied seasonally, since ingestion was greater in the warm season than the cold season. Grasses were greatly increased during the cold season, as was also greater diversity in the diet.

13.3.6.6 Behaviour

This species lives in herds of one or more adult males, several adult females with their young and various juveniles (Bertonatti and Pastore 1998). The older individuals are more vigilant, and at the slightest sign of danger induce fleeing of the herd, through vocalizations (snorts), stamping the ground with the front feet and rapid movements of the tail (Pastore et al. 1997). During oestrus, in large groups, subgroups form composed of one male and 2–3 females. Sometimes, juveniles are also in the group, and fights occur between males, usually with young males that try to separate females from an older male. These fights are not usually violent; the combatants lock antlers, push, and try to make their opponent retreat (sparring) (Pastore 2012). They have well defined bedding areas and trails, and the most frequently observed activities are feeding, rumination and locomotion, which can cover several kilometres (Pastore 2012).

13.3.6.7 Threats and Conservation

The principal threat affecting the taruca results from cattle ranching—trampling of the ground, over pasturing, transmission of diseases from cattle and fires set to improve cattle pasture. The fires degrade and consume large areas of natural pasture that are the taruca's habitat. Other threats include the growth of urban areas, roadkills, mining and illegal hunting for sport, cultural reasons and for subsistence (Bertonatti and Pastore 1998; Dellafiore and Maceira 1998; Guerra et al. 2019). In Jujuy, hunting associated with rituals occurs during Easter week, where the species is considered a symbol of the devil, hunted and offered as a sacrifice in Incan ruins (Díaz 1995; Bertonatti and Pastore 1998). The taruca is protected as a Natural National Monument (National Law No. 24,702/1996) and a Natural Provincial Monument in the province of Jujuy (Law 5405/2004). It was categorized on the national scale as Endangered (Guerra et al. 2019).

13.3.7 Blastocerus dichotomus

13.3.7.1 Common Names

Ciervo de los pantanos, guazú pucú, Cervo do Pantanal, Marsh Deer.

13.3.7.2 Distribution in Argentina

Formosa, Chaco, Corrientes, Santa Fe, Entre Ríos and Buenos Aires provinces. It became extinct in historical times from the provinces of Salta, Jujuy and Misiones (Díaz et al. 2000; Massoia et al. 2006).

13.3.7.3 Ecoregions

Humid Chaco, Iberá marshes, Delta and Islands of the Paraná and Uruguay rivers, Campos y Malezales, extinct in the Dry Chaco.

13.3.7.4 Habitat and Density

The species is dependent on wetlands (marshes, swamps, and lagoons) of the basins of the Paraná, Uruguay, and Paraguay rivers. The largest marsh deer population is found in the Iberá Marshes and adjacent areas (San Lucía, Miriñay, Aguapey, Riachuelo, and Batel marshes) in the province of Corrientes, with a population of nearly 10,000 individuals and densities of between 0.24 and 1.39 individuals/km^2 (Cano et al. 2012; De Angelo et al. 2018; Pereira et al. 2019). The second most important population inhabits wetlands and willow and poplar plantations on the islands of the Lower Delta of the Paraná and Uruguay rivers, in the Provinces of Buenos Aires and Entre Ríos. Although there is currently no density estimation, the Lower Delta population trend has been increasing in the last 15 years (Varela et al. in litt.). There are smaller populations in the northeast of Santa Fe, and eastern Chaco and Formosa, but their population status is uncertain (see Eberhardt et al. 2009; D'Alessio et al. 2018). Because marsh deer inhabit flooded environments, they are dependent on upland areas (levees, mounds, polders) and floating marshes to survive during extraordinary floods (Tomás et al. 2001; Varela 2003).

13.3.7.5 Diet and Feeding Ecology

The species has a herbivorous diet, browsing on palustrine, aquatic, and riverside species (Beccaceci 1996; Varela 2003). In Ibera, *B. dichotomus* consumes primarily C3 plants as determined by carbon isotope studies; this indicates that it preferentially feeds in flooded areas (Loponte and Corriale 2013). However, in the Paraná delta, it consumes leaves from commercial willow and poplar plantations, resulting in some conflict with local producers (Iezzi et al. 2018b). Detailed studies of its diet in the Paraná Delta are currently being carried out (Pereira et al. 2018).

13.3.7.6 Reproductive Biology

The reproductive biology of the marsh deer in Argentina is practically unknown. Females are polyoestrous with a mean oestrous cycle length of 24 days and a gestation period of 8½–9 months. Females can have a postpartum oestrus (Piovenzan et al. 2010; Polegato et al. 2018). Although the reproductive cycle does not show clear seasonality, with fawns being born all year round, births usually occur between April and October (Pereira et al. 2019).

13.3.7.7 Behaviour

The marsh deer is a cathemeral species. It is usually solitary or found in pairs, although it can be grouped during extreme flooding events (e.g. on floating marshes of the Paraná Delta (D'Alessio 2016). Mean home range size varies according to sex; males have home ranges double the size of females, and they probably also vary with the size of the inundated area (Piovenzan et al. 2010). The only estimates of home range for this species in Argentina come from the Paraná river Delta, where five individuals (three males and two females) were monitored by satellite telemetry between 2016 and 2017, and had home ranges between 228 and 396 ha (100% MCP) after 80–290 days of monitoring (J. Pereira, unpublished).

13.3.7.8 Threats and Conservation

The species's conservation status in Argentina is Vulnerable, but the subpopulation of the Paraná Delta river is considered as Endangered, and those of the Humid Chaco as Critically Endangered (Pereira et al. 2019). The main threats are wetland modification and poaching. Conversion of wetlands for cattle ranching, agriculture, and forestry has reduced and fragmented marsh deer habitat. In the Paraná Delta, for example, the species disappears in cattle ranching lands (Varela et al. 2017), which are also easily colonized by the *Axis* deer, an exotic invader (J. Pereira, unpublished). In Iberá marshes and Delta and Islands of Paraná river, extreme flooding has a strong impact on marsh deer in synergy with other factors such as poaching, diseases or habitat degradation (Orozco et al. 2013, 2018; Cruz and Courtalon 2017; Varela et al. 2017; Guillemi et al. 2019). Roadkill threat is increasing in the North of Corrientes (Balbuena et al. 2015) and could be a short- or medium-term issue in Buenos Aires, judging by the increased trend in roadkill records (Pereira et al. 2019). Finally, deaths of individuals due to domestic dog attacks and occasionally to fires have been reported (D'Alessio et al. 2018).

The species is not well represented in the system of protected areas of the country, although the recent creation of the Marsh Deer National Park (5588 ha) and the Ibera National Park (183.500 ha) help to protect the most important populations of the country. It has been declared a "Natural Provincial Monument" in the provinces of Buenos Aires (Law 12.209/1998), Entre Ríos (Res. 852/2015), Corrientes (Law 1.555/1992) and Chaco (Law 4.306/1996). It is listed in Appendix I of CITES.

13.3.8 Mazama americana

13.3.8.1 Common Names

Corzuela roja, Corzuela colorada, Pardo, Peñera, Venado, Veado-mateiro, Red Brocket.

13.3.8.2 Taxonomic Comments

The taxonomic identity of the red brocket is complex. Due to its high karyotypic polymorphism and wide geographic range, some authors consider *Mazama americana* a super-species that masks several possible cryptic species (see Duarte and

Merino 1997; González et al. 2016a, b); therefore, further taxonomic studies are necessary to identify the populations of Argentina. A recent study in Brazil indicates that the populations present in the Upper Parana Atlantic Forest may correspond to a new species called *Mazama rufa* (Luduvério 2018).

13.3.8.3 Distribution in Argentina

Humid or mountain forests of Misiones, Chaco, Formosa, Salta, Jujuy, and Tucuman provinces.

13.3.8.4 Ecoregions

Paranaense Forest, Yungas Forest, and Humid Chaco. The presence of the species in the Argentinean Dry Chaco and Campos and Malezales, is not confirmed.

13.3.8.5 Habitat and Density

The red brocket seems to have more specific habitat requirements than the gray brocket in that it prefers more conserved, closed habitats. In Yungas an altitudinal segregation with the gray brocket has been suggested, in which the red brockets use mainly the upper forest floors (Montane forest), above 900 masl. In winter they would descend to the open areas of the Pedemontane or transitional forests looking for tender pastures (Olrog 1979; Heinonen and Bosso 1994). However, possibly, this segregation depends more on topography than altitude in that this species uses more environments with steep slopes and ravines (de Bustos, obs. pers.). In the Paranaense Forest ecoregion, the red brocket prefers the best preserved humid lowland forest with low hunting pressures (Di Bitetti et al. 2008, Varela et al. in litt.). It is present in pine plantations close to large native forest remnants where hunting pressure is low (Varela et al. in litt.).

13.3.8.6 Diet and Feeding Ecology

There are no detailed studies on red brocket diet in Argentina, although there are casual observations. Its diet is composed mainly of fruits, seeds, mushrooms, shoots and tender leaves. As with *M. gouazoubira* in Yungas, more leaves, especially of grasses, were consumed during long periods of relative scarcity of fruits (Juliá and Richard 2001) In Misiones province, the consumption of aquatic plants, such as pondweed (*Potamogeton* sp.) has been mentioned (Massoia et al. 2006).

13.3.8.7 Behaviour

It is solitary although it occurs in pairs or with its young. In Yungas, it shows cathemeral behaviour but with less activity between dawn and midday (Reppucci et al. 2017).

13.3.8.8 Threats and Conservation

The loss and degradation of the forest are among the threats to this species, followed by poaching (Varela et al. 2019). In extremely fragmented landscapes it disappears or is found infrequently, probably due to more hunting and to harassment and predation by dogs. In Yungas, extensive cattle ranching seems to interfere or compete with *M. americana* (Reppucci et al. 2017). In Misiones various roadkills have been recorded (D. Varela obs. pers.). This species has been categorized nationally as Vulnerable (Varela et al. 2019).

13.3.9 Mazama gouazoubira

13.3.9.1 Common Names

Corzuela, Corzuela Parda, Guazuncho, Guazuncha, Virá, Guazu Virá, Viracho, venado, sacha cabra, Tabuka, Tánaganagá, Pité (Toba), Cahitá, Veado-catingueiro, Brown brocket, Grey brocket.

13.3.9.2 Distribution in Argentina

Catamarca, Chaco, Córdoba, Corrientes, Entre Ríos, Formosa, Jujuy, La Rioja, Mendoza, Misiones, Salta, San Luis, Santiago del Estero, Santa Fé and Tucumán (Juliá et al. 2019).

13.3.9.3 Ecoregions

Humid and Dry Chaco, Campos y Malezales, Iberá Marshes, Yungas Forest, Paranaense Forest, Espinal, Montes de Sierras y Bolsones (Juliá et al. 2019).

13.3.9.4 Habitat and Density

Gray brocket deer are widely distributed and use different ecosystems in different regions (Cartes Yegros 1999; Periago and Leynaud 2009; Periago 2017; Juliá et al. 2019). In the Argentinean Dry Chaco, habitat selection was studied in the arid portion of the region: in Los Llanos in La Rioja (Cartes Yegros 1999) and in the Chancaní Reserve (Periago and Leynaud 2009; Periago 2017). In the arid Chaco, gray brocket deer were positively associated with woody land-cover; and in the Chancaní Reserve, with the edges of primary forests located in high altitude areas (Periago and Leynaud 2009). These forests are more humid during summer. Periago (2017) studied habitat selection in the Chancaní Reserve. She found that intensity of habitat use decreases from more pristine environments to more anthropized areas: it uses mainly primary forests, followed by secondary forests, and Monte bushlands. Although gray brockets use the best conserved forests more intensively, they also use secondary forests with extensive ranching. There is controversy regarding its tolerance to areas completely transformed for intensive agriculture in the Dry Chaco. Periago (2017) did not detect it in these areas. However, they were present in forest-strips of landscapes dominated by intensive agriculture in Salta Province (Núñez-Regueiro et al. 2015) as in other regions (Juliá et al. 2019). To our knowledge, there is no density estimate for the whole Dry Chaco region. In the Chancaní Reserve Periago and Leynaud (2009) counted faecal pellet groups in 70 plots in summer and winter. They estimated that the density in this area was 4.41 individuals/km^2 in summer and 5.12 individuals/km^2 in winter. In the Mburucuyá National Park (Corrientes), using direct counts from transects, a density of 6.9 ± 0.9 individuals/km^2 was estimated, without differentiating between seasons. The species showed a greater affinity for mesophytic forests with respect to Yatay palm forests and mesophytic scrubland (Romero and Chatellenaz 2013). In the Natural Provincial Reserve of Multiple Uses (La Picada, Parana department, Entre Rios) density fluctuated between 0.19 individuals/km^2 in native forests, 0.08 individuals/km^2 in plantations and 0.21 individuals/km^2 in mixed forests, determined through footprints (Berduc et al. 2010). In El Rey National Park (Salta), it used more secondary forests than mature Yungas forests. In the Paranaense Forest, its presence is related to increased deforestation and fragmentation of the landscape. In northern Misiones, it appears only in highly fragmented areas and pine plantations (Iezzi et al. 2018a), where it seems to benefit from competition with its congeners *M. nana* and *M. americana* (Varela et al. in litt.). It is much more frequent in the forests of the south of Misiones province and in contact areas with the Campos and Malezales ecoregion.

13.3.9.5 Diet and Feeding Ecology

The diet varied depending on the habitat studied, and the diversity of items in the faeces was higher in bushlands than in other environments. The plant parts consumed include shoots, tender leaves, dead leaves, flowers and fruits of different types of plants such as bushes, trees, herbs, graminoids, vines and also mushrooms (Richard and Juliá 2001; Periago 2017). Additionally, coprophagy and geophagy have been observed (de Bustos obs. pers.), and the ingestion of small invertebrates, mainly ticks (Richard and Juliá 2001). Periago (2017) studied the diet in the Chancaní Reserve, analyzing 64 faecal samples collected in five different habitats: primary and secondary forests, bushlands, areas with selective tree extraction and

"Jarillal" (*Larrea divaricata*). As in the Dry Chaco, its diet included leaves of many different plants (e.g. *Castela coccinea*, *Maytenus spinosa*, *Ziziphus mistol*, *Acacia* spp., *Aspidosperma quebracho-blanco*) (Serbent et al. 2011; Periago 2017).

13.3.9.6 Reproductive Biology

Oestrus can occur at any time of year, the gestation period is 206–210 days, and a single spotted fawn is born. Births occur in all months except May and June. Spots last 3–4 months but begin to disappear after a month. The oestrous cycle has a mean duration of 26.9 days with a mean period of 2.3 days of receptivity (Black-Décima et al. 2010).

13.3.9.7 Behaviour

In various previous studies throughout its distribution the species has been found to be either crepuscular or to have significant activity during the day. More recent studies in the Argentinian Chaco and Yungas, based on camera traps, found a cathemeral behaviour, with significant morning activity but also considerable activity in the afternoon and at night (Periago and Leynaud 2009; Reppucci et al. 2017). It is basically solitary; the most common group is a female with a fawn. In high densities or under conditions of concentrated resources, two or more individuals can be observed foraging together. Social structure appears to be of solitary females occupying home ranges in which only core areas are defended, while males occupy territories that overlap more than one female (Black-Décima 2000; Black-Décima et al. 2010). Territorial marking is done by both males and females with males marking at a higher frequency. The most common and obvious marks are on trees, removing cortex with the incisors and then forehead rubbing, and the maintenance of numerous small latrines, which are monitored, re-marked regularly and counter-marked (Black-Décima and Santana 2011).

13.3.9.8 Threats and Conservation

In Argentina, indirect evidence suggests that populations may be decreasing throughout its distribution area as a result of the immense loss of habitat (Periago 2017). However, the species seems to tolerate large modifications in natural ecosystems, particularly the high deforestation rates of the Dry Chaco and Espinal, where ecosystems are completely transformed to intensive-agricultural lands with no forest cover (Baumann et al. 2016). These do not have negative effects on populations (Periago et al. 2014, Periago 2017) and deer reproduce in fragments and in the agricultural matrix (Núñez-Regueiro et al. 2015; Juliá et al. 2019). In Misiones, it is in expansion as a consequence of the fragmentation and degradation of the Paranaense Forest. However in Yungas, *M. gouazoubira* was found to segregate spatially and temporally from livestock, probably as a strategy to minimize competition for resources and as avoidance (Nanni 2015).

Hunting pressure is high (Altrichter 2005; Camino et al. 2018). In Argentina, researchers have observed that people come from cities and hunt many individuals that are then carried in cars or trucks with freezers (Camino 2016). In spite of this, the populations persist and recover rapidly when the pressure decreases (Juliá et al. 2019).

In Argentina, the species was categorized as Least Concern (Juliá et al. 2019).

13.3.10 Mazama nana

13.3.10.1 Common Names

Corzuela enana, Poca, Poquita, Veado-mao-curta, Veado-cambuta, Brazilian Dwarf Brocket.

13.3.10.2 Distribution

The Brazilian Dwarf Brocket is endemic to the Interior Atlantic Forest ecoregion. It is present in forest remnants of East Paraguay, Misiones province (Argentina) and inland Southeastern Brazil (Paraná, Santa Catarina, and Rio Grande do Sul states).

13.3.10.3 Distribution in Argentina

Only present in forests of Misiones province. The species is found in almost the whole Paranaense Forest ecoregion, although at its southern limit, in the ecotone with the Campos and Malezales ecoregion, the records of this species are often confused with those of *Mazama gouazoubira*.

13.3.10.4 Ecoregions

It is endemic to the Paranaense Forest ecoregion.

13.3.10.5 Habitat and Density

It inhabits semi-deciduous forests, Araucaria forests and ecotones, both in primary and secondary forests (Abril et al. 2010; Duarte et al. 2012), although it also uses commercial pine plantations (Varela et al. in litt.). *Mazama nana* is more frequent in secondary forests, fragmented, and less protected remnants than its relative *M. amer-*

icana (Chebez and Varela 2001, Di Bitetti et al. 2008, Varela et al. in litt.). Vogliotti (2008) showed a habitat partitionship between *M. nana* and *M. americana* in the Iguacu NP (Brazil), in which the former prefer forests with greater understory covers, fewer arboreal layers, and higher elevation areas.

13.3.10.6 Diet and Feeding Ecology

Like other *Mazama* species, Brazilian Dwarf Brockets are browsing herbivorous mammals, consuming leaves, flowers and fruits. There are no specific studies on the diet of *Mazama nana*.

13.3.10.7 Reproductive Biology

There are no studies on the reproduction of *Mazama nana*. However, they have a postpartum oestrus and only one offspring per year, after a gestation of about 7 months. In Misiones, females with offspring have been recorded with camera traps from November to March (D. Varela, pers. obs.). In captivity, the mating of *M. nana* with *M. gouazoubira* has produced infertile offspring (Abril et al. 2010); therefore, infertile hybridization cannot be discarded in overlapping range areas between the two species in southern Misiones (Argentina).

13.3.10.8 Behaviour

Mazama nana is a territorial and solitary species, although couples or females with offspring can be observed during the reproductive season. They have a cathemeral activity pattern, being more nocturnal in areas with higher anthropogenic pressures.

13.3.10.9 Threats and Conservation

The habitat of *Mazama nana* in Brazil and Paraguay suffered a dramatic process of deforestation due to the expansion of industrial agriculture, where over 80% of the original native forest has been lost. Currently, the largest remnants of habitat for *Mazama nana* are in the Paranaense Forest of Misiones, in Argentina. The species persists in fragmented and degraded native forests, even in pine plantations, when they are not overhunted (Varela et al. in litt.). Poaching is probably the main threat in Argentina. It is widely hunted for human consumption throughout the province of Misiones; however, it is more abundant in areas with high hunting pressure than its relative *Mazama americana* (Di Bitetti et al. 2008). Another threat is predation and persecution by domestic and feral dogs on the borders of protected areas (D. Varela, pers. obs.). In addition, there are roadkills on highways crossing protected natural areas. The Dwarf Brocket responds well to mitigation measures implemented along

Misiones roads, and it has been reported to use the wildlife underpassses and overpasses through monitoring with camera traps (Varela 2015). It is present in most of the public and private protected areas of Misiones. In Argentina, the species was categorized as Near Threatened (Varela 2019).

13.3.11 Ozotoceros bezoarticus

13.3.11.1 Common Names

Venado de las pampas, ciervo de las pampas, venado de campo, venado-campeiro, Pampas Deer.

13.3.11.2 Taxonomic Comments

Traditionally two subspecies of pampas deer have been recognized in Argentina: *Ozotoceros bezoarticus celer* (endemic) and *Ozotoceros bezoarticus leucogaster*, which were differentiated by morphological and craniometric characters, fur colour and geographic distribution (Chebez et al. 2008). However, an unpublished study using molecular and morphometric analysis indicates that there are only very slight differences between the four Argentine populations (Raimondi 2013).

13.3.11.3 Distribution in Argentina

Historically, the pampas deer had a wide distribution range in the grassland areas of the north and centre of Argentina, down to the Negro River in the south. The geographic distribution of *O. b. celer* covered the provinces of Buenos Aires, La Pampa, San Luis, Córdoba, Santa Fe and, marginally, Mendoza, Neuquén, Río Negro and probably Chubut. The geographic distribution of *O. b. leucogaster* covered the provinces of Entre Ríos, Santa Fe, Santiago del Estero, Chaco, Formosa, Salta, Tucumán, Corrientes and the south of Misiones (Chebez et al. 2008). However, at the beginning of the twentieth century, its populations began an enormous geographical and numerical reduction due principally to the loss and fragmentation of the habitat, hunting and, probably, competition with cattle for forage. At the present there are four isolated populations restricted to the northeast of Corrientes province, the northeast of Santa Fe province, the south-central area of San Luis province, and the coastal margin of Bahia Samborombon, Buenos Aires province (Miñarro et al. 2011). Recently some pampas deer have been found in north-central La Pampa province, which are presumed to have come from the San Luis population (Bilenca et al. 2018). Meanwhile, in Corrientes province, there is a project of translocation of individuals to repopulate Ibera National park, where deer went extinct at the beginning of the 1980s. So far this has produced two new groups (see Jiménez Pérez et al. 2009, 2016; Zamboni et al. 2017).

13.3.11.4 Ecoregions

Pampa, Espinal, Campos y Malezales, Iberá Marshes and Humid Chaco. According to Politis et al. (2011), it may have been also present, but not widely distributed, in the Dry Chaco. At the present, it is extinct in this ecoregion.

13.3.11.5 Habitat and Density

Ozotoceros bezoarticus is a native ungulate that inhabits the open grassland of South America (González et al. 2010). In Corrientes province a population of pampas deer in the Aguapey watershed has been recognized since the end of the 1980s (Jiménez Pérez et al. 2007). In this area the deer are found in grasslands subject to flooding known as "malezales" and "fofadales", where a diversity of grasses is found (*Andropogon lateralis*, *Sorghastrum setosum*, *Paspalum* spp., *Rhytachne subgibbosa*, etc.; Jiménez Pérez et al. 2007). While one of the reintroduced subpopulations prefers high mounds of grasses of *Elionurus muticus*, the other is found in grasslands of *Andropogon lateralis* and *Sorghastrum setosum* (Zamboni et al. 2014). At the end of the 1990s, population estimates in the region of Aguapey concluded that there were between 100 and 500 individuals (Merino and Beccaceci 1999; Parera and Moreno 2000). However, recent studies estimated a population of 1495 individuals with a population density varying between 0.74 and 1.84 individuals/km^2 (Zamboni et al. 2015). These data indicate that Corrientes province has the largest population of pampas deer in Argentina. It should be noted that the differences between estimates could be explained by differences in the design and analysis of the sample or by a real increase in abundance during the last 10 years (Zamboni et al. 2015). Finally, the two new subpopulations reintroduced into Iberá Marshes are growing (Zamboni et al. 2014, 2017). The least information is available for the population in Santa Fe province in Argentina. It is found in the "Submeridional Lowlands" in a localized area of 23,000 ha at extremely low densities (Pautasso et al. 2009). The deer occupy sites where the vegetation consists of grasses (*Spartina argentinensis*) and pastures of *Elionurus muticus*, and where the woody species are found in isolated mounds formed by chañares (*Geoffroea decorticans*), algarrobos or carobs (*Prosopis* spp.) and palo azul (*Cyclolepis genistoides*), among others (Pautasso et al. 2002, 2009). Although there are no density studies, the population is estimated to have no more than 50 individuals (Pautasso et al. 2002; González et al. 2010).

Turning now to the populations assigned to the subspecies *O. b. celer*, one is found in the coastal region of the Bahía Samborombón Wildlife Refuge, which covers about 244,000 ha of grasslands and wetlands in Buenos Aires province. This population has been declining and its distribution has been changing progressively throughout recent decades (Vila 2006; Pérez Carusi et al. 2009, 2017). The population in San Luis province inhabits the semiarid Pampean grasslands, a graminaceous steppe with small Chañar patches. Pampas deer are distributed over 400,000 ha, but the population is concentrated on 145,000 ha, which corresponds to areas where natural grasslands are in better condition (Dellafiore et al. 2003; Merino et al. 2011).

13.3.11.6 Diet and Feeding Ecology

The diet of *O. bezoarticus* has been described as "mixed", with preferences for grasses (Jackson and Giulietti 1988; Merino 2003). Although studies of diet have not been done on the populations discussed in this chapter, the population of Santa Fe has been mentioned to consume sprouting plants, after fires, of *Spartina argentinensis*, together with *Heliotropium curassavicum* and *Sarcocornia perennis* (Pautasso et al. 2002). Similar observations have been made on the populations of Corrientes province where deer have been seen selecting sprouts of several grasses (e.g. *Andropogon lateralis*, *Sorghastrum setosum*, *Paspalum durifolium*, *Rhytachne subgibbosa*) (S. Cirignoli obs. per.). Deer have also been reported to use and to feed on cultivations of soy, corn, sorghum, and sunflower, although this probably occurs only occasionally (Merino et al. 2009; Semeñiuk 2013).

13.3.11.7 Reproductive Biology

Although the number of births observed throughout the year varies with the different subpopulations (Merino et al. 2019), in the population translocated to Iberá Marshes, births occur in every month, but with a reproductive peak from the end of August until December (Zamboni et al. 2014). It is worthwhile noting that some females reach sexual maturity early (9 months of age) compared to that reported in the literature (e.g. González et al. 2010; Mattioli 2011). This could be due to good body condition, a high quality habitat and low population density (Zamboni et al. 2014). In Santa Fe does with fawns have been observed between September and February (Pautasso et al. 2009).

13.3.11.8 Behaviour

Pampas deer are active mainly during the day where they alternate periods of grazing and resting. They are essentially sedentary with a home range that can vary between 4 and 168 km^2 (González et al. 2010). The only reference to Argentine populations comes from the southern zone of Bahia Samborombon in Buenos Aires province. Here a mean home range of 898 ± 181 ha was found, with males occupying an area of 1422 ha and females only 523 ha (Vila et al. 2008).

The behaviour and social structure of *O. bezoarticus* seems to be complex and related to factors such as season, reproductive condition, age, sex and the availability of forage (González et al. 2010).

Deer populations in Argentina are characterized by low levels of sociability with solitary individuals or pairs being the most frequent social units. At the same time, there is a seasonal dynamic related to the size and composition of groups, since these are not constant throughout the year (Semeñiuk 2013). In Santa Fe province, deer tend to be solitary, and a maximum group size of four individuals has been observed (Pautasso et al. 2009). This could indicate the reduced population size. In the pastures of Corrientes a mean size of 1.91 individuals/group was calculated (Zamboni et al. 2015), which is the same as in the Buenos Aires population (Vila 2006), and less than the mean of 2.49 individuals/group in the San Luis population (Semeñiuk and Merino 2015).

13.3.11.9 Threats and Conservation

Currently, the pampas deer is classified as Near Threatened in light of an ongoing decline (González et al. 2016a, b). In turn, it is considered Endangered in Argentina, with some subpopulations Critically Endangered (Merino et al. 2019).

Without a doubt, the profound transformation of natural pastures for agriculture, cattle ranching and forestry, together with hunting, were the main causes of the drastic geographical reduction and population size that pampas deer have suffered, almost leading to its extinction in Argentina. In Corrientes province, the loss of natural pastures converted into forest plantations (mainly *Pinus* sp.) now occupy more than 24% of the habitat available for deer (Jiménez Pérez et al. 2009). Although *O. bezoarticus* can occupy young forests, the adult plantations are not an adequate habitat considering that it evolved in open habitats (González et al. 2010, S. Cirignoli pers. obs.). In the Santa Fe population, the advance of cultivation of sorghum forage could be affecting the species (Pautasso et al. 2009).

Poaching is a recurring threat in all subpopulations and has had a large impact on the small remnant in Santa Fe province (Pautasso et al. 2002, 2009). Extraordinary flooding as well as long droughts may have negatively affected this same subpopulation, although this cannot be definitively proved (Pautasso et al. 2009). Still, during droughts, dead deer have been found in water storage reservoirs (Pautasso et al. 2010).

The presence of dogs associated with cattle ranching or feral dogs represent another threat in the whole deer distribution range (Merino et al. 2019). Predation events have been recorded in the populations of Buenos Aires, Santa Fe and Corrientes (Vila 2006, Pautasso et al. 2005, S. Cirignoli pers. obs.).

Finally, negative interactions between pampas deer and introduced ungulates such as livestock and feral pigs have been proposed as among the main causes for its decline (Merino et al. 2019). Livestock have been considered a negative factor due to competition for food and habitat, and risk of disease transmission (Merino et al. 2019). At present, the grasslands have been drastically altered by overgrazing, plowing and exotic grass species introduction. Furthermore, cattle stocking rates have increased largely as a result of the introduction of exotic pastures and changes in cattle rearing practices (Merino et al. 2011). Pérez Carusi et al. (2017) found that the coexistence of pampas deer and cattle in the same pasture was possible only with a moderate number of heads of cattle (0.2–0.4 heads/ha), and deer were not found when densities were greater than 0.6 heads/ha.

In addition, negative interactions between pampas deer and feral pigs may exist, including inverse spatial relationships, possible predation of neonates, changes in the soil due to rooting behaviour and as a reservoir of pathogens (Carpinetti 1998; Carpinetti et al. 2017; Pérez Carusi et al. 2009, 2017). For the populations reintroduced into Ibera National Park, interaction with exotic ungulate populations is one of the principal threats, due to the size of the populations of *Sus scrofa* and *Axis axis* (see Cirignoli 2010; Ballari et al. 2015).

Pampas deer have been declared a Natural Provincial Monument in the provinces of Corrientes (Decree 1555/92), Buenos Aires (Provincial Law No. 11689/95), and Santa Fe (Provincial Law No. 12182/03). These laws provide the maximum legal protection possible for a species. They are declared of public interest in San Luis province, through a law for protection of pampas deer (Provincial Law No. 5499/04) which prohibits the hunting and capture of individuals. They are included in CITES Appendix I. In addition, Campos del Tuyú National Park has been created (3040 ha) to protect the pampas deer in Buenos Aires province.

13.4 Final Considerations

All of the seven ecoregions considered in this chapter begin at the northern border of Argentina and thus are extensions of more tropical ecoregions. As such, Argentine tropical ungulates are animals with more northern and tropical origins that continue their distributions south into favourable areas of the country; Argentina represents the southern limit of their distribution. Of the 11 species discussed in this chapter, four are classified as Endangered, four as Vulnerable, one as Near Threatened, and two as Least Concern at the national level. There is thus considerable concern with respect to the future of these species.

These species play a fundamental role in the ecological processes of ecosystems, either by determining the structure and composition of forests (herbivory, seed dispersion, etc.) or in the maintenance of trophic interactions with large carnivores, or because they directly influence the lives of local communities (Wilson and Mittermeier 2011).

However, in spite of their importance and the knowledge we have gained about them in recent years, basic taxonomic, ecological and conservation information is still scarce or absent for many of the species that inhabit Argentina. This is probably due to the difficulties in carrying out certain studies, the lack of financial support or of adequate governmental policies (Taber et al. 2016; Grotta-Neto and Duarte 2019). If, on top of this lack of knowledge, we add the accelerated transformation of their natural habitats into agricultural areas, livestock raising and forestry plantations, as well as the enormous hunting pressure to which they are submitted, the future of these populations is really uncertain.

Here we summarize the current conservation status of each species. The tapir's lower distribution limit is in northwest Argentina in the Yungas forests. It exists in low densities in the Paranaense and Yungas forests and in both Dry and Humid Chaco, occupying primary and secondary forests and even pine plantations where human interference is low. As a large animal with a long generation time and a low reproductive rate, it is very sensitive to the usual threats of habitat degradation and loss and hunting, and the large reduction in its distribution along with local extinctions document its problems. Classified as Vulnerable both by the IUCN and nationally, there is hope for its future as it is well represented in protected areas.

All three peccary species face different degrees of threat in Argentina. Two are classified as Endangered and one as Near Threatened nationally. The Chacoan peccary is classified as Endangered both by the IUCN and by Argentina. As a species endemic to the Gran Chaco it is especially vulnerable to the enormous deforestation of the Chaco (one of the highest in the world) and hunting. Effective conservation measures do not seem to be in place for this species and its future is uncertain. The collared peccary is classified as Least Concern by the IUCN but as Near Threatened in Argentina. Although it has a very wide distribution in the Americas, and seems to thrive in fragmented habitats, its abundance has diminished in certain parts of the Gran Chaco. In some cases there have been local extinctions due to the same factors that affect the Chacoan peccary. Formerly extinct in Corrientes province, it is now being reintroduced into Ibera National Park where four populational nuclei are now found with more than 100 animals. The white-lipped peccary is classified as Vulnerable by the IUCN but as Endangered in Argentina. It has a wide distribution in Central and South America but a much narrower one in Argentina. It exists at very low densities in the Chaco where its survival is uncertain. Its situation is better in remote parts of Yungas, where it maintains reasonable group sizes, and in the Paranaense forest, where it is recovering.

The guanaco is listed as Least Concern both by the IUCN and Argentina, but most of its populations are found either at high altitude in sub Andean areas or in Chile and Patagonia. The populations found in the Chaco would probably be listed as Critically Endangered as they are relicts and their survival uncertain.

The taruca is listed as Vulnerable by the IUCN and Endangered in Argentina. It is found principally in high altitude, non-tropical areas, but it extends into Yungas forests in Argentina. It is threatened by habitat loss and cattle ranching and is protected as a natural national and provincial monument by national and provincial laws.

The marsh deer is listed as Vulnerable both by the IUCN and by Argentina. It inhabits wetlands, including swamps and marshes. In Argentina there are four populations: the largest is found in Corrientes in Ibera and surrounding areas; the second, in the Parana delta; two smaller populations inhabit Santa Fe and Formosa and Chaco. The one inhabiting the Ibera wetlands is large, but the Parana delta population is considered Endangered and the two populations of the Humid Chaco, as

Critically Endangered. Its distribution has also diminished considerably with extinctions occurring in many provinces. It is threatened by wetland modification for productive purposes, poaching and flooding. It is not well represented in protected areas, but is found in two recent national parks, and has substantial legal protection as a provincial monument in four provinces. It is also on Appendix I of CITES.

Argentina has three species of brocket deer; two are listed as Vulnerable and one as Least Concern nationally. The red brocket deer is listed as Data Deficient by the IUCN, because it includes several cryptic species, and Vulnerable in Argentina. It occurs in the Yungas and Paranaense forests and also in Humid Chaco. It is very sensitive to degradation of forests and hunting and disappears from disturbed areas and areas with cattle ranching. The gray brocket is listed as Least Concern both by IUCN and nationally. It is very adaptable, capable of living in primary and secondary forests in various ecoregions and also in disturbed areas including areas dedicated to intensive agriculture, although at reduced densities. Its populations are decreasing due to loss of habitat and hunting, but are still healthy and capable of recovery. The Brazilian dwarf brocket is endemic to the Atlantic forest and its extension in Argentina, the Paranaense forest. It is listed as Vulnerable both by IUCN and in Argentina. It is found in primary and secondary forests and even pine plantations, but is vulnerable to habitat loss, dog predation, and hunting. It is found in most of the protected areas of Misiones.

The pampas deer was one of the most characteristic ungulates of Argentina, now suffering from drastically reduced populations. It is classified as Near Threatened by the IUCN but as Endangered in Argentina, as it has now been reduced to four isolated populations from its former extensive distribution throughout the grasslands area. It was decimated by the conversion of almost all of this area into production (agriculture and cattle ranching) and by extensive hunting. All of its populations now coexist with productive matrices in agricultural and ranching areas; however, in the case of two of these populations, Corrientes and San Luis, the populations are stable or increasing, indicating that productive activities can be compatible with deer conservation. The fact that pampas deer can maintain populations within productive areas managed with a sustainable approach to grasslands use (cattle load adjusted to the nutritional supply, rotational crops with a period of parcel rest and adequate distribution of watering stations), appears to be beneficial, at least for the survival of some populations. These factors seem to increase the availability of better-quality patches in the deer's habitat, and also directly affect habitat use. Consequently, this opens an important avenue for conservation that may complement the creation of protected areas (Merino et al. 2011). In Santa Fe province, protected areas on private property should also be established in order to preserve the last living populations in the Chaco environment. Finally, the successful reintroduction of deer into Ibera Park, in which one of the subpopulations now exceeds 120 individuals, should be a source of inspiration for this type of action in places where deer are in a critical situation (e.g. Santa Fe, Buenos Aires).

In conclusion, we are aware that many ungulate populations are found in the most productive regions of Argentina. This presents a challenge for conservation experts who must establish an equilibrium between the systems of production and the conservation of biodiversity.

We can see that the future of these species is linked to the creation of new protected areas, to enhance capacities of indigenous and rural communities for sustainable use of these species and their habitats, further investigation with adequate sources of funding and clear governmental policies with respect to land use. The principal threats that affect ungulate habitat in northern Argentina include agriculture and livestock expansion, poaching in habitat remnants, the impact of linear infrastructure and competition with exotic ungulates. Confronting these challenges will require innovative strategies for development at the local, national or regional level. Moreover, it would also be desirable to develop new rewilding projects both in the Chaco, in the Yungas jungles and the Atlantic forest.

Acknowledgements Many friends and colleagues have freely contributed information to this review. The authors especially thank Javi Pereira and Natalia Fracassi of the Pantano Project, Talia Zamboni and Sebastian Di Martino of the Conservation Land Trust and Mario Di Bitetti and Carlos De Angelo of the IBS. Thanks are also due to the projects Investigation and Conservation of the Tapir in the NOA and Jaguars at the Limit, to DRNOA of the National Parks Administration and the Secretary of the Environment of Salta. We thank the National Secretary of the Environment and Sustainable Development for funding (to DV and SC) in the program "Zero Extinction", and the Rufford Small Grant (20144-2), EDGE of existence Programme of the Zoological Society of London for funding (to MC) in Proyecto Quimilero. Finally we would like to thank Dario Podesta and Marcelo Ortiz for sharing their photographs of marsh deer, gray brocket, guanaco, tapir (DP) and taruca (MO).

References

Abril VV, Vogliotti A, Varela DM, JMB D, Cartes JL (2010) Brazilian dwarf brocket deer *Mazama nana* (Hensel 1872). In: Duarte JMB, González S (eds) Neotropical cervidology: biology and medicine of Latin American deer. Funep; IUCN, Jaboticabal; Gland, pp 160–165

Albanesi SA, Jayat JP, Brown AD (2016) Patrones de actividad de mamíferos de mediano y gran porte en el Pedemonte de Yungas del noroeste argentino. Mastozool Neotrop 23(2):335–358

Altrichter M (2005) The sustainability of subsistence hunting of peccaries in the Argentine Chaco. Biol Conserv 126:351–362

Altrichter M (2006) Wildlife in the life of local people of the semi-arid Argentine Chaco. Biodivers Conserv 15(8):2719–2736

Altrichter M, Boaglio GI (2004) Distribution and relative abundance of peccaries in the Argentine Chaco: associations with human factors. Biol Conserv 116:217–225

Altrichter M, Taber A, Noss A, Maffei L, Campos J (2015) *Catagonus wagneri*. The IUCN red list of threatened species 2015: e.T4015A72587993. https://doi.org/10.2305/IUCN.UK.2015-2.RLTS.T4015A72587993.en. Accessed 20 Dec 2018

Altrichter M, Desbiez A, Camino M, Decarre J (2016) Pecarí del Chaco o Taguá (*Catagonus wagneri*). Una estrategia para su conservación. Revisión de situación, análisis de viabilidad poblacional y aptitud del hábitat. UICN Grupo Especialista en Pecaríes, SSC, Guyra Paraguay, CCCI Paraguay, Asunción, 116 p

Aprile G (2016) Evaluación y diagnóstico del estado de situación del Proyecto Reintroducción de Guanacos del Parque Nacional Quebrada del Condorito (Córdoba, Argentina). Parque Nacional Quebrada del Condorito, Administración de Parques Nacionales, Córdoba

Balbuena PJ, Berdún JA, Bracho A, Faisal S, Ocampo EO, Contreras FR, Leiva L, Villalba R, Portela M, Benítez PP, Aguirre CA, Fraga VD, Holman B, Mestre LM, Chatellenaz ML, D'Angelo RC, Aued MB, Orozco MM, Varela D, Lartigau B (2015) Atropellamiento de mamíferos silvestres amenazados de extinción en la Ruta Nacional 12, al norte de la provincia de Corrientes. Un problema para la conservación de la fauna silvestre y la seguridad vial. Dirección Nacional de Fauna y Flora y Dirección de Recursos Naturales, Santa Fe, 49 p

Ballari SA, Cuevas MF, Cirignoli S, Valenzuela AE (2015) Invasive wild boar in Argentina: using protected areas as a research platform to determine distribution, impacts and management. Biol Invasions 17(6):1595–1602

Barbarán F (2017) Comercialización de cueros de pecarí Tayassuidae en la provincia de Salta, Argentina y evaluación de su política de conservación. Período 1973-2012. Rev Biodivers Neotrop 7(3):169–188

Bardavid S, de Bustos S, Politi N, Rivera LO (2019) Escasez de registros de pecarí labiado (*Tayassu pecari*) en un sector de alto valor de conservación de las Yungas australes de Argentina. Mastozool Neotrop 26(1):167–173

Barri FR (2016) Reintroducing guanaco in the upper belt of Central Argentina: using population viability analysis to evaluate extinction risk and management priorities. PLoS One 11(10):e0164806. https://doi.org/10.1371/journal.pone.0164806

Barri FR, Cufré M (2014) Supervivencia de guanacos (*Lama guanicoe*) reintroducidos con y sin período de preadapatación en el parque nacional Quebrada del Condorito, Córdoba, Argentina. Mastozool Neotrop 21(1):9–16

Barri FR, Falczuk V, Cingolani AM, Díaz S (2014) Dieta de la población de guanacos (*Lama guanicoe*) reintroducida en el Parque Nacional Quebrada del Condorito, Argentina. Ecol Austral 24(2):203–211

Baumann M, Piquer-Rodríguez M, Fehlenberg V, Pizarro GG, Kuemmerle T (2016) Land-use competition in the South American Chaco. In: Niewöhner J, Bruns A, Hostert P, Krueger T, Nielsen JØ, Haberl H, Lauk C, Lutz J, Müller D (eds) Land use competition. Ecological, economic and social perspectives. Springer International Publishing, Cham, pp 215–229

Baumann M, Gasparri I, Piquer-Rodríguez M, Gavier Pizarro G, Griffiths P, Hostert P, Kuemmerle T (2017) Carbon emissions from agricultural expansion and intensification in the Chaco. Glob Chang Biol 23(5):1902–1916

Beccaceci MD (1996) Dieta del ciervo de los pantanos (*Blastocerus dichotomus*), en la Reserva Iberá, Corrientes, Argentina. Mastozool Neotrop 3(2):193–198

Beck H (2005) Seed predation and dispersal by peccaries throughout the Neotropics and its consequences: a review and synthesis. In: Forget PM, Lambert JE, Hulme PE, Vander Wall SB (eds) Seed fate: predation, dispersal and seedling establishment. CABI Publishing, Wallingford, pp 77–115

Beck H, Thebpanya P, Filiaggi M (2010) Do Neotropical peccary species (Tayassuidae) function as ecosystem engineers for anurans? J Trop Ecol 26:407–414

Berduc A, Biering PL, Donello AV, Walker CH (2010) Lista actualizada y análisis preliminar del uso de hábitat de medianos y grandes mamíferos en un área natural protegida del Espinal con invasión de leñosas exóticas, Entre Ríos, Argentina. Revista FABICIB 14:9–27

Bertonatti C, Pastore H (1998) Libro Rojo: Taruca (*Hippocamelus antisensis*). Rev Vida Silv 61:21–22

Bilenca D, Codesido M, Abba A, Agostini MG, Corriale MJ, González Fischer C, Pérez Carusi L, Zufiaurre E (2018) Conservación de la biodiversidad en sistemas pastoriles. Buenas prácticas para una ganadería sustentable de pastizal. Kit de extensión para las Pampas y Campos. Fundación Vida Silvestre Argentina, Buenos Aires

Black-Décima P (2000) Home range, social structure and scent marking behavior in brown brocket deer in a large enclosure. Mastozool Neotrop 7(1):5–14

Black-Décima P, Santana M (2011) Olfactory communication and counter-marking in brown brocket deer *Mazama gouazoubira*. Acta Theriol 56:179–187

Black-Décima P, Rossi RV, Vogliotti A, Cartes JL, Maffei L, Duarte JMB, González S, Juliá JP (2010) Brown Brocket Deer *Mazama gouazoubira* (Fischer 1814). In: Duarte, J.M.B. and

González, S. (eds) Neotropical cervidology: biology and medicine of Latin American deer. Jaboticabal; Gland: Funep; IUCN, p 190-201

Cabrera AL (1971) Fitogeografía de la República Argentina. Bol Soc Arg Bot 14:1–42

Cabrera A, Yepes J (1960) Mamíferos Sud Americanos, 2nd edn. S. A. Editores, Buenos Aires

Camino M (2016) Selección de hábitat y ocupación de tres especies de pecaríes en el Chaco Semiárido Argentino. PhD dissertation, Facultad de Ciencias Exactas y Naturales de la Universidad de Buenos Aires, Argentina

Camino M, Cortez S, Cerezo A, Altrichter M (2016) Wildlife Conservation, perceptions of different co-existing cultures. Int J Conserv Sci 7(1):109–122

Camino M, Cortez S, Altrichter M, Matteucci SD (2018) Relations with wildlife of Wichi and Criollo people of the Dry Chaco, a conservation perspective. Ethnobiol Conserv 7:11. https://doi.org/10.15451/ec2018-08-7.11-1-21

Camino M, Cirignoli S, Varela D, Barri F, Aprile G, Periago ME, de Bustos S, Quiroga V, Torres R, Di Martino S (2019) *Pecari tajacu*. In: Categorización del estado de conservación de los mamíferos de Argentina 2019. Lista Roja de los mamíferos de Argentina. SAYDS - SAREM (eds.). Online versión.

Cano PD, Cardozo HG, Ball HA, D'Alessio S, Herrera P, Lartigau B (2012) Aportes al conocimiento de la distribución del ciervo de los pantanos (*Blastocerus dichotomus*) en la provincia de Corrientes, Argentina. Mastozool Neotrop 19(1):35–45

Carmanchahi P, Panebianco A, Leggieri L, Barri F, Marozzi A, Flores C, Moreno P, Schroeder N, Cepeda C, Oliva G, Kin M, Gregorio P, Ovejero R, Acebes P, Schneider C, Pedrana J, Taraborelli P (2019) *Lama guanicoe*. In: Categorización del estado de conservación de los mamíferos de Argentina 2019. Lista Roja de los mamíferos de Argentina. SAYDS - SAREM (eds.). Online versión.

Carpinetti BN (1998) Spatio-temporal variation in a pampas deer (*Ozotocerus bezoarticus celer*) population: influence of habitat structure and sympatric ungulates. MSc thesis, University of Kent at Canterbury, Canterbury, United Kingdom

Carpinetti B, Castresana G, Rojas P, Grant J, Marcos A, Monterubbianesi M, Sanguinetti HR, Serena MS, Echeverría MG, Garciarena M, Aleksa A (2017) Determinación de anticuerpos contra patógenos virales y bacterianos seleccionados en la población de cerdos silvestres (*Sus scrofa*) de la Reserva Natural Bahía Samborombón, Argentina. Analecta Vet 37(1):21–27

Cartes Yegros JL (1999) Distribución y uso de hábitat de la corzuela parda en los Llanos de la Rioja, Argentina. MSc thesis, Programa de Posgrado en Manejo de Vida Silvestre. Facultad de Ciencias Exactas, Físicas y Naturales, Universidad Nacional de Córdoba

Casco SL, Neiff M, Neiff JJ (2005) Biodiversidad en ríos del litoral fluvial. Utilidad del software PULSO. In: Temas de la Biodiversidad del Litoral fluvial argentino II, Miscelánea, vol 14. INSUGEO, Tucumán, pp 419–434

Chalukian S, Merino ML (2006) Orden Perissodactyla. In: Barquez RM, Díaz MM, Ojeda RA (eds) Mamíferos de Argentina. Sistemática y distribución. Sociedad Argentina para el Estudio de los Mamíferos, Mendoza, pp 113–114

Chalukian SC, de Bustos S, Lizárraga L, Saravia M, Garibaldi JF (2004) Uso de hábitat del tapir en relación a la presencia de ganado, en el Parque Nacional El Rey, Salta, Argentina. In: II Simposio Internacional de Tapir. Panamá. January 2004, pp 10–16

Chalukian SC, de Bustos S, Lizárraga L, Paviolo A, Varela D, Quse V (2009) Plan de acción para la conservación del tapir (*Tapirus terrestris*) en Argentina. Wildlife Conservation Society, Tapir Specialist Group-UICN, Dirección de Fauna-Secretaría de Ambiente y Desarrollo Sustentable de la Nación

Chalukian SC, de Bustos MS, Lizárraga RL (2013) Diet of lowland tapir (*Tapirus terrestris*) in El Rey National Park, Salta, Argentina. Integr Zool 8:47–55

Chebez JC (1994) Los que se van. Especies argentinas en Peligro. Albatros, Buenos Aires

Chebez JC, Varela D (2001) La corzuela enana. In: Dellafiore C, Maceira N (eds) Los ciervos autóctonos de la Argentina y la acción del hombre. Grupo Abierto de Comunicaciones, Buenos Aires, pp 51–56

Chebez JC, Johnson A, Pautasso A (2008) Venado de las Pampas. In: Chebez JC (ed) Los que se van. Fauna argentina amenazada, Mamíferos, vol 3. Albatros, Buenos Aires, pp 222–242

Cirignoli S (2010) El peligro de la fauna silvestre invasora en el Iberá: El enemigo fantasma. Boletín de los Esteros 8:8–10

Coley PD (1983) Herbivory and defensive characteristics of tree species in a lowland tropical forest. Ecol Monogr 53(2):209–229

Coppolillo P, Gómez H, Maisels F, Wallace R (2003) Selection criteria for suites of landscape species as a basis for site-based conservation. Biol Conserv 115:419–430

Costa T, Barri F (2018) *Lama guanicoe* remains from the Chaco ecoregion (Córdoba, Argentina): an osteological approach to the characterization of a relict wild population. PLoS One 13(4):e0194727. https://doi.org/10.1371/journal.pone.0194727

Cruz P (2012) Densidad, uso del hábitat y patrones de actividad diario del tapir (*Tapirus terrestris*), en el Corredor Verde de Misiones, Argentina. Licentiate thesis, University of Buenos Aires, Argentina

Cruz DP, Courtalon P (2017) Usos y percepciones de la fauna silvestre por pobladores de dos barrios aledaños a la Reserva Natural Otamendi, Campana, Argentina. Ecol Austral 27(2):242–251

Cruz P, Paviolo A, Bó RF, Thompson JJ, Di Bitetti MS (2014) Daily activity patterns and habitat use of the lowland tapir (*Tapirus terrestris*) in the Atlantic Forest. Mamm Biol 79:376–383

Cuéllar E (2011) Ecology and conservation of the guanaco *Lama guanicoe* in the Bolivian Chaco: habitat selection within a vegetation succession. PhD dissertation, University of Oxford, Oxford, England

Cuellar E, Noss A (2003) Mamíferos del Chaco y de la Chiquitanía de Santa Cruz, Bolivia. FAN, Santa Cruz de la Sierra

Cuéllar ES, Segundo J, Banegas J (2017) El guanaco (*Lama guanicoe* Müller 1776) en el Gran Chaco boliviano: Una revisión. Ecol Boliv 52(1):38–57

D'Alessio S (2016) Evaluación de la presencia de embalsados en las islas del Bajo Delta del Paraná y su importancia para el Ciervo de los Pantanos (*Blastocerus dichotomus*) en períodos de inundación. Licentiate thesis, University of Buenos Aires, Buenos Aires, Argentina

D'Alessio S, Aprile G, Lartigau B, Herrera P, Cano D, Eberhardt A, Antoniazzi L, Jimenez Perez I, Di Giacomo A, Meyer N, Fracassi N, Cardozo H, Varela D, Parera A, Ball H, Aued MB, Figuerero C, Fernandez A, Lezcano Y, Cowper Coles P, Sosa M, Cowper Coles N, Ramirez G, Kees A (2018) Ciervo de los pantanos *Blastocerus dichotomus* (Illiger, 1815). In: Manejo de fauna silvestre en la Argentina. Programa de Conservación de Especies Amenazadas. Ministerio de Ambiente y Desarrollo Sustentable y Fundación de Historia Natural Félix de Azara, Buenos Aires, pp 249–280. Available at: https://goo.gl/j5bL5R

D'Angelo R (2012) Estudio de dieta de taruca (*Hippocamelus antisensis* D'Orbigny 1834) en el Cerro Santa Ana, Provinica de Jujuy, Argentina. Licenture thesis, University of Buenos Aires, Buenos Aires, Argentina

De Angelo C, Martínez Pardo J, Varela D, Cirignoli S, Di Giácomo A (2018) Modelado del hábitat remanente para el ciervo de los pantanos en el noreste Argentino. In: XIII Congreso Internacional de Manejo de Fauna Silvestre en la Amazonia y Latinoamérica. May 2018. Ciudad del Este, Paraguay, pp 338–339

de Bustos S, Alderete E (in press) Primeros registros de pecari de collar Pecari tajacu (Mammalia, Aartiodactyla) para Monte de Sierras y Bolsones y en la provincia de Catamarca, Argentina. Notas sobre Mamiferos Sudamericanos.

de Bustos S (2018) Dispersión y herbivoría del tapir (*Tapirus terrestris*) en un bosque de Yungas. PhD dissertation, Universidad Nacional de Tucumán, Tucumán, Argentina

de Bustos S, Chalukian SC, Alveira M, Saravia M, Saravia M, Rodríguez K (2009) Habitat use and activity of ungulates using camera traps in El Rey National Park, Proc 10th Internat Mammal Congr, Mendoza, Argentina, August 2009, p 159

de Bustos S, Lizárraga L, Maras G, Reppucci J, Caruso F, Alveira M, Perovic P (2018) Situación crítica del pecarí labiado *Tayassu pecari* en las Yungas de la Alta Cuenca del Río Bermejo, Argentina. In: XXXI Jornadas Argentinas de Mastozoología, La Rioja, SAREM, November 2018, p 151

de Bustos S, Varela D, Lizárraga L, Cirignioli S, Quiroga V, Chalukian S, Giombini M, Juliá JP, Quse V, Giraudo A, Di Martino S, Camino M, Albanesi S (2019a) *Tapirus terrestris*. In:

Categorización del estado de conservación de los mamíferos de Argentina 2019. Lista Roja de los mamíferos de Argentina. SAYDS - SAREM (eds.). Online versión.

de Bustos S, Varela D, Lizárraga L, Camino M, Quiroga V (2019b) *Tayassu pecari*. In: Categorización del estado de conservación de los mamíferos de Argentina 2019. Lista Roja de los mamíferos de Argentina. SAYDS - SAREM (eds.). Online versión.

Dellafiore CM, Maceira NO (1998) Problemas de conservación de los ciervos autóctonos de la argentina. Mastozool Neotrop 5(2):137–145

Dellafiore CM, Demaría M, Maceira N, Bucher E (2003) Distribution and abundance of the pampas deer in San Luis Province, Argentina. Mastozool Neotrop 10(1):41–47

Di Bitetti MS, Paviolo A, Ferrari CA, De Angelo C, Di Blanco Y (2008) Differential responses to hunting in two sympatric species of brocket deer (*Mazama americana* and *M. nana*). Biotropica 40:636–645

Di Martino S, Jimenéz-Peréz I, Peña J (2015) Estrategia para la reintroducción de tapires (*Tapirus terrestris*) en la Reserva Natural Iberá (Corrientes, Argentina). The Conservation Land Trust Argentina, Buenos Aires. Available at: http://proyectoibera.org/

Díaz NI (1995) Antecedentes sobre la historia natural de la taruca (*Hippocamelus antisensis* d'Orbigny 1834) y su rol en la economía andina. Chungara Rev Antrop Chil 27(1):45–55

Díaz MM, Braun JK, Mares MA, Barquez RM (2000) An update of the taxonomy, systematics, and distribution of the mammals of Salta province, Argentina. Occ Pap Sam Noble Oklahoma Mus Nat Hist 10:1–52

Duarte JMB, González S (eds) (2010) Neotropical cervidology: biology and medicine of Latin American deer. Funep; IUCN, Jaboticabal; Gland

Duarte JMB, Merino ML (1997) Taxonomia e evolução. Biologia e conservacao de Cérvidos Sul-americanos: *Blastocerus*, *Ozotoceros* e *Mazama*. FUNEP, Jaboticabal

Duarte JMB, Abril VV, Vogliotti A, dos Santos ZE, de Oliveira ML, Tiepolo LM, Rodrigues LF, de Almeida LB (2012) Avaliação do risco de extinção do veado-cambuta *Mazama nana* Hensel, 1872, no Brasil. Biodiversidade Brasileira 2(3):59–67

Eberhardt A, Antoniazzi L, Kees A, Herrera P, Lartigau B, D'Alessio S (2009) Distribución y conservación del Ciervo de los Pantanos (*Blastocerus dichotomus*) en la provincia de Santa Fe, Argentina. Resultados Preliminares. Revista Biológica 10:43–53

Edge of Existence (2018). http://www.edgeofexistence.org/species/chacoan-peccary/

Ferraz KMPB, Silva Angelieri CC, Altrichter M, Desbiez A, Yanosky A, Campos Krauer JM, Torres R, Camino M, Cabral H, Cartes J, Cuellar RL, Gallegos M, Giordano AJ, Decarre J, Maffei L, Neris N, Saldivar Bellassai S, Wallace R, Lizarraga L, Thompson J, Velilla M (2016) Predicting the current distribution of the Chacoan peccary (*Catagonus wagneri*) in the Gran Chaco. Suiform Soundings 15:53–63

Flores CE, Cingolani AM, von Müller A, Barri FR (2013) Habitat selection by reintroduced guanacos (*Lama guanicoe*) in a heterogeneous mountain rangeland of central Argentina. Rangel J 34(4):439–445

Gasparini GM, Ubilla M, Tonni EP (2013) The Chacoan peccary, *Catagonus wagneri* (Mammalia, Tayassuidae), in the late Pleistocene (northern Uruguay, South America): paleoecological and paleobiogeographic considerations. Hist Biol 25(5-6):679–690

Geisa MG, Dottori N, Cosa MT (2018) Dieta de guanaco (*Lama guanicoe*) en el Chaco Árido de Córdoba, Argentina. Mastozool Neotrop 25(1):59–80

Giombini MI, Bravo SP, Martínez MF (2009) Seed dispersal of the palm *Syagrus romanzoffiana* by tapirs in the semi-deciduous Atlantic Forest of Argentina. Biotropica 41(4):408–413

Giombini MI, Bravo SP, Tosto DS (2016) The key role of the largest extant Neotropical frugivore (*Tapirus terrestris*) in promoting admixture of plant genotypes across the landscape. Biotropica 48:499–508

Gomez Vinassa ML, Beatríz Nuñez M (2016) Dieta estacional de guanacos (*Lama guanicoe*) y burros ferales (*Equus asinus*) en un ambiente semiárido de San Luis, Argentina. Ecol Austral 26(2):178–188

González S, Cosse M, Goss Braga F, Vila A, Merino ML, Dellafiore C, Cartes JL, Maffei L, Gimenez-Dixon M (2010) Pampas deer *Ozotoceros bezoarticus* (Linnaeus 1758). In Duarte,

J.M.B. and González, S. (eds) Neotropical cervidology: biology and medicine of Latin American deer. Jaboticabal; Gland: Funep; IUCN, p 119-132

González S, Duarte JMB, Cosse M, Repetto L (2016a) Conservation genetics, taxonomy, and management applications in Neotropical deer. In: Aguirre AA, Sukumar R (eds) Tropical conservation: perspectives on local and global priorities. Oxford University Press, New York

González S, Jackson JJ, Merino ML (2016b) *Ozotoceros bezoarticus*. The IUCN red list of threatened species 2016: Version 2016.1. Available at: http://www.iucnredlist.org. Accessed 12 Dec 2018

Grau HR, Torres R, Gasparri NI, Blendinger PG, Marinaro S, Macchi L (2015) Natural grasslands in the Chaco. A neglected ecosystem under threat by agriculture expansion and forest-oriented conservation policies. J Arid Environ 123:40–46

Grotta-Neto F, Duarte JMB (2019) Movements of neotropical forest deer: what do we know? In: Rafael Reyna-Hurtado R, Chapman CA (eds) Movement ecology of neotropical forest mammals. Focus on social animals. Springer Nature, Cham

Guerra I, Pastore H, Varela D, de Bustos S (2019) *Hippocamelus antisensis*. In: Categorización del estado de conservación de los mamíferos de Argentina 2019. Lista Roja de los mamíferos de Argentina. SAYDS - SAREM (eds.). Online versión.

Guillemi EC, Orozco MM, Argibay HD, Farber MD (2019) Evidence of *Ehrlichia chaffeensis* in Argentina through molecular detection in marsh deer (*Blastocerus dichotomus*). Int J Parasitol Parasites Wildl 8:45–49

Hansen MC, Potapov PV, Moore R, Hancher M, Turubanova SA, Tyukavina A, Thau D, Stehman SV, Goetz SJ, Loveland TR, Kommareddy A, Egorov A, Chini L, Justice CO, Townshend JRG (2013) High-resolution global maps of 21st-century forest cover change. Science 342(6160):850–853

Heinonen S, Bosso A (1994) Nuevos aportes para el conocimiento de la mastofauna del Parque Nacional Calilegua (Provincia de Jujuy, Argentina). Mastozool Neotrop 1(1):51–60

Hibert F, Sabatier D, Andrivot J, Scotti-Saintagne C, González S, Prévost MF, Grenand P, Chave J, Caron H, Richard-Hansen C (2011) Botany, genetics and ethnobotany: a crossed investigation on the elusive tapir's diet in French Guiana. PLoS One 6(10):e25850

Huemul Task Force (2012) Reassessment of morphology and historical distribution as factors in conservation efforts for the Endangered Patagonian Huemul Deer *Hippocamelus bisulcus* (Molina 1782). J Threat Taxa 4:3302–3311

Hurtado Martinez CM (2017) Reintroduction success and ecological aspects of reintroduced peccaries (*Pecari tajacu*) in the Ibera Natural Reserve, Corrientes, Argentina. Master thesis, Towson University, Vancouver, Canada

Hurtado CM, Beck H, Thebpanya P (2018) From exploration to establishment: activity changes of the first collared peccary (*Pecari tajacu*) group reintroduced in South America. Hystrix 29(2):229. https://doi.org/10.4404/hystrix-00058-2018

Iezzi ME, Cruz P, Varela D, De Angelo C, Di Bitetti MS (2018a) Tree monocultures in a biodiversity hotspot: impact of pine plantations on mammal and bird assemblages in the Atlantic Forest. For Ecol Manage 424:216–227

Iezzi ME, Fracassi NG, Pereira JA (2018b) Conservation of the largest cervid of South America: interactions between people and the Vulnerable marsh deer *Blastocerus dichotomus*. Oryx 52(4):654–660

Jackson JE, Giulietti J (1988) The food habits of pampas deer Ozotoceros bezoarticus celer in relation to its conservation in a relict natural grassland in Argentina. Biol Conserv 45(1):1–10

Jiménez Pérez I, Delgado A, Drews W, Solis G (2007) Estado de conservación de la última población de venado de las pampas (*Ozotocerus bezoarticus*) en Corrientes: reflexiones y recomendaciones. The Conservation Land Trust Argentina, Buenos Aires. Available at: http://proyectoibera.org/download/venado/estudio_venado_corrientes_07.pdf. Accessed 10 Jan 2019

Jiménez Pérez I, Delgado A, Srur M, Heinonen S (2009) Proyecto de conservación, rescate y restauración del venado de las pampas en la provincia de Corrientes. The Conservation Land Trust Argentina, Buenos Aires. Available at: http://proyectoibera.org/download/venado/proyecto_restauracion_venado.pdf. Accessed 10 Jan 2019

Jiménez Pérez I, Abuin R, Antúnez B, Delgado A, Massat M, Pereda I, Pontón F, Solís G, Spørring K, Zamboni T, Heinonen S (2016) Re-introduction of the pampas deer in Iberá Nature Reserve, Corrientes, Argentina. In: Soorae PS (ed) Global re-introduction perspectives: 2016. Case-studies from around the globe. IUCN/SSC Reintroduction Specialist Group and Abu Dhabi; Environment Agency-Abu Dhabi, Gland; Abu Dhabi, pp 221–227

Juliá JP, Richard E (2001) La corzuela colorada. In: Dellafiori C, Maceira N (eds) Los ciervos autóctonos de la Argentina y la acción del hombre. Secretaría de Desarrollo Sustentable y Política Ambiental. Ministerio de Desarrollo Social y Medio Ambiente. Buenos Aires, Argentina, pp 27–34

Juliá JP, Varela D, Periago ME, Cirignoli S, Muzzachiodi N, Camino M, Barri F, Iezzi ME, de Bustos S (2019) *Mazama gouazoubira*. In: Categorización del estado de conservación de los mamíferos de Argentina 2019. Lista Roja de los mamíferos de Argentina. SAYDS - SAREM (eds.). Online versión.

Ledesma NR (1992) Caracteres de la semiaridez en el Chaco Seco. Anales de la Acad Nac de Agron y Vet 46:21–32

López L, Cámara H (2007) Senderos de la Selva Misionera. Gobierno de la Provincia de Misiones, Alto Paraná S.A

López de Casenave JL, Pelotto JP, Protomastro J (1995) Edge-interior differences in vegetation structure and composition in a Chaco semi-arid forest, Argentina. For Ecol Manage 72(1):61–69

Loponte DM, Corriale MJ (2013) Isotopic values of diet of *Blastocerus dichotomus* (marsh deer) in Paraná Basin, South America. J Archaeol Sci 40(2):1382–1388

Luduvério DJ (2018) O status taxonômico de *Cervus rufus* illiger, 1811 e sua caracterização genética e morfológica. Master's thesis. Universidade Estadual Paulista (UNESP), Jaboticabal, Sao Paulo, Brasil

Manso Hernández N, Casertano SA, Garibaldi JF, Barrios Caro LX, Herrera JR, Correa Plasencia F (2010) Plan de manejo del parque provincial Puerto Península. Planificación de las Áreas Protegidas del Núcleo Norte de la Provincia de Misiones. Proyecto Araucaria XXI. AECID-APN MERNRyT, Misiones

Massoia E, Chebez JC, Bosso A (2006) Los mamíferos silvestres de la Provincia de Misiones, Argentina. Editorial L.O.LA, Buenos Aires

Matteucci SD (2018a) Ecorregión Campos y Malezales. In: Morello J, Matteucci SD, Rodríguez AF, Silva M et al (eds) Ecorregiones y complejos ecosistémicos Argentinos, 2nd edn. Orientación Gráfica Editora, Buenos Aires, pp 285–306

Matteucci SD (2018b) Ecorregión Esteros del Iberá. In: Morello J, Matteucci SD, Rodríguez AF, Silva M et al (eds) Ecorregiones y complejos ecosistémicos Argentinos, 2nd edn. Orientación Gráfica Editora, Buenos Aires, pp 331–351

Matteucci SD (2018c) Ecorregión Delta e Islas de los Ríos Paraná y Uruguay. In: Morello J, Matteucci SD, Rodríguez AF, Silva M et al (eds) Ecorregiones y complejos ecosistémicos Argentinos, 2nd edn. Orientación Gráfica Editora, Buenos Aires, pp 501–545

Matteucci SD, Morello J, Rodríguez AF, Mendoza N (2004) El alto Paraná Encajonado argentino-paraguayo. Ediciones FADU-UNESCO, Buenos Aires

Mattioli S (2011) Familia Cervidae. In: Wilson DE, Mittermeier RA (eds) Handbook of the mammals of the world, Hoofed mammals, vol 2. Lynx Edicions, Barcelona

Medici EP (2010) Assessing the viability of lowland tapir populations in a fragmented landscape. PhD dissertation, Philosophy in Biodiversity Management, Durrell Institute of Conservation and Ecology, University of Kent, Canterbury, United Kingdom

Merino ML (2003) Dieta y uso de hábitat del venado de las pampas, *Ozotoceros bezoarticus celer* Cabrera 1943 (Mammalia-Cervidae) en la zona costera de Bahía de Samborombón, Buenos Aires, Argentina. Implicancias para su conservación. PhD dissertation, Facultad de Ciencias Naturales y Museo, Universidad Nacional de La Plata

Merino ML, Beccaceci MD (1999) *Ozotocerus bezoarticus* (Artiodactyla, Cervidae) en Corrientes, Argentina: distribución, población y conservación. Iheringia Sér Zool 87:87–92

Merino ML, Semeñiuk MB, Olocco Diz MJ, Meier D (2009) Utilización de un cultivo de soja por el venado de las pampas, *Ozotoceros bezoarticus* (Linnaeus, 1758), en la provincia de San Luis, Argentina. Mastozool Neotrop 16(2):347–354

Merino ML, Semeñiuk MB, Fa JE (2011) Effect of cattle breeding on habitat use of Pampas deer *Ozotoceros bezoarticus* celer in semiarid grasslands of San Luis, Argentina. J Arid Environ 75:752–756

Merino ML, Cirignoli S, Perez Carusi L, Varela D, Kin M, Pautasso A, Demaria M, Beade M, Uhart M (2019) *Ozotoceros bezoarticus*. In: Categorización del estado de conservación de los mamíferos de Argentina 2019. Lista Roja de los mamíferos de Argentina. SAYDS - SAREM (eds.). Online versión.

Miñarro FO, Li Puma MC, Pautasso AA (2011) Plan nacional de conservación del Venado de las pampas (*Ozotoceros bezoarticus*). Resolución SAYDS 340/11. Dirección de Fauna Silvestre. Secretaría de Ambiente y Desarrollo Sustentable de la Nación. Jefatura de Gabinete de Ministros, Buenos Aires, 91 p

Morello J (2012) Ecorregión Chaco Húmedo. In: Morello J, Matteucci SD, Rodríguez AF, Silva M et al (eds) Ecorregiones y complejos ecosistémicos Argentinos, 1st edn. Orientación Gráfica Editora, Buenos Aires, pp 205–223

Nanni AS (2015) Dissimilar responses of the Gray brocket deer (*Mazama gouazoubira*), Crab-eating fox (*Cerdocyon thous*) and Pampas fox (*Lycalopex gymnocercus*) to livestock frequency in subtropical forests of NW Argentina. Mamm Biol 80(4):260–264

Neiff JJ (1999) El régimen de pulsos en ríos y grandes humedales de Sudamérica. In: Malvárez I (ed) Tópicos sobre humedales subtrópicales y templados de Sudamérica. MAB, UNESCO, Buenos Aires, pp 99–149

Neiff JJ (2004) El Iberá… ¿en peligro? Fundación Vida Silvestre Argentina, Buenos Aires

Neiff JJ (2005) Bosques fluviales de la cuenca del Paraná. In: Arturi MF, Frangi JL, Goya JF (eds) Ecologia y manejo de los bosques de Argentina. Facultad de Ciencias Agrarias y Forestales, Facultad de Ciencias Naturales y Museo, Universidad Nacional de La Plata, La Plata, pp 1–26

Neiff JJ, Patiño CAE, Casco SL (2005) Atenuación de las crecidas por los humedales del Bajo Paraguay. In: Peteán J, Capatto J (eds) Humedales fluviales de América del Sur. Hacia un manejo sustentable. Proteger Ediciones. IUCN, Santa Fe, pp 261–276

Nicolossi G (2004) Mortandad de vertebrados terrestres en canales de Empresa Ledesma. Informe técnico inédito. Parque Nacional Calilegua, Administración de Parques Nacionales

Nicolossi G, Baldo J (2011) Impacto del Sistema de riego sobre la fauna silvestre en el área de amortiguamiento del Parque Nacional Calilegua. Informe Técnico No 02/2011. Parque Nacional Calilegua, Administración de Parques Nacionales, Buenos Aires

Nigro NA, Lodeiro Ocampo N (2009) Atropellamiento de fauna silvestre en las rutas de la provincia de Misiones, Argentina: análisis y propuestas preliminares para mitigar su impacto. Rep Tigreros Ser Conserv 2:1–12. Available at: http://redyaguarete.org.ar/

Nori J, Torres R, Lescano JN, Cordier JM, Periago ME, Baldo D (2016) Protected areas and spatial conservation priorities for endemic vertebrates of the Gran Chaco, one of the most threatened ecoregions of the world. Divers Distrib 22(12):1212–1219

Novaro AJ, Redford KH, Bodmer RE (2000) Effect on hunting in source–sink system in the neotropics. Conserv Biol 14:713–721

Núñez-Regueiro M, Branch L, Fletcher RJ Jr, Marás GA, Derlindati E, Tálamo A (2015) Spatial patterns of mammal occurrence in forest strips surrounded by agricultural crops of the Chaco region, Argentina. Biol Conserv 187:19–26

O'Farrill G, Galetti M, Campos-Arceiz A (2013) Frugivory and seed dispersal by tapirs: an insight on their ecological role. Integr Zool 8:4–17

Oakley LJ, Prado D, Adámoli J (2005) Aspectos Biogeográficos del Corredor Fluvial Paraguay-Paraná. Temas de la Biodiversidad del Litoral fluvial argentino II. Miscelánea INSUGEO 14:1–14

Olrog CC (1979) Los mamíferos de la selva húmeda, Cerro Calilegua, Jujuy. Act Zool Lilloa 33(2):9–14

Orozco MM, Marull C, Jiménez Pérez I, Gürtler R (2013) Mortalidad invernal de ciervo de los pantanos (*Blastocerus dichotomus*) en humedales del Noreste de Argentina. Mastozool Neotrop 20:163–170

Orozco MM, Argibay H, Sotelo V, Müller G, Losada P, Mestres J, Ruiz Díaz G, Morales M, Moreira A, Ortiz H, Sosa D, Leiva P, Rodríguez P, Pérez P, Paszko L, Holman B (2018) Situación actual de *Blastocerus dichotomus* en Argentina – reseña sobre episodios de mortalidad. In: XIII Congreso Internacional de Manejo de Fauna Silvestre en la Amazonía y Latinoamérica, Ciudad del Este, Paraguay. May 2018, pp 346–347

Painter RLE (1998) Gardeners of the forest: plant-animal interactions in a Neotropical forest ungulate community. PhD disseertation, University of Liverpool, UK

Parera A, Moreno D (2000) El venado de las pampas en Corrientes, diagnóstico de su estado de conservación y propuestas de manejo. Publicación Especial Fundación Vida Silvestre Argentina, Buenos Aires

Pastore H (2012) Relevamiento biológico de la Finca El Puesto, Volcán, Provincia de Jujuy. Informe Inédito. Universidad Nacional del Comahue. Programa Conservación del Huemul, Administración de Parques Nacionales

Pastore H, Fernández C, Bertonatti C (1997) El venado andino: Monumento Natural de los Argentinos. Fundación Vida Silvestre Argentina, Buenos Aires

Pautasso AA, Peña MI, Mastropaolo JM, Moggia L (2002) Distribución y conservación del venado de las Pampas (*Ozotoceros bezoarticus leucogaster*) en el norte de Santa Fe, Argentina. Mastozool Neotrop 9(1):64–69

Pautasso AA, Chersich D, Peña MI, Mastropaolo JM, Fandiño B, Senn A, Raimondi VB (2005) El venado de las pampas (*Ozotoceros bezoarticus leucogaster* Cabrera 1943) en la fracción norte de los bajos submeridionales de la provincia de Santa Fe, Argentina. El venado de las pampas (Ozotoceros bezoarticus L. 1758) en la provincia de Santa Fe, Argentina. Situación Terminal. Com Mus Provincial Cienc Nat Florentino Ameghino 10(2):16–124

Pautasso AA, Fandiño B, Raimondi VB, Senn AI (2009) Avances sobre las Acciones Prioritarias del Plan Provincial para la Conservación del Venado de las Pampas en Santa Fe. Período 2006-2007. Rev Biol 9:8–24

Pautasso AA, Raimondi VB, Bierig PB, Leiva LA (2010) Mortalidad de venado de las pampas (*Ozotoceros bezoarticus*) y aguara guazu (*Chrysocyon brachyurus*) en represas de almacenamiento de agua en los bajos submeridionales de Santa Fe, Argentina. Notulas Faunisticas 52:1–6

Paviolo A (2010) Densidad de yaguareté (*Panthera onca*) en la Selva Paranaense: Su relación con la disponibilidad de presas, presión de caza y oexistencia con el puma (*Puma concolor*). PhD dissertation Ciencias Biológicas, Universidad Nacional de Córdoba

Paviolo A, De Angelo C, Di Blanco Y, Agostini I, Pizzio E, Melzew R, Ferrari C, Palacio L, Di Bitetti MS (2009) Efecto de la caza y el nivel de protección en la abundancia de los grandes mamíferos del Bosque Atlántico de Misiones. In: Carpinetti B, Garciarena M, Almirón M (eds) Parque Nacional Iguazú, conservación y desarrollo en la selva Paranaense de Argentina. Administración de Parques Nacionales, Buenos Aires, pp 243–260

Pereira JA, Fergnani D, Fernández V, Fracassi NG, González V, Lartigau B, Marín V, Tellarini J, Varela D, Wolfenson L (2018) Introducing the "Pantano Project" to conserve the southernmost population of the marsh deer. IUCN/Deer Specialist Group Newsletter 30:15–21

Pereira JA, Varela D, Aprile G, Cirignoli S, Orozco M, Lartigau B, De Angelo C, Giraudo A (2019) *Blastocerus dichotomus*. In: Categorización del estado de conservación de los mamíferos de Argentina 2019. Lista Roja de los mamíferos de Argentina. SAYDS - SAREM (eds.). Online versión.

Pérez Carusi LC, Beade MS, Miñarro F, Vila AR, Giménez Dixon M, Bilenca DN (2009) Relaciones espaciales y numéricas entre venados de las pampas (*Ozotoceros bezoarticus* celer) y chanchos cimarrones (*Sus scrofa*) en el Refugio de Vida Silvestre Bahía Samborombón, Argentina. Ecol Austral 19:63–71

Pérez Carusi LC, Beade MS, Bilenca DN (2017) Spatial segregation among pampas deer and exotic ungulates: a comparative analysis at site and landscape scales. J Mammal 98(3):761–769

Periago ME (2017) Efectos de los cambios en el uso de la tierra sobre la dieta y la selección de hábitat de mamíferos en el Chaco árido. PhD dissertation, Facultad de Ciencias Exactas, Físicas y Naturales de la Universidad Nacional de Córdoba. Córdoba, Argentina

Periago ME, Leynaud GC (2009) Density estimates of *Mazama gouazoubira* (Cervidae) using the pellet count technique in the arid Chaco (Argentina). Ecol Austral 19(1):73–77

Periago ME, Chillo V, Ojeda RA (2014) Loss of mammalian species from the South American Gran Chaco: empty savanna syndrome? Mammal Rev 45(1):41–53

Periago ME, Tamburini DM, Ojeda RA, Cáceres DM, Díaz S. (2017). Combining ecological aspects and local knowledge for the conservation of two native mammals in the Gran Chaco. J Arid Environ 147:54–62

Perovic P, de Bustos S, Reppucci J, Maras G (2017) Relevamiento de yaguareté (*Panthera onca*) en el Chaco Salteño. Informe técnico. DRNOA Administración de Parques Nacionales, Secretaria de Ambiente de Salta y Proyecto Jaguares en el Límite, Buenos Aires

Piovenzan U, Tiepolo LM, Tomas WM, Duarte JMB, Varela D, Marinho Filho JS (2010) Marsh Deer *Blastocerus dichotomus* (Illiger 1815) Duarte, J.M.B. and González, S. (eds) Neotropical cervidology: biology and medicine of Latin American deer. Jaboticabal; Gland: Funep; IUCN p 66-76

Polegato BF, Zanetti EDS, Duarte JMB (2018) Monitoring ovarian cycles, pregnancy and post-partum in captive marsh deer (*Blastocerus dichotomus*) by measuring fecal steroids. Conserv Physiol 6(1):cox073

Politis GG, Prates L, Merino ML, Tognelli MF (2011) Distribution parameters of guanaco (*Lama guanicoe*), pampas deer (*Ozotoceros bezoarticus*) and marsh deer (*Blastocerus dichotomus*) in Central Argentina: archaeological and paleoenvironmental implications. J Archaeol Sci 38(7):1405–1416

Puechagut PB, Politi N, Ruiz de los Llanos E, Lizarraga L, Bianchi CL, Bellis LM, Rivera LO (2018) Association between livestock and native mammals in a conservation priority area in the Chaco of Argentina. Mastozool Neotrop 25(2):407–418

Raimondi V (2013) Genética aplicada a la conservación de especies amenazadas y su hábitat. Estudio del aguará guazú (*Chrysocyon brachyurus*) y del venado de las pampas (*Ozotoceros bezoarticus*). PhD dissertation, Universidad de Buenos Aires, Argentina

Regidor HA, Costilla M (2003) Un mapa de distribución para la taruka (*Hippocamelus antisensis*) en el Noroeste Argentino. Memorias: manejo de fauna silvestre en Amazonia y Latinoamérica. CITES, Asunción, pp 266–268

Regidor HA, Garrido D, Ragno R (1997) Unidades de paisaje aptas para taruca *Hippocamelus antisensis* en Salta, Argentina. Parte I - Una Clasificación no supervisada. Manejo de Fauna. Univ Nacl Salta Publ Téc 8:21–23

Reppucci J, Perovic PG, de Bustos S, Maras G, Sillero C (2017) Patrones de actividad de mamíferos grandes y medianos de las Yungas argentinas. In: XXXI Jornadas Argentinas de Mastozoología, Bahía Blanca, Bs. As. Nov 2017, p 226

Richard E, Juliá JP (2001) La corzuela parda. In: Dellafiori C, Maceira N (eds) Los ciervos autóctonos de la Argentina y la acción del hombre. Secretaría de Desarrollo Sustentable y Política Ambiental. Ministerio de Desarrollo Social y Medio Ambiente. Buenos Aires, Argentina, pp 27–34

Rivera L, Martinuzzi S, Politi N, Bardabid S, de Bustos S, Radeloff V, Pidgeon A (2019) Influence of national parks on habitat use by lowland tapirs *Tapirus terrestris* in the Southern Yungas of Argentina, Oryx 18-A-0364

Rodríguez AF, Silva M (2012a) Ecorregión Selva Paranaense. In: Morello J, Matteucci SD, Rodríguez AF, Silva M et al (eds) Ecorregiones y complejos ecosistémicos Argentinos, 1st edn. Orientación Gráfica Editora, Buenos Aires, pp 225–245

Rodríguez AF, Silva M (2012b) Ecorregión de las selvas de Yungas. In: Morello J, Matteucci SD, Rodríguez AF, Silva M et al (eds) Ecorregiones y complejos ecosistémicos Argentinos, 1st edn. Orientación Gráfica Editora, Buenos Aires, pp 129–149

Romero VL, Chatellenaz ML (2013) Densidad de *Mazama gouazoubira* (Artiodactyla, Cervidae) en un Parque Nacional del nordeste de Argentina. Acta Zool Mex 29(2):388–399

Sabattini RA, Lallana VH (2007) Aquatic Macrophytes. In: Iriondo MH, Paggi JC, Parma MJ (eds) The Middle Paraná River: limnology of a subtropical wetland. Springer, Berlin, pp 205–226

Saldivar SS (2014) Status and threats to persistence of the Chacoan peccary (*Catagonus wagneri*) in the Defensores del Chaco National Park, Paraguay. MSc thesis, College of Environmental Science and Forestry, Syracuse, New York

Sallenave A (2009) Frugivory and seed dispersal by tapirs (*Tapirus terrestris*) in Atlantic forest of NE Argentine. Proc 10th Intern Mammal Cong, Mendoza, Argentina, Aug 2009, p 282

Sanderson EW, Redford KH, Vedder A, Coppolillo PB, Ward SE (2002) A conceptual model for conservation planning based on landscape species requirements. Landsc Urban Plan 58:41–56

Semeñiuk MB (2013) Ecología espacial y estructura social del venado de las pampas (*Ozotoceros bezoarticus* Linnaeus, 1758) en los pastizales semiáridos de la provincia de San Luis, Argentina: relaciones con el uso de la tierra. PhD dissertation, Universidad Nacional de La Plata, Buenos Aires

Semeñiuk MB, Merino ML (2015) Pamps deer (*Ozotoceros bezoarticus*) social organization in semiarid grasslands of San Luis, Argentina. Mammalia 79:131–138

Serbent MP, Periago ME, Leynaud GC (2011) *Mazama gouazoubira* (Cervidae) diet during the dry season in the arid Chaco of Córdoba (Argentina). J Arid Environ 75(1):87–90

Soler R (2005) Uso de hábitat del tapir (*Tapirus terrestris*) en tres rangos de altura, en el Parque Nacionale El Rey, Salta, Argentina. Licenciate Thesis dissertation Biodiversidad, Universidad Nacional del Litoral, Santa Fe

Taber AB, Doncaster CP, Neris NN, Colman FH (1993) Ranging behavior and population dynamics of the Chacoan peccary, *Catagonus wagneri*. J Mammal 74(2):443–454

Taber AB, Chalukian S, Minkowski K, Sanderson E, Lizárraga L, Rumiz D, Ventincinque E, Moraes E, De Angelo C, Antúnez M, Ayala G, Bodmer R, Boher S, Cartes JL, de Bustos S, Eaton D, Emmons L, Estrada N, de Oliveira LF, Fragoso JM, García R, Gómez C, Gómez H, Keuroghlian A, Ledesma K, Lizcano D, Lozano C, Montenegro O, Neris N, Noss A, Palacio Vieira JA, Paviolo A, Perovic P, Portillo H, Radachowsky J, Reyna-Hurtado R, Rodriguez Ortiz J, Salas L, Sarmiento Duenas A, Sarria Perea JA, Schiaffino K, de Thoisy B, Tobler M, Utreras V, Varela D, Wallace RB, Zapata Ríos G (2009) El destino de los arquitectos de los bosques neotropicales: Evaluación de la distribución y el estado de conservación de los pecaríes labiados y los tapires de tierras bajas. Tapir Specialist Group IUCN-Wildlife Conservation Society-Wildlife Trust, New York, 182 pp

Taber A, Beck H, González S, Altrichter M, Duarte JMB, Reyna-Hurtado R (2016) Why neotropical forest ungulates matter. The case for deer, peccary, and tapir conservation. In: Aguirre AA, Sukumar R (eds) Tropical conservation: perspectives on local and global priorities. Oxford University Press, New York

Tálamo A, Caziani SM (2003) Variation in woody vegetation among sites with different disturbance histories in the Argentine Chaco. For Ecol Manage 184:79–92

Tavarone EG, Ruíz L, Monguillot J, Ramírez D (2007) Proyecto piloto de reintroducción del guanaco (*Lama guanicoe*) en el Parque Nacional Quebrada del Condorito. Administración de Parques Nacionales, Córdoba. Available at: http://sib.gov.ar

Teta P, Abba AM, Cassini GH, Flores DA, Galliari CA, Lucero SO, Ramírez M (2018) Lista revisada de los mamíferos de Argentina. Mastozool Neotrop 25(1):163–198

Tomás WM, Salis SM, Silva MP, Miranda Mourão G (2001) Marsh deer (*Blastocerus dichotomus*) distribution as a function of floods in the Pantanal wetland, Brazil. Stud Neotrop Fauna Environ 36(1):9–13

Torres R, Camino M (2019) *Parachoerus wagneri*. In: Categorización del estado de conservación de los mamíferos de Argentina 2019. Lista Roja de los mamíferos de Argentina. SAYDS - SAREM (eds.). Online versión.

Torres R, Tamburini D, Lescano J, Rossi E (2017) New records of the endangered Chacoan peccary *Catagonus wagneri* suggest a broader distribution than formerly known. Oryx 51(2):286–289

Torres R, Tamburini D, Boaglio G, Decarre J, Castro L, Lescano J, Barri F (2019) New data on the endangered Chacoan peccary (*Catagonus wagneri*) link the core distribution with its recently discovered southern population. Mammalia 83:357. https://doi.org/10.1515/mammalia-2018-0105

Varela D (2003) Distribución, abundancia y conservación del ciervo de los pantanos (*Blastocerus dichotomus*) en el bajo delta del Río Paraná, provincia de Buenos Aires, Argentina. Licentiature thesis, University of Buenos Aires. Available at: https://goo.gl/V7RU4x

Varela D (2015) Ecología de rutas en Misiones. Evaluación de la efectividad de los pasafaunas y ecoductos. Informe con resultados para el período 2011-2014. Technical report. Dirección Provincial de Vialidad de Misiones and Conservación Argentina. Available: https://goo.gl/DV3xBt

Varela D, Lartigau B, Pereira JA (2017) Efectos de las inundaciones extraordinarias (Diciembre 2015–Agosto 2016) sobre la población de Ciervo de los Pantanos del Bajo Delta del Río Paraná. In: Informe técnico preparado al Ministerio de Ambiente y Desarrollo Sustentable de la Nación (Programa Extinción Cero). Proyecto Pantano. Centro de Investigaciones del Bosque Atlántico (CeIBA), Puerto Iguazú, 43pp

Varela D, de Bustos S, Cirignoli S, Di Bitetti MS (2019) *Mazama americana*. In: Categorización del estado de conservación de los mamíferos de Argentina 2019. Lista Roja de los mamíferos de Argentina. SAYDS - SAREM (eds.). Online versión.

Varela D (2019) *Mazama nana*. In: Categorización del estado de conservación de los mamíferos de Argentina 2019. Lista Roja de los mamíferos de Argentina. SAYDS - SAREM (eds.). Online versión.

Vila AR (2006) Ecología y conservación del venado de las pampas (*Ozotoceros bezoarticus celer*, Cabrera 1943) en la Bahía Samborombón, Provincia de Buenos Aires. PhD dissertation, Universidad de Buenos Aires, Ciudad Autónoma de Buenos Aires, Argentina

Vila AR, Beade MS, Barrios Lamuniere D (2008) Home range and habitat selection of pampas deer. J Zool 276:95–102

Vogliotti A (2008) Partição de habitats entre os cervídeos do Parque Nacional do Iguaçu. PhD dissertation, Universidade de São Paulo, São Paulo, Brazil

Wilson DE, Mittermeier RA (2011) Handbook of the mammals of the world, Hoofed mammals, vol 2. Lynx Edicions, Barcelona

Zamboni T, Jiménez-Pérez I, Abuín R, Antúnez B, Pereda I, Peña J, Massat M (2014) Proyecto de recuperación del venado de las pampas en la Reserva Natural Iberá y los bañados de Aguapey: informe de resultados y actividades (Año 2014). The Conservation Land Trust, Buenos Aires. Available at: http://www.proyectoibera.org/download/venado/informe_proyecto_venados_corrientes_2014.pdf

Zamboni T, Delgado A, Jiménez-Pérez I, De Angelo C (2015) How many are there? Multiple-covariate distance sampling for monitoring pampas deer in Corrientes, Argentina. Wildl Res 42(4):291–301

Zamboni T, Di Martino S, Jiménez-Pérez I (2017) A review of a multispecies reintroduction to restore a large ecosystem: the Iberá Rewilding Program (Argentina). Perspect Ecol Conser 15(4):248–256

Zamboni T, Yablonsky F, Ortiz H (2018) El pecarí de collar sigue recuperando terreno en el Parque Iberá: el inicio de dos nuevas poblaciones en Carambola y San Nicolás. Boletín CLT 3:3–4

Zarrilli AG (2004) Historia y economía del bosque chaqueño: la mercantilización de los recursos forestales (1890-1950). Anuario IEHS 19:255–283

Part II
Specific Topics of Tropical Ungulated Species

Chapter 14
The Mule Deer of Arid Zones

Sonia Gallina-Tessaro, Luz A. Pérez-Solano, Luis García-Feria, Gerardo Sánchez-Rojas, Dante Hernández-Silva, and Juan Pablo Esparza-Carlos

Abstract *Odocoileus hemionus* species occurs throughout western North America and its southernmost distribution reaches central Mexico, where the historical boundary is not very clear. In this chapter, we present information about its origin and geographic distribution, mainly concerning to mule deer lineage (*O. h. crooki*, *eremicus*, *fuliginatus*, *peninsulae*, *cerrosensis* and *sheldoni*) that are distributed in the arid zones of Mexico. Many biological and ecological aspects of mule deer have been widely studied in the southwestern USA, but in Mexico few studies have been done in arid lands (e.g. its diet, population, genetics), so we present a compilation of the information that is currently available for the species in these regions. A synthesis about the characteristics of the three most important habitat requirements (food, water and cover) that can become limiting factors for mule deer populations in arid zones is presented here; these habitat components have only been estimated in some occasions in Mexico. The quality and availability of these requirements are highly affected by overgrazing, and it has put the survival of many deer populations at risk. The mule deer is classified as one of the species of least concern in terms of conservation, although its populations and distribution areas have been reduced, even with some local extinctions. Mule deer provide an excellent example of the need to evaluate species at the regional and national levels to arrive at a more realistic classification of their conservation status.

S. Gallina-Tessaro (✉) · L. A. Pérez-Solano · L. García-Feria
Red de Biología y Conservación de Vertebrados, Instituto de Ecología, A.C., Xalapa, México
e-mail: sonia.gallina@inecol.mx; luis.garcia@inecol.mx

G. Sánchez-Rojas
Centro de Investigaciones Biológicas, Universidad Autónoma del Estado de Hidalgo, Pachuca, México
e-mail: ganchez@uaeh.edu.mx

D. Hernández-Silva
Research and Management, Wild Forest Consulting, S.C., Huitchila, México
e-mail: dante_hernandez@uaeh.edu.mx

J. P. Esparza-Carlos
Departamento de Ecología y Recursos naturales, Universidad de Guadalajara, Guadalajara, México
e-mail: juan.esparza@cucsur.udg.mx

S. Gallina-Tessaro (ed.), *Ecology and Conservation of Tropical Ungulates in Latin America*, https://doi.org/10.1007/978-3-030-28868-6_14

14.1 The Origin of the Mule Deer

The Cervidae family originated in Eurasia 20 million years ago (Ma.). The migration of the group to the Americas via the Bering Strait dates to the end of the Miocene (ca. 5 Ma.), with *Eocoileus gentryorum* the oldest recorded species (Gilbert et al. 2006). At the end of the Pliocene, during the Great American Interchange, some of the species of the Odocoileini Tribe migrated quite early towards South America, where they reached their maximum evolutionary radiation (mainly the genera *Blastocerus*, *Hippocamelus*, *Pudu*, and *Mazama*), as a result of colonization and recolonization, and by taking refuge and persisting during the glaciation events of the Pleistocene (Flagstad and Røed 2003; Gilbert et al. 2006).

Currently, there is quite some controversy over the taxonomic relationships of the deer of the New World (Odocoileini) owing to the high degree of contrast between the traditional classification—based primarily on morphological methods—and the newer classification based on molecular methods. With molecular tools, it has been found that there is no monophyly among its genera, and so further research is needed to determine their relationships (Gutiérrez et al. 2017). Within the genus *Odocoileus*, the phylogenetic relationships based on the mitochondrial cytochrome-b gene place this genus in a paraphyletic clade with respect to *Mazama pandora*, with *O. hemionus columbianus* and *O. h. sitkensis* a haplogroup (that could represent a valid species) sister to *M. pandora*, while the other nine mule deer subspecies together with the subspecies of the white-tailed deer (*Odocoileus virginianus*) appear to be more closely related (Gutiérrez et al. 2017).

Also, within the *O. hemionus* group, the cytochrome-b and control region markers indicate that the species has three well defined lineages: that of the Sitka black-tailed deer *O. h. sitkensis* (Kodiak Island, Yukon and northern coast of British Columbia) and the Columbian black-tailed deer *O. h. columbianus* (southern British Columbia), and the third lineage includes the nine subspecies of the mule deer (*O. h. hemionus*, *crooki*, *eremicus*, *inyoensis*, *californicus*, *fuliginatus*, *peninsulae*, *cerrosensis*, and *sheldoni*) that in equal measure correspond to previously identified morphological differences and the predictions based on the glacial cycles of the Pleistocene (Latch et al. 2009, 2014).

These glacial cycles exerted selective pressure on the biota distributed in the temperate zone of the Northern Hemisphere. During glacial advance, the species of temperate ecosystems were forced to survive in isolated refugia, where population size decreased allowing for genetic drift and inbreeding, as well as some restrictions to gene flow that led to the formation of the lineages by the process of vicariance (Latch et al. 2009, 2014). The lineages of black-tailed deer can be explained by these biogeographical phenomena, given their limited genetic structure. The opposite occurs with the mule deer, which lost and gained space during the different glaciations, and now exhibits *panmixia*, reflected in the great genetic diversity of its populations (Latch et al. 2014).

The oldest of the modern-day deer belonging to the genus *Odocoileus* is the white-tailed deer (*O. virginianus*), which appeared 3.5 Ma. in the Pliocene (Gilbert et al. 2006). It is possible that the *O. hemionus* group is the most recent and that it developed 2 Ma. during the glacial maximum since the oldest reported fossil record dates to the Irvingtonian (between 1.8 Ma. and 300,000 years ago) (Bell et al. 2004). There is a hypothesis that the mule deer originated from the white-tailed deer on two occasions: in the Early Pleistocene, branching from the primitive white-tailed deer and developing as the black-tailed deer of the West, and a second time in the Post-Pleistocene, as a result of the fusion of male black-tailed deer with female white-tailed deer some 11,000–9000 years ago (Geist 1998). Thus, the species multiplied, radiated and hybridized. The mule deer, as a new form, adapted to more extreme climates and open landscapes, it diverged in ethological signals and ornamental structures to attract a mate. Hybridization between male white-tailed deer and female mule deer occurred, allowing in a few cases, the introgression of the genes of one species into the other (Geist 1998).

The fossils of *O. hemionus* have been found in 19 localities in the Canadian provinces of British Columbia and Alberta, the US states of Alaska, California, Colorado, New Mexico, Texas, and Utah; however for Mexico in particular, the fossil record for this family is scarce. Recently, for the Early Pliocene cranial and post-cranial remains have been reported for at least three animals assigned to the genus *Odocoileus*, which were found in the states of Sonora, Nuevo León, San Luis Potosí, and Hidalgo (Aguilar et al. 2012).

One of the most relevant characteristics between the two current species of *Odocoileus* is that the *hemionus* group has more complex, larger antlers that have a dichotomous forking pattern (Gustafson 1985). Additionally, they are ecologically segregated by their different feeding adaptations and habits, and by their antipredator behavior, which makes it possible to predict their habitat preferences. For example, instead of using a fast gallop to escape like *O. virginianus* does, *O. hemionus* escapes in great leaps and bounds (stotting or pronking), which also allows it to gain enough speed to escape, and to rapidly climb steep slopes, all of which are formidable obstacles to any predator. *O. hemionus* can change direction and trajectory at will, unpredictably and instantly. It has the advantage in rough and mountainous terrain and in dense shrubby vegetation (Geist 1998).

14.2 Distribution

Odocoileus hemionus occurs throughout western North America from Alaska and Western Canada through the Rocky Mountains and the Western Plains states of the United States of America, south to the Peninsula of Baja California, Cedros Island, Tiburón Island, and Northwestern Mexico. Its southernmost distribution reaches central Mexico, but the historical boundary is not very clear (Sánchez Rojas and Gallina-Tessaro 2016). The historical distribution of populations of mule deer in the Chihuahuan Desert of Mexico, include the state of Coahuila, the extreme southwest of the state of Nuevo León, the northern region of San Luis Potosí and Zacatecas, a small region in western Tamaulipas, the desert region of the states of Chihuahua and Durango, and the northeastern region of Sonora (Fig. 14.1).

In Mexico, six subspecies of mule deer are recognized and these are distributed in arid zones: in Sonora, *O. h. eremicus*; on the southern Baja California Peninsula, *O. h. peninsulae* and to the north, *O. h. fuliginatus*; two island subspecies, *O. h. sheldoni* on Tiburón Island and *O. h. cerrosensis* on Cedros Island; and *O. h. crooki* in the Chihuahuan Desert (Anderson and Wallmo 1984) (Fig. 14.1). That being said, the type specimen *O. h. crooki* belonged to a hybrid between *O. virginianus* and *O. h. eremicus* (Heffelfinger 2000), though in spite of this it has recently been recognized as a valid subspecies, *O. h. crooki* (Latch et al. 2014).

14.3 Description

The mule deer grows to a total body length of 1300–2600 mm; it is reddish brown or lion yellow in color in the summer, and dark brown or brownish gray in the winter. The brown coloring extends to the face close to the eyes on the front of and to the sides of the nose, the rest of the face is white or gray. The tail is 115–190 mm long with a black tip and tones from white to black on the dorsal surface. Males can reach 64–114 kg, while females are lighter at 45–75 kg (Leopold 2000). One of the most obvious characteristics of this deer are its very large ears, which are three quarters the length of the face with dark edges and white interiors. There is a deep lacrimal fossa and a metatarsal gland longer than 25 mm (Anderson and Wallmo 1984). Males have antlers with a short sub-basal protuberance on the main trunk

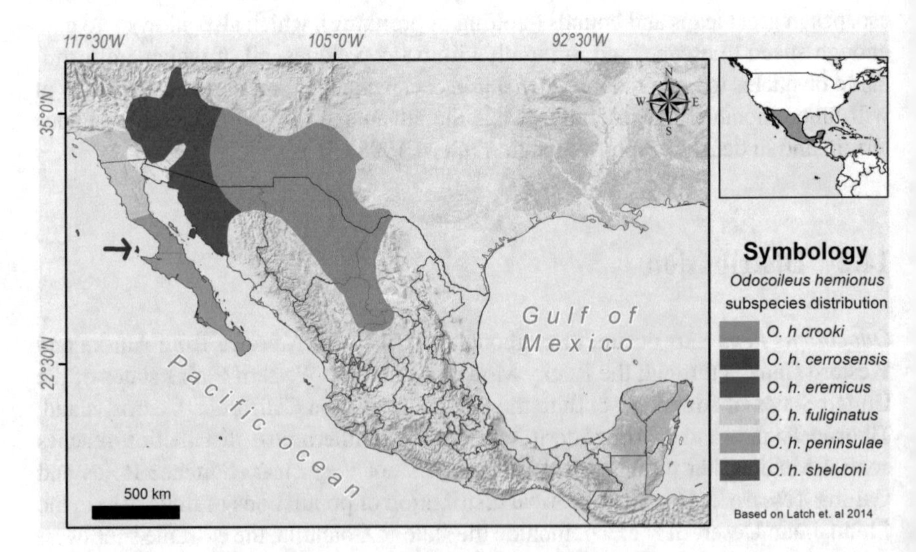

Fig. 14.1 Estimated geographic distribution of mule deer subspecies in Mexico (Modified by Adriana Sandoval Comte from Latch et al. 2014)

that curves upwards and outwards, and the points fork with divisions of the same size (Fig. 14.2). Females conceive during their second year of life, rarely in the first (Fig. 14.3). Fawns are born with mottling on their fur that disappears at 83–87 days, when it becomes grayish as they approach adulthood (Anderson and Wallmo 1984).

14.4 Diet

The diet of mule deer has been widely studied in the southwestern USA, but in Mexico few studies have been done in the desert. Desert mule deer forage on plant species that are widely distributed throughout their range (Alcalá-Galván and Krausman 2012), and like other herbivores deer are opportunistic, and their diet varies notably with place, elevation, season, and year (Heffelfinger 2006; Alcalá-Galván and Krausman 2012). Mule deer have the ability to consume a variety of different species, according to their availability, and the quality of natural and human disturbed vegetation. However, the composition of their diet differs as a function of varying availability among plant species in different habitats (Heffelfinger

Fig. 14.2 The mule deer's most defining characteristic is its large ears, photo taken in rancho Valle Colombia, from the Finan family, in the Chihuahuan Desert, Coahuila (Photo Alexander Peña-Peniche)

Fig. 14.3 Doe with fawn (Photo Luz A. Pérez-Solano)

2006; Haverson et al. 2014). This variation also is influenced by climate, which results in differences in plant quality and growth (Heffelfinger 2006).

In the Chihuahuan Desert during the rainfall seasons (summer and winter), the diet of mule deer is complemented with forbs (board-leaved weeds), cacti and shrub/tree fruits, which are consumed when abundant (Heffelfinger 2006). In the Sonoran Desert, for some years and in some areas during the rainy season, forbs represent 21% of the total diet (Alcalá-Galván and Krausman 2012) and as much as 60% of their diet in some areas of the Chihuahuan Desert (Olivas-Sánchez et al. 2015, 2018). During the dry season, bushes and succulents are the basic diet. Succulent plants are important in the diet of mule deer in the Chihuahuan Desert (Urness 1981; Guth 1987; Esparza-Carlos et al. 2011), and contribute 22–70% of the forage consumed throughout of the year: prickly pear (*Opuntia rastrera*), lechugilla (*Agave lechuguilla*) and candelilla (*Euphorbia antisyphilitica*), which are all important sources of water (Urness 1981; Guth 1987; Heffelfinger 2006).

The diet of the mule deer has been studied in the northern Chihuahuan Desert, but in Mexico, there are only two diet studies published for this region (Fig. 14.4). There is also a study for the Sonoran Desert. The first study of mule deer diet in the Mapimí Biosphere Reserve, central Chihuahuan Desert, was done by Guth (1987) who found that mule deer feed on 65 species of plants. Of these 43% were shrubs, 34% herbs (*Zephyranthes longifolia* 5.7%, *Acourtia parryi* 5.2%, and *Tidestromia gemmata* 3.52%), and 13% were succulents (candelilla *Euphorbia antisyphilitica*, range between months 10–24%; *Opuntia rastrera*, 2.9–18.7%), and 10% grass

Fig. 14.4 Mule deer foraging in the Chihuahuan Desert, Durango (Photo: Luz A. Pérez-Solano)

(Guth 1987). By chance, the study was conducted in the fifth rainiest year recorded to that date (1985, 378 mm; mean precipitation = 287 mm, range 122–504 mm; precipitation data from 1979 to 2006; *Estación Climática Laboratorio del Desierto*, INECOL). Thus, we suspect that during this year forbs were more abundant and consumed in greater quantities by mule deer than in other years.

The second study of mule deer diet was carried out by Olivas-Sánchez et al. (2015, 2018) near Aldama, Chihuahua, during the rainy seasons of 2 years. Diet was found to consist of 25 plant species. The species most frequently found in fecal pellets were *Ditaxis serrata* (herbaceous, 36–39%), *Opuntia* sp. (succulent, 12.4%) and *Flourensia cernua* (shrub, 5–15%), *Yucca elata* (succulent, 8%), along with *Prosopis glandulosa* and *Chilopsis linearis* (trees, 5%) and *Leucophyllum frutescens* (shrubs, 5%). Differences in plant community composition on the terrain results in mule deer feeding differentially; that is, on hills with slopes of 10–25°, 13% of diet was succulents (47% available), 32% was trees and bushes (42%), 53% was herbs (9%), and 2% grasses (2%). In flat areas, the diet consisted of 8% grasses (40% available), 44% herbs (24%), 26% succulents (23% available), and 22% trees and shrubs (13% available) (Olivas-Sánchez et al. 2015, 2018).

Seasonality also defines the consumption of plants by mule deer, that is, in the cold-dry season (December–March) trees and shrubs were the most frequently consumed (36%), followed by grasses and herbs (28.5%) and succulents (28%). During the temperate-dry season (April–June) succulent plants (*Opuntia* sp.) were the main component (87%) of the diet, followed by 5.2% of trees and shrubs and 6.4% for herbs and grasses, while during humid-temperate season (July–November) the herbs and grasses were the most frequently consumed with the highest value (53%), followed by trees and shrubs (30%) and succulents at 12% (Olivas-Sánchez et al. 2015, 2018).

There has been third study in Mexico, done in the Sonoran Desert. Mule deer consumed 96 species: 45 of which were bushes and trees, 33 forbs, 12 grasses and six succulents. The diet is highly variable between sites and seasons. Bushes and trees constitute 77–88% of the plants consumed. The most frequently consumed bushes and trees are mesquite (*Prosopis juliflora*), ironwood (*Olneya tesota*), jojoba (*Simmondsia chinensis*), bird of paradise (*Caesalpinia pumila*), and fairy duster (*Calliandra eriophylla*), which each contributed ≥10% of the diet. Forbs comprised 5–10% of the diet, and the most frequently consumed was Moradia (*Glandularia delticola*; 4–9%) (Alcalá-Galván and Krausman 2012). The variation in the consumption of forbs is related to spatial and temporal variation in precipitation regimes. Generally, forb consumption is greater during or following the rainy season in summer or winter (0–21%) (Alcalá-Galván and Krausman 2012). Grasses are the least consumed type of forage (1–9%), though in certain locations grasses account for 32% of the forage consumed in summer, mainly native species (Alcalá-Galván and Krausman 2012). The consumption of succulents is high, representing 1–12% of consumed forage, but varies (1–28%) between areas and seasons. The most consumed succulent is Chain Fruit choya (*Opuntia fulgida*), the consumption of which varies 1–26% between years and sites (Alcalá-Galván and Krausman 2012).

In Sonora, mule deer and cattle do not appear to compete for forage. Even though desert mule deer and cattle shared 45–77 forage species in central Sonora, diets only overlapped in springtime. Trees and bushes are the most important type of forage for mule deer (Alcalá-Galván and Krausman 2012).

14.5 Reproduction

Throughout the mule deer's distribution, the reproduction peak occurs from the end of November to mid-December (Anderson and Wallmo 1984), though in Mexico in the La Laguna mountain range in Baja California Sur, mating occurs between December and February (Gallina et al. 1992, 2000). The reproductive season of the subspecies *O. h. fuliginatus* is 1 or 2 months out of phase with that of the northern populations (Sierra de Juárez mountain range, April to May), and those of the south (El Rosario, February to April) on the Baja California Peninsula (Ayala-Cano and Peraza-Perales 2014).

Mule deer have a polygynous mating system in which the dominant male displaces the other males and follows the females that are in estrous to copulate with and fertilize them when they allow it (Hungerford et al. 1981; Geist 1998). After a gestation period of 200–208 days, the peak in births occurs between July and August, though on Cedros Island it has been reported to occur in April (Pérez-Gil

1981). In general, fawns weigh 2.7–4 kg at birth. Mean number of births per female is 1.14–1.85, depending on the quality of her nutrition during the pre-reproductive season (Geist 1998).

14.6 Behavior

In Mexico, data on natural populations has only been published for deer in the Mapimí Biosphere Reserve (Sánchez-Rojas and Gallina 2000a, b; Esparza-Carlos et al. 2011, 2016; Pérez-Solano et al. 2016, 2017), in Aldama Chihuahua, in the Cañón de Santa Helena Flora and Fauna Refuge, Chihuahua (Laundré et al. 2009; Olivas-Sánchez et al. 2015, 2018) and for the Sierra de La Laguna, Baja California Sur (Gallina et al. 1991; Álvarez-Cárdenas et al 1994, 1999a, 1999b), as well as sites where the species has been reintroduced, such as Coahuila (Ortega-Sánchez 2013).

The degree of socialization of the mule deer depends on the seasonality of the region, sex, the population and subspecies (Mackie et al. 2003). Feeding habits are similar between the sexes, which live apart for most of the year. Females with off-spring of different generations tend to make family groups that grow and shrink in

Fig. 14.5 Females mule deer tend to make family groups, which vary in size depending on seasonality (Photo: Sonia Gallina)

size depending on seasonality (Fig. 14.5, Hungerford et al. 1981, Geist 1998). There may also be small groups of adult males or groups of young males with wandering females (Fig. 14.6). The home ranges of these groups may overlap those of other groups, especially in feeding sites or where there are sources of water (Mackie et al. 2003).

14.7 Habitat Requirements

The main habitat requirements of mule deer are water, food, and cover.

14.7.1 Water

In arid lands, water distribution is a key factor for deer. Home ranges are established close to water sources, and some animals will even leave their established home range to drink water and then return to their usual range. But the influence of water

Fig. 14.6 Male mule deer in rancho Valle Colombia, from the Finan family, which is part of a Grassland Priority Conservation Area in the Chihuahuan Desert, Coahuila (Photo: Alexander Peña-Peniche)

availability within the home range is small (Marshal et al. 2006b). The benefits of additional water sources is controversial; sometimes they have a minor influence on habitat use and other times a negative effect, mainly when the site becomes a successful kill site for predators (Heffelfinger 2006), or because the deer want to avoid cattle and/or people, both of which tend to make greater use of the areas around permanent water sources (Loft et al. 1991; Galindo-Leal and Weber 1998; Esparza-Carlos et al. 2011). Desert animals have physiological adaptations to mitigate water loss, such as adjusting their activity patterns and obtaining water from forage, as well as consuming succulents (Krausman 1978; Short 1981; Urness 1981; Heffelfinger 2006).

14.7.2 Food

Mule deer are primarily browsers, and they select the most nutritious plants and their most nutritious parts (Watkins et al. 2007). However, precipitation pattern influences feeding behavior and the use deer make of their desert habitat. For example, in extremely dry years, mule deer gather around hills or mountains, because the flat zone is not habitable (Esparza-Carlos et al. 2011). In this case, the deer depend heavily on browsing and succulents (Marshal et al. 2005; Heffelfinger 2006). Both the home ranges of mule deer and their movement within these areas are influenced by food availability. During the dry season the displacement of mule deer is less than it is in the wet season (Pérez-Solano et al. 2017). In the rainy season the home range of mule deer increases by several kilometers (Pérez-Solano et al. 2016) and they use the flatlands where forbs become abundant (Esparza-Carlos et al. 2011).

14.7.3 Cover

Mule deer cover requirements fall into two categories: (1) vertical cover, which is related to predation risk: (a) Foraging in dense-intermediate cover, facilitates surprise attacks by stalking predators such as puma (Esparza-Carlos et al. 2016, 2018; Davies et al. 2012); (b) Hiding cover (also known as escape or security cover) is dense-to impenetrable cover that makes it more difficult to detect immobile fawns or resting adults (Cossío-Bayúgar and Sisto-Burt 2011; Davies et al. 2012), thereby diminishing predation risk by chase predators, such as coyotes (Lingle 2002, Lingle and Pellis 2002). (2) Horizontal cover, which is related to thermal cover, is very important in extreme environments. While cover can be critical to mule deer survival, the tradeoff between food resources and vegetation cover has been only been estimated on a few occasions (Sánchez-Rojas and Gallina 2000a; Pérez-Solano et al. 2016, 2017).

14.8 Habitat Use

The combined population-level consequences of site fidelity and philopatry are behavioral adaptations found in many species but these have received little attention despite their importance for understanding spatial patterns in connectivity and population dynamics. Bose et al. (2017) used an integral approach to explore the effects of fidelity and philopatry on the fine-scale genetic structure of black-tailed deer (*Odocoileus hemionus columbianus*). They assessed philopatry from mitochondrial DNA (mtDNA) haplotypes and found very small movements and seasonal home ranges, together with high site fidelity. The molecular data indicated multigenerational periods of philopatric behavior by black-tailed deer, suggesting that matrilineal groups might best serve as the basic units of conservation and management (Bose et al. 2017).

It is likely that philopatry in the black-tailed deer is a function of habitat quality since this species requires sites with specific characteristics for ruminating, resting, protecting its offspring and protecting itself from the hostile climate as a strategy for conserving its energy (Hungerford et al. 1981; Lang and Gates 1985; Huegel et al. 1986; Pollock et al. 1994). In the region of Mapimí, Sánchez-Rojas and Gallina (2000a) detected that the population of mule deer was particularly structured, and its most suitable habitats were located in the foothills or *bajadas*. Groups of deer select these areas because they have more heterogeneous terrain and are close to bodies of water during the dry season.

Two factors appear to result in the greater presence of this species: vegetation type, with their preferred vegetation composed of tall woody plants, and the availability of cover and food (Guth 1987; Álvarez-Cárdenas et al. 1999a), and with woody habitats that increase horizontal and vertical cover (Pérez-Solano et al. 2017). Areas with steeper slopes have been found to be preferred by the species, as they allow the deer to more effectively employ their antipredator tactics (Geist 1998). However, in Mexico under drier conditions, deer habitat use was explained primarily by food resource variables, followed by variables related to predation risk. During wetter years, food resources become unimportant while cover and visibility explain deer habitat use (Esparza-Carlos et al. 2011; Pérez-Solano et al. 2017).

Trade-offs are made between obtaining food and predation risk: the deer take a chance that they will be able to avoid predation by seeking out safe habitats for foraging. Mule deer integrate variables at different scales to make their foraging decisions (Esparza-Carlos et al. 2016). Mule deer forage less in risky habitats in the Chihuahuan Desert, than in other ecosystems (Esparza-Carlos et al. 2016). For example, mule deer avoid tall shrubs in the riskiest habitat close to hills, probably because the height of the bushes provides ambush cover for puma (*Puma concolor*). Mule deer perceive sites near hills at the macro-habitat scale and taller bushes at the micro-habitat scale as risky habitats (Esparza-Carlos et al. 2016).

Pérez-Solano et al. (2016) reported a home range of 14.70 km^2 for females and 18.05 km^2 for a male in the Mapimí Biosphere Reserve. In general, mule deer have larger home ranges in open habitats where the topography and vegetation are not complex, and that have more variable environmental conditions; this in comparison with the small home ranges that are characteristic in closed, more diverse habitats with stable environments, where the home ranges of males are larger than those of the females (Mackie et al. 2003). In semidesert and desert regions mule deer tend to inhabit home ranges at higher elevations during the summer, moving in autumn to even higher locations, and then moving to lower land during winter and spring, which is related to the time of the year when the females give birth (July–August; Marshal et al. 2006a, b). Mule deer home range size was larger in the arid environments of western Sonora (27.3 km^2 ± 2.6 [SE]) than in central Sonora (14.5 km^2 ± 2.0 [SE]). In contrast, the average size of the mule deer's home range in west-central Texas was 2.47 ± 0.29 km^2, with yearly variation: in spring it was 3.9 ± 0.32 km^2 and in summer it was 2.82 ± 0.32 km^2. The mean core area was 0.73 ± 0.10 km^2 and 0.61 ± 0.09 km^2, respectively (Brunjes et al. 2006).

In the coniferous forests of the San Pedro Mártir mountain range, Baja California Norte, Mexico, mule deer prefer plains with 20–40% plant canopy cover, abundant herbaceous plants, that are lightly rocky and near water (Ahumada-Cervantes 2000). However, they also inhabit sites high in the mountains when there is no snow (Mellink 2005). In the Sierra de La Laguna mountains, in the southern part of the Baja California peninsula, they prefer sites with slopes greater than 30%, rocky soils and a high density of shrubs (Gallina et al. 1991; Álvarez-Cárdenas et al. 1999a).

Habitat selection and use is mainly influenced by vegetation height and proximity to water bodies (Esparza-Carlos et al. 2011, Pérez-Solano et al. 2017). This determines the home range and core areas of the animals, as well as the number or density of animals that we can find at a given site (Begon et al. 2006). Based on the sites where populations of mule deer are found, density changes with the potential of sustenance offered by the site or with environmental characteristics (Mandujano and Gallina 2015). For example, in Aldama and Gemelos, Chihuahua, mean mule deer density was 2.2 ind/km^2 (± 0.7 SE), while in Cuevitas it was 0.9 ind/km^2, both for spring and summer (Vital-García et al. 2016). Studies done in Coahuila and Nuevo León report that for some sites the absence of mule deer is related to the presence of cattle and pasture deterioration from overgrazing. The density in these sites was 0.95 and 13.3 ind/km^2 (Martínez-Muñoz et al. 2003).

There is great variation in the density of mule deer throughout their distribution. For *O. h. hemionus* in the Rocky Mountains of Utah and Wyoming low densities have been reported from >1 to 2 ind/km^2, intermediate densities range from 4 to 6 ind/km^2 and high densities are 10–12 ind/km^2 (Brown 2009). Further south, in Tucson, Arizona, low densities have been reported for *O. h. eremicus*: 0.9 ind/km^2 and 2.5 ind/km^2 during summer and winter respectively (Koenen et al. 2002), while in western Texas mean density was 2.4 ind/km^2 (Brunjes et al. 2006). In southwestern California the composition of the population of *O. h. eremicus* was 41–74% females, 6–31% males, and 6–34% juveniles, with a density of 0.5–0.13 ind/km^2. This low density was attributed to low precipitation that was highly unpredictable resulting in unpredictable forage availability (Marshal et al. 2006a).

In Coahuila, the estimated density for *O. h. eremicus* was 11 ind/km^2 during the dry season, 4.7 ind/km^2 during the early rainy season and 1.8 ind/km^2 in the late rainy season (Lozano-Cavazos et al. 2015). In the Municipality of Pitiquito, Sonora, very close to Tiburón Island, density ranged from 0.5 to 2.5 ind/km^2 (Serra-Ortíz et al. 2008).

On islands there are different subspecies of mule deer. A density of 13.32 ind/km^2 has been estimated for *O. h. sheldoni* in the desert matorral scrub of Tiburón Island (Servín et al. 2010). On Santa Catalina Island, California, for introduced *O. h. californicus* Stapp and Guttilla (2006) reported a density of 5.41 ind/km^2, with a range of 3.14 ind/km^2 in January and March, to 7.85 ind/km^2 in July.

14.9 Sexual Segregation

Sexual segregation has been widely studied in many groups of animals (Ruckstuhl and Neuhaus 2002; Alves et al. 2013) and is recognized as a pattern of separation of the males from the females along ecological and behavioral dimensions, in which males and females use different resources and perform activities in groups of the same sex and age during certain times. However, as detailed in other chapters (Gallina et al. 2019), the quantification of sexual segregation in ungulates is complex because behavior must be evaluated taking spatiotemporal scales into account (Bowyer et al. 1996; Barboza and Bowyer 2000; Bowyer 2004; Bowyer and Kie 2004) at both the individual (Perez-Barberia et al. 2005; Calhim et al. 2006) and population levels (Barboza and Bowyer 2000; Perez-Barberia and Gordon 2000; Perez-Barberia et al. 2002; Michelena et al. 2006; Siuta and Bobek 2006; Gallina et al. 2014).

Sexual segregation can also be estimated at four different levels: spatial, habitat, social, and diet (Hungerford et al. 1981; Conradt 1998; Alves et al. 2013; Gallina et al. 2019). In most cases, studies done at different levels generate hypotheses that are not mutually exclusive. Mule deer exhibit sexual segregation (Bowyer 1984; Bowyer and Kie 2004), and different hypotheses have been put forth to explain this phenomenon (Main and Coblentz 1996, Main et al. 1996). The most widely supported hypotheses to explain sexual segregation in mule deer are those of predation risk (Main and Coblentz 1996; Bowyer and Kie 2004; Pierce et al. 2004) and sexual dimorphism (Barboza and Bowyer 2000).

14.10 Factors that Affect Mule Deer Populations

As mentioned above, the three most important habitat requirements (food, water, and cover) can become limiting factors for mule deer populations in arid zones. However, overgrazing is a factor that affects the quality and availability of these three requirements. In most of the Chihuahuan Desert, extensive cattle ranching has expanded notably since the sixteenth century, resulting in the loss of natural grasslands and has altered the structure of desert vegetation (Barral 1991; Barral and

Hernández 2001). This has put the survival of many deer populations at risk (Martínez-Muñoz et al. 2003).

The presence of cattle has negative effects on mule deer populations, mainly in terms of water use and the probability of being preyed upon near watering sites (Loft et al. 1991; Galindo-Leal and Weber 1998; Esparza-Carlos et al. 2011). Although it has been suggested that mule deer and cattle make differential use of the habitat and thus are not competing for food (Cossío-Bayúgar 2015), it has also been documented that the cattle and activities associated with cattle ranching affect deer behavior, displacing them or modifying their activity patterns (Kie 1996; Loft et al. 1991; Galindo-Leal and Weber 1998; Chaikina and Ruckstuhl 2006; Heffelfinger et al. 2006; Esparza-Carlos et al. 2011).

14.10.1 Mortality Factors

The dynamic behavior of a population is determined by the rate of change rate in demographic parameters over time, such that the number of animals that die over a period of time is also a determinant of the rate of population growth (Sinclair et al. 2006). Among the causes of death of individuals in a population, are malnutrition, disease and predation. For mule deer, malnutrition and disease appear to be at least as important as predation in causing early fawn mortality. Also, there is evidence that with better nutritional status in pregnant females there is an increase in early survival rate of their fawns, and the improved nutritional status of the fawns can in turn significantly increase their survival rates even when predation is the primary cause of mortality (Bishop et al. 2005). Thus, when habitat conditions are favorable, mule deer are better nourished and have more cover, which reduces their susceptibility to predation (Ballard et al. 2001).

14.10.2 Predation Risk and Impact on Populations

The main predator of the adult mule deer in Mexico and southwestern USA is the puma (*Puma concolor*), however, coyotes (*Canis latrans*) will kill adults whose death is imminent owing to illness or old age. Although fawns are preyed upon by coyotes, bobcats (*Lynx rufus*), and golden eagles (*Aquila chrysaetos*), the effect of these losses on the population are minimal (Ballard et al. 2001; Heffelfinger 2006). When coyotes have been removed there is no effect (or only a marginal effect) on population density and fawn survival, however, when pumas are removed fawn survival increases slightly (Connolly 1981; Hurley et al. 2011).

The antipredator strategies for mule deer depend on predator type (stalking or chasing predators), and the responses of mule deer are different from the antipredator strategies of white-tailed deer (Lingle and Pellis 2002). Mule deer are more vulnerable to be hunted by chasing predators such as coyotes on low slopes or gently rolling terrain, while white-tailed deer are more vulnerable to predation on high slopes and rugged terrain. For this reason, mule deer stand high on slopes to avoid predation by coyotes, and when a mule deer is attacked by a coyote, the deer rarely run away, instead they group together to face the predator rather than trying to escape. This increases their probability of survival, even when there are fawns in the group, because the females will fend off the coyotes and defend both their own offspring and heterospecific fawns (Lingle 2002; Lingle and Pellis 2002; Lingle et al. 2005). If mule deer do flee when attacked, they are more likely to move upslope at a shallow angle (Lingle 2002). The strategy of fleeing uphill is used to run away from chasing predators such as canids but is not used with stalking predators like puma (Caro 2005; Geist 1981; Stankowich and Coss 2006). In contrast, white-tailed deer will run away and move downhill and away from slopes to escape (Lingle 2002; Lingle and Pellis 2002; Lingle et al. 2005).

Since the hunting style of pumas is stalking and ambushing, places with low visibility are associated with greater predation risk (Esparza-Carlos et al. 2018), because cover makes it difficult for the prey to sight stalking predators (Caro 2005) and makes it easier for pumas to hide, approach their prey undetected, and carry out a successful attack (Robinette et al. 1959, Beier 1995). Visual obstruction is provided by bushes, rocks, uneven terrain, etc. For this reason, mule deer avoid habitats and microhabitats where the structure provides ambush cover that makes it easier for puma to stalk and capture mule deer (Altendorf et al. 2001; Hernández et al. 2005; Esparza-Carlos et al. 2011, 2016).

Predation has indirect and direct effects on the population density and diet quality of mule deer. In enclosures free of predators, mule deer density has been observed to increase, while outside where predators are present, density is low (Carrera et al. 2015). Furthermore, the risk of predation has an important effect on the nutritional status of mule deer, likely because in the presence of predators, mule deer spend more time alert, watching out for predators and this reduces the time they can spend actually foraging (Ydenberg et al. 2007). In the absence of predators, mule deer select better forage, access all areas and maximize their energy intake, which improves their physical condition and health (Carrera et al. 2015).

14.10.3 Predation Risk Influence Habitat Use

In desert zones, the risk of predation differs with scale: (a) at the macro-habitat scale, the areas close to hills (<1.5 km) are dangerous, so mule deer feel more apprehensive and forage less because proximity to the hills increases the probability of encounters with pumas (Esparza-Carlos et al. 2011, 2016); (b) at the micro-habitat scale, the bottom of hills makes deer more vulnerable to stalking and ambush by predators (Esparza-Carlos et al. 2011, 2016, 2018). Low visibility owing to vegetation cover is related to increased predation risk, mainly by cougar which is the deer's main predator (Esparza-Carlos et al. 2011, 2016, 2018). In areas far from hills, at the macro-habitat scale, tall shrubs have no negative effect on mule deer apprehensiveness or foraging, probably owing to the fact that puma make less use of these areas (Esparza-Carlos et al. 2016). At the micro-habitat scale, mule deer make less use of areas where visibility is <35% owing to wider bushes (>2.5 m) (Esparza-Carlos et al. 2011). These scales correlate with the density of mule deer: when visibility is low (larger shrubs and tall plants) deer density is also low (Sánchez-Rojas and Gallina 2000a).

14.11 Conservation Prospects

The mule deer is listed as a species of least concern in conservation terms (Sánchez Rojas and Gallina-Tessaro 2016); however, data from Coahuila and Nuevo León, Mexico on poaching, habitat destruction (Martínez-Muñoz et al. 2003), and the expansion of cattle ranching in the northern part of the country (Galindo-Leal 1993; Sánchez-Rojas and Gallina 2006; Sánchez Rojas and Gallina-Tessaro 2016), have reduced both the populations and distribution areas of this species, with even some local extinctions (Martínez-Muñoz et al. 2003). Other data reflect the important dynamics of this species as a metapopulation in Mapimí Biosphere Reserve (Sánchez-Rojas and Gallina 2000a, b, 2007; Sánchez Rojas and Gallina-Tessaro 2016). Mule deer offer an excellent example of the need to evaluate the species at the regional and national scale in order to arrive at a more realistic classification of their conservation status. Once this criterion is applied, it is possible that the status of mule deer will be changed to vulnerable, at least in this region (Sánchez Rojas and Gallina-Tessaro 2016).

The long-term well-being of mule deer depends on the conservation status of their habitats. The habitat requirements of mule deer must be incorporated into land management plans (Watkins et al. 2007), including water management since this may be a key limiting factor for mule deer in many habitats (Heffelfinger et al. 2006). Also, it is critical to develop conservation strategies that include migratory landscapes (Jachowski et al. 2018), as well as altered habitats where deer track changing plant phenology (Sawyer and Kauffman 2011; Merkle et al. 2016).

Land management plans and the rewilding of abandoned landscapes (Gillson et al. 2011; Navarro and Pereira 2013) should promote the use of wilderness areas as refugia where natural ecological and evolutionary processes can operate with minimal human disturbance (Watson et al. 2016) and mule deer populations such as those of the Chihuahuan Desert can prosper.

Acknowledgments Adriana Sandoval-Comte provided support with the Geographic Information Systems for the map. Bianca Delfosse translated parts of the manuscript and revised the English.

References

Aguilar FJ, Arroyo-Cabrales J, Johnson E et al (2012) Distribución de los venados (Mammalia: Cervidae) durante el pleistoceno tardío en México. In: XIII Simposio sobre venados de México, Facultad de Medicina Veterinaria y Zootecnia, Universidad Nacional Autónoma de México, Toluca, Estado de México, pp 23–25

Ahumada-Cervantes R (2000) Propuesta de plan de manejo para el venado bura (*Odocoileus hemionus fuliginatus*) en la sierra de San Pedro Mártir. PhD Thesis. Universidad Autónoma de Baja California, Mexico

Alcalá-Galván C, Krausman P (2012) Diets of desert mule deer in altered habitats in the lower Sonoran Desert. Calif Fish Game 98:81–103

Altendorf KB, Laundré JW, López-González C (2001) Assessing effects of predation risk on foraging behavior of mule deer. J Mammal 82(2):430–439

Alves J, da Silva A, Soares AMV et al (2013) Sexual segregation in red deer: is social behaviour more important than habitat preferences? Anim Behav 85:501–509

Anderson AE, Wallmo OC (1984) *Odocoileus hemionus*. Mamm Species 219:1–9

Ayala-Cano SG, Peraza-Perales IA (2014) Estrategias de manejo y conservación del venado bura (*Odocoileus hemionus*), con base en sus características biológicas y los elementos estructurales de su hábitat, en diferentes UMAS de Baja California, México. In: XIV Simposio sobre venados de México, Facultad de Medicina Veterinaria y Zootecnia, Universidad Nacional Autónoma de México, Mérida, Yucatán, pp 19–21

Álvarez-Cárdenas S, Gallina S, Galina-Tessaro P (1994) Dinámica poblacional del venado bura de la Sierra de la Laguna, Baja California Sur, México. In: IV simposio sobre venados de México, Facultad de Medicina Veterinaria y Zootecnia, Universidad Nacional Autónoma de México, Nuevo Laredo, Tamaulipas, pp 114–135

Álvarez-Cárdenas S, Gallina S, Galina-Tessaro P et al (1999a) Habitat availability for the mule deer (Cervidae) population in a relictual oak-pine forest in Baja California Sur, Mexico. Trop Zool 12(1):67–78

Álvarez-Cárdenas S, Gallina S, Galina-Tessaro P et al (1999b) Population dynamics in a relictual oak-pine forest in Baja California Sur, Mexico. In: Ffolliott PF, Ortega-Rubio A (eds) Ecology and Management of forests, woodlands and shrublands in the dryland regions of the United States and Mexico: perspectives for the 21th Century, University of Arizona-CIBNOR, pp 197–210

Ballard WB, Lutz D, Keegan TW et al (2001) Deer-predator relationships: a review of recent North American studies with an emphasis on mule and black-tailed deer. Wildl Soc Bull 29:99–115

Barboza PS, Bowyer RT (2000) Sexual segregation in dimorphic deer: a new gastrocentric hypothesis. J Mammal 81:473–489

Barral H (1991) Bolsón de Mapimí, ayer y hoy. TRACE 19:53–58

Barral H, Hernández L (2001) Los ecosistemas pastoreados desérticos y sus diversas formas de aprovechamiento: análisis de tres casos. In: Hernández L (ed) Historia ambiental de la ganadería en México. Instituto de Ecología, Xalapa, pp 85–97

Begon M, Townsend CR, Harper JL (2006) Ecology: from individuals to ecosystems, 4th edn. Blackwell, Oxford

Bell CJ, Lundeliu EL, Barnosky AD et al (2004) The Blancan, Irvingtonian, and Rancholabrean mammal ages. In: Woodburne MO (ed) Late Cretaceous and Cenozoic mammals of North America: biostratigraphy and geochronology. Columbia University Press, New York, pp 232–314

Bishop CJ, Unsworth JW, Garton EO (2005) Mule deer survival among adjacent populations in southwest Idaho. J Wildl Manag 69(1):311–321

Bose S, Forrester TD, Brazeal JL et al (2017) Implications of fidelity and philopatry for the population structure of female black-tailed deer. Behav Ecol 28(4):983–990

Bowyer RT (1984) Sexual segregation in southern mule deer. J Mammal 65(3):410–417

Bowyer RT (2004) Sexual segregation in ruminants: definitions, hypotheses, and implications for conservation and management. J Mammal 85(6):1039–1052
Bowyer RT, Kie JG (2004) Effects of foraging activity on sexual segregation in mule deer. J Mammal 85(3):498–504
Bowyer RT, Kie JG, Van Ballenberghe V (1996) Sexual segregation in black-tailed deer: effects of scale. J Wildl Manag 60(1):10–17
Brown DE (2009) Effects of Coyote Removal on Pronghorn and Mule Deer Populations in Wyoming. Dissertation, Utah State University
Brunjes K, Ballard W, Humphrey M et al (2006) Habitat use by sympatric mule and white-tailed deer in Texas. J Wildl Manag 70(5):1351–1359
Beier P (1995) Dispersal of juvenile cougars in fragmented habitat. J. wildl. Manage 59(2):228–237
Calhim S, Shi JB, Dunbar RIM (2006) Sexual segregation among feral goats: testing between alternative hypotheses. Anim Behav 72:31–41
Caro T (2005) Antipredator defenses in birds and mammals. University of Chicago Press, Chicago
Carrera R, Ballard WB, Krausman PR et al (2015) Reproduction and nutrition of desert mule deer with and without predation. Southwest Nat 60(4):285–298
Chaikina NA, Ruckstuhl KE (2006) The effect of cattle grazing on native ungulates: the good, the bad, and the ugly. Rangelands 28:8–14
Connolly GE (1981) Limiting factors and population regulation. In: Wallmo OC (ed) Mule and black-tailed deer of North America. University of Nebraska Press, Lincoln, pp 245–285
Conradt L (1998) Measuring the degree of sexual segregation in group-living animals. J Anim Ecol 67(2):217–226
Cossío-Bayúgar A (2015) Uso del hábitat y su relación con la presencia-ausencia de parásitos en el venado bura (*Odocoileus hemionus*) de la Reserva de la Biosfera de Mapimí, México. PhD Thesis, Instituto de Ecología, A.C.
Cossío-Bayúgar A, Sisto-Burt AM (2011) Definición y medición del bienestar animal. In: Medina-Cruz M (ed) Clínica, cirugía y producción de becerras y vaquillas lecheras, vol 12. Editorial AC, Ciudad de México, pp 116–128
Davies NB, Krebs JR, West SA (2012) An introduction to behavioural ecology, 4th edn. Wiley-Blackwell, Chichester.
Esparza-Carlos JP, Laundré JW, Sosa VJ (2011) Precipitation impacts on mule deer habitat use in the Chihuahuan desert of Mexico. J Arid Environ 75:1008–1015
Esparza-Carlos JP, Laundré JW, Hernández L, Íñiguez-Dávalos LI (2016) Apprehension affecting foraging patterns and landscape use of mule deer in arid environments. Mamm Biol 81(6):543–550
Esparza-Carlos JP, Íñiguez-Dávalos LI, Laundré JW (2018) Microhabitat and top predator presence affects prey apprehension in a subtropical mountain forest. J Mammal 99:596–607
Flagstad Ø, Røed KH (2003) Refugial origins of reindeer (*Rangifer tarandus* L.) inferred from mitochondrial DNA sequences. Evolution 57(3):658–670
Galindo-Leal C (1993) Densidades poblacionales de los venados cola blanca, cola negra y bura en Norte América. In: Medellín RA, Ceballos G (eds) Avances en el estudio de los mamíferos de México, vol 1. Asociación Mexicana de Mastozoología A.C, Ciudad de México, pp 371–391
Galindo-Leal C, Weber M (1998) El venado de la Sierra Madre Occidental. Edicusa, México
Gallina S, Galina-Tessaro P, Álvarez-Cárdenas S (1991) Mule deer density and pattern distribution in the pine-oak forest at the Sierra de la Laguna in Baja California Sur, México. Ethol Ecol Evol 1(3):27–33
Gallina S, Galina-Tessaro P, Alvarez-Cárdenas S (1992) Hábitat y dinámica poblacional del venado bura. In: Ortega A, Arriaga L (eds) Uso y Manejo de los Recursos Naturales de la Sierra de La Laguna, Baja California sur, Mexico. Centro de Investigaciones Biológicas, Baja California Sur, pp 297–327
Gallina S, Alvarez-Cárdenas S, Galina-Tessaro P (2000) Artiodactyla: Cervidae. P. In: ST Alvarez Castañeda y JL Patton (eds.) Los Mamíferos del Noroeste deMexico II. CIBNOR p.793–815

Gallina S, Sánchez-Rojas G, Buenrostro-Silva A et al (2014) Comparison of faecal nitrogen concentration between sexes of White-tailed deer in a tropical dry forest in southern Mexico. Ethol Ecol Evol 27:103–115

Gallina S, Sánchez-Rojas G, Hernández-Silva D, Pérez-Solano LA, García-Feria L, Esparza-Carlos JP (2019) The mule deer of the mapimi biosphere reserve. Chapter 3. In: Gallina S (ed) Ecology and conservation of Latin American ungulates. Springer, New York, p XX

Geist V (1981) Behavior: adaptative strategies in mule deer. In: Wallmo C (ed) Mule and black-tailed deer of North America. Nebraska University Press, Lincoln, pp 157–223

Geist V (1998) Deer of the world: their evolution, behaviour, and ecology. Stackpole Books, Mechanicsburg

Gilbert C, Ropiquet A, Hassanin A (2006) Mitochondrial and nuclear phylogenies of Cervidae (Mammalia, Ruminantia): systematics, morphology, and biogeography. Mol Phylogenet Evol 40(1):101–117

Gillson L, Ladle RJ, Araújo MB (2011) Baselines, patterns and process. In: Ladle RJ, Whittaker RJ (eds) Conservation biogeography. Wiley-Blackwell, Oxford, pp 31–44

Gustafson EP (1985) Antlers of Bretzia and *Odocoileus* (Mammalia, Cervidae) and the evolution of New World Deer. Trans Nebraska Acad Sci XIII:83–92

Guth A (1987) Hábitos alimenticios del venado bura (*Odocoileus hemionus* Rafinesque 1817) en la Reserva de la Biosfera de Mapimí Dgo. PhD Thesis, Escuela Nacional de Estudios Profesionales Iztacala, Universidad Nacional Autónoma de México, Ciudad de México, México

Gutiérrez EE, Helgen KM, McDonough MM et al (2017) A gene-tree test of the traditional taxonomy of American deer: the importance of voucher specimens, geographic data, and dense sampling. ZooKeys 697:87–131

Haverson LA, Ortega AL, Alcalá GC (2014) Venado Bura. In: Valdez R, Ortega AJ (eds) Ecología y manejo de fauna silvestre en México. Colegio de Postgraduados, Guadalajara, pp 389–412

Heffelfinger JR (2000) Status of the name *Odocoileus hemionus crooki (Mammalia: Cervidae)*. Proc Biol Soc Wash 113:319–333

Heffelfinger J (2006) Deer of the Southwest: a complete guide to the natural history, biology, and management of southwestern mule deer and white. Texas A&M University Press, College Station

Heffelfinger JR, Brewer C, Alcalá-Galván CH, Hale B, Weybright DL, Wakeling BF, Carpenter LH, Dodd NL (2006) Habitat guidelines for mule deer: southwest deserts ecoregion. Mule Deer Working Group, Western Association of Fish and Wildlife Agencies, Arizona, pp 4–6

Hernández L, Laundre JW, Gurung M (2005) Use of camera traps to measure predation risk in a puma-mule deer system. Wildl Soc Bull 33(1):353–358

Huegel CN, Dahlgren RB, Gladfelter HL (1986) Bed site selection by white-tailed deer fawns in Iowa. J Wildl Manage 50:474–480

Hungerford CR, Burke MD, Folliott PF (1981) Biology and population dynamics of mule deer in Southwestern United States. In: Folliott PF, Gallina S (eds) Deer Biology, habitat requirements and management in western North America MAB. Instituto de Ecología, A. C. Jalapa, Veracruz, pp 109–132

Hurley MA, Unsworth JW, Zager P et al (2011) Demographic response of mule deer to experimental reduction of coyotes and mountain lions in southeastern Idaho. Wildlife Monogr 178:1–33

Jachowski DS, Kauffman MJ, Jesmer BR et al (2018) Integrating physiological stress into the movement ecology of migratory ungulates: a spatial analysis with mule deer. Conserv Physiol 6(1):coy054. https://doi.org/10.1093/conphys/coy054

Kie JG (1996) The effects of cattle grazing on optimal foraging in mule deer (*Odocoileus hemionus*). Forest Ecol Manag 88:131–138

Koenen KKG, DeStefano S, Krausmann PR (2002) Using distance sampling to estimate seasonal densities of desert mule deer in a semidesert grassland. Wildl Soc Bull 30(1):53–63

Krausman PR (1978) Forage relationships between two deer species in Big Bend National Park, Texas. J Wildl Manage 42(1):101–107

Lang BK, Gates JE (1985) Selection of sites for winter night beds by white-tailed deer in a hemlock-northern hardwood forest. Am Midl Natur 113:245–254

Latch EK, Heffelfinger RJ, Fike AJ et al (2009) Species-wide phylogeography of North American mule deer (*Odocoileus hemionus*): cryptic glacial refugia and postglacial recolonization. Mol Ecol 18:1730–1745

Latch EK, Reding DM, Heffelfinger RJ et al (2014) Range-wide analysis of genetic structure in a widespread, highly mobile species (*Odocoileus hemionus*) reveals the importance of historical biogeography. Mol Ecol 23(13):3171–3190

Laundré JW, Loredo-Salazar J, Hernández L (2009) Evaluating potential factors affecting puma Puma concolor abundance in the Mexican Chihuahuan Desert. Wildlife Biol 15:207–212

Leopold AS (2000) Fauna silvestre de México: aves y mamíferos de caza. Instituto Mexicano de Recursos Naturales Renovables, México

Lingle S (2002) Coyote predation and habitat segregation of white-tailed deer and mule deer. Ecology 83(7):2037–2048

Lingle S, Pellis S (2002) Fight or flight? Antipredator behavior and the escalation of coyote encounters with deer. Oecologia 131(1):154–164

Lingle S, Pellis SM, Wilson WF (2005) Interspecific variation in antipredator behaviour leads to differential vulnerability of mule deer and white-tailed deer fawns early in life. J Anim Ecol 74:1140–1149

Loft E, Menke J, Kie J (1991) Habitat shifts by mule deer: the influence of cattle grazing. J Wildl Manage 51(1):16–26

Lozano-Cavazos EA, Ortega-Santos A, Tarango-Arámbula LA et al (2015) Densidad y uso del hábitat por el venado bura (*Odocoileus hemionus eremicus* Rafinesque) en Coahuila, México. Agroproductividad 8(5):62–68

Mackie RJ, Kie JG, Pac DF et al (2003) Mule deer. *Odocoileus hemionus*. In: Feldhamer GA, Thompson BC, Chapman JA (eds) Wild mammals of North America. Biology, management, and conservation. Johns Hopkins University Press, Baltimore, pp 889–905

Main MB, Coblentz BE (1996) Sexual segregation in Rocky Mountain mule deer. J Wildl Manag 60:497–507

Main MB, Weckerly FW, Bleich VC (1996) Sexual segregation in ungulates: new directions for research. J Mammal 77(2):449–461

Mandujano S, Gallina S (2015) Conceptos y modelos para estimar la capacidad de carga del hábitat para ungulados. In: Gallina S (ed) Manual de técnicas del estudio de la fauna. Instituto de Ecología, A.C., Xalapa Veracruz, pp 145–160

Marshal JP, Krausman PR, Bleich VC (2005) Dynamics of mule deer forage in the Sonoran Desert. J Arid Environ 60:593–609

Marshal JP, Lesicka LM, Bleich VC et al (2006a) Demography of desert mule deer in southeastern California. Calif Fish Game 92(2):55–66

Marshal JP, Bleich VC, Krausman PR et al (2006b) Factors affecting habitat use and distribution of desert mule deer in an arid environment. Wildl Soc Bull 34:609–619

Martínez-Muñoz A, Hewitt DG, Valenzuela S et al (2003) Habitat and population status of desert mule deer in Mexico. Z Jagdwiss 49:14–24

Mellink E (2005) El venado bura de Baja California. In: Sánchez-cordero V, Medellín RA (eds) Contribuciones Mastozoológicas en Homenaje a Bernardo Villa. Universidad autónoma de México-CONABIO, Ciudad de México, pp 367–372

Merkle JA, Monteith KL, Aikens EO et al (2016) Large herbivores surf waves of green-up during spring. Proc R Soc B 283:20160456

Michelena P, Noel S, Gautrais J et al (2006) Sexual dimorphism, activity budget and synchrony in groups of sheep. Oecologia 148(1):170–180

Navarro LM, Pereira HM (2013) Rewilding abandoned landscapes in Europe. In: Pereira HM, Navarro LM (eds) Rewilding European landscapes. Springer Open, New York, pp 3–23

Olivas-Sánchez MP, Vital-García C, Flores-Márquez JP et al (2015) Cambios estacionales en la dieta del venado bura (*Odocoileus hemionus* crooki) en matorral desértico Chihuahuense. Agroproductividad 8(6):59–64

Olivas-Sánchez MP, Vital-García C, Flores-Márquez JP et al (2018) Mule deer forage availability and quality at the Chihuahuan Desert rangelands, Mexico after a severe three-year drought. Cogent Biol 4(1):1536315

Ortega-Sánchez A (2013) Evaluation of a translocated population of desert mule deer in the Chihuahuan desert of northern Coahuila, Mexico. PhD Thesis, Texas A&M University, Texas, United States of America

Perez-Barberia FJ, Gordon IJ (2000) Differences in body mass and oral morphology between the sexes in the Artiodactyla: evolutionary relationships with sexual segregation. Evol Ecol Res 2(5):667–684

Perez-Barberia FJ, Gordon IJ, Pagel M (2002) The origins of sexual dimorphism in body size in ungulates. Evolution 56(6):1276–1285

Perez-Barberia FJ, Robertson E, Gordon IJ (2005) Are social factors sufficient to explain sexual segregation in ungulates? Anim Behav 69:827–834

Pérez-Gil SR (1981) A preliminary study of the deer from Cedros Island, Baja California, México. PhD Thesis, University of Michigan, Michigan, United States of America

Pérez-Solano LA, Gallina-Tessaro S, Sánchez-Rojas G (2016) Individual variation in mule deer (Odocoileus hemionus) habitat and home range in the Chihuahuan Desert, Mexico. J Mammal 97:1228–1237

Pérez-Solano LA, García-Feria LM, Gallina-Tessaro S (2017) Factors affecting the selection of and displacement within core areas by female mule deer (*Odocoileus hemionus*) in the Chihuahuan Desert, Mexico. Mammal Biol 87:152–159

Pierce BM, Bowyer RT, Bleich VC (2004) Habitat selection by mule deer: forage benefits or risk of predation? J Wildl Manag 68(3):533–541

Pollock MT, Whittaker DG, Demarais S et al (1994) Vegetation characteristics influencing site selection by male white-tailed deer in Texas. J. Range Manage 47:235–239

Robinette WL, Gashwiler JS, Morris OW (1959) Food habits of the cougar in Utah and Nevada. J Wildl Manage 23(3):261–273

Ruckstuhl KE, Neuhaus P (2002) Sexual segregation in ungulates: a comparative test of three hypotheses. Biol Rev 77:77–96

Sánchez Rojas G, Gallina-Tessaro S (2016) *Odocoileus hemionus*. The IUCN red list of threatened species 2016: e.T42393A22162113. https://doi.org/10.2305/IUCN.UK.2016-1.RLTS.T42393A22162113

Sánchez-Rojas G, Gallina S (2000a) Factors affecting habitat use by mule deer (*Odocoileus hemionus*) in the central part of the Chihuahuan Desert, Mexico: an assessment with univariate and multivariate methods. Ethol Ecol Evol 12:405–417

Sánchez-Rojas G, Gallina S (2000b) Mule deer (*Odocoileus hemionus*) density in a landscape element of the Chihuahuan Desert, Mexico. J Arid Environ 44:357–368

Sánchez-Rojas G, Gallina S (2006) La metapoblación del venado bura en la reserva de la biosfera Mapimí, México: consideraciones para su conservación. Cuadernos de Biodiversidad 22:7–15

Sánchez-Rojas G, Gallina S (2007) Metapoblaciones el reto en la biología de la conservación: El caso del venado bura en el Bolsón de Mapimí. In: Sánchez-Rojas G, Rojas-Martínez A (eds) Tópicos en Sistemática, Biogeografía, Ecología y Conservación de Mamíferos. Universidad Autónoma del Estado de Hidalgo, Hidalgo, pp 115–124

Sawyer H, Kauffman MJ (2011) Stopover ecology of a migratory ungulate. J Anim Ecol 80:1078–1087

Serra-Ortíz MA, González-Saldivar FN, Cantú-Ayala C et al (2008) Evaluación del hábitat disponible para dos especies de cérvidos en el noroeste de México. Rev Mex Mastozool 12:43–58

Servín J, López-Pérez A, Huerta A et al (2010) Evaluación de la densidad poblacional del venado bura (*Odocoileus hemionus sheldoni*) de la isla Tiburón, Sonora, México. In: XII Simposio sobre Venados de México, Ing. Jorge G. Villarreal González, Facultad de Medicina Veterinaria y Zootecnia, Universidad Nacional Autónoma de México/Asociación Nacional de Ganaderos Diversificados/Consejo Estatal de Flora y Fauna de Nuevo León, A.C. Ciudad de México, pp 23–25

Short HL (1981) Nutrition and metabolism. In: Wallmo OC (ed) Mule and black-tailed deer of North America. University of Nebraska Press, Lincoln, pp 99–127

Sinclair ARE, Fryxell JM, Caughley G (2006) Wildlife ecology, conservation, and management. Blackwell Publishing, London

Siuta A, Bobek B (2006) Comparison of red deer stomachs in relation to different foraging habitats. Med Weter 62(1):32–35

Stankowich T, Coss RG (2006) Effects of risk assessment, predator behavior, and habitat on escape behavior in Columbian black-tailed deer. Behav Ecol 18(2):358–367

Stapp P, Guttilla DA (2006) Population density and habitat use of mule deer (*Odocoileus hemionus*) on Santa Catalina Island, California. Southwest Nat 51:576–582

Urness PJ (1981) Desert and chaparral habitats. Part 1. Food habits and nutrition. In: Wallmo OC (ed) Mule and black-tailed deer of North America. University of Nebraska Press, Lincoln, pp 347–365

Vital-García C, Olivas Sánchez MP, García-Acosta LL et al (2016) Estatus de las poblaciones del venado bura, *Odocoileus hemionus*, en el estado de chihuahua. In: Abstracts of XV Simposio sobre Venados de México. Ing. Jorge G. Villareal González, Departamento de Etología, Fauna Silvestre y Animales de Laboratorio de la Facultad de Medicina Veterinaria y Zootecnia, Universidad Nacional Autónoma de México/Asociación Nacional de Ganaderos Diversificados/ Consejo Estatal de Flora y Fauna de Nuevo León, A.C. Ciudad de México, pp 27–29

Watkins BE, Bishop CJ, Bergman EJ et al (2007) Habitat guidelines for mule deer: Colorado plateau shrubland and forest ecoregion. Mule Deer Working Group, Western Association of Fish and Wildlife Agencies, Colorado, p 72

Watson JEM, Shanahan DF, DiMarco M et al (2016) catastrophic declines in wilderness areas undermine global environment targets. Curr Biol 26:2929–2934

Ydenberg RC, Brown JS, Stephens DW (2007) Foraging: an overview. In: Stephens DW, Brown JS (comp) Foraging: Behavior and Ecology. University of ChicagoPress, Chicago, p 1–28.

Chapter 15
Recent Studies of White-Tailed Deer in the Neotropics

Sonia Gallina-Tessaro, Eva López-Tello, and Salvador Mandujano

Abstract The white-tailed deer is the species with the greatest distribution in the American continent. This deer is considered a species with high plasticity because it lives in different types of vegetation ranging from temperate forests to dry tropical forests, arid zones and secondary vegetation. The distribution and results of genetic studies about the subspecies are presented in this chapter. According to different studies about the diet, deer selects a high number of plant species from different families, but particularly consumes a greater percentage of few shrubs and trees. Analyzing the nutritional composition of the deer diet, it is found that the amount of crude protein and acid detergent fiber were lower in the winter diet compared to that of autumn and spring, the digestibility was higher in autumn than in the other seasons. Movement of white-tailed deer could be influenced by many ecological, environmental, and behavioral variables. The animal energetic requirements depend on the basal metabolism, behavior and physiological conditions. Each animal activity represents an energy cost. We presented knowledge about home range size and activity patterns of the species. Variation in the reproductive chronology of the white-tailed deer in the region has been linked to environmental variables associated with latitude such as photoperiods and food availability. Deer population density is one of the parameters most evaluated in different regions and habitat types. The population estimation of deer is one of the central themes in Latin America, mainly in Mexico, since this allows to make a sustainable use of the species from the hunting point of view. The most important diseases that affect the health of deer are gastrointestinal parasitosis caused mainly by helminths and protozoa. Deer selects specific sites to rest, ruminate and protect their young. Therefore, these sites should offer food, water, protection against predators and thermal cover that allows deer to minimize both the absorption of heat by exposure to the sun and the loss of water by transpiration. Some studies have compared deer preferences between conserved

S. Gallina-Tessaro (✉) · S. Mandujano
Red de Biología y Conservación de Vertebrados, Instituto de Ecología A.C., Xalapa, Veracruz, México
e-mail: sonia.gallina@inecol.mx

E. López-Tello
Posgrado, Instituto de Neuroetología, Universidad Veracruzana, Xalapa, Veracruz, México

S. Gallina-Tessaro (ed.), *Ecology and Conservation of Tropical Ungulates in Latin America*, https://doi.org/10.1007/978-3-030-28868-6_15

sites and sites disturbed by human activity. In relation to habitat quality, differences in use have been found. Although the species is highly adaptable and widely distributed, we can ensure that it has populations that are exploited in a sustainable manner, but some of the subspecies may be at risk from poaching and severe transformation of their habitat.

15.1 Distribution

The white-tailed deer is the species with the greatest distribution in the American continent; it is found from Canada to South American countries such as Venezuela, Colombia, Peru, and Brazil (Smith 1991; Gallina et al. 2010b; Mandujano et al. 2014). Thirty-eight subspecies have been described, of which 24 live in Latin America where their conservation status is often uncertain (Baker 1984; Weber and González 2003; Mandujano et al. 2010; Ortega et al. 2011). In Mexico, 14 subspecies have been recorded (*O. virginianus mexicanus*, *O. v. couesi*, *O. v. texanus*, *O. v. miquihuanensis*, *O. v. carminis*, *O. v. sinaloae*, *O. v. oaxacanensis*, *O v. acapulcensis*, *O. v. O. v. veraecrucis*, *O. v. thomasi*, *O. v. toltecus*, *O. v. nelsoni*, *O. v. truei*, and *O. v. yucatanensis*), in Colombia five subspecies (*O. v. goudotti*, *O. v. ustus*, *O. v. curassavicus*, *O. v. apurensis*, and *O. v. tropical*), in Panama three (*O. v. chiriquensis*, *O. v. truei*, and *O. v. rothschildi*), while in Costa Rica there are only two subspecies (*O. v. truei* and *O. v. chiriquensis*) (Ortega et al. 2011, Fig. 15.1).

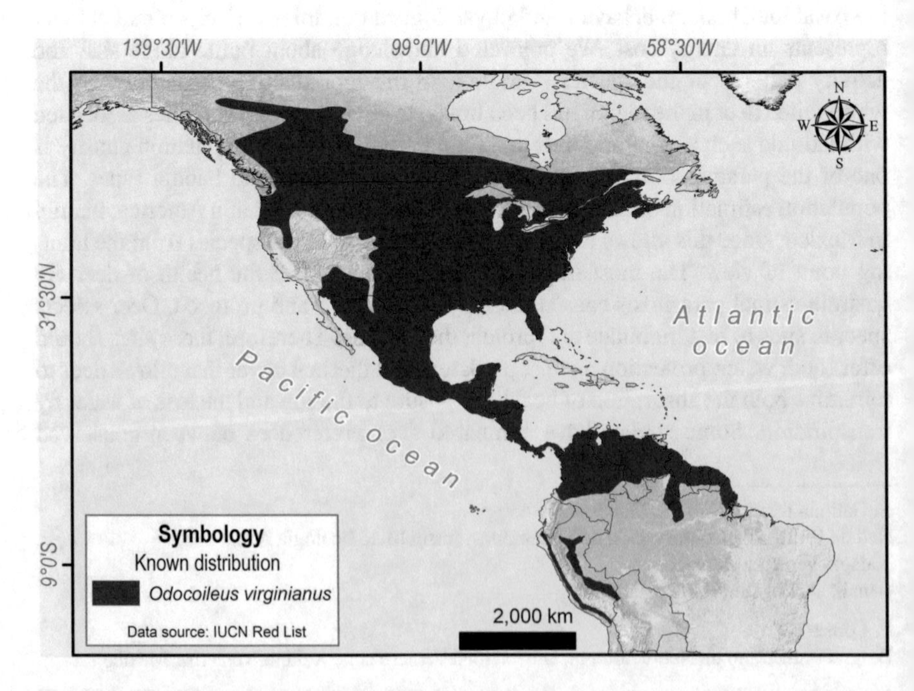

Fig. 15.1 Distribution map of white-tailed deer (*Odocoileus virginianus*) by Adriana Sandoval Comté

White-tailed deer is considered a species with high plasticity because it lives in different types of vegetation ranging from temperate forests to dry tropical forests, arid zones and secondary vegetation. In Mexico of the 14 subspecies, only two (*O. v. texanus* and *O. v. carminis*) do not inhabit tropical forests, while the other 12 subspecies include at least some of the tropical forests within their distribution areas. The subspecies with a greater part of their total distribution area occupied by tropical dry forests are *O. v. sinaloae* (Fig. 15.2), *O. v. mexicanus*, *O. v. acalpucensis*, and *O. v. yucatanensis*. In contrast, there are another subspecies *O. v. veraecrucis* (Fig. 15.3), *O. v. toltecus*, *O. v. thomasi*, *O. v. nelsoni*, and *O. v. truei* (Mandujano et al. 2010). In Colombia, the potential distribution of white-tailed deer includes more than one region that goes from the savannah to the open forest. In Costa Rica, Colombia, and Venezuela this species is distributed over most of its territory and inhabits different types of vegetation, mainly in savannahs (Ortega et al. 2011).

15.2 Genetics

White-tailed deer differ in body size, coat color, and other physical features throughout its range. De La Rosa-Reyna et al. (2012) sampled white-tailed deer from 13 sites in the range of five subspecies occurring in Mexico: *O. v. texanus*, *O. v. carminis*, *O. v. veraecrucis*, *O. v. sinaloae*, and *O. v. yucatanensis*. They estimated

Fig. 15.2 *Odocoileus virginianus sinaloae* in a dry tropical forest of Chamela, Jalisco, Mexico (Photo: Alberto González Gallina)

Fig. 15.3 Female of *Odocoileus virginianus veraecrucis* from Carrizal, Veracruz, México (Photo: Sonia Gallina- Tessaro)

genetic diversity and structure based on 12 microsatellite DNA loci and observed heterozygosity (HO) that was comparable to white-tailed deer in the USA (HO = 0.53–0.64), except for *O. v. yucatanensis* (HO = 0.41). These authors observed statistically significant genetic structure among all 13 sites (FST = 0.15). Analysis of molecular variance revealed that grouping sites by subspecies (FSC = 0.09) or geographic region (FSC = 0.13–0.14). Five loci with null alleles were found exclusively in *O. v. yucatanensis*. The frequency of null alleles in congeneric species tends to increase with increasing phylogenetic distance from the focal species for which the locus was derived. Deer in Yucatan were genetically differentiated from other subspecies and had comparatively lower genetic diversity, consistent with the biogeographic history of the region. Populations of white-tailed deer in Mexico are subject to a range of management challenges. Additional research is needed to understand the effect of management on the diversity and genetic structure of white-tailed deer.

Hernández-Llamas (2014) research was to determine and describe the genetic variability of the subspecies *O. v. miquihuanensis*, *O. v. couesi*, and *O. v. mexicanus* in the Center-North of Mexico by using a fragment of the control region (D-loop) of mitochondrial DNA. In the results of trees subspecies separation was observed *O. v. mexicanus* in one clade and the union of *O. v. miquihuanensis* with *O. v. couesi* in another one. Analyzes resulted high genetic differentiation among subspecies analyzed.

Comer et al. (2005) analyze 14 microsatellite DNA loci to calculate pairwise relatedness among individual deer and to assign doe pairs to putative relationship categories. Relationship categories differed in mean spatial distance, but only 60%

of first-degree-related doe pairs (full sibling or mother–offspring pairs) and 38% of second-degree-related doe pairs (half sibling, grandmother–granddaughter pairs) were members of the same social group based on spatial association (Comer et al. 2005).

In Colombia, Suárez et al. (2017) did a study to know the genetic variability of three groups of white-tailed deer in semi-captivity in the Department of Boyacá, by amplifying five microsatellites finding a low heterozygosity: 0.625, 0.700, and 0.444, lower than in other works carried out with the species: 0.64–078 with an allelic diversity of 6.8–8.3 and mean polymorphism information 0.59–0.73 (DeYoung et al. 2003), 0.860 (Miller et al. 2010).

15.3 Diet

The diet varies throughout the year depending on their nutritional and energy requirements, which change according to sex, age and season. According to different studies, deer selects a high number of species from different families, but particularly consumes a greater percentage of few species of shrubs and trees (González and Briones-Salas 2012; Granados et al. 2014; Vásquez et al. 2015). For example, in temperate forests of Mexico, it mainly consumes the following plant species: *Acalypha setosa*, *Smilax pringlei*, *Psidium sartorianum*, and *Dendropanax arborea*. Of the total species consumed, 79% are part of the cloud forest and 20.5% of the pine-oak forest (Aguilera-Reyes et al. 2013). In dry forests and scrub it has been reported that it consumes 139 species belonging to 51 families, mainly 40 herbaceous and 36 shrubby. The legumes represent 20.1%, followed by the Cactaceae 13.8%, grasses with 7.2% and Agavacea with 6.5%. It also consumes several plant species as a source of water, among which are *Ceiba parvifolia*, *Pachycereus weberi*, *Spondias purpurea*, and *Opuntia* sp. (Villarreal-Espino et al. 2011). In forests and fragmented forests, it consumes 25 species from 17 families, of which Asteraceae and Fabaceae are the most frequent (Clemente-Sánchez 2018).

González and Briones-Salas (2012) reported that in the temperate forests of northern Oaxaca, Mexico, diet deer is composed of 42 species belonging to 23 families, of which the most representative were Fagaceae, Asteraceae, Ericaceae, and Fabaceae. The preferred species were *Senecio* sp., *Sedum dendroideum*, *Arctostaphylos pungens*, and *Satureja macrostema*. While Granados et al. (2014) found in the tropical rain forest of Campeche, their diet is made up of 40 species from 15 botanical families. The highest richness was 29 species in the rainy season. The shrub species were preferred throughout the year, while the herbaceous species in the rainy season. Another study conducted in sites with tropical dry forest and crassicaule scrub reports that the annual diet of the deer is constituted by 83 species of 36 families, of these only 13 represent more than 50% of the diet. The main families in the rainy season were the Malvaceae, Commelinaceae, and Mirtaceae, while in the dry season were Fabaceae, Poaceae, Rubiaceae, Begoniaceae, and Malpighiaceae. The main species were *Bursera schlechtendalii*, *Caesalpinia* sp.,

Lysiloma divaricatum, *Opuntia lasiacantha*, *Agave macroacantha*, *Ceiba parvifolia*, *Euphorbia* sp., *Talinum paniculatum*, *Bursera fagaroides*, *Solanum lanceolatum*, *Senna wislizeni*, and *Karwinskia humboldtiana* (Vásquez et al. 2015).

In the Mixteca Poblana, a site dominated by tropical dry forest and xerophilous scrub, the deer consumes 133 species belonging to 50 families; legumes represent 19.5%, followed by cacti, 14.3%, grasses with 6.8%, and Asteraceae and Agavaceae each with 6%. Of the total of 76 species were consumed preferably in the rainy season (May–October), representing 57.1%, while in the dry season there were 46 species that contributed 34.6%. Of the total families six represented 68.2% of the diet: Agavaceae, Fagaceae, Mimosoideae, Fabaceae, Caesalpinioideae, and Sterculiaceae, all these families included 46 species but only 11 were the most representative. It emphasizes the importance of Fabaceae which contributed 37.1% (Villarreal-Espino et al. 2007, Fig. 15.4).

Some studies in Mexico have evaluated the nutritional composition of the deer diet and accounted that in the dry tropical forest of the State of Morelos, the composition varies depending on the time of year. According to this study, the amount of crude protein and acid detergent fiber were lower in the winter diet compared to that of autumn and spring, the digestibility was higher in autumn than in the other seasons. Of the five most consumed plants (*Sida* sp., *Randia aculeata*, *Begonia* sp.,

Fig. 15.4 White-tailed deer eating in the Tehuacán-Cuicatlán Biosphere Reserve, Mexico (Photo: Salvador Mandujano)

Malpighia mexicana, and *Bouteloua repens*), *Sida* sp. had the highest digestibility and crude protein content, followed by *Randia aculeata* (López-Pérez et al. 2012). In the Mixteca Poblana, the highest percentage of dry matter (66.6%) is provided by seven families: Agavaceae, Asteraceae, Fagaceae, Mimosoideae, Fabaceae, Caesalpinioideae, and Malpighiaceae (Villarreal-Espino et al. 2007).

There have also been some studies in captivity and semi-captivity to know the preferences of deer consumption, as well as the chemical composition of the diet. Plata et al. (2009), found that deer selected mainly the arboreal species (76.52%), followed by the shrubs (19.37%), herbaceous (5.81%) and in smaller proportion grasses (0.29%). *Leucaena leucocephala* and *Brosimum alicastrum* constituted 51.20% of the total diet, which had a composition of 56.97% of dry matter, 17.64% of crude protein and 42.53% of cell walls. In Guatemala, Vaides-Arrue (2017) reported that the main forage plants were grasses because they exist in greater availability (75%), followed by shrubs (22%) and finally tree species (3%). The diet of the white-tailed deer is mainly constituted by shrub species (*Hibiscus rosa-sinensis*, *Pyrus communis*, and *Bauhinia* sp.), and grasses (*Andropogon bicornis*, *Panicum sellowii*, and *Paspalum conjugatum*).

In Nicaragua, a comparative study was made to know the livestock preferences (cattle, goats and sheep) and deer, on four species of common forage trees in the Central American tropics (*Acacia pennatula*, *Guazuma ulmifolia*, *Gliricidia sepium*, *Enterolobium cyclocarpum*). The preferred plant species for the three species of cattle was *A. pennatula*, while the deer preferred *G. ulmifolia*. The authors suggest that *A. pennatula* is more apt for livestock uses and that there is no possible competition between cattle and deer due to the difference in their preferences (García-Caballero et al. 2016).

15.4 Behavior

Movement of white-tailed deer could be influenced by many ecological, environmental, and behavioral variables such as hunger, reproduction, physiological condition, habitat, and predators or human activity. In the southern portion of the distribution of the species, the information about the movement ecology is practically unknown, especially in tropical areas (Contreras-Moreno et al. 2019a).

In northeastern Mexico, a number of studies have compared the behavior of males and females of the subspecies *O. v. texanus* (Gallina et al. 2003) by analyzing the use of habitat between years, sexes and reproductive periods (Bello et al. 2001a, 2003), home range (Bello et al. 2001b, c), movements in relation to precipitation (Bello et al. 2004), distances traveled (Bello et al. 2006), energy expenditure (Gallina and Bello 2010), and activity (Gallina and Bello 2014). It has been observed that in the northeastern arid lands of Mexico, the daily traveled distances of the white-tailed deer depended on the environmental temperature and precipitation (Delfín-Alfonso et al. 1998; Gallina et al. 2003).

Water is a crucial factor for wildlife in arid and semiarid regions of Mexico and the USA. Management strategies include installation of artificial water sources. However, there is limited information on its effect on wildlife behavior. Bello et al. (2001c) during a long-term study of white-tailed deer (*O. virginianus*) ecology in semiarid northeastern Mexico, estimated differences between sexes, seasons and years of: (1) home range and core area sizes, (2) number of water sources in each activity area, and (3) distance of deer to these sources. Their study site was a fenced area with great water availability, located between Lampazos, Nuevo León and Progreso, Coahuila, Mexico. There were 32 artificial water sources (troughs) and three small dams, with a density of 3.4 sources per km^2. Mean home range size was larger, during reproductive season (230 ha), and smaller during fawning season (166 ha). Sex, season, or year had no effect on core area size or minimum distance to water sources. Abundant water at the site could explain the lack of sex- and year-related differences, even in dry years. They concluded that water management activities at this site influence deer behavior, meanwhile differences between seasons in home ranges could be influenced by individual differences in physiological requirements.

The current knowledge about pregnancy and fawning is limited. It is considered that home-range size and habitat use directly influence the deer population dynamics (Gallina et al. 1998; Green et al. 2017), reproductive patterns (gestation, offspring, time of parturition, fawn survival, size of mothers and fawns), and the behavior of mothers (Green et al. 2017). There is little knowledge about the behavior strategies used by white-tailed deer (*O. v. texanus*) females with fawns. In the same study area, in the northeast of Mexico, Soto-Werschitz et al. (2018) analyzed the variation in home range and the use of vegetation types by eight females during the breeding season in a xeric shrubland in northeastern Mexico and compared the strategies of females with and without fawns. The home range was significantly larger in females with fawns.

The animal energetic requirements depend on the basal metabolism, behavior and physiological conditions. Each animal activity represents an energy cost. It is unknown how deer confront the thermal conditions in arid and semiarid areas where the air temperature is higher than 40 °C and precipitation is low than 400 mm annually as in northern Mexico (Gallina and Bello 2010). So, it is important to know the energetic cost between sexes, in different physiological conditions, and in different years to understand the response of individuals to the environmental changes, mainly on precipitation quantity and distribution that influence habitat quality as food availability and cover protection. Gallina and Bello (2010) found significant differences between sexes in terms of total expenditure. The males spent more energy (1726 ± 38 kcal/day/individual) than females (1556 ± 35 kcal/day/individual). Between the 4 years of study they also found significant differences, with lower energy expenditure in the driest year and much higher in years with greater precipitation. Indicating that deer have a behavioral strategy for try to save its energy expenditure when environmental conditions are unfavorable.

In the tropics, deer face excesses such as floods that occur annually in certain ecosystems. In addition, with climate change, not only droughts are more prolonged, but rainfall is becoming more intense, which is causing the flood time to continue (Contreras-Moreno and Torres-Ventura 2018). Bodmer et al. (2014) recognize that in the Amazon the events of hydrological changes have caused an impact on the populations of land ungulates of the floodplain forests. This has forced the fauna to take refuge in smaller spaces and with fewer available food resources, which increases competition, not only intraspecific, but also with other species, which leads to a deterioration of the corporal condition of terrestrial ungulates and a higher mortality.

Average home range estimated (ha) with Mixed Kernel Density, for white-tailed deer during the monitoring period (May 2016–May 2017), in the Laguna de Términos, Campeche, Southeastern Mexico, were 16.06 ± 7.63 km^2 during the rainy season, and 17.62 ± 4.72 km^2 during dry season (Contreras-Moreno et al. 2019a). In that study the mean distance traveled was 1044 m (SD ± 501 m), with a minimum of 302 m and maximum 4831 m, being in the dry season when they travel greater distances. The breeding season that occurs in the southwest of Campeche from February to June, as well as the scarcity (availability) of resources, is what causes these main movements in the dry season. The smallest distances in the rainy season from July to January, is due to floods in the area that reach 0.5 m. Also, in the tropical dry forests of the Pacific coast of the country, the daily traveled distances were longer during the wet season of the year (June-November) corresponding to the fawning season (Sánchez-Rojas et al. 1997).

The activity patterns of animals are related to their need to meet their basic requirements for food, movement, social interaction, and rest (Beier and McCullough 1990). These patterns can vary as a function of intrinsic characteristics to the animal such as sex, age, physiological state, and as a function of external factors such as forage availability and habitat quality (Marchinton and Hirth 1984; Beier and McCullough 1990). White-tailed deer are known to carry out most activities in the crepuscular hours (sunrise and sunset) when climate conditions are favorable and energy loss owing to cold temperatures can be avoided, as can water loss resulting from high temperatures (Beier and McCullough 1990; Gallina et al. 2005). Little is known about the activity patterns of deer in tropical dry forest.

The objective of the study of López-Tello et al. (2015) was to document deer activity, both daily and during reproductive season for four sites in the Tehuacán-Cuicatlán Biosphere Reserve (TCBR) in Puebla-Oaxaca, Mexico. The Tehuacán-Cuicatlán Biosphere Reserve (TCBR) covers 490,186 ha. It is in the southeastern part of the state of Puebla and the northeastern part of the state of Oaxaca in Mexico (17°32′24.00″–18°52′55.20″ N and 96°59′24.00″–97°48′43.20″ W). To record deer activity, digital camera traps with motion detectors were used, from September 2011 to January 2013. With a total sampling effort of 11,204 trap-days, 798 independent records were obtained. To determine daily activity peaks, the records were grouped into 3-h intervals because of the low number of records obtained at two locations. Activity during reproductive season was divided into three stages of equal length: the rut (November to February), gestation (March to June), and fawning (July to October). The peak activity time for the deer was in the morning from 0900 to 1159 h, and in the afternoon, there was another peak in activity from 1500 to

1759 h and least active from 2400 to 0259 h. Activity was highest during the mating season (November to February) and lowest during gestation (López-Tello et al. 2015).

In a Cloud Forest at the biosphere Reserve of Manantlán, Jalisco, Mexico, six females were radiotracked, finding a bimodal daily activity cycle (González-Pérez 2003): 0900–1200 and 1900–0000 h. Deer in arid zones were more active during crepuscular hours: 1700–2100 h and early in the morning 0500–0900 h. During the breeding season, deer were active at all hours of the day.

Correa-Viana (1995) reported that deer in the paramo Mucubaji, Venezuela, had two activity peaks: at 0900 h and in the sunset. In the Paramo region of Colombia, deer spent 55% of their time for feeding in the morning hours, 22.8% ruminating, and 21.6% resting. Deer used only 0.1% of the time for drinking (Mora and Mosquera 2000).

15.5 Reproduction

In the tropics, breeding can be year-round with peaks in certain seasons (Geist 1998). The fawning period for *O. v. gymnotis* occurs mainly from July to November (rainy season), but a second peak occurs during February and March (dry season). The fawning peak of *O. v. apurensis* occurs mainly from November to February, perhaps in adaptations to heavy rains and widespread flooding on the low llanos. In Peru, fawning varies regionally but generally occurs from January to March (Brokx 1984). McCoy-Colton and Vaughan-Dickhaut (1985) reported the fawning season in April and May in Guanacaste, Costa Rica, at the beginning of the rainy season. Blounch (1987) reported that, in the Colombian Llanos, the fawning season was from September to March, with a peak in December. The same pattern was observed in the high lands, but with a high frequency of births in December, January and February (Blanco and Zabala 2005).

Contreras-Verteramo et al. (2016) identified the relationship between the fluctuations in the levels of testosterone in feces, the cycle of the antlers and some environmental characteristics of the white-tailed deer habitat of four Management Units for Conservation of Wildlife (UMAs) in the Huasteca region of the states of Veracruz and San Luis Potosí, Mexico. The vegetation of the four sampling sites is mainly composed of induced pastures with different degree of presence of native species typical of low spiny forests such as *Piscidia communis*, *Ebenopsis ebony*, *Acacia farnesiana*, *A. pringlei*, *A. cornigera*, *A. gregii*, *Xanthoxylum fagara*, *Sabal mexicana*, *Phyllostylon brasiliense*, *Bursera simaruba*, and *Prosopis glandulosa*, among others. The results for the quantification of testosterone in feces indicate that, from July to February, the highest levels of the hormone testosterone are present. On the other hand, the photographs shown that, in that same period, the antlers of the deer are ossified and free of velvet (clean). In contrast, the lowest concentrations of testosterone were detected between March and June, a season in which the deer lose their antlers and another growth cycle begins. In this study it was possible to determine that the white-tailed deer of the region Huasteca present a very broad fertile period. These results contrast significantly with what can be observed for northern Mexico, where deer show clean antlers from October to February (Villarreal 1999) and even more at higher latitudes where the period of highest testosterone concentration is October to December (Bubenik et al. 1990). They also differ from the lack of periodicity that occurs in equatorial latitudes (Halls 1984).

In Southeastern Mexico, from 2009 to 2017, Contreras-Moreno et al. (2019b) performed camera trap surveys, in which they obtained 10,134 independent photographs of deer. They classified deer bucks in three stages: velvet, hard antler, and no antler. They found that deer with velvet occur from January to July; hard antlers occur from May to October; no antlers occur from the second part of November to the first part of February; and, fawning occurs from February to June. The fawning season in Campeche and Tabasco was different from what occurs in the north, central and western portions of Mexico, where the fawning season goes from July to September and the hard antlers are present from mid-November to the end of January (Villarreal-Espino 2000; Gallina et al. 2010b; Ditchkoff 2011; Mandujano and Gallina 2014; Mandujano et al. 2014).

Variation in the reproductive chronology of the white-tailed deer in that region has been linked to environmental variables associated with latitude such as photoperiods and food availability (Demarais and Strickland 2011). However, the observed fawning season from February to May (with most of the fawn observations in April and May) matches the dry season in the area, a time characterized by a severe water deficiency and daily extreme temperatures exceeding 50 °C, and when the forage availability for females is likely reduced due to the deciduousness of the tropical forests of the region. Fawns that born during this period are approximately 3 months old when water begins rising with the arrival of the rainy season in June, as occurs with the white-tailed deer populations of the Everglades in Florida, an area with a similar flooding season (MacDonald-Beyers and Labisky 2005).

Deer evolved a system of scent and visual cues as the rut. The sexual behavior leaving a scent that attract the opposite sex, and visual signposts, contributes to increment the exit of mating during the breeding season. Males during the pre-reproductive season used to rubbing the antlers against shrubs and trees. This behavior to shed the velvet of the antlers (died skin), serve to establish areas for dominance between males, and for communication with scent, auditive, and visual cues to females. In the ejido El Limón de Cuachichinola, in the Biosphere Reserve Sierra de Huautla, in the State of Morelos, Mexico, during October–November, the rubbing process indicates the beginning of mating season (Buenrostro-Silva et al. 2008). In this study they registered the shrubs and trees with rubbing along transects and found a total of 25 rubbings. Bucks rub frequently in shrubs and trees with a diameter between 5 and 10 cm, and nine species of trees were used by males. Deer selected *Lysiloma divaricata* (40%) and *Conzattia multiflora* (16%).

Positive relationships between age, sexually selected traits, and male reproductive success have been reported for several polygynous ungulates; however, relatively little is known about the factors influencing male reproductive success in ungulate species whose mating system is characterized by tending-bond behaviors.

Newbolt et al. (2017) investigated male reproductive success in white-tailed deer across a range of sex ratios and age structures using a known population of deer. Their supported model indicated that annual body size and antler size of the individual were positively associated with annual male breeding success; however, the effects of annual antler size were sensitive to changes in mean male age of the herd.

15.6 Population Dynamics

The population estimation of deer is one of the central themes in Latin America, mainly in Mexico (Table 15.1), since this allows to make a sustainable use of the species from the hunting point of view. Population density is one of the parameters most evaluated in different regions and habitat types. However, there is a considerable variation in the estimates due to differences in sampling procedures depending on the method used (direct and indirect counts). One of the most used methods is the

Table 15.1 White-tailed deer density estimation in Mexico

Deer/km²	Method	Vegetation type	State	Reference
2.1	Fecal group counts	Pine-oak forest	Hidalgo	Sánchez-Rojas et al. (2009)
5.5	Direct counts in strip transect	Secondary vegetation, tropical evergreen forest, tintales, and savannah of palms	Quintana Roo	González-Marín et al. (2008)
6.7	Fecal group counts	Tropical dry forest and secondary vegetation	Michoacán	Mandujano et al. (2013)
2.74	Fecal group counts	Temperate forest	Morelos	Flores-Armillas et al. (2011)
2.36	Camera traps	Rosette desert scrub and grassland	Sonora	Lara-Díaz et al. (2011)
0.49–1.30	Fecal group counts	Temperate forest	Oaxaca	Piña and Trejo (2014)
5–11.6	Fecal group counts	Oak forest, subtropical scrub and induced grassland	Aguascalientes	Medina-Torres et al. (2015)
0.63–1.30	Fecal group counts	Secondary vegetation	Tabasco	Contreras-Moreno et al. (2015a)
5.64 (wet) 6.35 (dry)	Fecal group counts	Oak forest, mixed forest and scrub-grassland	Puebla	Ocaña-Parada et al. (2014)
8.75 (Bajo Balsas) 2.03 (RBTC)	Fecal group counts	Tropical dry forest	Michoacán Oaxaca	Yáñez-Arenas and Mandujano (2015)
2.0–6.3 (wet) 0.8–12 (dry)	Camera trap	Pine-oak forest	Estado de México	Soria-Díaz and Monroy-Vilchis (2015)

counting of fecal groups, however there is also great variation in the estimates due to the daily defecation rate (DDR) used. For example, in Honduras it was estimated that the average DDR is 12.5 fecal groups per individual per day, but the authors recommend using the highest rate of 15 fecal groups to avoid over estimates (Portillo et al. 2010). In Mexico, rates of 15.2, 17, and 20.9 have been reported, while in Venezuela it was 14 fecal groups per day (Mandujano 2014). Therefore, it is not always possible to make comparisons between sites and/or regions.

In Mexico it has been reported that deer density can vary depending on the characteristics of the habitat, among the most important variables are climatic, vegetation type and human activities (Flores-Armillas et al. 2011; Mandujano et al. 2013; Ocaña-Parada et al. 2014; Medina-Torres et al. 2015; Yáñez-Arenas and Mandujano 2015). For example, in the Sierra El Laurel, Aguascalientes, Mexico, the highest density was found in the subtropical scrubland, followed by the oak forest (*Quercus* sp.) and the induced pasture (Medina-Torres et al. 2015). In the Parque Estatal Lázaro Cárdenas del Río "Flor del Bosque," Puebla, Mexico, there was greater density in the mixed forest, both in the rainy season and in the dry season (Ocaña-Parada et al. 2014). In the Mixteca Poblana the population density varies from 2 to 7 ind/km^2 (Villarreal-Espino et al. 2011). In different studies it has been reported that the sites with the highest deer density and abundance are characterized by being conserved, presenting low human activity, as well as irregular topography, dense coverage in the understory and the tree stratum. While sites with fragmentation, poaching, high density of livestock and agricultural activities, have low density and abundance (Flores-Armillas et al. 2011; Mandujano et al. 2013; Contreras-Moreno et al. 2015a; Clemente-Sánchez 2018). There have also been several studies on the population structure, for example Contreras-Moreno et al. (2015a) estimated the proportion of sexes and age classes in two localities of Tabasco, Mexico, dominated by acahuales (secondary forest). In the Caudillo a male-to-female ratio of 1:1.25 and a female-to-fawn ratio of 1: 0.64 were estimated, while in San Joaquín a male-to-female ratio of 1:2.75 and a female-to-fawn ratio of 1:0.5 were recorded; these authors consider that the habitat conditions are favorable for a good population structure.

15.7 Mortality Factors

Among the most important diseases that affect the health of deer are gastrointestinal parasitosis caused mainly by helminths and protozoa. Different species of parasites present in domestic livestock can also infect different species of wild mammals. Currently the high overlap in the use of habitat between white-tailed deer and domestic livestock, as well as their phylogenetic closeness, produces a potential transmission of parasites and diseases (Romero-Castañón et al. 2008, Fig. 15.5). One of the genera present in tropical and subtropical regions is *Babesia*. In a study conducted in the northern region of Mexico, a high prevalence of *Babesia bovis* (75%) and *Babesia bigemina* (53%) was found in deer (Cantú-Martínez et al. 2008).

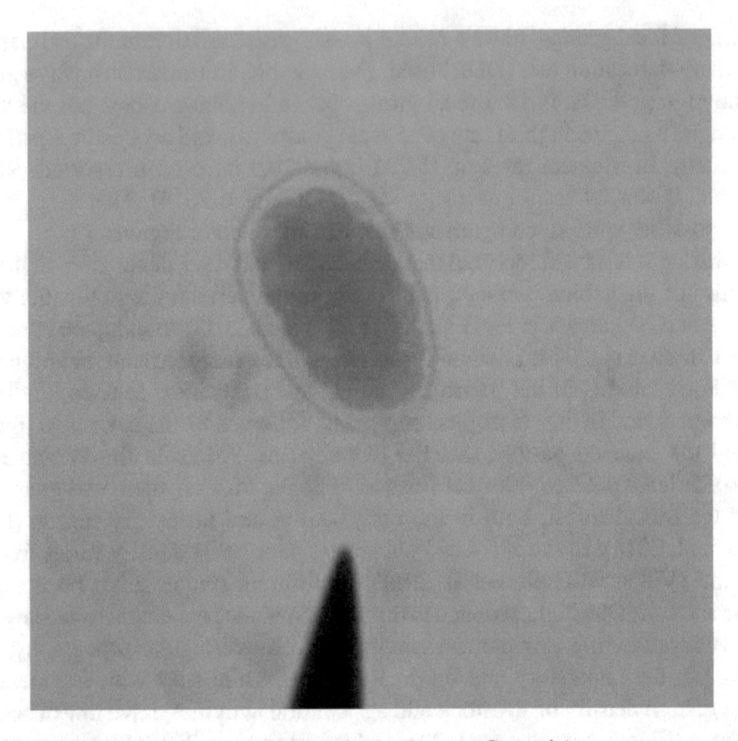

Fig. 15.5 *Trichostrongylus* sp. egg (Photo Salvador Romero-Castañón)

Also, infections by nematodes of the suborder Strongylida and by ectoparasites of the genus *Lipoptena* and *Amblyomma* have been reported.

In a study made in the Natural Reserve of the Paratroopers Brigade, San Jose, Guatemala, 15 deer were captured, which were found parasitized with *Rhipicephalus microplus*, 13.3% with *Amblyomma* sp., 6.67% with *Amblyomma parvum* and 6.67% with *Dermacentor variabilis* (Álvarez-Robles et al. 2018). In the dry tropical forest of the state of Yucatan, Mexico, it was reported that four individuals captured were parasitized with gastrointestinal nematodes of the suborder Strongylida, fly genus *Lipoptena* and a species of tick (*Amblyomma cajennese*) (Mukul-Yerves et al. 2014a). Mukul-Yerves et al. (2014b) collected 64 fecal samples in five intensive management units for wildlife conservation (UMAs), located in the state of Yucatan, Mexico. Their results indicated that 15 individuals presented parasitic load. The parasites that they identified belonged to the suborder Strongylida (Fig. 15.6), and the order Eucoccodiidae and the genera *Strongyloides* and *Trichuris*. The suborder Strongylida was the most frequent and the genus *Trichuris* the least frequent.

Romero-Castañón et al. (2008), compared the presence and prevalence of parasites in wild and domestic ungulates that live in the Selva Lacandona, Chiapas, Mexico. They identified 18 species of nematodes and seven species of protozoa (only in domestic ungulates) in feces, three species of nematodes, two species of trematodes and one species of cestode during postmortem examination. The infec-

Fig. 15.6 White-tailed deer with a bad physical condition infested by ticks from Sonora, México (Photo Carlos López-González)

tion by *Paramphistomum cervi* was prevalent and intense in the deer, reason why the authors consider that it represents a potential risk for the populations of these ungulates. Of the 16 ectoparasites that they identified, 15 were ticks of the family Ixodidae and one was a dipterous of the family Hippoboscidae. According to the similarity index of Morisita, the greatest similarity in the composition of the endoparasite and ectoparasite community was between *Bos taurus* and *O. virginianus*.

15.8 Habitat Requirements

Deer selects specific sites to rest, ruminate and protect their young (Ortega et al. 2011). Therefore, these sites should offer food, water, protection against predators, and thermal cover that allows deer to minimize both the absorption of heat by exposure to the sun and the loss of water by transpiration (Gallina 1994; Rothley 2001; Serra-Ortiz et al. 2008; Delfín-Alfonso et al. 2009; Plata et al. 2013). The cover is of greater importance in arid and semiarid zones, where water loss should be avoided as a result of high environmental temperatures (>40 °C) and limited water availability (Gallina 1994; Gallina et al. 2010a). Also, the slope, altitude and orientation are considered important variables for the presence of the deer (Gallina 1994; Rothley 2001; Delfín-Alfonso et al. 2009).

According to Ortega-Santos and Fulbright (2014), the ideal habitat for white-tailed deer in northern Mexico should include patches of dense and diverse shrub vegetation, interspersed in the landscape to provide quality coverage for males and females with fawns. The presence of cactus (*Opuntia* sp.) and other succulents is important especially in the most arid habitats. The coverage of medium and high height grasses is important near the shrub cover to provide protection for the offspring, a proper grazing management of domestic animals is important to achieve this condition. Shrub vegetation interspersed with open areas dominated by diverse herbaceous plants is essential for feeding areas. The streams or drainage areas are important because of the diversity of the vegetation as they accumulate humidity and serve as corridors. The surface water sources and the mentioned habitat characteristics must be juxtaposed with each other to produce the forage classes necessary to cover the nutritional requirements of the deer.

In the Sierra del Laurel, Aguascalientes, Mexico, the habitat attributes that influence the likelihood of being used to a greater or lesser extent by deer were evaluated. The results indicated that the type of vegetation, the slope, the altitude and the distance to the bodies of water are the most important characteristics, likewise the scrub was the type of vegetation most used (Medina-Torres et al. 2008). Meanwhile, in the temperate forests of the state of Morelos, Mexico, of the nine types of vegetation available, the deer preferred to use the pine-oak and mountain cloud forests and avoided the pine forest (Flores-Armillas et al. 2011). It has also been reported that in the dry tropical forest of San Luis Potosí, Mexico, it prefers sites consisting mainly of forest and minimum pasture areas (88.4% of forests, 6.6% of crops and 5% of pastures) (García-Marmolejo et al. 2015).

Some studies have compared deer preferences between conserved sites and sites disturbed by human activity. For example, in the state of Tabasco, Mexico, the deer prefers the subperennifolia medium forest and the savannah pasture, while avoiding the low-flood forest, which may be due to the density of livestock in the study area (Contreras-Moreno et al. 2015b). In the temperate forests of Chichinautzin, Morelos, Mexico, there is a greater presence of deer in the cloud and pine-oak forest, which are characterized by higher values in the number of shrubs, coverage, volume and average tree height. While the sites with less presence were those with less coverage, height, tree volume and less diversity of shrub species. Therefore, it is suggested that the deer uses more conserved sites with low anthropic activity (Flores-Armillas et al. 2011, 2013). However, two studies conducted in Mexico, one in the Calakmul region, Campeche and the other in Lacandona, Chiapas, report that deer uses more than expected deforested, agricultural and secondary vegetation sites, while the other types of vegetation uses them in relation to their availability and/or less than expected (Weber 2008; Tejeda-Cruz et al. 2009).

In semiarid zones the thermal cover is important for the survival of deer, therefore, it is necessary to have information about the characteristics of the rest sites. For example, Gallina et al. (2010a) carried out a characterization of these sites in a xerophilous thicket in northeastern Mexico. Their results indicated that the males selected shrubs with greater thermal coverage, volume and height than the females, which prefer sites with mesquites (*Prosopis glandulosa*) and huizaches (*Acacia farnesiana*). These sites are closer to the dams and are used mainly from 11:00 to

14:00 when the temperatures are generally higher. They also found seasonal differences in the selection, during the post-breeding season, they used sites with bushes of greater height and volume, but in fewer species. While in the breeding season the females selected sites with greater coverage of protection against predators.

In relation to habitat quality, differences in use have been found. In the center of Veracruz, Mexico, an evaluation of the quality of the deer's habitat was carried out. In which it was obtained that only 29.75% of the area presents a relative quality of medium to high, while more than 50% of the territory is of low quality to inappropriate. Within the biomes considered as potential habitat are temperate forests, tropical forests, scrub and the combination at the edges of these three biomes, which together comprise only 27% of the total area of the study area, while the transformed biomes (induced pasture, cultivated pasture, agricultural and urban areas) cover the majority of the area (Delfín-Alfonso et al. 2009). In the same region, Bolívar-Cimé and Gallina (2013) found a significant difference in the use of habitat according to its quality, those of low quality were avoided, and those of medium and high quality were used. The oak and the dry tropical forest were preferred while grassland was avoided.

In the Balsas River depression, Puebla, Mexico, habitat quality was evaluated in two UMAs with three main vegetation types: dry tropical forest, xerophytic scrub and oak forest. The results indicated that the bushes were the type of vegetation with the highest protection coverage (42.5% and 34.9%), followed by the tropical dry forest (32.3% and 34.6%) and the oak forest (19.9% and 25.5%). However, the oak forest presents a higher percentage of bare soil (54% and 51.4%) and organic matter (27% and 23%) which is considered as a feature that prevents soil erosion. Therefore, the oak forest presents the best quality habitat for the deer, followed by the scrub, while the dry tropical forest is the one with the lowest quality (Villarreal-Espino et al. 2013).

15.9 Conservation Prospects

Although the species is highly adaptable and widely distributed, we can ensure that it has populations that are exploited in a sustainable manner for both hunting and subsistence hunting, but that some of the subspecies may be at risk from poaching and severe transformation of their habitat. In Mexico the problem that has arisen is that the subspecies *O. v. texanus* is the most appreciated as a trophy and has been introduced in sites outside its distribution (which is the northwest of the country), putting at risk the genetic diversity of the other subspecies. Social behavior of white-tailed deer can have important management implications. The formation of matrilineal social groups among female deer has been documented, and management strategies have been proposed based on this well-developed social structure. As can be seen in the chapter, it is one of the most studied species of ungulates in both Mexico and Latin America.

Acknowledgements Thanks to Adriana Sandoval Comte for the distribution map.

References

Aguilera-Reyes U, Sánchez-Cordero V, Ramírez-Pulido J et al (2013) Hábitos alimentarios del venado cola blanca *Odocoileus virginianus* (Artiodactyla: Cervidae) en el parque natural sierra Nanchititla, Estado de México. Rev Biol Trop 61:243–253. https://doi.org/10.15517/rbt.v61i1.11059

Álvarez-Robles E, Fuentes-Rousselin H, Meoño-Sáncez E et al (2018) Aproximación al estudio de los parásitos externos del venado cola blanca (*Odocoileus virginianus*) en la reserva natural de la Brigada de Paracaidistas de San José, Escuintla Guatemala. Revista Electrónica de Veterinaria 19:1–11

Baker RH (1984) Origin, classification and distribution. In: Halls LK (ed) White-tailed deer: ecology and management. Stackpole, Harrisburg, pp 1–18

Beier P, McCullough DR (1990) Factors influencing white-tailed deer activity patterns and habitat use. Wildl Monogr 109:1–51

Bello J, Gallina S, Equihua M (2001a) Characterization and habitat preferences by white-tailed deer in Mexico. J Range Manage 54:537–545. https://doi.org/10.2307/4003582

Bello J, Gallina S, Equihua M et al (2001b) Activity areas and distance to water sources by white-tailed deer in Northeastern Mexico. Vida Silvestre Neotropical 10:30–37

Bello J, Gallina S, Equihua M et al (2001c) Home range, core area and distance to water sources by white tailed deer in northeastern Mexico. Vida Silvestre Neotropical 10:30–37

Bello J, Gallina S, Equihua M (2003) El venado cola blanca: uso del hábitat en zonas semiáridas y con alta disponibilidad de agua del Noreste de México. In: Polanco-Ochoa R (ed) Manejo de fauna silvestre en Amazonía y Latinoamérica: selección de trabajos V congreso internacional. CITES, Fundación Natura, Bogotá, pp 67–76

Bello J, Gallina S, Equihua M (2004) Movements of the white-tailed deer and their relationship with precipitation in Northeastern Mexico. Dermatol Int 29:357–361

Bello J, Gallina S, Equihua M (2006) Distancias de desplazamiento del Venado Cola Blanca y su relación con factores ambientales en el Noreste de México. In: Memorias del VI congreso internacional sobre manejo de fauna silvestre en la Amazonía y Latinoamérica, pp 146–151

Blanco L, Zabala AI (2005) Recopilación del conocimiento local sobre el venado cola blanca (*Odocoileus virginianus*) como base inicial para su conservación en la zona amortiguadora del Parque Nacional Natural Pisba, en los municipios de Tasco y Socha. BSc thesis, Universidad Pedagógica y Tecnológica de Colombia

Blounch RA (1987) Reproductive seasonality of the white-tailed deer on the Colombia llanos. In: Wemmer C (ed) Biology and management of the cervidae. Smithsonian, Washington, pp 339–343

Bodmer RE, Fang GT, Puertas PE et al (2014) Cambio climático y fauna silvestre en la Amazonía peruana, Fundación Latinoamericana para el Trópico Amazónico-Fundamazonia, Iquitos, Perú

Bolívar-Cimé B, Gallina S (2013) An optimal habitat model for the white-tailed deer (*Odocoileus virginianus*) in central Veracruz, Mexico. Anim Prod Sci 52:707–713. https://doi.org/10.1071/AN12013

Brokx PA (1984) White-tailed deer of South America. In: Halls LK (ed) Ecology and management of the white-tailed deer. Stackpole, Harrisburg, pp 525–546

Bubenik GA, Brown RD, Schams D (1990) The effect of latitude on the seasonal pattern of reproductive hormones in the male white-tailed deer. Comp Biochem Physiol 97:253–257

Buenrostro-Silva A, Gallina S, Sánchez-Rojas G (2008) Los talladeros de machos de venado cola blanca *Odocoileus virginianus mexicanus* (Gmelin, 1788) y su ubicación para definir los sitios reproductivos. In: Lorenzo C, Espinoza E, Ortega J (eds) Avances en el estudio de los mamíferos de México II. Asociación Mexicana de Mastozoología A. C., México, pp 219–238

Cantú-Martínez MA, Salinas-Meléndez JA, Zarate-Ramos JJ et al (2008) Prevalence of antibodies against *Babesia bigemina* and *B. bovis* in white-tailed deer (*Odocoileus virginianus texanus*) in farms of northeastern Mexico. J Anim Vet Adv 7:121–123

Clemente-Sánchez F (2018) Dieta, población y capacidad de carga del venado cola blanca (*Odocoielus virginianus*) en dos condiciones de hábitat en Tlachichila, Zacatecas, México. Agroproductividad 11:15–23

Comer CE, Kilgo JC, D'Angelo GJ et al (2005) Fine-scale genetic structure and social organization in female white-tailed deer. J Wildl Manag 69:332–344. https://doi.org/10.2193/0022-541x(2005)069<0332:fgsaso>2.0.co;2

Contreras-Moreno F, Torres-Ventura Y (2018) El cambio climático y los ungulados silvestres F. Desde el herbario CICY 10:144–150

Contreras-Moreno FM, Zúñiga-Sánchez J, Bello-Gutiérrez J (2015a) Parámetros poblacionales de *Odocoileus virginianus* (Cervidae) en dos comunidades de Tabasco, México. Revista Latinoamericana de Conservación 4:7–13

Contreras-Moreno FM, Zúñiga-Sánchez S, Bello-Gutiérrez J (2015b) Preferencia de hábitat de *Odocoileus virginianus thomasi* merriam en dos ejidos ganaderos del sureste de México. Agroproductividad 8:49–55

Contreras-Moreno F, Hidalgo-Mihart MG, Contreras-Sánchez WM (2019a) Daily traveled distances by the white-tailed deer in relation to seasonality and reproductive phenology in a tropical lowland of southeastern Mexico. In: Reyna-Hurtado R, Chapman C (eds) Movement ecology of neotropical forest mammals. Springer, Cham, pp 111–123

Contreras-Moreno F, Hidalgo-Mihart MG, Cruz J et al (2019b) Seasonal antler cycle in white-tailed deer in south-eastern mexico. Eur J Wildl Res 65:53

Contreras-Verteramo A, Gallina S, Alvarado-Sánchez B, Rangel-Lucio JA (2016) Patrón Reproductivo de Los machos de venado cola blanca en la región Huasteca de los Estados de Veracruz y San Luis Potosí. In: Memorias del XXXIII Simposio sobre Fauna Silvestre, Facultad de Medicina Veterinaria y Zootecnia, UNAM, México, 7-9 noviembre 2016

Correa-Viana M (1995) Distribución y estado actual del venado de páramo en el parque Nacional Sierra nevada, Mérida, Venezuela, Universidad Nacional Experimental de los Llanos Occidentales Ezequile Zamora

De la Rosa-Reyna XF, Calderón-Lobato DR, Parra-Bracamonte GM et al (2012) Genetic diversity and structure among subspecies of white-tailed deer in Mexico. J Mammal 93:1158–1168

Delfín-Alfonso C, Mandujano S, Gallina S et al (1998) Patrones de desplazamiento del venado cola blanca en un rancho con manejo de agua en el Noreste de México. In: Memorias del VI Simposio sobre Venados de México, Universidad Nacional Autónoma de México, Asociación Nacional de Ganaderos Diversificados Criadores de Fauna (ANGADI), Xalapa, Veracruz, México

Delfín-Alfonso C, Gallina S, López-González C (2009) Evaluación del hábitat del venado cola blanca utilizando modelos espaciales y sus implicaciones para el manejo en el centro de Veracruz, México. Trop Conserv Sci 2:215–228. https://doi.org/10.1177/194008290900200208

Demarais S, Strickland B (2011) Antlers. In: Hewitt D (ed) Biology management of white tailed deer. CRC Press, Boca Raton, pp 107–145

DeYoung RW, Demarais S, Honeycutt RL et al (2003) Evaluation of a DNA microsatellite panel useful for genetic exclusion studies in white-tailed deer. Wildl Soc Bull 31:220–232

Ditchkoff SS (2011) Anatomy and physiology. In: Hewitt D (ed) Biology and management of white-tailed deer. CRC Press, Boca Raton, pp 43–73

Flores-Armillas VH, Gallina S, García-Barrios JR et al (2011) Selección de hábitat por el venado cola blanca *Odocoileus virginianus mexicanus* (Gmelin, 1788) y su densidad poblacional en dos localidades de la región centro del Corredor Biológico Chichinautzin, Morelos, México. Therya 2:263–277. https://doi.org/10.12933/Therya-11-31

Flores-Armillas VH, Botello F, Sánchez-Cordero V et al (2013) Caracterización del hábitat del venado cola blanca (*Odocoileus virginianus mexicanus*) en los bosques templados del Corredor Biológico Chichinautzin y modelación de su hábitat potencial en Eje Transvolcánico Mexicano. Therya 4:377–393. https://doi.org/10.12933/Therya-13-118

Gallina S (1994) Uso del hábitat por el venado cola blanca en la Reserva de la Biosfera La Michilía, México. In: Vaughan C, Rodríguez M (eds) Ecología y manejo del venado cola blanca en México y Costa rica. Universidad de Costa Rica, Costa Rica, pp 299–314

Gallina S, Bello J (2010) El gasto energético del venado cola blanca (*Odocoileus virginianus texanus*) en relación a la precipitación en una zona semiárida de México. Therya 1:9–22. https://doi.org/10.12933/Therya-10-1

Gallina S, Bello J (2014) Patrones de actividad del venado cola blanca en el noreste de México. Therya 5:423–436. https://doi.org/10.12933/Therya-14-200

Gallina S, Pérez-Arteaga A, Mandujano S (1998) Patrones de actividad del venado de cola blanca (*Odocoileus virginianus texanus*) en un matorral xerófilo de México. Boletín de la Sociedad de Biología, Concepción, Chile 69:221–228

Gallina S, Corona P, Bello J (2003) El venado cola blanca: comportamiento en zonas semiáridas del Noreste de México. In: Memorias del V Congreso Internacional en Manejo de fauna silvestre en Amazonía y Latinoamérica. CITES, Fundación Natura, Bogotá, pp 165–173

Gallina S, Corona-Zárate P, Bello J (2005) El comportamiento del venado cola blanca en zonas semiáridas del Noreste de México. In: Sánchez-Cordero V, Medellín RA (eds) Contribuciones mastozoológicas en homenaje a Bernardo Villa. Instituto de Biología UNAM, Instituto de Ecología UNAM, CONABIO, México, pp 193–203

Gallina S, Bello J, Contreras-Verteramo C, Delfín-Alfonso C (2010a) Daytime bedsite selection by the texan white-tailed deer in xerophyllous brushland, North-eastern Mexico. J Arid Environ 74:373–377. https://doi.org/10.1016/j.jaridenv.2009.09.032

Gallina S, Mandujano S, Bello J et al (2010b) White-tailed deer *Odocoileus virginianus* (Zimmermann 1780). In: Barbanti-Duarte B, González S (eds) Neotropical cervidology: biology and medicine of Latin American deer. Funep; IUCN, Jaboticabal; Gland, pp 101–118

García-Caballero A, Barragán-Portillo B, Querol-Carranza D (2016) El silvopastoralismo como herramienta de conservación: ejemplificado con el venado cola blanca en Nicaragua. BSc Thesis, Universidad Autónoma de Barcelona

García-Marmolejo G, Chapa-Vargas L, Weber M, Huber-Sannwald E (2015) Landscape composition influences abundance patterns and habitat use of three ungulate species in fragmented secondary deciduous tropical forests, Mexico. Global Ecol Conserv 3:744–755. https://doi.org/10.1016/j.gecco.2015.03.009

Geist V (1998) Deer of the world. Their evolution, behavior and ecology. Stackpole, Harrisburg

González G, Briones-Salas M (2012) Dieta de *Odocoileus virginianus* (Artiodactyla: Cervidae) en un bosque templado del norte de Oaxaca: México. Rev Biol Trop 60:447–457

González-Pérez GE (2003) Uso del hábitat y área de actividad del venado cola blanca (*Odocoileus virginianus sinaloae* J. Allen) en la Estación Científica Las Joyas, Reserva de la Biosfera de Manantlán, Jalisco. MSc thesis, Universidad Autónoma de México

González Marín R, Gallina S, Mandujano S, Weber M (2008) Densidad y distribución de ungulados silvestres en la Reserva Ecológica El Edén, Quintana Roo, México. Acta Zoológica Mexicana (n.s.) 24:73–93

Granados D, Tarango L, Olmos G et al (2014) Dieta y disponibilidad de forraje del venado cola blanca *Odocoileus virginianus thomasi* (Artiodactyla: Cervidae) en un campo experimental de Campeche, México. Rev Biol Trop 62:699–710

Green ML, Kelly AC, Satterthwaite-Phillips D et al (2017) Reproductive characteristics of female white-tailed deer (*Odocoileus virginianus*) in the Midwestern USA. Theriogenology 94:71–78. https://doi.org/10.1016/j.theriogenology.2017.02.010

Halls LK (1984) White-tailed deer: ecology and management. Stackpole, Harrisburg

Hernández-Llamas ÁR (2014) Variabilidad genética y relación filogeográfica de tres subespecies de venado cola blanca (*Odocoileus virginianus*) en la región Centro-Norte de México. MSc thesis, Colegio de Postgraduados

Lara-Díaz NE, Coronel-Arellano H, González-Bernal A, Gutiérrez-González C, López-González CA (2011) Abundancia y densidad de venado cola blanca (Odocoileus virginianus couesi) en Sierra de San Luis, Sonora, México. Therya, 2(2): 125–137.

López-Pérez E, Serrano-Aspeitia N, Aguilar-Valdés BC, Herrera-Corredor A (2012) Composición nutricional de la dieta del venado cola blanca (*Odocoileous virginianus ssp. mexicanus*) en Pitzotlán, Morelos. Revista Chapingo Serie Ciencias Forestales y del Ambiente 18:219–229. https://doi.org/10.5154/r.rchscfa2011.01.006

López-Tello E, Gallina S, Mandujano S (2015) Activity patterns of white-tailed deer in the Tehuacán-Cuicatlán Biosphere Reserve, Puebla-Oaxaca, Mexico. Deer Spec Group IUCN Newslett 27:32–43

MacDonald-Beyers K, Labisky RF (2005) Influence of flood waters on survival, reproduction, and habitat use of white-tailed deer in the Florida Everglades. Wetlands 25:659–666. https://doi.org/10.1672/0277-5212(2005)025[0659:iofwos]2.0.co;2

Mandujano S (2014) PELLET: an Excel®-based procedure for estimating deer population density using the pellet-group counting method. Trop Conserv Sci 7:308–325

Mandujano S, Gallina S (2014) Conceptos y cálculo de la capacidad de carga de venados. In: Gallina S, Mandujano S, Villareal-Espino OA (eds) Monitoreo y manejo del venado cola blanca, conceptos y métodos. Instituto de Ecología A. C., Benemérita Universidad Autónoma de Puebla, Xalapa, pp 111–120

Mandujano S, Delfín-Alfonso C, Gallina S (2010) Comparison of geographic distribution models of white-tailed deer *Odocoileus virginianus* (Zimmermann, 1780) subspecies in Mexico: biological and management implications. Therya 1:41–68. https://doi.org/10.12933/Therya-10-5

Mandujano S, Yáñez-Arenas CA, González-Zamora A, Pérez-Arteaga A (2013) Habitat-population density relationship for the white-tailed deer *Odocoileus virginianus* during the dry season in a Pacific Mexican tropical dry forest. Mammalia 77:381–389

Mandujano S, Gallina S, Ortega A (2014) Venado Cola Blanca en México. In: Valdéz R, Ortega AJ (eds) Ecología y manejo de fauna silvestre en México. Colegio de Postgraduados, Texcoco, pp 399–420

Marchinton RL, Hirth DH (1984) Behavior. In: Halls LK (ed) White-tailed deer: ecology and management. Stackpole, Harrisburg, pp 129–168

McCoy-Colton MB, Vaughan-Dickhaut C (1985) Resultados preliminares del estudio del venado cola blanca (*Odocoileus virginianus*) en Costa Rica. In: Guier E (ed) Investigaciones sobre fauna silvestre de Costa Rica. Universidad Estatal a Distancia, San José

Medina-Torres SM, García-Moya E, Márquez-Olivas M et al (2008) Factores que influyen en el uso del hábitat por el venado cola blanca (*Odocoileus virginianus couesi*) en la Sierra del Laurel, Aguascalientes, México. Acta Zoológica Mexicana 24:191–212

Medina-Torres SM, García-Moya E, Márquez-Olivas M et al (2015) Relación hábitat-densidad de *Odocoileus virginianus couesi* (Coues and Yarrow, 1875) en la Sierra del Laurel, Aguascalientes. Revista Mexicana de Ciencias Forestales 6:17–36

Miller BF, DeYoung RW, Campbell TA et al (2010) Fine-scale genetic and social structuring in a central Appalachian white-tailed deer herd. J Mammal 91:681–689. https://doi.org/10.1644/09-MAMM-A-258.1.Key

Mora CA, Mosquera A (2000) Estudio preliminar del comportamiento alimenticio del venado cola blanca (*Odocoileus virginianus goudoti*) en ecosistemas de subpáramo y páramo del Parque Nacional Natural Chingaza, en Cundinamarca-Meta, Colombia. BSc thesis, Universidad Nacional de Colombia

Mukul-Yerves J, Pereira-Hoíl A, Rodríguez-Vivas R, Montes-Pérez R (2014a) Parasitosis gastrointestinal en venados cola blanca (*Odocoileus virginianus yucatanensis*) y temazate (*Mazama temama*) en condiciones de cautiverio en Yucatán, México. Bioagrociencias 7:33–37

Mukul-Yerves JM, Zapata-Escobedo M, Montes-Pérez R et al (2014b) Parásitos gastrointestinales y ectoparásitos de ungulados silvestres en condiciones de vida libre y cautiverio en el trópico mexicano. Revista Mexicana de Ciencias Pecuarias 5:459–469

Newbolt CH, Acker PK, Neuman TJ et al (2017) Factors influencing reproductive success in male white-tailed deer. J Wildl Manag 81:206–217. https://doi.org/10.1002/jwmg.21191

Ocaña-Parada CJ, Jiménez-García D, López-Olguín J et al (2014) Population density of white-tailed deer (*Odocoileus virginianus*), in a natural protected area State of Puebla, Mexico. World J Zool 9:52–58. https://doi.org/10.5829/idosi.wjz.2014.9.1.82259

Ortega JA, Mandujano S, Villarreal J et al (2011) Managing white tailed deer: Latin American. In: Hewitt DG (ed) Biology and management of white-tailed deer. CRC Press, Boca Raton, pp 565–597

Ortega-Santos JA, Fulbright T (2014) Capacidad de carga, planeación y manejo de la cosecha del venado cola blanca. In: Gallina S, Mandujano S, Villareal-Espino O (eds) Monitoreo y manejo del venado cola blanca, conceptos y métodos. Instituto de Ecología A. C., Benemérita Universidad Autónoma de Puebla, Xalapa, pp 137–160

Piña E, Trejo I (2014) Densidad poblacional y caracterización de hábitat del venado cola blanca en un bosque templado de Oaxaca, México. Acta Zoológica Mexicana (n.s.), 30(1): 114–134.

Plata FX, Ebergeny S, Resendiz JL et al (2009) Palatabilidad y composición química de alimentos consumidos en cautiverio por el venado cola blanca de Yucatán (*Odocoileus virginianus yucatanensis*). Archivos de Medicina Veterinaria 41:123–129. https://doi.org/10.4067/S0301-732X2009000200005

Plata FX, Martínez JA, Mendoza GD et al (2013) Incorporación de la cobertura de escape en un modelo de capacidad de carga para venado de cola blanca. Archivos de Medicina Veterinaria 45:91–97. https://doi.org/10.4067/S0301-732X2013000100015

Portillo H, Hernández J, Elvitt F et al (2010) Estimación de la tasa de defecación del venado cola blanca (*Odocoileus virginianus*) en cautividad en Honduras. Mesoamericana 14:55–57

Romero-Castañón S, Ferguson BG, Guiris D et al (2008) Comparative parasitology of wild and domestic ungulates in the Selva Lacandona, Chiapas, Mexico. Compar Parasitol 75:115–126. https://doi.org/10.1654/4267.1

Rothley KD (2001) Manipulative, multi-standard test of a white-tailed deer habitat suitability model. J Wildl Manag 65:953–963

Sánchez-Rojas G, Gallina S, Mandujano S (1997) Área de actividad y uso del hábitat de dos venados cola blanca (*Odocoileus virginianus*) en un bosque tropical de la costa de Jalisco, México. Acta Zoológica Mexicana 72:39–54

Sánchez Rojas G, Aguilar Miguel C, Hernández Cid E (2009) Estudio poblacional y uso de hábitat por el venado cola blanca (Odocoileus virginianus) en un bosque templado de la Sierra de Pachuca, Hidalgo, México. Tropical Conservation Science 2:204–214.

Serra-Ortiz MA, González-Zaldívar FN, Cantú-Ayala C et al (2008) Evaluación del hábitat disponible para dos especies de cérvidos en el noroeste de México. Revista Mexicana de Mastozoología 12:43–58

Smith WP (1991) *Odocoileus virginianus*. Mamm Species 388:1–13

Soto-Werschitz A, Mandujano S, Gallina S (2018) Home-range analyses and habitat use by white-tailed deer females during the breeding season. Therya 9:1–6

Soria-Díaz L, Monroy-Vilchis O (2015) Monitoring population density and activity pattern of white-tailed deer (Odocoileus virginianus) in Central Mexico, using camera trapping. Mammalia, 79(1): 43–50.

Suárez D, Arrieta L, Gaona L (2017) Variabilidad genética de *Odocoileus virginianus* (Zimmerman, 1970) en Boyacá-Colombia: Reporte de Caso. Revista UDCA Actualidad & Divulgación Científica 20:479–484

Tejeda-Cruz C, Naranjo E, Cuarón A et al (2009) Habitat use of wild ungulates in fragmented landscapes of Lacandon Forest, Southern Mexico. Mammalia 73:211–220. https://doi.org/10.1515/MAMM.2009.044

Vaides-Arrue SJ (2017) Características generales del comportamiento del venado cola blanca (*Odocoileus virginianus*) en semicautiverio en Cobán, Alta Verapaz, Guatemala. Revista Ciencia Multidisciplinaria CUNORI 1:83–84

Vásquez Y, Tarango L, López-Pérez E et al (2015) Variation in the diet composition of the white tailed deer (*Odocoileus virginianus*) in the Tehuacán-Cuicatlán Biosphere Reserve. Revista Chapingo Serie Ciencias Forestales y del Ambiente 22:58–67. https://doi.org/10.5154/r.rchscfa.2015.04.012

Villarreal J (1999) Venado cola blanca: manejo y aprovechamiento cinegético, Segunda edición. Unión Ganadera Regional de Nuevo León, Nuevo Léon

Villarreal-Espino OA (2000) El aprovechamiento sustentable del venado cola blanca mexicano (*Odocoileus virginianus mexicanus*): Una alternativa para el uso del suelo en la región de la Mixteca poblana. In: Memorias del VII Simposio sobre venados de México, Universidad Nacional Autónoma de México, Asociación Nacional de Ganaderos Diversificados Criadores de Fauna (ANGADI), pp 127–152

Villarreal-Espino OA, Guevara R, Cortes I et al (2007) Alimentación del venado cola blanca mexicano (*Odocoileus virginianus mexicanus*) en el sur de Puebla, México. Deer Spec Group News 22:21–24

Villarreal-Espino OA, Plata-Pérez FX, Camacho-Ronquillo JC et al (2011) El Venado Cola Blanca en la mixteca poblana. Therya 2:103–110. https://doi.org/10.12933/Therya-11-25

Villarreal-Espino OA, Ortega-Aguilar B, Hernández-Hernández JE et al (2013) Condition of quality the white tail deer habitat, in Río Balsas depression-México. Int J Plant Anim Environ Sci 3:210–221

Weber M (2008) Un especialista, un generalista y un oportunista: uso de tipos de vegetación por tres especies de venados en Calakmul, Campeche. In: Lorenzo C, Espinoza-Medinilla E, Ortega J (eds) Avances en el estudio de los mamíferos de México II. Asociación Mexicana de Mastozooología A. C., El Colegio de la Frontera Sur, San Cristóbal de las Casas, Chiapas, pp 579–592

Weber M, González S (2003) Latin American deer diversity and conservation: a review of status and distribution. Ecosci 10:443–454

Yáñez-Arenas C, Mandujano S (2015) Evaluating the relationship between white-tailed deer and environmental conditions using spatially autocorrelated data in tropical dry forests of central Mexico. Trop Conserv Sci 8:1126–1139. https://doi.org/10.1177/194008291500800418

Chapter 16
Brocket Deer

Sonia Gallina-Tessaro, Luz A. Pérez-Solano, Rafael Reyna-Hurtado, and Luis Arturo Escobedo-Morales

Abstract Brocket deer (*Mazama*) are one of the least known members of the family Cervidae in the world, but important advances have been reached recently. We present here information about its geographic distribution and phylogenetic relationships. Studies based on molecular data have noted the polyphyly of genus, but there is not consensus about the specific position among *Mazama* species, even taxonomic status for some of their species still needs to be defined. Biological traits of the species within this genus, like evasive and mostly nocturnal, and low population densities as well as accelerated habitat loss rates, put most of them in high extinction risk. A synthesis about their reproduction, activity patterns, food habits, habitat requirements, and main threats to the persistence of brocket deer species face is presented here. More research efforts in systematics, population genetics, and ecology for these species are required to assess the human impact on their populations and effectively revert the negative trend that some populations of these species are showing through its distributional range.

16.1 Distribution and Species

Brocket deer (*Mazama*) are one of the least known members of the family Cervidae in the world (Weber and González 2003; Grubb 2005). They are endemic to the Neotropics and it distribution includes most of the humid and subhumid tropical

S. Gallina-Tessaro (✉) · L. A. Pérez-Solano
Red de Biología y Conservación de Vertebrados, Instituto de Ecología, A.C., Xalapa-Enríquez, Veracruz, Mexico
e-mail: sonia.gallina@inecol.mx

R. Reyna-Hurtado
Conservación de la Biodiversidad, El Colegio de la Frontera Sur, Lerma, Campeche, Mexico
e-mail: rreyna@ecosur.mx

L. A. Escobedo-Morales
Universidad Nacional Autónoma de México, Mexico City, Mexico
e-mail: luis.escobedo@st.ib.unam.mx

S. Gallina-Tessaro (ed.), *Ecology and Conservation of Tropical Ungulates in Latin America*, https://doi.org/10.1007/978-3-030-28868-6_16

forest from NE Mexico through Central and South America (Grubb 2005; Merino and Rossi 2010; Groves and Grubb 2011), occupying a wide variety of environmental and elevation conditions, from lowland and montane tropical and subtropical forests, cloud forest, open lowland grassland, and paramos (Emmons and Feer 1997; Eisenberg 1998; Eisenberg and Redford 1999). There is also a great variation in the size of their geographic range, with some of them inhabiting most of tropical South America, such as red brocket deer (*Mazama americana*) and other ones with highly restricted areas, for example the Merida brocket deer (*M. bricenii*) inhabiting only some highlands of Venezuela or the Yucatan brown brocket (*M. pandora*) that lives only in the Yucatán Península in Southern Mexico (Fig. 16.1).

Mazama is also one of the more diverse genera among deer with ten recognized species (Weber and González 2003; Grubb 2005; Groves and Grubb 2011), however this number is under review and their phylogenetic relationships are complex and even controversial. Allen (1915) made one of the earliest and more detailed studies based on an exhaustive examination of cranial and morphological characters, and he recognized 18 species divided on two groups, red and brown brockets. Besides pelage coloration, length of the nose and body size, Allen (1915) noted the difficulty to find diagnostic characters to clearly differentiate among all these forms. Other authors have reviewed the validity of those taxa using biogeographic, morphological, cytogenetic or genetic criteria adding new species or renaming some species (e.g. Cabrera 1960; Jorge and Benirschke 1977; Czernay 1987; Duarte 1996; Medellín et al. 1998; Gutiérrez et al. 2015).

A completely resolved phylogeny of Neotropical deer is not available until now, but important advances have been reached recently. Studies based on molecular data

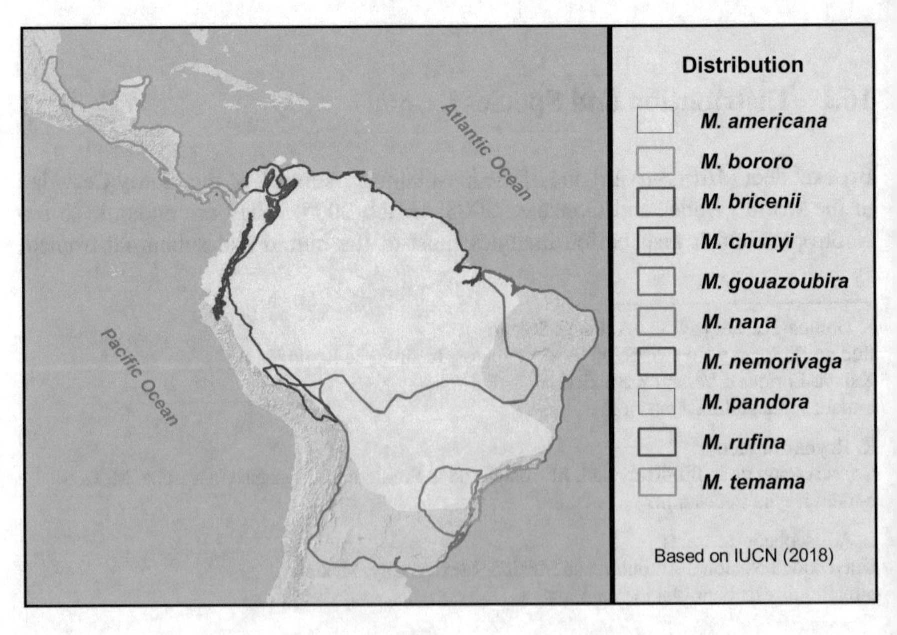

Fig. 16.1 Distribution range for *Mazama* species

have noted the polyphyly of genus *Mazama* (Smith et al. 1986; Gilbert et al. 2006; Duarte et al. 2008; Hassanin et al. 2012; Escobedo-Morales et al. 2016; Gutiérrez et al. 2017). Eventhough there is no consensus in the specific position of all members, these studies agree in pointing a close relationship of gray brocket deer *M. gouazoubira* and *M. nemorivaga* with South American genera *Blastocerus*, *Hippocamelus*, *Ozotoceros* and *Pudu;* named as the "gray clade" (Duarte et al. 2008). Recently Heckeberg et al. (2016) and Gutiérrez et al. (2017) include *M. chunyi* within the gray clade. On the other side, *M. americana*, *M. bororo*, *M. bricenii*, *M. nana*, *M. pandora* and *M. temama* are grouped altogether with genus *Odocoileus*, comprising the "red clade" (Duarte et al. 2008; Escobedo-Morales et al. 2016; Gutiérrez et al. 2017). Recently, van Roosmalen (2015) described a new species, *M. tienhoveni*, from the lower Rio Aripauanã Basin, Central Amazonia based on a reduced specimen's sample and a small fragment (less than 300 bp) of the mitochondrial gene cytochrome b. His phylogenetic tree presents low nodes support and should be taken with caution until new evidence is available. A more detailed discussion about the taxonomic implications of the relationships between *Mazama* species with the other Neotropical deer can be consulted in Escobedo-Morales et al. (2016), Heckeberg et al. (2016), and Gutiérrez et al. (2017), that propose to exclusively use the name *Mazama* for the red clade and *Nanelaphus* for gray clade brocket deer, as well as the pertinence to include Yucatan brocket deer *M. pandora* (Fig. 16.2) into the genus *Odocoileus*.

Fig. 16.2 *Mazama pandora* (Photo: Rosa María González Marín)

These recent contributions have enhanced the knowledge about the interspecific relationships of Neotropical deer; however, they have not documented the level of intraspecific differentiation within these species, mainly because of samples scarcity since data are available for only a few *Mazama* species. The presence of cryptic species is another unsolved question mainly in those with broad distribution. Duarte et al. (2008) found two lineages for *M. americana* that diverged approximately 2 mya from an ancestral form, and probably low ecological plasticity promoted isolated populations restricted to wet tropical forests. Abril et al. (2010) confirmed the existence of these two lineages from west and east Amazonia respectively and proposed that they should be considered as separated species. Figueiredo (2014) reviewed inter and intraspecific genetic variation in *M. gouazoubira* and *M. nemorivaga* using mitochondrial and nuclear genes and found a clear phylogeographic structure for both of them. For Amazonian *M. nemorivaga*, she suggested reviewing the taxonomic status of *M. n. rondoni*, and to raise it to species level. Similar conclusions are reached for *M. gouazoubira*, with five lineages, also showing phylogeographic structure related to presence of different biomes (Figueiredo 2014). Other species with extensive geographic, for example Central American red brocket deer *M. temama*, distributed from NE Mexico, Central America, through N Colombia, with three valid subspecies (Grubb 2005; Bello-Gutiérrez et al. 2010) probably presents similar phylogeographic patterns or even being a species complex. On the other size, taxonomic status for sister species with allopatric distribution has been questioned. This is the case for Andean *M. bricenni* and *M. rufina*, occurring in Cordillera de Mérida (Venezuela) and Cordillera Central (Colombia), respectively. Gutiérrez et al. (2015) included specimens of both species and did not found significant differences in cranial morphology or high genetic distance to consider them as independent lineages. They suggested that *M. bricenii* should be considered as a synonym of *M. rufina*.

The use of complementary approaches such as ecological niche modelling (Peterson et al. 2011), traditional and geometric morphometrics (González et al. 2018), cytogenetics and reproductive characters (Cursino et al. 2014; Carranza et al. 2017), new sequencing techniques (Gutiérrez et al. 2017) would provide more evidence allowing elucidate evolutionary patterns and its causes for this fascinating Neotropical deer.

16.2 Reproduction

The reproductive parameters of *Mazama* spp. in the wild are almost unknown. Most of the information about their breeding behaviour are based in opportunistic observations (Fig. 16.3) that suggest that in general, adults are solitaries that meet only during courtship and mating (Hurtado-Gonzales and Bodmer 2006). There are also reports that their breeding occurs along the year but have birthing peaks (Bisbal 1994).

Fig. 16.3 Male and female of *Mazama pandora* In the Calakmul Biosphere Reserve, México. (Photo: Rafael Reyna-Hurtado)

Apparently, variation in food availability is the main factor responsible for the seasonal reproductive pattern (Goodman 1999).

The red brocket deer (*Mazama americana*) is a medium-sized artiodactyl, weighing between 20 and 30 kg at maturity. In the Amazon region, the female red brocket deer is considered to be nonseasonal polyestric, apparently breeding year-round (Hurtado-Gonzales and Bodmer 2006) but exhibits clear peaks of conceptions and births. This species has a mean gestation length of 210 days (Muller and Duarte 1992) and litter size is usually one, but occasionally twins may occur (Branan and Marchinton 1987; Redford and Eisenberg 1992; Bisbal 1994; Hurtado-Gonzales and Bodmer 2006).

On the other hand, in the Northeastern Peruvian Amazon, conceptions occurred year-round in the red brocket deer but there were peaks in the rate of conception. Estimated yearly reproductive production was 0.76–0.82 young per adult female. In the Amazon region, the lack of large seasonal variation in the availability of resources in the tropical environment could result in an enough food supply for deer to sustain a pregnancy regardless of the month of the year (Mayor et al. 2011).

Fig. 16.4 Female of *Mazama pandora* with fawn in the Calakmul Biosphere Reserve, México (Photo: Rafael Reyna-Hurtado)

16.3 Behaviour

Brocket deer are small, shy animals that live in well conserved forest rarely seen in perturbed areas (Reyna-Hurtado and Tanner 2005) (Fig. 16.4). These features make them difficult to study, so there are few studies that exist regarding their behaviour.

In five sympatric brocket deer species throughout Brazil, it has been suggested that their circadian activity (diurnal or nocturnal activity) are differentiated by the clade to which they belong, rather than any other ecological influence. These species belong to two different clades (the red clade which include *M. americana*, *M. nana* and *M. bororo* and the gray clade by *M. gouazoubira* and *M. nemorivaga*) and show strong morphological and life style convergent features, but currently inhabit different habitats (Leite de Oliveira et al. 2016).

The red clade species displayed a more nocturnal pattern, with *M. americana* showing some activity in the early morning in the Amazon Forest. In contrast, the grey clade species presented a concentration of activity during daylight, with *M. nemorivaga* activity occurring more clearly confined to late morning and *M. gouazoubira* more active in the early morning and late afternoon. The high overlap between *M. americana* and *M. nana* in their activity peaks suggests that another niche axis (e.g. diet, fine-scale habitat use), instead of the time axis, allows them to coexist. The opposite situation was observed between *M. americana* and

M. nemorivaga in the Amazon Forest, where little overlap occurred (Leite de Oliveira et al. 2016).

Red brocket deer are mainly nocturnal with a bimodal pattern of activity: 18.00–22:00 h and 01.00–05:00 h but they were diurnal in the best-protected areas. Dwarf brockets also show a bimodal pattern, but during the morning 06:00–11:00 h and evening and early night 18.00–20:00 h (Leite de Oliveira et al. 2016).

16.4 Diet

Weber (2005) in a detailed study in feeding habits found that the gray brocket deer (*M. pandora*) was a generalist in terms of the species that feed on while the red brocket deer (*M. temama*) was a specialist with more fruits consumed than the gray brocket deer and the white-tailed deer. This author found that gray brocket deer switch from frugivorous to browsers during the year while red brocket deer specialized in fruits all year around (Weber 2014).

Serbent et al. (2011) studied the gray brocket deer (*M. gouazoubira*) diet and estimated the food availability in the arid Chaco, of Cordoba Province, Argentina. Thirty-three plant species were identified in the faecal samples. The diet samples analyzed were woody plants (67.7%), succulents (11.2%), and herbs (6.7%), and grass consumption was very low, reaching only 2.0%. The availability of food items in the area was dominated by woody plants with 62.1% cover, followed by grasses (26.3%), herbs (7.5%), succulents (2.8%) and the pteridophyte, *Selaginella sellowii* (1.3%). The most consumed species were *Castela coccinea* (23.1%), *Maytenus spinosa* (10.2%), *Condalia microphylla* (9.5%), *Schinus fasciculatus* (8.5%) and *Ximenia americana* (7.4%), which represent more than 50% of the plants ingested. The succulents corresponded to species in the Cactaceae and Bromeliaceae families. *M. gouazoubira*'s ability to feed in a selective manner (on woody and succulent plants) as well as in an opportunistic manner (from herbs), and to adjust to variations in resource availability, can explain its extensive geographic distribution. In this study, fruits were the second most consumed item, which is consistent with the fruting pattern of the woody plants consumed (Varela 2003), with a peak in the spring season. The selection of succulent plants and cacti detected in this study is possibly related to the deer's water requirements (Stalling 1984; Sombra and Mangione 2005), which seems consistent, as the study was conducted during the dry season when water was limited.

In the Calakmul Biosphere Reserve *M. temama* was reported as forest specialist species because it consumes at least 20 species of forest plants. Its diet was composed by 80% of fruits and the rest were leaves and some flowers (Weber 2005). The diet of *M. temama* was also studied in a cloud forest by direct observation and microhistologic methods and it was found that 48 species were consumed by the species and that shrubs and herbs were the preferred ones while grasses and acorns were least preferred (Villarreal-Espino-Barros et al. 2008).

16.5 Habitat Requirements

The presence of brocket deer species along their distribution has been considered as an indicator of well-preserved tropical forests (Eisenberg 1998). The deer species of the genus *Mazama* are adapted morphologically for life in forest habitats and in a variety of vegetation types throughout their large Neotropical distribution, extending from southern Mexico to central Argentina, and including the whole of Brazil (Duarte 1996; Eisenberg and Redford 1999; Weber and González 2003) (Table 16.1).

All the *Mazama* species occur in areas and habitats with different characteristics, even those that are distributed in the same region. *M. americana* is widely distributed through most forests of South America from Colombia to Northern Argentina (Varela et al. 2010). *M. nana* is associated with the mixed ombrophilous forest in southern Brazil and part of Paraguay and Argentina (Abril et al. 2010), while *M. bororo* is found in a restricted area of the southern Brazilian coast where dense ombrophilous forest occurs (Vogliotti and Duarte 2010). *M. gouazoubira* is associated with the forested areas of the Brazilian Cerrado but can also occur in more open lands such as the Pantanal, while its clade mate *M. nemorivaga* is distributed throughout the Amazon Forest (Black-Décima et al. 2010; Rossi et al. 2010). All these habitats show differing features: dominant vegetation, types of food, main predators or human disturbances. Although detailed information on these potential differences between habitats is scarce, the effect of the habitat component on activity patterns is to be expected.

The red brocket deer (*M. americana*) and the gray brocket deer (*M. gouazoubira*) are sympatric species and present several ecological similarities in their diet and habitat use, although their interspecific competition is poorly understood. *M. americana* is generally thought to be dependent on mature forest and it has been related with the large trees density (diameter at breast height > 50 cm) and the edge forest distance, while *M. gouazoubira* is more flexible and uses a wider variety of habitats than *M. americana*. Even *M. gouazoubira* has been found in sparse vegetation, including grasslands, and disturbed habitats such as agricultural crops (Andrade-Núñez and Aide 2010; Ferreguetti et al. 2015).

Red brocket deer (*M. temama* Kerr 1792) inhabit two of the most threatened ecosystems in Mexico, the cloud forest and tropical evergreen forest; due to the lack of studies little is known about their current condition. It has been classified as a habitat specialist (Weber 2008), this species could use areas covered with secondary growth vegetation and even crops (Bello 2004; Lira-Torres and Naranjo 2003). In Mexico, it can also use areas covered with croplands such as the seasonal use of coffee agroecosystems, due to the fructification of coffee plants (Lira-Torres and Naranjo 2003; Pérez-Solano et al. 2016), where the surrounding forests presents high-quality habitats (Lira-Torres and Naranjo 2003; Bello 2004). In general, it has been considered that a high-quality habitat for a brocket deer is characterized by the availability of water sources, a high vegetation cover, and low human hunting pressure (Weber 2008).

Table 16.1 Species characteristics of the genus *Mazama* (see details in the complete book of Duarte and González 2010)

Mazama species	Common name	Height (cm)	Length (cm)	Weight (kg)	Reproduction (births)	Habitat	Home range (ha)	Behaviour	Conservation status
M. americana	Red brocket deer	58–80	90–145	30–38		Lowland and montane tropical and subtropical forests including mature, secondary and gallery forest and savannas	52.2–66.7	Mostly during dusk and night (18:00–06:00 h)	In IUCN red list it is classified as data deficient, are the main subsistence hunting deer in neotropical forests
M. nana	Brazilian dwarf brocket deer	45		15	September and February	Mixed ombrophilus Forest and ecotones adjacent to semideciduous forest and Cerrado			It appears as deficient data but it is the most threatened deer of Brazil and possibly of the neotropical region, in Argentina is considered vulnerable
M. temama	Central american red brocket deer	60–70	90–120	12–32	Throughout the year	Perennial and subperennial forests, cloud forests, low-dry forests andlow flooded forests			Data deficient

(continued)

Table 16.1 (continued)

Mazama species	Common name	Height (cm)	Length (cm)	Weight (kg)	Reproduction (births)	Habitat	Home range (ha)	Behaviour	Conservation status
M. bororo	Small red brocket deer	50	80–85	25	Throughout the year	Brazilian Atlantic forests	24.8–63.9	Crepuscular and nocturnal between 18:00–22:00 h	Vulnerable for IUCN, but could be considered one of the most threatened cervid of the world for it is endemic to the Brazilian Atlantic forest and the intense destruction of this ecosystem
M. rufina	Dwarf red brocket deer	45	85–90	10–15		Montane Forest, tropical montane cloud forest and paramos		Nocturnal	Vulnerable, but a near threatened species in Colombia and Ecuador
M. bricenii	Merida brocket deer	45–50	85–90	8–13	December	Remnant mountain forest patches, paramos and Polylepsis woods		Nocturnal	Vulnerable for the habitat loss. It is considered an endangered species in Venezuela
M. chunyi	Peruvian dwarf brocket deer	37	60			Lowland and subAndean Amazonian forests to the montane cloud forests		Day and night mostly crepuscular	Vulnerable for habitat destruction

Mazama species	Common name	Height (cm)	Length (cm)	Weight (kg)	Reproduction (births)	Habitat	Home range (ha)	Behaviour	Conservation status
M. gouazoubira	Brown brocket deer	50–65	88–106	11–25	Almost all months in Argentina, and in Bolivian Chaco from August to April	Low secondary woodlands and open grasslands, Brazilian Cerrado in bushy vegetation	25–35 in Bolivian Chaco and 37–348 in Cerrado Brazil	Active the entire day, peaks of activity 05:00–10:00 and 17:00–22:00	Is one of the most hunted (for sport and subsistence) but is considered Least concern for the IUCN IN 2008
M. nemorivaga	Amazonian brown brocket deer	49–50	76–102	14–15.5	January to March and July to October and December	Tropical and subtropical broadleaf moist forests		Throught the entire day	Least concern for IUCN-habitat destruction is the main type of threat
M. pandora	Yucatan brown brocket deer			16–25	Almost all months	Tropical evergreen forests (tall and medium)			Is not in risk in the Mexican list but is considered by IUCN as vulnerable inferred for habitat loss

Fig. 16.5 *Mazama temama* fawn in a cloud forest of the Reserve La Otra Opción, Los Tuxtlas, Mexico (Photo: Alejandro González Gallina)

In the Calakmul region of Mexico, Ramírez (2016) found Yucatan brocket deer (*Mazama pandora*) to be more abundant in dry tropical forest than *M. temama* that was an inhabitant of more humid tropical forest.

In the mountain regions in Mexico, where the cloud forests are the main vegetation type, *M. temama* (Fig. 16.5) relative abundance have a positive linear relationship with the density of tree coverage (0.4 trees/m^2), 60% of canopy cover, 70% of protected cover for fawns and 50% for adults, and shrub richness (> 4 species) (Muñoz-Vazquez and Gallina 2016). This species selects sites with high vegetation cover and intermediate plant species richness, and the use of these sites is associated with the proximity of patches of vegetation in good condition and avoided secondary growth vegetation (Pérez-Solano et al. 2012, 2016). These results suggest that *M. temama* behaves like a forest specialist, as their abundance and welfare depend on food availability, habitat quality in terms of forest cover and the remoteness of settlements, which are generally the best preserved and most inaccessible fragments (Muñoz-Vazquez and Gallina 2016). In these mountains of Mexico, the potential habitat for the species has been dramatically reduced by about 50% due to human activity, mainly deforestation, agriculture and settlement. Moreover, the potential area for distribution that is under protection remains minimal (7.9%) (Pérez-Solano and Mandujano 2013) (Fig. 16.6).

Fig. 16.6 In Mexican mountains the habitat for *Mazama temama* has been dramatically reduced mainly deforestation, agriculture and settlement, and their area under federal protection is minimal (Photo: Pérez-Solano et al. 2012)

16.6 Population Estimates

Population estimates of brocket deer are scarce and when available is a mix of the two species given the almost no possibility of differentiating the tracks of one species of the other. In the Calakmul Biosphere Reserve, Reyna-Hurtado and Tanner (2007) found that brocket deer' signs (the combination of tracks of *M. pandora* and *M. temama*) (Fig. 16.7) were the most abundant of all ungulate species and that track 'relative abundance did not vary between a set of hunted sites and the Calakmul protected area. The same authors also found that tracks of brocket deer were the most abundant tracks in dry forest of the protected area and that dry forest were used more than expected for these species, while in the hunted sites the low-flooded forest was used more than expected (Reyna-Hurtado and Tanner 2005). Using camera traps, Ramírez (2016) found that *M. temama* was more abundant in a community forest than in Calakmul Biosphere Reserve and the opposite occurred for *M. pandora* that was more abundant in the protected area than in the community forest. The vegetation types of both sites can explain these findings. In Calakmul Biosphere Reserve, the dry forests are the favourite habitat of *M. pandora*, while the more humid community forests are preferred by *M. temama* (Bello-Gutiérrez et al. 2010).

Fig. 16.7 *Mazama pandora* (Photo: Rafael Reyna-Hurtado)

Some estimates of density using observations in transects are between 0.90 and 1.5 deer/km^2 for both species of *Mazama* for the Calakmul region (Weber 2005). Information that is highly needed for these two species includes the impact of hunting activities and deforestation in the population of the two species and the home range size and movement patterns for both species would be exciting research with conservation implications.

16.7 Hunting and Other Uses

Brocket deer are highly appreciated by their meat and skin (Escamilla et al. 2000) (Fig. 16.8) and somtimes are hunted because they crop-raid bean, maize and pepper fields (Mendez and Bello 2005). In addition to the meat, people use hooves and antlers as handcraft gifts (Fig. 16.9), and in some regions the blood is used to treat some diseases (González 2003; Santos-Fita et al. 2012), the liver to improve blood content in sick people and the marrow to cure teethache (Ávila-Najera et al. 2011; Santos-Fita et al. 2012). In Mexico, brocket deer are also very important species for subsistence hunters, especially in the south areas of the Yucatan peninsula around the Calakmul Biosphere Reserve (Escamilla et al. 2000; Weber 2000) and the red

Fig. 16.8 Skin's red brocket deer in a Nahuatl indigenous territory, within the Zongolica mountain range, Veracruz, México (Photo: Luz A. Pérez-Solano)

brocket deer have been subjected to sport hunting in the last years in some communities where sport hunting is allowed under the UMA (Units for Wildlife Management and Conservation) scheme (Weber et al. 2006; R. Reyna-Hurtado pers. obs.).

On the other hand, Di-Bitetti et al. (2008) examined the effect of hunting on the abundance and activity patterns of sympatric *M. americana* and *M. nana*, in three areas within the Atlantic Forest of Misiones, Argentina, that differ in protection and hunting pressure. More frequently recorded were red brocket deer than dwarf brocket deer, and higher records found in areas with better protection, increasing with the distance to main accesses used by poachers (used as a proxy to hunting pressure). This study indicates that the response to hunting is species specific and while one brocket deer is negatively affected (*M. americana*) the other (*M. nana*) is unaffected or may even benefit for some level of hunting. Red brockets are heavier (30–50 kg contrasting with the dwarf brockets that are much lighter 10–13 kg) (Weber and González 2003) and hunters would prefer this species.

Fig. 16.9 Coat rack made with hooves of red brocket deer in Zongolica, Veracruz, México (Photo: Luz A. Pérez-Solano)

16.8 Conservation Prospects

Between 2000 and 2012, 2.3 million km^2 of forest was lost globally, and in the tropics, forest loss increased each year (Hansen et al. 2013). Tropical forests in the Neotropics face high rates of habitat loss (FAO 2016), threatening the persistence of all *Mazama* species, and some concerns occur about its conservation status. The International Union for Conservation of Nature and Natural Resource (IUCN 2018) consider six species as Vulnerable (*M. bororo*, *M. bricenni*, *M. chunyi*, *M. nana*, *M. pandora* and *M. rufina*) while the other members of the genus are catalogued as Data deficient (*M. americana* and *M. temama*) or Least Concern (*M. gouazoubira* and *M. nemorivaga*). The Convention on International Trade in Endangered Species of Wild Fauna and Flora (CITES 2018) includes Guatemalan populations of *M. temama ceracina* into its Appendix III.

We need to increase the efforts on research of the ecological aspects of these species and the human impact on their populations before it is too late to revert the trend that some populations of these species are showing. Conserving these interesting species of tropical deer, we will assure that they will continue playing the important ecological roles they play such as seed dispersers, herbivores, seed predators and also prey of carnivores and in that way, they will benefit tropical ecosystems in the same way they have done for thousands of years.

References

Abril VV, Carnelossi EAG, González S, Duarte JMB (2010) Elucidating the evolution of the red brocket deer *Mazama americana* complex (Artiodactyla; Cervidae). Cytogenet Genome Res 128:177–187. https://doi.org/10.1159/000298819

Allen JA (1915) Notes on American deer of the genus Mazama. Bull Am Mus Nat Hist 34:521–553

Andrade-Núñez MJ, Aide TM (2010) Effects of habitat and landscape characteristics on medium and large mammal species richness and composition in northern Uruguay. Fortschr Zool 27:909–917. https://doi.org/10.1590/S1984-46702010000600012

Ávila-Najera DM, Rosas-Rosas OC, Tarango-Arámbula LA, Martínez-Montoya JF, Santoyo-Brito E (2011) Conocimiento, uso y valor cultural de seis presas del jaguar (Panthera onca) y su relación con éste, en San Nicolás de los Montes, San Luis Potosí, Mexico. Revista Mexicana de Biodiversidad 82:1020–1028

Bello J (2004) Mamíferos del estado de Tabasco: Diversidad y especies amenazadas. Revista Kuxulka'b 12:1–12

Bello-Gutiérrez J, Reyna-Hurtado R, Wilham J (2010) Central American red brocket deer *Mazama temama* (Kerr, 1992). In: Duarte JMB, González S (eds) Neotropical cervidology. Biology and medicine of Latin American deer. Funep/IUCN, Jaboticabal/Gland, pp 166–171

Bisbal FJ (1994) Biología poblacional del venado matacán (Mazama spp.) (Artiodactyla: Cervidae) en Venezuela. Rev Biol Trop 42:305–313

Black-Décima P, Rossi RV, Vogliotti A et al (2010) Brown brocket deer *Mazama gouazoubira* (Fischer 1814). In: Neotropical cervidology. Biology and medicine of Latin American deer. Funep/IUCN, Jaboticabal/Gland, pp 190–201

Branan WV, Marchinton RL (1987) Reproductive ecology of white-tailed and red brocket deer in Suriname. In: Wemmer C (ed) Biology and management of the Cervidae. Smithsonian Institution, Washington DC, pp 344–351

Cabrera A (1960) Catálogo de los mamíferos de América del Sur. Revista Museo Argentino Bernardino Rivadavia 4:309–732

Carranza J, Roldán M, Peronia EFC, Duarte JMB (2017) Weak premating isolation between two parapatric brocket deer species. Mamm Biol 87:17–26

CITES (2018) Convención Sobre El Comercio Internacional De Especies Amenazadas De Fauna Y Flora Silvestres. Apéndices I, II y III. UNEP

Cursino MS, Salviano MB, Abril VV, Zanetti EDS, Duarte JMB (2014) The role of chromosome variation in the speciation of the red brocket deer complex: the study of reproductive isolation in females. BMC Evol Biol 14:40. https://doi.org/10.1186/1471-2148-14-40

Czernay S (1987) Spiesshirsche und Pudus. Die Neue Brehm Bucherei 581:1–84

Di-Bitetti MS, Paviolo A, Ferrari CA, De Angelo C, Di Blanco Y (2008) Differential responses to hunting in two sympatric species of brocket deer (*Mazama americana* and *M. nana*). Biotropica 40:636–645

Duarte JMB (1996) Guia de identificação de cervídeos brasileiros. Funep, Jaboticabal

Duarte JMB, González S (eds) (2010) Neotropical cervidology. Biology and medicine of Latin American deer. Funep/IUCN, Jaboticabal/Gland

Duarte JMB, González S, Maldonado JE (2008) The surprising evolutionary history of South American deer. Mol Phylogenet Evol 49:17–22. https://doi.org/10.1016/j.ympev.2008.07.009

Eisenberg JF (1998) Mammals of the neotropics, The Northern Neotropics: Panama, Colombia, Venezuela, Guyana, Suriname, French Guiana, vol 1. The University of Chicago Press, Chicago

Eisenberg JF, Redford KH (1999) Mammals of the neotropics, The central neotropics, vol 3. University of Chicago Press, Chicago, IL

Emmons LH, Feer F (1997) Neotropical rain forest mammals, a field guide, 2nd edn. The University of Chicago Press, Chicago

Escamilla A, Sanvicente M, Sosa M, Galindo-Leal C (2000) Habitat mosaic, wildlife availability, and hunting in the tropical forest of Calakmul, Mexico. Conserv Biol 14:1592–1601

Escobedo-Morales LA, Mandujano S, Eguiarte LE, Rodríguez-Rodríguez MA, Maldonado JE (2016) First phylogenetic analysis of Mesoamerican brocket deer *Mazama pandora* and *Mazama temama* (Cetartiodactyla: Cervidae) based on mitochondrial sequences: Impli- cations on Neotropical deer evolution. Mamm Biol 81:303–313. https://doi.org/10.1016/j.mambio.2016.02.003

FAO (2016) El Estado Mundial de la Agricultura y la alimentación: cambio climático, agricultura y seguridad alimentaria. Organización de las Naciones Unidas para la Alimentación y la Agricultura, Roma

Ferreguetti ÁC, Tomás WM, Bergallo HG (2015) Density, occupancy, and activity pattern of two sympatric deer (*Mazama*) in the Atlantic Forest, Brazil. J Mammal 96:1245–1254. https://doi.org/10.1093/jmammal/gyv132. Published online August 11, 2015

Figueiredo MG (2014). Filogenia e taxonomia dos veados cinza (Mazama gouazoubira e M. nemorivaga). Doctoral thesis, Universidade Estadual Paulista Júlio de Mesquita Filho, Faculdade de Ciências Agrárias e Veterinárias

Gilbert C, Ropiquet A, Hassanin A (2006) Mitochondrial and nuclear phylogenies of Cervidae (Mammalia, Ruminantia): systematics, morphology, and biogeography. Mol Phylogenet Evol 40:101–117. https://doi.org/10.1016/j.ympev.2006.02.017

González JA (2003). Patrones generales de caza y pesca en comunidades nativas y asentamientos de colonos aledaños a la Reserva Comunal Yanesha, Pasco, Perú. R. Polanco-Ochoa (ed). Manejo de fauna silvestre en Amazonía y Latinoamérica. Selección de trabajos V Congreso Internacional. CITES, Fundación Natura, Bogota, Colombia, pp 89-102

González S, Mantellato AMB, Duarte JMB (2018) Craniometrical differentiation of gray brocket deer species from Brazil. Rev Mus Argentino Cienc Nat, ns 20:179–193

Goodman R (1999) Seasonal reproduction, mammals. In: Knobil E, Neill JD (eds) Encyclopedia of reproduction, vol 4. Academic Press, San Diego, CA, pp 341–352

Groves C, Grubb P (2011) Ungulate taxonomy. The Johns Hopkins University Press, Baltimore, MD, p 309

Grubb P (2005) Order artiodactyla, 637-722. In: Wilson DE, Reeder DM (eds) Mammal species of the world. A taxonomic and geographic reference, vol 2, 3rd edn. Johns Hopkins University Press, Baltimore, p 142

Gutiérrez EE, Maldonado JE, Radosavljevic A, Molinari J, Patterson BD, Martínez-C JM, Rutter AR, Hawkins MTR, Garcia FJ, Helgen KM (2015) The taxonomic status of *Mazama bricenii* and the significance of the Táchira depression for mammalian endemism in the cordillera de Mérida, Venezuela. Plos one 10:e0129113. https://doi.org/10.1371/journal.pone.0129113

Gutiérrez EE, Helgen KM, McDonough MM, Bauer F, Hawkins MTR, Escobedo-Morales LA, Patterson BD, Maldonado JE (2017) A gene-tree test of the traditional taxonomy of American deer: the importance of voucher specimens, geographic data, and dense sampling. ZooKeys 697:87–131. https://doi.org/10.3897/zookeys.697.15124

Hansen MC, Potapov PV, Moore R, Hancher M, Turubanova SA, Tyukavina A, Thau D, Stehman SV, Goetz SJ, Loveland TR, Kommareddy A, Egorov A, Hini L, Justice CO, Townshend JRG (2013) High resolution global maps of 21st-century forest cover change. Science 342:850–853

Hassanin A, Delsuc F, Ropiquet A, Hammer C, van Vuuren BJ, Maththee C, Ruiz-Garcia M, Catzeflis F, Areskoug V, Nguyen TT, Couloux A (2012) Pattern and timing of diversification of Cetartiodactyla (Mammalia, Laurasiatheria), as revealed by a comprehensive analysis of mitochondrial genomes. C R Biol 335:32–50

Heckeberg NS, Erpenbeck D, Wörheide G, Rössner GE (2016) Systematic relationships of five newly sequenced cervid species. PeerJ 4:e2307. https://doi.org/10.7717/peerj.2307

Hurtado-Gonzales JL, Bodmer RE (2006) Reproductive biology of female Amazonian brocket deer in northeastern Peru. Eur J Wildl Res 52:171–177

International Union for Conservation of Nature (IUCN) (2018) The IUCN red list of threatened species. Version 2018-2. http://www.iucnredlist.org/. Accessed 15 Dec 2018

Jorge W, Benirschke K (1977) Centromeric heterochromatin and G-banding of the red brocket deer, *Mazama americana temama* (Cervoidea, Artiodactyla) with a probable non-Robertsonian translocation. Cytologia 42:711–721

Leite de Oliveira M, de Faria Peres PH, Vogliotti A et al (2016) Phylogenetic signal in the circadian rhythm of morphologically convergent species of Neotropical deer. Mamm Biol 81:281–289

Lira-Torres I, Naranjo E (2003) Abundancia, preferencias de hábitat e impacto del ecoturismo sobre el puma y dos de sus presas en la reserva de la biósfera El Triunfo, Chiapas, México. Revista mexicana de Mastozoología 7:20–39

Mayor P, Bodmer RE, Lopez-Bejar M, López-Plana C (2011) Reproductive biology of the wild red brocket deer (Mazama americana) female in Peruvian Amazon. Anim Reprod Sci 128:123–128

Medellín RA, Gardner AL, Aranda JM (1998) The taxonomic status of the Yucatan brown brocket, *Mazama pandora* (Mammalia, Cervidae). Proc Biol Soc Washington 111:1–14

Mendez SM, Bello J (2005) Daños a los cultivos de frijol, por mamíferos silvestres, en el ejido Agua Blanca, Tacotalpa, Tabasco, México. In: T. Ramon (ed) Semana de divulgación y video científico, UJAT 2005. UJAT, Villahermosa, Tabasco

Merino ML, Rossi RV (2010) Origin, systematics and morphological radiation, 2–11. In: Neotropical cervidology: biology and medicine of Latin American deer. Jaboticabal

Muller E, Duarte JMB (1992) Utilizacao da citología vaginal efolativa para monitoracao do ciclo estral em veadocatingueiro. In: XVI Congresso da sociedade de Zoologicos do Brasil Anais

Muñoz-Vazquez B, Gallina S (2016) Influence of habitat fragmentation on abundance of Mazama temama at different scales in the cloud forest. Therya 7:77–87

Pérez-Solano L, Mandujano S (2013) Distribution and loss of potential habitat of red brocket deer (*Mazama temama*) in Sierra Madre Oriental, México. Deer Specialist Group News, pp 12–18

Pérez-Solano L, Mandujano S, Contreras-Moreno F, Salazar J (2012) Primeros registros del temazate rojo Mazama temama (Kerr 1792) en áreas aledañas a la Reserva de la Biosfera de Tehuacán-Cuicatlán, México. Revista Mexicana de Biodiversidad 83:875–878

Pérez-Solano L, Hidalgo-Mihart MG, Mandujano S (2016) Preliminary study of habitat preferences of red brocket deer (*Mazama temama*) in a mountainous región of Central Mexico. Therya 7:197–203

Peterson AT, Soberón J, Pearson RG et al (2011) Ecological niches and geographic distributions. Princeton University Press, Princeton

Ramírez L (2016) Abundancia relativa y patrones de actividad por venados en dos sitios de la región de Calakmul, Campeche, México. Tesis de Licenciatura, Universidad Autónoma de Campeche, San Francisco de Campeche, Campeche, México

Redford KH, Eisenberg JF (1992) Mammals of the Neotropics: Chile, Argentina, Uruguay, Paraguay. The Southern Cone. University of Chicago Press, Chicago

Reyna-Hurtado R, Tanner GW (2005) Habitat preferences of ungulates in hunted and nonhunted areas in the Calakmul Forest, Campeche, Mexico. Biotropica 37:676–685

Reyna-Hurtado R, Tanner GW (2007) Ungulate relative abundance in hunted and non-hunted sites in Calakmul Forest (Southern Mexico). Biodivers Conserv 16:743–756

van Roosmalen MGM (2015) Hotspot of new megafauna found in the Central Amazon (Brazil): the lower Rio Aripuanã Basin. Biodiversity Journal 6:219–244

Rossi RV, Bodmer R, Barbanti JM, Trovati RG (2010) Amazonian brown brocket deer Mazama nemorivaga (Cuvier 1817). In: Neotropical cervidology: Biology and Medicine of Latin American Deer. FUNEP/IUCN, Jaboticabal/Gland, pp 202–210

Santos-Fita D, Naranjo EJ, Rangel-Salazar JL (2012) Wildlife uses and hunting patterns in rural communities of the Yucatan Peninsula, Mexico. J Ethnobiol Ethnomed 8:38

Serbent MP, Periago ME, Leynaud GC (2011) *Mazama gouazoubira* (Cervidae) diet during the dry season in the arid Chaco of Córdoba (Argentina). J Arid Environ 75:87–90

Smith MH, Branan WV, Marchinton RL et al (1986) Genetic and morphologic comparisons of red brocket, brown brocket, and white-tailed deer. J Mammal 67:103–111

Sombra MS, Mangione AM (2005) Obsessed with grasses? The case of mara Dolichotis patagonum (Caviidae: Rodentia). Revista Chilena de Historia Natural 78:401e408

Stalling JR (1984) Notes on feeding habits of Mazama gouazoubira in the Chaco Boreal of Paraguay. Biotropica 16:155e157

Varela RO (2003) Frugivoría y dispersión de semillas por 13 especies de vertebrados del Chaco salteño, Argentina. Tesis doctoral, Facultad de CienciasExactas, Físicas y Naturales, Universidad Nacional de Córdoba

Varela DM, Trovati RG, Guzmán KR, Rossi RV, Duarte JMB (2010) In: Duarte JMB, Gonzalez S (eds) Red brocket deer Mazama americana (Erxleben 1777). Neotropical cervidology. Funep/IUCN, Jaboticabal, pp 151–159

Villarreal-Espino-Barros OA, Campos-Armendia LE, Castillo-Martínez TA et al (2008) Composición botánica de la dieta del venado temazate rojo (Mazama temama), en la sierra nororiental del estado de Puebla. Universidad y ciencia 24:183–188

Vogliotti A, Duarte JMB (2010) Small red brocket deer–Mazama bororo. In: Duarte JMB, Gonzalez S (eds) Neotropical cervidology, biology and medicine of Latin American deer. Funep/IUCN, Jaboticabal/Gland, pp 218–227

Weber M (2000) Effects of hunting on tropical deer populations in Southeastern México. M.Sc. Thesis, Royal Veterinary College, University of London, London, UK, 80 pp

Weber M (2005) Ecology and conservation of sympatric tropical deer populations in the Greater Calakmul Region, Mexico. School of Biological and Biomedical Sciences PhD. Sc. dissertation, University of Durham, Durham, UK, 245 pp

Weber M (2008) Un especialista, un generalista y un oportunista: uso de tipos de vegetación por tres especies de venados en Calakmul, Campeche. In: Lorenzo C, Espinoza E, Ortega J (eds) Avances en el Estudio de los Mamíferos de México, vol 2. Asociación Mexicana de Mastozoología, A. C., México, D. F.

Weber M (2014) Temazates y venados cola blanca tropicales. In: Valdéz R, Ortega-S. JA (eds) Ecología y manejo de fauna silvestre en México. Colegio de Posgraduados, Texcoco, México, pp 421–452

Weber M, González S (2003) Latin American deer diversity and conservation: a review of status and distribution. Écoscience 104:443–454

Weber M, García-Marmolejo G, Reyna-Hurtado R (2006) The tragedy of the commons Mexican style: a critique to the Mexican UMAs concept as applied to wildlife management and use in South-Eastern Mexico. Wildl Soc Bull 34:1480–1488

Chapter 17
Recent Studies on White-Lipped Peccary and Collared Peccary in the Neotropics

Salvador Mandujano and Rafael Reyna-Hurtado

Abstract In this chapter we synthesize the published information of almost 100 papers of scientific journals, books and other sources on the collared peccary (*Pecari tajacu*) and the white-lipped peccary (*Tayassu pecari*) in the Neotropical region. Studies in the Neotropical region include among the different techniques: the direct observation, the counting of tracks and the use the of camera-traps. Studies of peccaries include diverse topics as population estimation, group size, modeling distribution and abundance, social behavior, home range and habitat use, diet and foraging patterns, predation, competitive interactions, genetics, physiology, parasites, ecosystem engineering, subsistence hunting and conservation. However, the continuous increase in fragmentation and hunting pressure observed in all Neotropical forests make it urgent to monitor populations of peccary species all over its range.

17.1 Introduction

The present chapter details published information of almost 100 papers of scientific journals, books and other sources on the collared peccary (*Pecari tajacu*) and the white-lipped peccary (*Tayassu pecari*). To obtain the information Scopus and Google-Academic databases were consulted using as keywords the scientific name and common name of the two species. The searches were restricted to the period between 2000 and 2018 with the aim to compile the most recent information. Despite this intensive searching is possible that some publications have been ignored

S. Mandujano (✉)
Red de Biología y Conservación de Vertebrados, Instituto de Ecología A.C., Veracruz, México
e-mail: salvador.mandujano@inecol.mx

R. Reyna-Hurtado
Departamento de Conservación de la Biodiversidad, El Colegio de la Frontera Sur (ECOSUR), Unidad Campeche, Campeche, México
e-mail: rreyna@ecosur.mx

S. Gallina-Tessaro (ed.), *Ecology and Conservation of Tropical Ungulates in Latin America*, https://doi.org/10.1007/978-3-030-28868-6_17

involuntarily. In addition, this chapter focus only in the Neotropical region, from Southern Mexico to Peru, excluding the Amazonian area of Brazil and northern Argentina (to consult more details about life history of these two species please see: Reyna-Hurtado et al. 2014; Reyna-Hurtado et al. 2018; Beck et al. 2018).

17.2 Geographical Distribution

Peccaries are restricted to the New World. Of the three recognized species, two, the white-lipped peccary (*Tayassu pecari*) and the collared peccary (*Pecari tajacu*), occur in Mexico. The white-lipped peccary occurs from northern Argentina through southeastern Mexico, and the collared peccary occurs from northern Argentina to the southwestern USA. The third species, the Chacoan peccary (*Catagonus wagneri*), known only from fossil specimens until the 1970s, was documented as an extant species in the dry forest of the Chaco region in Paraguay, Argentina, and Bolivia (Wetzel et al. 1975; Campos-Krauer et al. 2012) (Fig. 17.1). A fourth species, the giant peccary (*Pecari maximus*), from the Amazon region, was proposed by Van Roosmalen et al. (2007); however, it is still undetermined whether it is a subspecies of collared peccary (Gongora et al. 2007).

Despite the size and conspicuity of the peccaries and the existing knowledge of their distribution (Reyna-Hurtado et al. 2018; Beck et al. 2018), there are still new records of these species in areas where they have not been previously reported; for example, the collared peccary has been reported for the Cordoba region of Colombia (Humanez-López et al. 2016), for the Morelos state in Central Mexico (Mason-Romo et al. 2008); and there are some records of a relatively stable population of this species for Guanajuato, also in Central Mexico (Charre-Medellín et al. 2018).

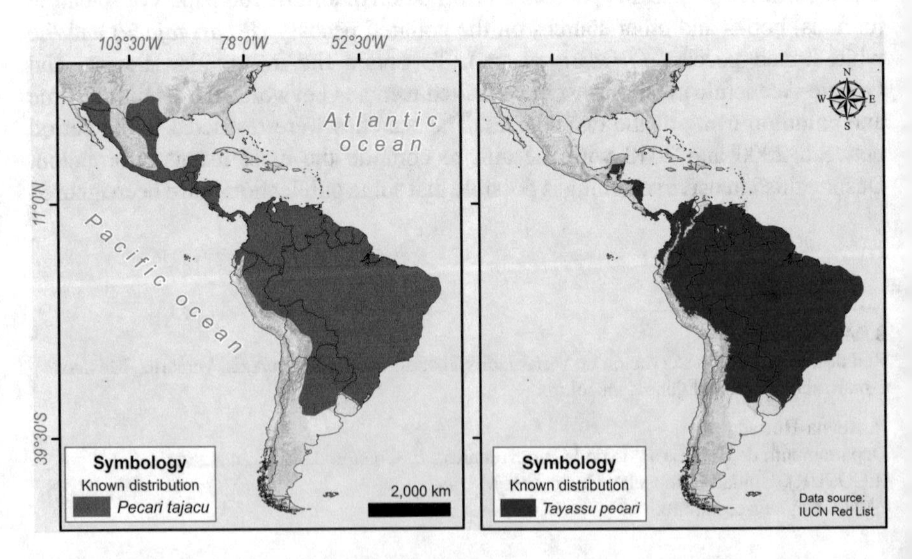

Fig. 17.1 Peccary distribution maps by Adriana Sandoval Comte

Naranjo et al. (2015) evaluated the presence, relative abundance and threats for tapir and white-lipped peccary in ten sites of five states in southern Mexico. The areas verified for white-lipped peccary in Mexico were in Campeche, Chiapas, Oaxaca, Quintana Roo, Veracruz and Yucatán. However, the areas more important for the species due to the large extension and habitat quality are Calakmul-Balamkú-Balam-Kin (Campeche), Montes Azules-Lacantún (Chiapas), Los Chimalapas (Oaxaca), and Sian Ka'an and communal forest lands (ejidos) (Quintana Roo). The main threats were habitat loss and fragmentation, uncontrolled hunting, fires, and diseases. The new records of these species in Mexico are incentives to work for the conservation of peccaries and to increase research efforts about them.

17.3 Studies on Population and Group Size of Peccaries

Studies in the Neotropical region include among the different techniques the direct observation, the counting of tracks and the use of automated cameras (photo-trapping). These studies allow for a glimpse of group size, abundance and occupancy rate of the two species of peccaries. For Oaxaca state in Southern Mexico, Pérez-Irineo and Santos-Moreno (2016) provide information about relative abundance, herd size, activity pattern, and occupancy of ungulates at Los Chimalapas region of Oaxaca State in Southern Mexico. There, white-lipped peccary was found to be highest in relative abundance, while collared peccaries' abundance was lowest. The herd size was smaller compared to other populations of *P. tajacu* and *T. pecari*; in particular, the occupancy probability of *T. pecari* was low. Lira-Torres (2006) evaluated relative abundance, population density and habitat preferences in la Tuza de Monroy, Santiago Jamiltepec. Relative abundance and density of *P. tajacu* was 0.55 tracks/km and 1.98 ind/km^2 while secondary vegetation was the preferred habitat for this species. Cortés-Marcial and Briones-Salas (2014) used camera traps and tracks to estimate the diversity, relative abundance and activity patterns of mammals in a tropical deciduous forest located in the Isthmus of Tehuantepec, Oaxaca. One of the most abundant species was *Pecari tajacu*. Lira-Torres and Briones-Salas (2012) studied mammals populations using camera traps in the Chimalapas region of same state and one of the most abundant species was *P. tajacu*. Similar results were obtained by same authors in the Selva Zoque (Lira-Torres and Briones-Salas 2011).

In Chiapas, Mexico, García-Marmolejo et al. (2015) evaluated the potential role of secondary forests for wildlife conservation, three ungulate species: *Mazama temama*, *Odocoileus virginianus*, and *Pecari tajacu*, at two spatial scales: Local, where three different successional stages of tropical deciduous forest were compared, and Landscape, where available habitats were compared in terms of landscape composition. They found, that *P. tajacu* and *O. virginianus* mainly used early successional stages, while *M. temama* used all successional stages in similar proportions. *P. tajacu* and *O. virginianus* also selected landscapes covered essentially by forests, although they required smaller percentages of forest. All ungulate species avoided landscape fragments covered by pastures. They concluded that landscape is the fundamental scale for ungulate management, and that secondary forests are potentially important landscape elements for ungulate conservation.

Chávez-Hernández et al. (2011) evaluated relative abundance of three ungulates populations in La Sepultura Biosphere Reserve of Chiapas, Mexico and found collared peccary with low abundance in comparison with tapir (*Tapirus bairdii*) or white-tailed deer (*Odocoileus virginianus*). Bolaños and Naranjo (2001) estimated relative abundance and distribution of collared peccary and white-lipped peccary among other ungulates in the Lacantun river basin and on the Montes Azules Biosphere Reserve of Chiapas and found a density of 3.79 ind/km^2 for the white-lipped peccary and of 2.34 ind/km^2 for the collared peccary.

In Yucatan Peninsula, also in Mexico, different authors studied peccary populations alone or together with other ungulates species. For example, in Calakmul Biosphere Reserve, Reyna-Hurtado and Tanner (2007) estimated relative abundance of six species of ungulates through counting track abundance in hunted sites and in the southern area of the protected area. Collared peccary relative abundance remained similar among sites while white-lipped peccary was significantly higher in the protected area than in the hunted sites. In the same area Reyna-Hurtado and Tanner (2005) determined that white-lipped peccary is an inhabitant of medium semi-perennial forest while collared peccary is more a habitat generalist. In the same sites but 10 years later, Briceño-Méndez et al. (2014) evaluated habitat preferences for white-lipped peccary and determined that the species is frequently visitor of medium semi-perennial forest but preferred the Low-flooded forest and its presence is relate with the presence of *Manilkara zapota* and *Brosimum alicastrum* fruits.

González-Marín et al. (2008) estimated abundance and distribution of wild ungulates in several vegetation types of El Eden Ecological Reserve in northern Quintana Roo state of Mexico. The density of the collared peccary was 1.9 ± 0.8 ind/km^2 and 0.2 ± 0.1 groups/km^2. The preferred vegetation type was the secondary vegetation.

In South America, particularly in the Tuparro National Park and Puinawai National Reserve of Colombia, Gómez-Valencia et al. (2016) estimated and compared relative abundance of the two species of peccaries using three methods. Walking transects, counting tracks and recording presence/absence of the species. They found a variation in the collared peccary relative abundance but not for the white-lipped peccary. The flooded forest was positively correlated with white-lipped peccary relative abundance while the distances to the villages with the collared peccary. In the Ucayal Departamento of Peru, Lleellish et al. (2003) evaluated the population size of both species and found a density of 2.5 ind/km^2 and 9.2 ind/km^2 for the collared peccary and the white-lipped peccary respectively.

Wallace and Painter (2000) compared relative abundances of both species of collared and white-lipped peccaries from several survey sites in lowland Amazonian Bolivia. In general, survey sites with high biotic diversity are characterized by relatively low population densities of a number of large mammalian species, in contrast, the forests of the Beni Biosphere Reserve alluvial plain in Bolivia, which are characterized by a relatively low biotic diversity include high densities of several larger mammalian species including the peccaries. Given that in several subsistence hunting sites peccary population densities are apparently stable, the authors suggest that wildlife management initiatives within these extensive tracts of the Beni alluvial plain should form a critical part of a national strategy for the conservation of peccaries.

Desbiez et al. (2010) examined the distribution of densities and metabolic biomass of medium- to large-sized nonvolant mammals in forest, cerrado, and floodplain landscapes, in an area with low anthropogenic influence, in the central area of the Brazilian Pantanal during a prolonged drought. In addition, at the time of the study, frugivores were found to have higher energy consumption than browser/grazers across the three landscapes; most fruits are produced in forested areas stressing their importance.

17.4 Modeling Abundance and Distribution

With the advance of the analytical methods of modeling species distribution and abundance using algorithms developed to that purposes many studies have modeled peccary's population distribution. For example, using a dataset of 36 studies, Mandujano and Naranjo (2010) evaluated variation in ungulate biomass across a rainfall gradient using polynomial models, aiming to: (1) compare Neotropical and Paleotropical dry and wet forests as well as African savannas; and (2) evaluate the usefulness of polynomial models to predict ungulate biomass at Neotropical sites using data from a dry forest of Chamela-Cuixmala Biosphere Reserve (CCBR) and a wet forest of Montes Azules Biosphere Reserve (MABR) in Mexico. Rainfall predicted the expected ungulate biomass in Neotropical ecosystems relative to that of Paleotropical ones under similar precipitation regimes but did not correctly predict the observed ungulate biomass at local level if data outside the Neotropics are included in the model. This was more evident when we compared observed biomass against predicted biomass in the tropical dry forest of CCBR, while some polynomial models successfully predicted the observed biomass for the tropical wet forest of MABR. Factors such as Pleistocene extinctions and the absence of large, native grazers (i.e., Bovidae) that have kept ungulate richness and standing biomass relatively low in Neotropical forests should be accounted for when comparing data sets from different regions. In particular, collared peccary and white-lipped peccary contribute with important proportion of the estimate's ungulate biomass in the Neotropical forests.

For collared peccary there are several studies that have used the ecological niche modeling approach to draw current distribution and predict potential areas where the species may be distributed. For example, in the Biosphere Reserve of Zicuirán-Infiernillo (RBZI) in the state of Michoacan (Mexico) the possible effects of changes in land use on the distribution of this ungulate were evaluated. The potential distribution covered an area of 5762 km^2 (Yañez-Arenas et al. 2012). Considering the small patches sizes to maintain minimum viable populations, the authors suggest that the best strategy for the conservation of these species in this region is a scheme based on source–sink dynamics, taking as a source the RBZI and as a sink the management units (UMA) that are already established on the periphery of the area. Also, in the Tehuacán-Cuicatlán Biosphere Reserve (TCBR), located between the states of Oaxaca and Puebla in Mexico, the potential distribution of collared peccary was estimated to cover 78% of the 4906 km^2 of this reserve (Ortíz-García et al. 2012).

Ortíz-García and Mandujano (2011) developed habitat suitability index to evaluate the habitat quality of collared peccary in the TCBR, Mexico. They considered three habitat variables and one anthropogenic variable in their analyses. They also classified the total area (490,186 ha) in four categories of habitat quality (very low, low, medium and high). They found that 40.3% of the area contains medium to high quality habitat while 59.7% low to very low habitat quality. They found that peccaries prefer high quality habitat and that anthropogenic activities are having a high impact on the habitat. These maps are currently being used to elaborate conservation actions for the species in the reserve.

In other study in the northernmost American tropical forests of eastern Mexico the remaining potential area of all three species is fragmented and has apparently been nearly lost in the lowlands with only less than 14% remaining) (García-Marmolejo et al. 2016). In particular, the distribution model allowed detecting an important location in the western portion of their study area, which may function as a large biological corridor in the Sierra Madre Oriental biogeographic province, a region heavily transformed by land use change.

Recently, Martínez-Gutiérrez et al. (2018) evaluated whether geographic variation in population abundance of the collared peccary is related to its location with respect to the centroid of its ecological niche or to the centroid of its geographic range. They modeled the ecological niche of the species using occurrence and environmental data and created spatial models of distance to the niche centroid (DNC) and to the geographic centroid (DGC); and tested the relationships between population abundance and DNC and between abundance and DGC. Results show a negative relationship between abundance and DNC and a nonsignificant relationship between abundance and DGC. Population abundances are associated with the structure of the ecological niche, especially the maximum abundance expected in an area. Thus, the DNC approach can be useful in obtaining a spatial approximation of potential abundance patterns at biogeographic extents. To achieve a better prediction of realized abundances, it is critical to consider the human influence.

17.5 Social Behavior, Home Range and Habitat Use

Mayor et al. (2007) analyzed reproductive parameters resulting from 74 parturitions in a population of captive collared peccaries in the eastern Amazon, Belém, State of Pará, Brazil. Parturitions were homogeneously distributed throughout the year. The average litter size was 1.85 newborns per parturition. Collared peccary females had a newborn sex ratio of 52.6% females and 47.4% males. The mean age at first parturition was 639 days, although, the earliest first parturition occurred at age of

381 days. Estimated parturition-conception interval was 58 days. Mean farrowing interval was 196 days and mean production was 1.03 litters and 1.86 newborns per year per female. Mortality rate in newborns less than 2 months old was 26.9% of the total newborn population. Most identified causes of death were abandonment by the mother (22.2%) and trauma due to aggression (13.9%). Newborn deaths occurred mainly during the first 2 days of life.

Along the same lines, Nogueira et al. (2010) evaluated the effects of space allowance on the occurrence of social behavior patterns on farmed collared peccary. Enclosure size had a significant effect on agonistic patterns of peccaries during feeding, in that more agonistic behavior was observed in smaller spaces. They also found that shelter usage increased as space decreased. Differing space allowances, however, did not influence the occurrence of positive interactions that were more frequent before compared to during feeding. They recommend that animal welfare can be improved by adopting at least 187 m^2 per peccary. In addition, their study also confirmed the importance of shelter areas in collared peccary husbandry.

Radiotelemetry studies are very important to determine the area that peccaries need to fulfill their daily requirements. For example, in the Calakmul Biosphere Reserve, a semidry tropical forest of Southern Mexico, Reyna-Hurtado et al. (2009) determined the home range of four groups of white-lipped peccary that they followed for 18 months, and found some of the largest home range reported for this species, where some groups need 100 and even 121 km^2 every year to find water and food. They also found that during the dry season their movements were constrained near to the few water sources available during that season. Compositional analyses (Aebischer et al. 1993), on white-lipped peccary use of habitat determined that ponds, medium forest and flooded forest were preferred, and they avoid dry forest types (Reyna-Hurtado et al. 2009).

Moreira-Ramírez et al. (2019) followed three groups of white-lipped peccaries over the Calakmul Biosphere Reserve in southern Mexico and one group outside the reserve and found larger home ranges 140 km^2 in the group living in the hunted site as this group always keeps moving. Authors conclude that this group was behaving in that way due to hunting pressure.

Meyer et al. (2019) conducted the first study of white-lipped peccary movements in the Darien National Park in Panama and found consistent home ranges size for a group that followed for 15 months. The group always stayed in relatively small home range of 59 km^2. The authors conclude that this group finds all they need in this mature well-conserved forest of the Darien National Pak.

Keuroghlian et al. (2004) conducted a long-term radio telemetry study of area use by white-lipped and collared peccaries in a fragmented of semideciduous Atlantic forest in southeastern Brazil. The population of 150 ± 52 white-lipped peccaries was divided among 3–4 subherds. They observed a high frequency of switching of individuals among subherds and documented periodic subherd fusion. Seven to 22 collared peccary herds (mean herd size, 9) persisted in the main forest fragment. Average home range size for the white-lipped peccaries was 1871 ha (90% HM). This was similar to the few reliable estimates available from continuous tropical forests. Despite supposed area restrictions, white-lipped peccaries maintained distinct seasonal ranges with a minimum of overlap. Within seasons, subherds overlapped spatially but were separated temporally.

In order to understand how white-lipped peccaries use forest nested in a matrix of agricultural land, Jorge et al. (2019) analyzed WLP movement (i.e., linear distances moved) at varying temporal intervals (3, 6, 12, 24, 168, and 720 h) and monthly herd ranges (MCP 30%, 50%, 70%, and 90%) in two agricultural regions of Central Brazil. Short and long-term movement did not show variation across months or seasons. Yet, long-term movement and ranges positively correlated with the diversity of available fruits and negatively correlated with the percent of forest cover. Furthermore, the negative relationship between ranges and forest cover was more pronounced during the wet season, with herds in areas with less forest cover having ranges twice as large as those in areas with more forest cover.

In French Guiana, Richard-Hansen et al. (2019) followed two groups of white-lipped peccaries and found stable home ranges of 50–70 km^2 of groups moving in pristine mature tropical forest while doing occasional trips to coastal vegetation strips.

Average home range sizes of two radtracked collared peccary herds were 305 and 123 ha. Home range boundaries and seasonal ranges of collared peccary herds shifted minimally, and herd subgrouping was short-lived. Spatial overlap between the two peccary species varied seasonally with white-lipped herd movements.

Rumiz and Venegas (2006) analyzed GPS data obtained from radiotelemetry collars affixed to white-lipped peccaries at three WCS study sites in Bolivia. Each site represented a different ecoregion, including pre-Andean wet forest, Chiquitano semideciduous forest, and Chaco dry forests/shrublands. Locations from the tall pre-Andean forest were very scarce compared to those of the Chiquitano forest, and Chaco forest and shrubland.

Reyna-Hurtado and Tanner (2005) and Reyna-Hurtado (2007) investigated habitat preference in the community of ungulates in the Calakmul Forest of Southern Mexico through systematically counting the tracks of six species. Brocket deer (*Mazama americana* and *Mazama pandora*) preferred low-dry forest in the nonhunted area and low-flooded forest in the hunted areas. Collared peccary was a generalist species in the nonhunted area, whereas in the hunted areas, it preferred the subperennial forest. White-lipped peccary was found only in subperennial forest in the nonhunted area and favored low-flooded forest in the hunted areas. White-tailed deer preferred low-flooded forest in the hunted areas, while it was a generalist in the nonhunted area. Tapir preferred low-flooded forest in the hunted areas. The most evident habitat difference among hunted and nonhunted areas was a major use of low-flooded forest in the hunted areas for the species.

Tejeda-Cruz et al. (2009) studied the habitat use of five ungulate species in two fragmented sites in the Lacandon Forest, Southern Mexico. They found that Baird's tapir and the white-lipped peccary are rainforest specialists avoiding disturbed habitats in the Lacandon Forest, while the red brocket deer, the collared peccary, and the white-tailed deer show different degrees of success in using habitat mosaic derived from human activities. Given that fragmentation patterns have an effect on habitat use of ungulate species in their study area, more information about the effects of landscape configuration on the distribution and abundance of those species is needed. Bello-Gutiérrez et al. (2004) characterized the micro-habitat of collared peccary as well as other ungulates of the mountains of the southern state of Tabasco, Mexico. They found collared peccary to account with the majority of records of the ungulate community, above the white-tailed deer and the brocket deer. The collared peccary favored sites with tall trees and high diversity of tree species and collared peccary used perturbed sites but with tree coverage higher than 80%.

Martínez-Kú et al. (2008) tested the hypothesis that larger density of large and medium mammals are found near ponds in the Calakmul Biosphere Reserve in Southern Mexico and they found that peccaries as well as other ungulates accounted for 84% of the records found of all mammal species in the periphery of the ponds.

17.6 Diet and Foraging Patterns

López et al. (2006) determined the potential nutritional levels in 25 species of plants, and in earthworms, that constitute part of the diet of white-lipped peccary in Corcovado National Park, Costa Rica. The highest content of fat and energy was found in seeds of the Myristicacea family. In vegetative parts of *Dieffenbachia* spp., they found the highest content of calcium. Nutritious contents differed among plant parts (seeds, fruits, stems and leaves). Fat and energy content were larger in seeds and fruits, whereas the largest content of protein was found in fruits and leaves. Mineral content also differed among plant parts. Calcium, potassium and magnesium were higher in leaves whereas copper and zinc were higher in seeds. They found several species with higher fat and energy content than palms, which can explain the low consumption of palm seeds in Corcovado. It is possible that the regular consumption of stems and leaves of some species is related to their high mineral content. Seasonality of reproduction in Corcovado seems to be related not only to fruit availability but also to the nutritional quality of food. Altrichter et al. (2001) studied fruit availability, diet and habitat use by white-lipped peccary in Corcovado National Park, southwest Costa Rica. The results show that the availability of important fruits for the white-lipped peccary differs between habitats and climatic seasons. Fruit availability was highest in the primary forest than secondary and coastal forest. There was a period of shortage of fruits to ends of the wet season, during which the consumption of not seasonal resources like leaves and shafts increased. The important fruits during this period of shortage were *Ficus* sp. and *Licania operculipetala*. The several types of forest were used according to the fruit availability, and it was a direct relation between the consumption and the fruit availability.

Experimental studies about diet of captive collared peccaries have been also conducted by Montes-Pérez et al. (2012b) in Yucatan, Mexico, where they evaluated the voluntary consumption of four species of plants (*Leucaena leucocephala*, *Guazuma ulmifolia*, *Brosimum alicastrum* and *Pennisetum purpureum*). They found that *G. ulmifolia* was the preferred species followed by *L. leucocephala* and *B. alicastrum*, while *P. purpureum* was the less consumed. Montes-Pérez et al. (2018) evaluated also the effect of some food types on the ovaric activity of collared peccary females that were free to consume these items for 35 days. They found that food mixed with 55% of *Brosimum alicastrum*, 40% of *Pennisetum purpureum* and 5% of molasses was the favorite and all females that feed on it increase twofold their progesterone levels.

Perez-Cortez and Reyna-Hurtado (2008) determined the main components and their seasonal variation of the diet of the two species of peccaries in the Calakmul Biosphere Reserve of Southern Mexico. They found that fruits (57.9%) and leaves (30.1%) were the most frequent elements of the 22 stomachs of collared peccary and that of 37 species of food plants, the most frequent ones were *Brosimum alicastrum*, *Manilkara zapota*, *Piper amalago*, *Zea mays*, and an unidentified one. Nine-stomach content and 16 feces of white-lipped peccaries indicated 41 plant species with fruits accounted for most of the parts consumed with 81.2% and 66.8% in stomach and feces respectively. The main species were *B. alicastrum*, *Chamaedorea* sp., *M. zapota*, *Mimosa* sp., and *P. amalago*. Both species of peccaries shared 32 species of plants and their ecological niche did not overlapped significantly.

Reyna-Hurtado et al. (2012) studied the searching strategies of a highly social mammal, the white-lipped peccary, in Calakmul Biosphere Reserve, Mexico. They attempted to determine what theoretical searching model best explained the movement patterns of groups of white-lipped peccaries, including short-tailed, long-tailed, and scale-free distributions. They found that the only distribution that was well supported by the data was a zero-inflated lognormal distribution; this implies a general pattern of normally short-range intensive searching with occasional long-distance directed movements taking the animals away from previously searched areas. They also found that groups concentrated foraging activities around sources of water during the dry season, behaving as central-place foragers while occasionally searching distant areas. They discuss the potential adaptive values of such behavioral strategies for social species living in highly heterogeneous environments.

17.7 Peccaries as Prey

In the Calakmul Biosphere Reserve of Southern Mexico, Aranda (2002) studied the content of 37 jaguar feces and found the collared peccary as the main species being present in 42% of the samples followed by coatimundi (*Nasua narica* 18%) and armadillos (*Dasypus novemcinctus* 12%). Aranda (2002) also contrasted these results with prey availability and found that collared peccary was consumed in higher proportion than expected based in its availability. Aranda has also propounded an interesting hypothesis based in the almost perfect overlap of peccary species distribution and jaguar, where he argued that they evolved together, and jaguars develop hunting strategies to hunt peccaries as they were and are the favorite prey (Aranda 2002). Avila-Najera et al. (2018) analyzed jaguar diet and puma (*Puma concolor*) in Quintana Roo, a southern state in Mexico. Using molecular techniques identified the feces of both species and the species that these cats feed on. At the same time, they estimated prey availability using photo-trapping techniques. They found that both species of felines preferred collared peccary and brocket deer. Finally, Hernández-Saint Martín et al. (2015) analyzed the food habits of the jaguar and puma in a protected area surrounded by a fragmented landscape in the Biosphere Reserve Sierra del Abra-Tanchipa, northeastern Mexico. The diet of both felids was comprised mainly of wild artiodactyls; the collared peccary contributed 35.2% to the biomass of jaguar diets and white-tailed deer contributed 51.9% to puma diets.

17.8 Competitive Interactions

The introduction of a species into an ecosystem with species already occupying a similar trophic level is predicted to lead to a high degree of niche overlap. The feral pig (*Sus scrofa*), one of the world's worst invasive species, was introduced to the Pantanal about 200 years ago and is thought to compete with the native white-lipped peccary and collared peccary. In South America, the invasive feral pig has become established in Argentina, Uruguay, Paraguay and in a wide range within Brazil, along the southern half of the Atlantic Forest, in the Cerrado (savanna) and in the Pantanal wetland. The geographical ranges of the two most common South American native peccary overlap almost entirely, and the feral pig now co-occurs with them in several areas. Because feral pig, white-lipped and collared peccary are considered ecological equivalents, there has been much speculation about possible competitive interactions among them (Oliveira-Santos et al. 2011). Fox example, to understand why wildlife hunting is not a major conservation issue in the Pantanal an exploratory survey, Desbiez et al. (2011) realized semistructured interviews, skull collection and tooth wear analysis of feral pig, white-lipped peccary and collared peccary were conducted, and hunting registers distributed, in the central region of the Pantanal. The results showed that feral pigs are the main hunting target. Feral pigs are effectively acting as a replacement species for hunting of native wildlife because the pigs provide a constant, culturally acceptable, readily available and free source of meat and oil to remote ranches.

Hidalgo-Mihart et al. (2014) presented the first record of feral pig in the Southeast Mexico, specifically on the area of Laguna de Terminos Protected Area. Hernández-Peréz (2019) determined that this feral population of pigs coexists with collared peccary and while the feral pigs inhabit wetlands and African palm oil plantations, the collared peccary favored more the remaining patches of original vegetation of the area.

Desbiez et al. (2008) studied partitioning between the three species including analysis of fruit items and plants in fecal samples as well as encounter rates in different habitats, to help generate hypotheses about competitive interactions among the species. Overlaps in food resources and habitat use between feral pigs and peccaries were found to be lower than expected. In fact, niche overlap was highest between the native species. Results indicate that currently, feral pigs are not a direct threat to the native peccaries in the study area. Differences in morphology and behavior indicate possible mechanisms of niche partitioning between the species. Feral pigs may, nevertheless, impact the wildlife community in other ways as predators of eggs, by destruction of vegetation through rooting, or by functioning as disease reservoirs. Cattle-ranching activities may favor feral pigs and the current anthropogenic changes in the landscape could lead to changes in competitive dynamics between feral pigs and native species.

17.9 Genetics

Ruiz-García et al. (2015) sequenced the mitochondrial DNA (mtDNA) control region of 59 peccaries (44 white-lipped peccaries, and 15 collared peccaries). Also genotyped three DNA microsatellites from 78 white-lipped peccaries representing the four putative morphological subspecies (i.e., *spiradens*, *aequatoris*, *pecari*, and *albirostris*) present in northwestern South America (i.e., Colombia, Ecuador, Peru, and Bolivia). Results showed: (1) the estimated diversity of the mtDNA control region in the *T. pecari* population was extremely high, whereas the average genetic diversity for the microsatellites was medium to high and similar to that observed in European pig breeds; (2) there was no significant genetic heterogeneity among the quoted putative morphological subspecies at the mitochondrial marker, but they did detect significant (although relatively small) genetic heterogeneity using microsatellites, indicating that *T. pecari albirostris* is a uniquely differentiated group; and (3) the phylogenetic mtDNA trees showed that haplotypes were intermixed independent of their "a priori" subspecies classification. In addition, the microsatellite assignation analyses yielded low percentages of well-classified individuals when the analysis considered the geographic morphology of the subspecies. Thus, the molecular results do not support the putative morphological subspecies of *T. pecari* in northwestern South America. Finally, their results did not detect clear historical demographic changes using the mtDNA control region sequences.

DNA (Gongora et al. 2006) and morphological (Grives and Grubb 2011) data show that collared peccary may be split in two species, one in North/Central and one in South America. In addition, it is probable that there is a third species in Central America, however, these are preliminary conclusions that need to be confirmed in the future with more genetic studies (see further details in Gongora et al. 2018).

17.10 Physiology

Castelo et al. (2010) evaluated the influence of the thawing rate on the quality of frozen-thawed (cryopreserved in Tris-based extenders) semen obtained from collared peccaries. There were no significant differences between the two extenders after extension, chilling, or glycerol addition. In conclusion, semen from collared peccaries was successfully cryopreserved in Tris-based extenders and thawed with two protocols (37 °C for 1 min or 55 °C for 7 s). Costa et al. (2010) investigated testis structure, spermatogenic cycle length, Sertoli cell efficiency, and spermatogenic efficiency, in collared peccary. Testis weight and gonadosomatic index were 23.7 + 1.8 g and 0.2 + 0.1%, respectively. Seminiferous tubule volume density was 77.4 + 1.7%. Leydig cells occupied 12.8 + 1.8% of the testis parenchyma and presented a peculiar cytoarchitecture in the periphery of the seminiferous tubule lobes. The premeiotic, meiotic, and postmeiotic stage frequencies were very similar to those found for wild and domestic boars. The spermatogenic cycle and entire spermatogenic process (based on 4.5 cycles) lasted approximately 12.3 + 0.2 days and 55.1 + 0.7 days, respectively. Daily sperm production per gram of testis in the collared peccary was approximately $23.4 \pm 2 \times 10^6$, which is similar to that of domestic and wild boars. Romero-Solorio (2005) investigated urea and creatinine values on captive white-lipped peccaries of Lima, Peru and found values in the range of 27.4–2.7 mg/dl and 2.8–0.2 mg/dl respectively. Costa and Henry (2004) analyzed morphologically and functionally the spermatogenesis in adult collared peccary. The mitosis efficiency coefficient in collared peccary was 22.54, the meiotic profile, 1:2.7, and spermatogenesis general profile was 1:64. It was concluded that spermatogenesis in peccary is very similar to that in pigs. Souza et al. (2009) compared the effects of acepromazine-tiletamine-zolazepam and propofol used in anesthetic protocols for semen collection by electroejaculation from captive collared peccaries. Conclusion suggests the use of propofol for anesthetic restraint of collared peccaries enhanced collection of semen by electroejaculation.

Mayor et al. (2004) described the morphological characteristics of external genitals and vaginal epithelia during the estrous phases in females of collared peccary. This study showed a predominant presence of superficial and intermediate cells (values above 60%) during a period of 4.4 ± 2.6 days. This predominance of cells occurred 2.3 ± 1.5 days before the estradiol peak. They concluded that changes in vaginal cytology and changes in external genitals have a predictive value of 86.4% and 88.9% respectively. Guerra-Centeno (2007) took blood samples of 66 captive white-lipped peccary of Guatemala zoos. The blood chemistry values provided a reliable reference for the health and nutritional stage of the species in captivity.

In order to optimize breeding management of captive born collared peccaries in semiarid conditions, Maia et al. (2014) described and correlated the changes in the ovarian ultrasonographic pattern, hormonal profile, vulvar appearance, and vaginal cytology during the estrus cycle in this species. According to hormonal dosage, six estrous cycles were identified as lasting 21.0 days, being on average 6 days for the estrogenic phase and 15 days for the progesterone phase. Estrogen presented mean peak values of 55.6 pg/mL. During the luteal phase, the high values for progesterone were 35.3 ng/mL. The presence of vaginal mucus, a reddish vaginal mucosa and the separation of the vulvar lips were verified in all animals during the estrogenic peak. Through ultrasonography, ovarian follicles measuring 0.2 cm were visualized during the estrogen peak. Corpora lutea presented hyperechoic regions measuring 0.4 cm identified during luteal phase. No significant differences between proportions of vaginal epithelial cells were identified when comparing estrogenic and progesterone phases. In conclusion, female collared peccaries, captive born in semiarid conditions, have an estral cycle that lasts 21.0 days, with estrous signs characterized by vulvar lips edema and hyperemic vaginal mucosa, coinciding with developed follicles and high estrogen levels. Villa et al. (2013) estimated age and wear and growth lines of teeth in white-lipped peccary and collared peccary skulls obtained from subsistence hunters in Tierra Comunitaria de Origen Takana-I, La Paz, Bolivia. They established age between 1 and 12 years with 12% of samples not being determined or showing mixed results. Age lines in teeth showed ages of 1–18 years old. Both methods were better to predict age between 2 and 10 years old for both species of peccaries.

García et al. (2011) studied the estrous cycle of collared peccary by vaginal cytology. The estrous cycle period for this specie was 28.45 ± 5.45 days. They observed a significant difference between the cell types in the same phase of the estrous cycle. During the proestrus phase, the exfoliative cytology arrangement showed high frequency of intermediate and superficial cells. The estrus phase had elevated rates of superficial cells, when compared to the other types, and an absence of leukocytes. During this phase, the external genitalia were reddish, tumescent; and the cervical mucus was evident. In the metaestrus phase, they observed a decrease of superficial cells, as well as the highest indices of intermediate cells, an increase of leukocytes and the presence of metestrus and foam cells. In the diestrus phase, the intermediate cells increased, and leukocytes decreased.

Montes-Peréz et al. (2009) evaluated stress levels produced at two different densities of captive collared peccary and found a correlation between density, sex and stress levels. Later, Montes-Pérez et al. (2012a) estimated cortisol levels and progesterone in captive collared peccary and conclude that groups of this species with a single male in areas larger than 5 m^2 per individual do not have increase in the stress level.

17.11 Parasites

Herrera et al. (2008) reported infections of *Trypanosoma evansi* and *Trypanosoma* cruzi in the sympatric suiformes—collared peccary, white-lipped peccary, and feral pig by parasitological, serological, and molecular tests in the Brazilian Pantanal. The results show that peccaries and feral pigs play an important role on the maintenance of both *T. evansi* and *T. cruzi* in the Brazilian Pantanal. Health impairment was observed only in the white-lipped peccary infected with *T. evansi*. Despite presenting low *T. evansi* parasitemia, all infected white-lipped peccaries displayed low hematocrit values and marked leucopenia. The hematological values showed that the *T. evansi* infection is more severe in young white-lipped peccaries. The presented data show that feral pigs and peccaries are immersed in the transmission net of both trypanosome species, *T. cruzi* and *T. evansi*, in the Pantanal region. Tantaleán et al. (2008) identified and determined frequency of helmints present in wild white-lipped peccaries of Madre de Dios Department of Peru. Of 81.81% of positives samples they found 81.81% and 12.12% of individuals with nematode and trematode eggs respectively. Nematode eggs found were of *Ascaris* sp. (51.51%), *Ancylostomatidae* (33.33%), and *Spirurida* types (6.06%), and the trematodes were all *Paragonimus* sp. (12.12%).

Mukul-Yerves et al. (2014) identified gastrointestinal parasites in collared peccary and white-tailed and brocket deer of captivity in southern Mexico and found nematodes of Strongylida order and protozoans of Eucoccidiorida order. They found collared peccary have *Oesophagostomum* sp., *Eimeria* sp., and *Isospora* sp. All ungulate species have *Strongyloides* spp. and ectoparasites such as flea *Pulex irritans* and the louse *Gliricola porcelli* and the mite *Amblyomma* spp.

17.12 Ecosystem Engineers

The concept of ecosystem engineering has catalyzed novel approaches and models for non-trophic interactions and ecosystem functions. Ecosystem engineers physically modify abiotic and biotic environments, then creating new habitats that can be colonized by a new suite of species. Beck (2006) studied the relationships of peccaries with palms (Arecaceae) that are a dominant element within the neotropical plant community. He argued that the evolution of a strong mastication apparatus, unique interlocking canines, patterns of movement, and foraging ecology in peccaries are viewed as adaptations to exploit hard seeds, particularly palm seeds. But how strong are the interactions between peccaries and palms, and what are the ecological ramifications? To respond to this question Beck (2006) analyzed and synthesized results over 76 papers, published between 1917 and 2004, which revealed that peccaries consumed fruits from 46 palm species, 73% of whose seeds were destroyed after ingestion. Furthermore, peccaries disperse palm seeds; eat flowers, seedlings, and roots; and trample seedlings. Thus, peccaries affect the spatiotemporal distribution and demography of palms. Local extinction of peccaries results always in dramatic changes in the forest ecology. Keuroghlian and Eaton (2009) conducted fruit removal and medium-to-large-sized mammalian exclusion experiments to: (1) quantify seasonal fruit consumption from high-density patches beneath parent trees by *T. pecari* and other consumers, and (2) measure impacts of *T. pecari* rooting and foraging activities on seedling dynamics in *E. edulis* stands. During the dry season, when *S. romanzoffiana* palms provided 68% of fruit dry weight in the fragment, *T. pecari* consumed significantly greater amounts than other consumers, and along with *Pecari tajacu* and *Tapirus terrestris*, were potential seed dispersers. More than

95% of *E. edulis* fruit removal was due to seed predation by *T. pecari*. Intense removal during the dry season was closely linked with previously documented range shifts and habitat preferences by *T. pecari*. Exclusion plot experiments in *E. edulis* (palmito) stands showed that the number and proportion of nonpalmito species seedlings increased dramatically in the absence of *T. pecari* rooting and foraging activities that disturbed soil and thinned seedlings.

In the Peruvian Amazonas, Beck et al. (2010) tested why peccaries function as ecosystem engineers by creating and maintaining wallows. Such wallows are critical aquatic habitats and breeding sites for anuran species during dry seasons. Wallows had a significantly higher density of tadpoles, metamorphs and adult anurans, as well as higher ß-diversity and species richness than ponds. This not only provides the first systematic evidence of the ecosystem engineering processes of peccaries, but also reveals positive consequences of such for anuran. Michel et al. (2014) tested two hypotheses: (1) insectivorous birds and bats initiate trophic cascades in tropical rain-forest understory; and (2) the native, omnivorous collared peccary (*Pecari tajacu*) negates these cascades via non-trophic effects in northeastern Costa Rica. Excluding birds and bats increased total arthropod densities by half, both with and without peccaries. Bird/bat exclosures increased Diptera density by 28% and leaf damage by 24% without peccaries, consistent with a trophic cascade. However, bird/bat exclosures decreased Diptera density by 32% and leaf damage by 34% with peccaries, a negation of the trophic cascade. Excluding peccaries increased leaf damage by 43% on plants without birds and bats. This is the first study, to our knowledge, to demonstrate that the non-trophic activity of an omnivorous ungulate can reverse a trophic cascade.

Galetti et al. (2015) used a natural experiment in the Brazilian Atlantic rainforest to investigate the ecological responses of rodents to the functional extinction of a dominant terrestrial mammal, the white-lipped peccary. They detected a 45% increase in the abundance and a decrease in diversity of rodents in defaunated forests. Two of these species (*Akodon montensis* and *Oligoryzomys nigripes*) are important hosts of Hantavirus, a lethal virus for humans. Stable isotope ratios derived from the hair of rodents and peccaries and their food resources indicate that at least two rodent species shifted to a diet more similar to peccaries in the defaunated forest.

Reider et al. (2013) studied the effects of peccaries in leaf removal and reptiles and amphibian densities and considered that peccary loss could have important consequences for litter amphibians and reptiles that depend entirely upon the litter for shelter, foraging and reproduction sites, and thermoregulation. Results demonstrate that peccaries should be viewed not just as seed predators or ecosystem engineers for palms and pond-breeding amphibians, but also as important agents that affect leaf litter structure and abundance of terrestrial amphibians and reptiles.

Serna-Motta et al. (2008) studied the role of collared peccaries as seed dispersers and predators of ten fleshy-fruited species, including both native and naturalized exotic plants to the Atlantic rainforest remnants of Southern Bahia, Brazil. Most seeds were killed and only guava seeds had improved germination after passage through peccary guts when compared to the control.

17.13 Subsistence Hunting and Conservation

Wild animals have been a source of food and income through subsistence hunting by forest-dwelling people in Neotropical countries in spite of the fact that sometimes hunting appears to be unsustainable as it leads to the depletion of wild fauna. Nogueira and Nogueira-Filho (2011) reviewed and discusses the implications for tropical forest integrity and rural population dependency on forest resources. Conclude that establishing captive management programs for peccaries is an effective way of avoiding wild stock depletion, deforestation, and guaranteeing the livelihood of forest dwellers in the Neotropics. However, it is essential that governmental and/or nongovernmental agencies be involved in providing subsides to establish peccary farms, provide technical assistance, and introducing peccary captive breeding centers to supply founder stock. Noss and Leny-Cuéllar (2008) presented data from Bolivia confirming that peccaries are being over-exploited by subsistence hunters. However, a finer examination within the Isoso region in Bolivia indicates that in certain zones current hunting rates may be sustainable, either because these animals are rarely encountered and therefore offtakes are virtually nil, or because productivity is relatively high perhaps reflecting source–sink dynamics between the Isoso and the neighboring Parque Nacional Kaa-Iya del Gran Chaco of Bolivia.

Wallace and Painter (2000) compared general mammalian relative abundances from a number of survey sites in lowland Amazonian Bolivia. In general survey sites with high biotic diversity are characterized by relatively low population densities of several large mammalian species including both species of Amazonian peccary. In contrast, the forests of the Beni alluvial plain, which are characterized by a relatively low biotic diversity include high densities of several larger mammalian species including the peccaries. Despite being a potentially critical region for the conservation of peccaries and other mammalian fauna the Beni alluvial plain is currently represented by just one relatively small protected area; the Beni Biosphere Reserve. Nevertheless, the Beni alluvial plain has many indigenous territorial demands.

Ramos et al. (2016) examined the relationship between hunting pressure and age structure in the white-lipped peccary, analyzing the distribution of age classes at seven sites in the region Terra do Meio in the Brazilian Amazon. Results indicated that the white-lipped peccary was the most frequently hunted terrestrial animal in the region. Fishing, followed by hunting, provided the main sources of animal protein. Their data suggest there was no relationship between age structure and hunting at the study sites. The social structure and mobility of white-lipped peccaries seem to minimize the effects of hunting on age structure. Their results, similar to previous studies, show that the age structure of the white-lipped peccary is robust to hunting impacts. Other factors may have stronger effects on age structure than subsistence hunting. They suggest that deforestation may explain the prevalence of older individuals in peccary populations to the north of their study sites.

Fang et al. (2006) described the implementation of a pilot project on peccary pelt exportation certification in communities along the Yavari-Miri River region of

Loreto Department in Peru. The goal of the project is to encourage a legal trade of peccary pelts through sustainable hunting rates, with communities that follow the rules being certified to sell their products at higher value and obtained conservation incentives. Several communities participated and have been sustainable hunting peccaries for several years now (Fang et al. 2006). In the Ucayali River region of Peru Lleellish et al. (2003) evaluated the population size of both species of peccaries and found an average of 2.5 ind/km^2 and 9.2 ind/km^2 for collared peccary and white-lipped peccary respectively. The hunting pressure was estimated in 0.57 hunted ind/km^2 and 1.14 hunted ind/km^2 of collared and white-lipped peccaries respectively. According to the authors of this study, hunting of peccaries never surpassed more than 5% of the population productivity; therefore, they conclude it was sustainable.

In México, Briceño-Méndez et al. (2011) described the hunting rates of collared peccaries in Tzucacab, Yucatán. They recorded 93 hunting events in which 22 collared peccaries were hunted with a biomass of 374 kg and an additional 17-wounded peccaries that were not killed. There were not differences in the sex preferences of the hunted animals. Of 30 hunters of the community, 32% do hunting trips daily, 28% weekly and 40% monthly. There were three main hunting techniques the drive hunting, waiting in trees and nocturnal searching with powerful lights. They conclude that collared peccary hunting is associated with cultural and social aspects of the communities in addition to provide meat for their families. The drive hunting methods, the most frequent technique but has more negative effects on the population of collared peccary; therefore, alternatives must be find to make the hunting rates more sustainable in the region. Avila-Nájera et al. (2011) obtained information about rural hunters in southern Mexico and the knowledge they have about jaguar and their preys, among them the two peccaries. According to these hunters, both peccaries are important prey for jaguars.

17.14 Status in the Wild

The continuous increase in fragmentation and hunting pressure observed in all Neotropical forests make it urgent to monitor populations of peccary species all over its range, especially in those areas were distinct taxonomic units may exist. Peccaries are very important species that ecologically play critical roles as seed predators and dispersers (Beck 2005, Beck 2006), ecosystem engineers, and soil and pond modifiers (Beck et al. 2010). They are also an important prey for large predators such as jaguar and pumas and help maintain local livelihoods throughout its geographical distribution. Efforts must be made to assure that peccaries are conserved in all areas of its original range and to reduce threats such as forest fragmentation, hunting pressure, introduced exotic species, and disease transmission. Native mature forests serve as a reservoir of animal and plant populations, but they can also be incorporated into community protected areas that link existing wildlife reserves with movement and dispersal corridors. The continued existence of peccaries will depend on the successful implementation of wildlife conservation and management programs that will benefit both wildlife and human communities.

Acknowledgements The first author appreciates the support that Conacyt provided in the project (No. CONACYT CB-2015-01-256549 "*Evaluación de métodos para estimar el tamaño poblacional del venado cola blanca y del ganado en libre pastoreo en la Reserva de la Biosfera Tehuacán-Cuicatlán*"). TRRH appreciate the support of Conacyt through the project Conacyt Ciencia-Basica (Project Number: 182386-12) and the sabbatical year supported by Conacyt to RRH. Thank you to Paola Trillo who compiled the references for this chapter and to Adriana Sandoval Comte for the distribution maps.

References

Aebischer N, Robertson PA, Kenward RE (1993) Compositional analysis of habitat use from animals radio-tracking data. Ecology 74:1313–1325

Altrichter M, Carrillo E, Sáenz J, Fuller T (2001) White-lipped peccary (*Tayassu pecari*, artiodactyla: Tayassuidae) diet and fruit availability in a Costa Rican rain forest. Rev Biol Trop 49(3-4):1183–1192

Aranda M (2002) Importancia de los pecaríes para la conservación del jaguar en méxico. l jaguar en el nuevo milenio (RA Medellın, ed.). Universidad Nacional Autónoma de México and Wildlife Conservation Society, New York, pp.101–105

Avila-Nájera DM, Rosas-Rosas OC, Tarango-Arámbula LA, Martínez-Montoya JF, Santoyo-Brito E (2011) Conocimiento, uso y valor cultural de seis presas del jaguar (*Panthera onca*) y su relación con éste, en San Nicolás de los Montes, San Luis Potosí, México. Rev Mex Biodivers 82(3):1020–1028

Avila-Najera D, Palomares X, Chávez C, Tigar X, Mendoza E (2018) Jaguar (*Panthera onca*) and puma (*Puma concolor*) diets in Quintana Roo, Mexico. Anim Biodivers Conserv 41:257

Beck H (2005) Seed predation and dispersal by peccaries throughout the Neotropics and its consequences: a review and synthesis. In: Forget P-M, Lambert JE, Hulme PE, Vander Wall SB (eds) Seed fate: predation, dispersal and seedling establishment. CABI Publishing, Wallingford, pp 77–115

Beck H (2006) A review of peccary-palm interactions and their ecological ramifications across the neotropics. J Mammal 87(3):519–530

Beck H, Thebpanya P, Filiaggi M (2010) Do neotropical peccary species (Tayassuidae) function as ecosystem engineers for anurans? J Trop Ecol 26(4):407–414

Beck H, Keuroghlian A, Reyna-Hurtado R, Altrichter M, Gongora J (2018) White-lipped peccary *Tayassu pecari* (link 1795). In: Melleti M, Meijaard E (eds) Ecology, conservation and management of wild pigs and peccaries of the world. Cambridge University Press, Cambridge

Bello-Gutiérrez J, Guzmán-Aguirre C, Chablé-Montero C (2004) Caracterizacióndel habitat de tres especies de artiodáctilos en un área fragmentada de Tabasco México. Technical report

Bolaños JE, Naranjo E (2001) Abundancia, densidad y distribución de las poblaciones de ungulados en la cuenca del río Lacantún, Chiapas, México. Rev Mex Mastozool 5:45

Briceño-Méndez M, Montes-Péres R, Aguilar-Cordero W, Pool-Cruz A (2011) Cacería del pecarí de collar (*Pecari tajacu*) (Artiodactyla: Tayassuidae) en tzucacab, Yucatán, México. Rev Mexi Mastozool 1:1

Briceño-Méndez M, Reyna-Hurtado R, Calmé S, García-Gil G (2014) Preferencia de hábitat y abundancia relativa de tayassu tajacu en un área con cacería en la región de calakmul, campeche, México. Rev Mex Biodivers 85:242

Campos-Krauer JM, Benitez IK, Robles V, Meritt DA (2012) A range-wide survey to determine the current distribution and population status of the Chacoan peccary in the Paraguayan Chaco. Suiform Soundings 16:38

Castelo T, Bezerra F, Souza A, Moreira M, Paula V, Oliveira M, Silva A (2010) Influence of the thawing rate on the cryopreservation of semen from collared peccaries (*Tayassu tajacu*) using tris-based extenders. Theriogenology 74(6):1060–1065

Charre-Medellín J, Rangel-Rojas J, Magaña-Cota G, Monterrubio-Rico T, Charre-Luna J (2018) Evidence for current presence of a collared peccary (*Pecari tajacu*) in Guanajuato, Mexico. West North Am Nat 78:106

Chávez-Hernández C, Moguel-Acuña J, González-Galván M, Guiris-Andrade D (2011) Abundancia relativa de tres ungulados en la reserva de la biosfera "La Sepultura" Chiapas, México. Therya 2:111

Cortés-Marcial M, Briones-Salas M (2014) Diversidad, abundancia relativa y patrones de actividad de mamíferos medianos y grandes en una selva seca del Istmo de Tehuantepec, Oaxaca, México. Rev Biol Trop 62(4):1433

Costa DS, Henry MR (2004) Spermatogenesis in the collared peccary (*Pecari tajacu*). Arq Bras Med Vet Zootec 56(1):46–51

Costa G, Leal M, Silva J, Cassia A, Ferreira S, Guimaraes DA, Francca L (2010) Spermatogenic cycle length and sperm production in a feral pig species (collared peccary, *Tayassu tajacu*). J Androl 31(2):221–230

Desbiez AJ, Santos SA, Keuroghlian A, Bodmer RE (2008) Niche partitioning among white-lipped peccaries (*Tayassu pecari*), collared peccaries (*Pecari tajacu*), and feral pigs (*sus scrofa*). J Mammal 90(1):119–128

Desbiez A, Bodmer R, Tomas W (2010) Mammalian densities in a neotropical wetland subject to extreme climatic events. Biotropica 42(3):372–378

Desbiez A, Keuroghlian A, Piovezan U, Bodmer R (2011) Invasive species and bushmeat hunting contributing to wildlife conservation: the case of feral pigs in a neotropical wetland. Oryx 45(1):78–83

Fang T, Ríos C, Bodmer R (2006) Implementación de un programa piloto de certificación de pieles de pecaríes (*Tayassu tajacu* y *T. pecari*) en la comunidad de Nueva Esperanza, Río Yavarí-Mirín. Rev Electrónica Manejo Fauna Silv Latinoamérica 1:1–15

Galetti M, Guevara R, Neves C, Rodarte R, Bovendorp R, Moreira M, Hopkins J III, Yeakel J (2015) Defaunation affects the populations and diets of rodents in neotropical rain- forests. Biol Conserv 190:2–7

García SC, Pendu Y, Albuquerque N (2011) Determination of the estrous cycle in collared peccary *Pecari tajacu*: colpocytological and clinical aspects. Acta Amazon 41(4):583–588

García-Marmolejo G, Chapa-Vargas L, Weber M, Huber-Sannwald E (2015) Landscape composition influences abundance patterns and habitat use of three ungulate species in frag- mented secondary deciduous tropical forests, Mexico. Global Ecol Conserv 3:744–755

García-Marmolejo G, Weber M, Rosas-Rosas OC, Chapa-Vargas L, Huber-Sannwald E, Martínez-Calderas J (2016) Potential distributional patterns of three wild ungulate species in a fragmented tropical region of northeastern Mxico. Trop Conserv Sci 6(4):539

Gómez-Valencia B, Montenegro O, Sánchez-Palomino P (2016) Variación en la abundancia de ungulados en dos áreas protegidas de la Guyana Colombiana estimadas con modelos de ocupación. Therya 7(1):89–106

Gongora J, Morales S, Bernal JE, Moran C (2006) Phylogenetic divisions among collared peccaries (Pecari tajacu) detected using mitochondrial and nuclear sequences. Mol Phylogen Evol 41:1–11

Gongora J, Taber A, Keuroghlian A, Altrichter M, Bodmer RE, Mayor P, Moran C, Damayanti CS, González S (2007) Re-examining the evidence for a "new" peccary species, "Pecari maximus," from the Brazilian Amazon. PPHSG. Newsletter 7:19–26

Gongora J, Groves C, Meijaard E (2018) Evolutionary relationships and taxonomy of Suidae and Tayassuidae. In: Melleti M, Meijaard E (eds) Ecology, conservation and management of wild pigs and peccaries of the world. Cambridge University Press, Cambridge

González-Marín RM, Gallina S, Mandujano S, Weber M (2008) Densidad y distribución de ungulados silvestres en la Reserva Ecológica El Edén, Quintana Roo, México. Acta Zool Mex 24(1):73–93

Grives CP, Grubb P (2011) Ungulate Taxonomy. John Hopkins University Press, Baltimore MD
Guerra-Centeno D (2007) Valores de referencia para química sérica del pecarí de labios blancos: efectos del sexo, edad y población. RECVET 2:1–15
Hernández-Peréz E (2019) Distribución y relaciones ecológicas entre cerdos ferales (*Sus scrofa*) y poblaciones nativas de peccaries. Master thesis. El Colegio de la Frontera Sur. Campeche, Mexico
Hernández-Saint Martín A, Rosas-Rosas O, Palacio-Núñez J, Tarango-Arambula L, Clemente-Sánchez F, Hoogesteijn A (2015) Food habits of jaguar and puma in a protected area and adjacent fragmented landscape of northeastern Mexico. Nat Areas J 35:308
Herrera HM, Abreu UG, Keuroghlian A, Freitas TP, Jansen AM (2008) The role played by sympatric collared peccary (*Tayassu tajacu*), white-lipped peccary (*Tayassu pecari*), and feral pig (*Sus scrofa*) as maintenance hosts for trypanosoma evansi and trypanosoma cruzi in a sylvatic area of brazil. Parasitol Res 103(3):619–624
Hidalgo-Mihart M, Pérez-Hernández D, Pérez-Solano L, Contreras-Moreno F, Angulo-Morales J, Hernández-Nava J (2014) Primer registro de una población de cerdos asilvestra- dos en el área de la Laguna de Términos, Campeche, México. Rev Mex Biodivers 83:990
Humanez-López E, Racero-Casarrubia J, Arias-Alzate A (2016) Anotaciones sobre distribución y estado de conservación de los cerdos de monte *Pecari tajacu* y *Tayassu pecari* (Mammalia: Tayassuidae) para el departamento de córdoba, Colombia. Notas Mastozool 3:24
Jorge ML, Keuroghlian A, Bradham J, Emi J, Oshima F, Cezar-Ribeiro M (2019) White-Lipped Peccary Movement and Range in Agricultural Lands of Central Brazil. In: Reyna-Hurtado R, Chapman CA (eds) Movement Ecology of Neotropical Forest Mammals. Springer Nature, Switzerland
Keuroghlian A, Eaton DP (2009) Removal of palm fruits and ecosystem engineering in palm stands by white-lipped peccaries (*Tayassu pecari*) and other frugivores in an isolated Atlantic forest fragment. Biodivers Conserv 18(7):1733
Keuroghlian A, Eaton DP, Longland WS (2004) Area use by white-lipped and collared peccaries (Tayassu pecari and Tayassu tajacu) in a tropical forest fragment. Biol Conserv 120:411–425
Lira-Torres I (2006) Abundancia, densidad, preferencia de hábitat y uso local de los vertebrados en la Tuza de Monroy, Santiago Jamiltepec, Oaxaca. Rev Mex Mastozool 10(1):41–66
Lira-Torres I, Briones-Salas M (2011) Impacto de la ganadería extensiva y cacería de sub- sistencia sobre la abundancia relativa de mamíferos en la selva Zoque, Oaxaca, México. Therya 2(3):217–244
Lira-Torres I, Briones-Salas M (2012) Abundancia relativa y patrones de actividad de los mamíferos de los Chimalapas, Oaxaca, Mexico. Acta Zool Mex 28(3):566–585
Lleellish M, Amanzo J, Hooker Y, Yale S (2003) Evaluación poblacional de pecaríes en la región del Alto Purús. Alto Purús. Biodiversidad, Conservación y Manejo. Center for Tropical Conservation. Nicholas School of the Environment, Lima, pp 137–145
López MT, Altrichter M, Sáenz J, Eduarte E (2006) Valor nutricional de los alimentos de *Tayassu pecari* (Artiodactyla: Tayassuidae) en el Parque Nacional Corcovado, Costa Rica. Rev Biol Trop 54(2):687–700
Maia KM, Peixoto GC, Campos LB, Bezerra JA, Ricarte A, Moreira N, Oliveira MF, Silva AR (2014) Estrus cycle monitoring of captive collared peccaries (*Pecari tajacu*) in semiarid conditions. Pesquisa Vet Brasileira 34(11):1115–1120
Mandujano S, Naranjo EJ (2010) Ungulate biomass across a rainfall gradient: a comparison of data from neotropical and palaeotropical forests and local analyses in Mexico. J Trop Ecol 26(1):13–23
Martínez-Gutiérrez PG, Martínez-Meyer E, Palomares F, Fernández N (2018) Niche centrality and human influence predict rangewide variation in population abundance of a widespread mammal: The collared peccary (pecari tajacu). Divers Distribut 24(1):103–115
Martínez-Kú DH, Escalona-Segura G, Vargas-Contreras JA (2008) Importancia de las aguadas para los mamíferos de talla mediana y grande en Calakmul, Campeche, México. In: Lorenzo C, Espinoza E, Ortega J (eds) Avances en el estudio de los mamíferos de México II/México. Asociación Mexicana de Mastozoología, Colegio de la Frontera del Sur, México, pp 449–468

Mason-Romo ED, Villa-Mendoza EP, Rendón-Alquicira G, Valenzuela-Galván D (2008) Primer registro del pecarí de collar (*Pecari tajacu*) en el estado de Morelos. Rev Mex Mastozool 12(1):170–175

Mayor P, Gálvez H, Guimaráes DA, López-Béjar M (2004) Características del estro de la hembra de pecarí (*Tayassu tajacu*) del este Amazónico. Arch Zootecnia 63:393

Mayor P, Guimaraes DA, Le Pendu Y, Da Silva JV, Jori F, Lopez-Béjar M (2007) Reproductive performance of captive collared peccaries (*Pecari tajacu*) in the eastern Amazon. Anim Reprod Sci 102(1-2):88–97

Meyer N, Moreno R, Martinez-Morales MA, Reyna-Hurtado R (2019) Spatial ecology of a large and endangered tropical mammal, the white-lipped peccary in Darién Panama. In: Reyna-Hurtado R, Chapman CA (eds) Movement Ecology of Neotropical Forest Mammals. Springer Nature, Switzerland

Michel N, Sherry T, Carson W (2014) The omnivorous collared peccary negates an insectivore-generated trophic cascade in costa rican wet tropical forest understorey. J Trop Ecol 30(1):1–11

Montes-Peréz R, Solís-Sosa A, Yokoyama-Kano J, Segura-Correa J (2009) Evaluación de estrés en el *Pecari tajacu* sometido a dos densidades de población. Arch Zootecnia 58:463

Montes-Pérez RC, Kuri ML, Mukul-Yerves J, Segura-Correa JC, Centurión-Castro FG (2012a) Effects of two space allowances on cortisol levels, agonistic behavior and their relationship with the ovarian cycle of the collared peccary (pecari tajacu) in captivity. Arch Latinoamericanos Prod Anim 20:77

Montes-Pérez RC, Mora-Camacho O, Mukul-Yerves JM (2012b) Forage intake of the collared peccary (*pecari tajacu*). Rev Colombiana Cienc Pecuarias 25(4):586–591

Montes-Pérez R, Borges-Ventura D, Solorio-Sánchez F, Sarmiento-Franco L, Magaña-Monforte J (2018) Preferencia del consumo de ensilado y su efecto sobre la actividad ovárica del *pecari tajacu*. Abanico Veterinario 8:47

Moreira-Ramírez JF, Reyna-Hurtado R, Hidalgo-Mihart M, Naranjo EJ, Ribeiro MC, García-Anleu R, McNab R, Radachowsky J, Mérida M, Briceño-Méndez M, Ponce-Santizo G (2019) White-lipped peccary home range in the Maya Forest of Guatemala and Mexico. In: Reyna-Hurtado R, Chapman CA (eds) Movement Ecology of Neotropical Forest Mammals. Springer Nature, Switzerland

Mukul-Yerves J, Zapata-Escobedo R, Montes-Pérez R, Rodríguez-Vivas I, Torres-Acosta J (2014) Parásitos gastrointestinales y ectoparásitos de ungulados silvestres en condiciones de vida libre y cautiverio en el trópico mexicano. Rev Mex Ciencias 5:459

Naranjo E, Amador-Alcalá S, Falconi-Briones F, Reyna-Hurtado R (2015) Distribución, abundancia y amenazas a las poblaciones de tapir centroamericano (*Tapirus bairdii*) y pecarí de labios blancos (*Odocoileus virginianus*) en méxico. Therya 6:227

Nogueira SC, Nogueira-Filho SLG (2011) Wildlife farming: an alternative to unsustain- able hunting and deforestation in neotropical forests? Biodivers Conserv 20(7):1385–1397

Nogueira S, Silva M, Dias S, Pompeia S, Cetra M, Nogueira-Filho S (2010) Social behaviour of collared peccaries (*Pecari tajacu*) under three space allowances. Anim Welf 19(3):243–248

Noss AJ, Leny-Cuéllar R (2008) La sostenibilidad de la cacería de *tapirus terrestris* y de *Tayassu pecari* en la tierra comunitaria de origen isoso: el modelo de cosecha unificado. Mastozool Neotrop 15(2):241–252

Oliveira-Santos L, Dorazio R, Tomas W, Mourao G, Fernandez F (2011) No evidence of interference competition among the invasive feral pig and two native peccary species in a neotropical wetland. J Trop Ecol 27(5):557–561

Ortíz-García AI, Mandujano S (2011) Modelando la calidad del hábitat para el pecarí de collar en una reserva de biosfera de México. Suiform Soundings 11:14–27

Ortíz-García AI, Ramos-Robles MI, Pérez-Solano LA, Mandujano S (2012) Distribución potencial de los ungulados silvestres en la Reserva de Biosfera de Tehuacán-Cuicatlán, México. Therya 3(3):334–348

Perez-Cortez S, Reyna-Hurtado R (2008) La dieta de los pecaries (*Pecari tajacu* y *Odocoileus virginianus*) en la región de Calakmul, Campeche, México. Rev Mex Mastozool 12(1):17–42
Pérez-Irineo G, Santos-Moreno A (2016) Abundance, herd size, activity pattern and occu- pancy of ungulates in southeastern Mexico. Anim Biol 66(1):97–109
Ramos RM, Pezzuti JCB, Vieira EM (2016) Age structure of the vulnerable white-lipped peccary *Tayassu pecari* in areas under different levels of hunting pressure in the amazon forest. Oryx 50(1):56–62
Reider K, Carson W, Donnelly M (2013) Effects of collared peccary (*Pecari tajacu*) exclusion on leaf litter amphibians and reptiles in a Neotropical wet forest, Costa Rica. Biol Conserv 163:90–98
Reyna-Hurtado R (2007) Social ecology of the white-lipped peccary in Calakmul forest, Campeche, Mexico. PhD Thesis. University of Florida. Gainesville
Reyna-Hurtado R, Tanner GW (2005) Habitat preferences of ungulates in hunted and nonhunted areas in the Calakmul forest, Campeche, Mexico. Biotropica 37:676
Reyna-Hurtado R, Tanner GW (2007) Ungulate relative abundance in hunted and non- hunted sites in Calakmul forest (southern Mexico). Biodivers Conserv 16(3):743–756
Reyna-Hurtado R, Rojas-Flores E, Tanner GW (2009) Home range and habitat preferences of white-lipped peccary groups in a seasonal tropical forest of the Yucatan Peninsula, Mexico. J Mammal 90(5):1199
Reyna-Hurtado R, Chapman CA, Calme S, Pedersen EJ (2012) Searching in heterogeneous and limiting environments: foraging strategies of white-lipped peccaries. J Mammal 93(1):124–133
Reyna-Hurtado R, March I, Naranjo E, Mandujano S (2014) Pecaríes en México. Colegio de Postgraduados Chap 14:339–361
Reyna-Hurtado R, Keuroghlian A, Altrichter M, Beck H, Gongora J (2018) Collared peccary *Pecari* spp (Linneaus, 1758). In: Melleti M, Meijaard E (eds) Ecology, conservation and management of wild pigs and peccaries of the world. Cambridge University Press, Cambridge
Richard-Hansen C, Berzins R, Petit M, Rux O, Goguillon B, Clément L (2019) Movements of white-lipped peccary in French Guiana. In: Reyna-Hurtado R, Chapman CA (eds) Movement Ecology of Neotropical Forest Mammals. Springer Nature, Switzerland
Romero-Solorio M (2005) Niveles referenciales séricos de urea y creatinina en huanganas *Tayassu pecari* mantenidas en cautiverio en el zoológico patronato parque de la leyendas. Rev Investigaciones Vet Perú 19:79
Ruiz-García M, Pinedo-Castro M, Luengas-Villamil K, Vergara C, Rodriguez JA, Shostell JM (2015) Molecular phylogenetics of the white-lipped peccary (*Tayassupecari*) did not confirm morphological subspecies in northwestern south america. Genet Mol Res 14(2):5355–5378
Rumiz DI, Venegas C (2006) Exploración de datos de radio-collares gps en pecaríes de labios blancos. Qué podemos aprender de una muestra de n= 4. Suiform Soundings, newletter of the IUCN/SSC Wild Pig Specialist Group. Peccary Spec Group Hippo Spec Group 6:38–44
Serna-Motta TC, Giné G, Nogueira SC, Nogueira-Filho SL (2008) Digestive seed dispersion and predation by collared peccaries in the southern bahian atlantic forest, brazil. Suiform Soundings 8(1):45–52
Souza AL, Castelo TS, Queiroz JP, Barros IO, Paula VV, Oliveira MF, Silva AR (2009) Evaluation of anesthetic protocol for the collection of semen from captive collared peccaries (*Pecari tajacu*) by electroejaculation. Anim Reprod Sci 116(3-4):370–375
Tantaleán M, Nancy C, Leguía VG, Alcázar PG, Donadi R (2008) Frecuencia de helmintos en huanganas silvestres (*Tayassu pecari* Link, 1795) residentes en áreas protegidas del departamento de Madre de Dios, Perú. Neotrop Helminthol 2(2):48–53
Tejeda-Cruz C, Naranjo E, Cuarón A, Perales H, Cruz-Burguete L (2009) Habitat use of wild ungulates in fragmented landscapes of Lacandon forest, southern Mexico. Mammalia 73:211
Van Roosmalen MGM, Frenz L, van Hooft P, de Iong HH, Leirs H (2007) A new species of living peccary (Mammalia: Tayassuidae), from the Brazilian Amazon. PPHSG Newslett 7:15–19

Villa M, Miranda-Chumacero G, Wallace R (2013) Estimación de edades mediante análisis en tales individuos de *Tayassu pecari* y *Pecari tajacu* (Artiodactyla: Tayassuidae). Rev Mex Biodivers 84(4):1167–1178

Wallace RB, Painter RL (2000) Conservación de pecaríes en la Amazonía boliviana: Biodiversidad vs viabilidad poblacional. In: Fang T, Valqui M, Bodmer R (eds) Manejo de fauna silvestre en Amazonía y América Latina. Memorias del X Congreso Internacional de fauna silvestre en américa latina, Salta, argentina, 14 a 18 de mayo, 2012, pp 263–271

Wetzel RM, Dubos RE, Martin RL, Myers P (1975) Catagonus, an "extinct" peccary, alive in Paraguay. Science 189:379–381

Yañez-Arenas C, Mandujano S, Martínez-Meyer E, Pérez-Arteaga A, González-Zamora A (2012) Modelación de la distribución potencial y el efecto del cambio de uso de suelo en la conservación de los ungulados silvestres del Bajo Balsas, México. Therya 3(1):67–80

Chapter 18
Tapirs of the Neotropics

Eduardo J. Naranjo

Abstract This chapter presents a concise review of available information on the ecology and conservation of the three Neotropical tapir species: the lowland tapir (*Tapirus terrestris*), the mountain tapir (*T. pinchaque*), and Baird's tapir (*T. bairdii*). Tapirs play important roles in the dynamics of tropical ecosystems as browsers, seed dispersers and seed predators, and they have been used as food sources in rural communities of the Neotropics for centuries. A considerable number of research projects have been conducted on these species during the last 25 years in most range countries. Research shows evident declines in population sizes and distributions of the three species throughout Central and South America, primarily due to habitat fragmentation, habitat loss, and poaching. Habitat and population management practices coupled with ecological research and social involvement are essential for tapir conservation across the Neotropics. Participation of local communities is necessary to ensure tapir populations' persistence in both protected and unprotected areas.

18.1 Description

Tapirs (Perissodactyla: Tapiridae) are odd-toed ungulates that originated over 40 million years ago in Asia and later spread into the Americas (Eisenberg 1981; Hershkovitz 1954). These mammals were distributed across North America and Eurasia three million years ago, when they colonized South America (Eisenberg 1989, 1997). Tapirs had disappeared from most of North America by the end of the Pleistocene, 12,000 years ago (Feldhamer et al. 1999). The global distribution of tapirs has been shrinking since that time primarily due to human transformation of ecosystems and overhunting (Brooks et al. 1997). These mammals are currently distributed in some tropical rainforests of Southeast Asia (malayan tapir, *Tapirus indicus*), in extensive tropical and subtropical forest tracts of Central America

E. J. Naranjo (✉)
El Colegio de la Frontera Sur, San Cristobal de Las Casas, Chiapas, Mexico
e-mail: enaranjo@ecosur.mx

S. Gallina-Tessaro (ed.), *Ecology and Conservation of Tropical Ungulates in Latin America*, https://doi.org/10.1007/978-3-030-28868-6_18

(Baird's tapir, *T. bairdii*) and South America (lowland tapir, *T. terrestris*), and throughout the paramo and montane forest ecosystems of the northern Andes (mountain tapir, *T. pinchaque*) (Brooks et al. 1997). A fifth living tapir species (*T. kabomani*) from western Amazonia has been recently described (Cozzuol et al. 2013). However, its discovery is still disputed in the scientific community and more evidence is needed to clarify if this is actually a new species rather than a subspecies of *T. terrestris* (Voss et al. 2014). Tapirs play important roles in the dynamics of tropical ecosystems as browsers, seed dispersers, and seed predators, and they have been used as food sources in rural communities of the Neotropics since pre-Columbian times (Eisenberg 1989; O'Farrill et al. 2006; Naranjo 2009; Fragoso et al. 2016).

Tapir bodies are robust and cylindrical, their tails are short, and their heads are remarkably large, with a long upper lip forming a flexible proboscis (Eisenberg 1989). Limbs are thick and short, with four toes on the forefeet and three toes on the hind feet. All toes end in wide, thick black nails that form hooves (Reid 1997). Their thick, short body hair varies from deep black in *T. pinchaque* to dark brown or dark grey in adults of *T. terrestris* and *T. bairdii*, the last two with light grey tones on their chest, throat, and ear tips. Mountain tapirs have distinctively white lips. Young tapirs are reddish brown with white stripes and spots on their back and sides. Maximum adult length may be about 2 m, while biomass ranges between 150 and 300 kg (Eisenberg 1989; March and Naranjo 2005). Females appear to be slightly larger and heavier than males (Padilla and Dowler 1994; Naranjo 2002). Well-maintained captive tapirs can live up to 30 years (Matola et al. 1997). However, factors such as disease, parasites, predators, and stress reduce the average longevity of Baird's tapir to probably less than 20 years in the wild.

18.2 Distribution

Tapirs were once widely distributed throughout Central and South America in pre-Columbian times (Eisenberg 1989). Nevertheless, the expansion of human presence and activity has reduced tapir habitat considerably, especially during the last 50 years. The lowland tapir lives in low altitudes of northern and central South America, including tropical forests and wetlands of Argentina, Bolivia, Brazil, Colombia, Ecuador, French Guiana, Guyana, Paraguay, Peru, Suriname, and Venezuela (Naveda et al. 2008; Fig. 18.1). This species has disappeared from most tropical dry forests and grasslands of northeastern and southeastern Brazil, northern Argentina, and southern Paraguay. The species has also been extirpated from the inter-Andean valleys of central and northern Colombia, and western Venezuela (Naveda et al. 2008).

The mountain tapir is now extinct in much of its former range, including the Venezuelan Andes and the northern Colombian Andes. It currently occurs in wild areas of the Andes of southern Colombia, Ecuador, and northern Peru at altitudes between 1400 and 4500 m (Downer 1997; Lizcano et al. 2016; Fig. 18.1). Most of

Fig. 18.1 Current distribution of Neotropical tapir species. Modified from Naveda et al. (2008), García et al. (2016), Lizcano et al. (2016), and Naranjo (2018)

the remaining mountain tapir populations are restricted to protected areas around volcanoes above 2000 m. In Colombia, this species is still present in the southern part of the eastern and central Andes, but has entirely disappeared from the Cordillera Occidental as well from the northern parts of the Andes, the Sierra Nevada de Santa Marta, and Serrania de la Macarena (Lizcano et al. 2016). The distribution of mountain tapirs in Ecuador includes scattered forest and paramo patches along the central and eastern Andes, and a few isolated localities of the western Andes (Lizcano et al. 2016; Tapir Specialist Group 2010). In Peru, mountain tapirs have been recorded in parts of Cajamarca, Lambayeque, and Chota provinces (Lizcano et al. 2016).

Current distribution of Baird's tapir includes parts of southern Mexico, Belize, Costa Rica, Guatemala, Honduras, Nicaragua, Panama, and northwestern Colombia (Matola et al. 1997; García et al. 2016; Fig. 18.1). It has been extirpated from El Salvador, and its current or historical presence has not been confirmed in Ecuador (Matola et al. 1997; Tapir Specialist Group 2010). Baird's tapir populations are present in the southeastern Mexican states of Campeche, Chiapas, Oaxaca, Quintana Roo, Veracruz, and probably Tabasco and Yucatan (Naranjo 2009, 2018). Populations of this mammal remain in all districts of Belize, in the Guatemalan Departments of Izabal, Petén, Alta Verapaz, and Quiché, in the Departments of Atlántida, Cortés, El Paraíso, Gracias a Dios, Olancho, and Yoro in Honduras. In Nicaragua, Baird's tapirs are restricted to the tropical rainforests and swamps along the Caribbean Coast, while in Costa Rica they are present in the largest protected areas throughout the country (Naranjo 1995; García et al. 2016). This ungulate has been recorded in the Panamanian provinces of Bocas del Toro, Chiriquí, Coclé, Colon, Darién, Kuna Yala, Panamá, and Veraguas (García et al. 2016). The only recent records of Baird's tapirs in Colombia come from the northwestern Los Katíos National Park and Serranía del Darién (Ministerio de Ambiente, Vivienda y Desarrollo Territorial 2005). Human activities such as farming, cattle ranching, logging, and overhunting have considerably reduced tapir species distributions throughout their ranges in the Neotropics (see a description mortality factors below).

18.3 Life History

18.3.1 Diet

Tapirs are browsers and frugivores feeding on leaves, shoots, bark, fruits and flowers of many plant species (Eisenberg 1989; Bodmer 1990a). Their large body mass and their habitat preferences require them to spend over 70% of their active hours feeding or searching for food (Terwilliger 1978; Foerster and Vaughan 2015). Although their main food items are shoots and leaves, fruits constitute their most important sources of calories, and these are actively sought throughout the forest floor (Naranjo 1995; O'Farrill et al. 2006). Up to 264 plant species have been recorded in mountain tapirs´ diet, with high consumption frequency of Araliaceae, Asteraceae, Fabaceae (particularly *Lupinus* sp.), Gunneraceae, Melastomataceae, Pteridaceae, and Solanaceae (Downer 2001; Ministerio de Ambiente, Vivienda y Desarrollo Territorial 2005; Tapir Specialist Group 2010). Foerster and Vaughan (2015), and Naranjo (2009) listed 126 and 98 plant species in Baird's tapir diet in Costa Rica and Mexico, respectively. Plant families frequently consumed were Araceae, Asteraceae, Fabaceae, Lauraceae, Moraceae, Rubiaceae, and Solanaceae (Naranjo 2009; Foerster and Vaughan 2015). Lowland tapirs feed on at least 120 plant species primarily belonging to the families Arecaceae, Mimosaceae, Moraceae, and Piperaceae, among others (Salas and Fuller 1996; Ministerio de Ambiente, Vivienda y Desarrollo Territorial 2005; Tobler 2008).

Tapirs are important seed dispersers and seed predators, depending on the plant species (Janzen 1981; Olmos 1997; Galetti et al. 2001). For instance, bulky seeds such as those of *Licania platypus* and *Pouteria sapota* are chewed and spat, but not ingested by Baird's tapirs. In contrast, smaller seeds such as those of *Bactris balanoidea*, *Brosimum alicastrum*, *Ficus* spp., *Manilkara zapota*, and *Spondias mombin* are either ingested and later defecated without damage, or destroyed by chewing or digestion (Janzen 1981; Naranjo 2009). Some of the seeds consumed by tapirs maintain or increase their germinative power after passing through their digestive system. Oily fruits and seeds of trees such as *Attalea butyracea*, *Bactris balanoidea*, *Euterpe edulis, Licania platypus*, *Manilkara zapota*, *Mauritia flexuosa*, *Maximiliana maripa*, *Raphia taedigera*, and *Pouteria sapota* are among the most frequently consumed by tapirs (Bodmer 1990a; Naranjo 1995, 2009; Fragoso 1997; O'Farrill et al. 2006).

Tapirs probably are the last potential dispersers of plant species with large seeds that were formerly dispersed by large mammals that are now extinct (O'Farrill et al. 2013). O'Farrill et al. (2013) compiled evidence from numerous studies showing that tapirs have key roles as seed dispersers and seed predators. Tapirs influence plant diversity by seed predation or seed removal over long distances (Galetti et al. 2001; Fragoso et al. 2003). Depending on the availability of food items, tapirs can shift their foraging strategy among habitat types and seasons. The most noticeable changes in proportions of food items ingested by tapirs throughout the year are those related to fruit consumption. Fruit (2–33%) usually constitutes a smaller proportion of tapir food than leaves and other fiber sources (67–98%) (Bodmer 1990a; Olmos 1997; Naranjo 2009; Foerster and Vaughan 2015).

18.3.2 Reproduction

Many studies on the reproductive behavior of captive tapirs have been conducted (e.g., Álvarez del Toro 1966; Barongi 1993). However, their reproductive habits remain poorly known in the wild. Tapir gestation lasts between 390 and 400 days (about 13 months; Álvarez del Toro 1966). Females are able to mate when they are 3–4 years old, and they have estrous periods of about 48 h every 50–80 days (Barongi 1993). The minimum period between litters is 17 months. They give birth to one young weighing around 10 kg, which stays with its mother for about a year (Leopold 1959; Eisenberg 1989). During courtship, males engage in intensive active behaviors such as walking, running, swimming, biting females on their feet and back, and repeatedly trying to copulate with them. Once a female has accepted a male, she may copulate several times (Naranjo 2014). Captive newborn tapirs gain over 2 kg per week (Wilson and Wilson 1973). Juveniles reach their adult size at around 18 months of age. All the milk teeth erupt within the first month, and all the permanent teeth erupt after about 29 months (Matola et al. 1997).

18.3.3 Behavior

Tapirs tend to be solitary and may be active day and night. However, they travel much longer distances during the first hours of darkness (Foerster 1998). In areas close to human settlements and under hunting pressure, tapirs usually become almost completely nocturnal to avoid contact with humans and domestic animals (Brooks et al. 1997; Naranjo 2009). Tapirs are excellent swimmers and often travel along watercourses. On dry land, they frequently move through a well-defined trail system in search of food or water. Although these ungulates are shy and try to avoid encounters with humans, in extreme situations they may attack, especially when females with offspring are chased (March and Naranjo 2005). The main predators of tapirs are humans, jaguars, pumas, and large crocodiles (Eisenberg 1989; Brooks et al. 1997; Naranjo 2009).

An interesting aspect of habitat use by tapirs relates to its defecation habits. Tapirs frequently defecate in shallow water, although they may use particular sites on dry land, forming latrines (Fragoso 1997; Naranjo 2009). In areas with abundant streams and ponds, most defecation (>90%) occurs in permanent or seasonal water bodies (Naranjo 2009). In areas where water is scarce, a significant proportion of latrines have been observed on dry land, especially along mountain crests. Tapirs probably use these high places as both marking sites and corridors to move between different habitat types (Naranjo and Cruz 1998; Tobler et al. 2006). Tapirs have very well developed senses of smell and hearing, which facilitate nonvisual communication. Acoustic communication consists primarily of high-pitched vocalizations similar to whistles, as well as clicking noises and nasal snorts, depending on the situation (Hunsaker and Hahn 1969). Tapirs also produce olfactory signals by spraying urine over their territories and defecating in either water bodies or dry latrines (Kuehn 1986; Naranjo 2009).

18.4 Population Dynamics

18.4.1 Density and Home Range

A substantial number of density estimates obtained through a variety of techniques ranging from line transect sampling to camera-trapping have been published during the last decade for both the lowland tapir and Baird's tapir. Yet, very few data on this matter are available for the mountain tapir. Lowland tapir densities have been estimated at 0.07–3.5 ind/km^2 in different range countries and habitat types (Passos-Cordeiro 2004; Trolle et al. 2008; Cruz 2012; Noss et al. 2012; Tobler et al. 2014; Ferreguetti et al. 2017). Baird's tapir density estimates range from 0.03 to 2.9 ind/km^2 (Naranjo 2009, 2018; González-Maya et al. 2012). The density of mountain tapirs has been estimated so far at 0.17–0.25 ind/km^2 (Acosta et al. 1996; Downer 1996; Lizcano and Cavelier 2000). These noticeable variations in density estimates within and among species may result from considerable differences in topography, altitude, cover type, food and water availability, hunting, human presence, and methodologies, among other factors (Naranjo 2009). In general terms, densities tend to be higher in lower altitudes, plain terrain, heavily forested areas, unhunted sites, and far from human settlements (Brooks et al. 1997; Ministerio de Ambiente, Vivienda y Desarrollo Territorial 2005; Tobler 2008; Tapir Specialist Group 2010; Naranjo 2018). Both growth and mortality rates are usually very slow in tapir populations given their limited reproductive outputs (Eisenberg 1989). Consequently, there are no noticeable fluctuations in their population sizes unless their habitats suffer rapid fragmentation processes or hunting pressure increases within short periods. Such conditions very likely will result in decreasing population sizes.

Tapir home ranges have been assessed through radiotelemetry and camera-trapping in several study areas across Central and South America. Available home range estimates for the lowland tapir fluctuate between 1 and 4.8 km^2 (Tobler 2008; Medici 2010), while those for Baird's tapir range from 0.67 through 1.8 km^2 (Williams 1984; Foerster and Vaughan 2002; Naranjo and Bodmer 2002). An atypically large home range size (23.9 km^2) was observed for a male Baird's tapir followed for 4 years in Calakmul, Mexico by Reyna et al. (2016). The only published estimate of Mountain tapir home range is 3.5 km^2 (Lizcano and Cavelier 2004).

Variations in tapir movements and home range sizes may be related to differences in habitat quality (e.g., food and water availability) (Naranjo 2009; Lira et al. 2014). Freshwater bodies usually are important components of their home ranges. Foerster and Vaughan (2002) found variations among individual movements (379–720 m per night), and home ranges (1.6 km^2 for males, and 1.02 km^2 for females), with significant overlap among individuals.

18.4.2 Population Structure

Sex ratios have not been thoroughly analyzed in wild tapir populations. In southern Mexico, Naranjo (2002) observed that females and males represented 66.7% and 33.3% of individuals that could be identified. Unhunted tapir populations generally have a large proportion of adults, while juveniles and young occur in smaller percentages (Montenegro 1999). An example of this was observed by Naranjo (2009) in an unhunted site of the Lacandon Forest, Mexico, where adults, juveniles, and young tapirs comprised 85.8%, 7.1%, and 7.1% of the population, respectively. The low numbers of juveniles and young in tapir populations may be explained by their very low growth rates, considerable longevity, and low natural mortality rates (Bodmer et al. 1997; Brooks et al. 1997).

18.5 Habitat Requirements

Studies on tapir habitat use have shown that these mammals prefer: (1) A high density of permanent water sources; (2) a diverse and dense understory (which implies abundant food); (3) large extensions of native vegetation (usually forest); (4) low fire incidence; and (5) low hunting pressure and minimal human presence (Salas and Fuller 1996; Fragoso 1997; Tobler et al. 2006; Naranjo 2009, 2018). Lowland and Baird's tapirs frequently use lowland, mature tropical evergreen forests as well as palm swamps, which offer abundant high-quality food and aquatic habitats, in addition to resting cover (Bodmer 1990b; Medici 2010; Naranjo 2018). High-quality habitat for tapirs usually includes rivers, streams, lagoons, ponds, and swamps, where they may spend several hours a day for thermal regulation, as well as for relief from biting and stinging insects (Brooks et al. 1997; March and Naranjo 2005).

Aquatic habitats are also important as escape cover from potential predators such as humans and large felines, but that is not the case to escape from crocodiles. Mountain and lowland tapirs frequently use natural salt licks (Ministerio de Ambiente, Vivienda y Desarrollo Territorial 2005; Tapir Specialist Group 2010). This has been documented in the Peruvian Amazon (Acosta et al. 1994; Montenegro 1998, 1999), in the Sierra de la Macarena and in the Colombian Andes (Ministerio de Ambiente, Vivienda y Desarrollo Territorial 2005), especially at night during the dry season (Montenegro 1998).

Most tapir habitats across the Neotropics currently occur within protected areas, and good-quality habitat remaining outside protected areas is increasingly being transformed for human use because of its potential for agriculture, cattle grazing, and timber extraction (Naveda et al. 2008; García et al. 2016; Lizcano et al. 2016). In countries like Mexico, loggers, farmers, and ranchers continue to receive government subsidies to expand habitat-destructive activities and increase the removal of native vegetation (Naranjo et al. 2015; Naranjo 2018). To reduce these impacts, managers of protected areas should invest time and resources in outreach and cooperative programs focused on avoiding further deforestation and overhunting by neighboring communities.

18.6 Mortality Factors

The low productivity of tapirs is explained by their long period of sexual immaturity, small litter size (one young in most cases), and long gestation (13 months) and lactating (8–10 months) periods (Eisenberg 1989; Brooks et al. 1997). Survival rates are high for tapir populations inhabiting high-quality habitats without human disturbance. Besides humans (their main predator), jaguars, pumas, Andean bears, and large crocodiles are significant predators, especially of young and juvenile tapirs (Eisenberg 1989; Ministerio de Ambiente, Vivienda y Desarrollo Territorial 2005; Tapir Specialist Group 2010).

The most relevant mortality factors for tapirs throughout the Neotropics are persistent hunting, deforestation, forest fragmentation, droughts, floods, forest fires, and road collisions (Naveda et al. 2008; García et al. 2016; Lizcano et al. 2016; Naranjo 2018). Tapir subsistence hunting persists in the majority of range countries, where a diversity of practices has been documented (Ministerio de Ambiente, Vivienda y Desarrollo Territorial 2005; Naranjo 2009; Fragoso et al. 2016). In particular, farmers around protected areas sheltering tapir populations face more frequent crop damages by these ungulates passing through their cultivars or consuming crops (Serrano-MacGregor 2017). Consequently, these mammals have a greater risk of being shot down by farmers without notice to authorities. Sustainable harvests seem unlikely for tapir populations (Bodmer et al. 1997). Therefore, hunting restrictions should be enforced in protected areas and their surroundings to allow recovery particularly in isolated habitat patches. On the other hand, road collisions involving tapirs have become more common on many roads across Central and South America

(Contreras-Moreno et al. 2013; Naranjo et al. 2015; Poot and Clevenger 2018). As traffic becomes heavier on the roads running through and around tapir habitat, more road kills would be expected over time (Poot and Clevenger 2018).

18.7 Conservation Prospects

Millions of hectares of Neotropical forests have been cleared or fragmented during the last 20 years (Keenan et al. 2015). Many studies have documented declines of tapir populations as their habitats are progressively transformed into farming and grazing areas, or lost (Naranjo and Bodmer 2007; Naveda et al. 2008; García et al. 2016; Lizcano et al. 2016). Effective habitat management and protection is probably the most important tools to maintain viable tapir populations in the long term. The creation and maintenance of protected area networks and corridors among them would be of utmost importance to improve habitat availability for tapirs across the Neotropics. Law enforcement is needed to mitigate harmful practices (e.g., poaching, extensive cattle grazing, and burning) threatening tapir populations in most range countries, particularly within protected areas (Brooks et al. 1997; Naranjo 2009). Isolation of tapir populations is especially prevalent in the shrinking forests of southern Mexico and Central America, as well as in the Colombian, Ecuadorean, and Peruvian Andes (García et al. 2016; Lizcano et al. 2016). Further isolation of these populations can be mitigated by encouraging connectivity among forest fragments through cooperative programs to establish community reserves, and promoting environmentally friendly agricultural and land use practices such as agroforestry systems (e.g., organic cacao and shade coffee plantations), sustainable cattle grazing, and sustainable harvests of native plant species around protected areas (Naranjo 2009, 2014). In addition, tapirs can be focal species for ecotourism. However, excessive numbers of tourists in protected areas may be counterproductive for tapirs and other wildlife. Nevertheless, when properly planned and managed, ecotourism can be a non-consumptive alternative for promoting the conservation and sustainable management of tapirs and other vulnerable species (Alvarez-Loaiza et al. 2017).

Habitat and population management practices coupled with ecological research and social involvement are essential for tapir conservation across the Neotropics. Participation of local communities is necessary to ensure tapir populations' persistence in both protected and unprotected areas. Tapirs cannot sustain commercial hunting, but the sustainable harvest of more productive species such as armadillos, deer, and peccaries may help fulfill the needs of communities for subsistence hunting in comanagement systems (Bodmer and Puertas 2000; Naranjo 2009). Threats to tapirs and their habitats in Latin America are complex and challenging. Local economic development programs, such as the training and hiring of community residents as guards in protected areas, environmental educators, and field assistants in research and monitoring projects, may be considerably more effective than other strategies. Local or national government programs encouraging agroforestry and ecotourism projects in tapir distribution areas may also help residents' economy.

Finally, an essential element in fostering wise management and conservation of tapirs and their habitats is the training of professionals. In this sense, the growing number of conservation-oriented graduate and undergraduate programs and courses available in academic institutions throughout Latin America is encouraging. Environmental education and communication programs should also be strongly supported and encouraged by authorities in both urban and rural areas, particularly those near tapir habitats.

References

Acosta H, Cabrera JA, Miraña J (1994) Aportes al conocimiento de *Tapirus terrestris* en el Parque Natural Cahuinarí (Amazonas-Colombia). Fundación Natura-Colombia, Bogotá

Acosta H, Cavelier J, Londoño S (1996) Aportes al conocimiento de la biología de la danta de montaña *Tapirus pinchaque* en los Andes Centrales de Colombia. Biotropica 28:258–266

Álvarez del Toro M (1966) A note on the breeding of Baird's tapir at Tuxtla Gutiérrez Zoo. Int Zoo Yearbook 6:196–197

Alvarez-Loaiza PJ, Abrigo PA, Vite FM, Trelles DA, Espinoza AC, Yánez P (2017) El tapir de montaña (*Tapirus pinchaque*) como especie bandera en los Andes del sur del Ecuador. INNOVA Res J 2:86–103

Barongi RA (1993) Husbandry and conservation of tapirs. Int Zoo Yearbook 32:7–15

Bodmer RE (1990a) Fruit patch size and frugivory in the lowland tapir (*Tapirus terrestris*). J Zool 222:121–128

Bodmer RE (1990b) Responses of ungulates to seasonal inundations in the Amazonian floodplain. J Trop Ecol 6:191–201

Bodmer RE, Puertas PE (2000) Community-based co-management of wildlife in the Peruvian Amazon. In: Robinson JG, Bennett EL (eds) Hunting for sustainability in tropical forests. Columbia University Press, New York, pp 395–412

Bodmer RE, Eisenberg JF, Redford KH (1997) Hunting and the likelihood of extinction of Amazonian mammals. Conserv Biol 11:460–466

Brooks DM, Bodmer RE, Matola S (eds) (1997) Tapirs: status survey and conservation action plan. IUCN/SSC Tapir Specialist Group, Gland/Cambridge

Contreras-Moreno FM, Hidalgo-Mihart MG, Pérez-Solano LA, Vásquez-Maldonado YA (2013) Nuevo registro de tapir centroamericano (*Tapirus bairdii*) atropellado en el noroeste del estado de Campeche, México. Tapir Conserv 20:22–25

Cozzuol MA, Clozato CL, Holanda EC, Rodrigues FHG, Nienow S, de Thoisy B, Redondo RAF, Santos FR (2013) A new species of tapir from the Amazon. J Mammal 94:1331–1345

Cruz MP (2012) Densidad, uso del habitat y patrones de actividad diaria del tapir (*Tapirus terrestris*) en el Corredor Verde de Misiones, Argentina. BSc thesis, Universidad de Buenos Aires, Argentina

Downer C (1996) The mountain tapir, endangered flagship species of the high Andes. Oryx 30:45–58

Downer C (1997) Evaluación del estado y plan de acción para el tapir andino (*Tapirus pinchaque*). In: Brooks DM, Bodmer RE, Matola S (eds) Tapirs: status survey and conservation action plan. IUCN/SSC, Tapir Specialist Group, Gland/Cambridge, pp 75–88

Downer C (2001) Observations on the diet and habitat of the mountain tapir (*Tapirus pinchaque*). J Zool 254:279–291

Eisenberg JF (1981) The mammalian radiations: an analysis of trends in evolution, adaptation, and behavior. University of Chicago Press, Chicago

Eisenberg JF (1989) Mammals of the Neotropics, The northern neotropics, vol 1. University of Chicago Press, Chicago

Eisenberg JF (1997) Introduction. In: Brooks DM, Bodmer RE, Matola S (eds) Tapirs: status survey and conservation action plan. IUCN/SSC Tapir Specialist Group, Gland/Cambridge, pp 1–2

Feldhamer GA, Drickamer LC, Vessey SH, Merritt JF (1999) Mammalogy: adaptation, diversity, and ecology. WCB McGraw-Hill, New York

Ferreguetti AC, Tomás WM, Bergallo HG (2017) Density, occupancy, and detectability of lowland tapirs, Tapirus terrestris, in Vale natural reserve, southeastern Brazil. J Mammal 98:114–123

Foerster CR (1998) Ecología de la danta centroamericana (Tapirus bairdii) en un bosque lluvioso tropical de Costa Rica. MSc thesis, Universidad Nacional de Costa Rica, Heredia

Foerster CR, Vaughan C (2002) Home range, habitat use, and activity of Baird's tapir in Costa Rica. Biotropica 34:423–437

Foerster CF, Vaughan C (2015) Diet and foraging behavior of a female Baird's tapir (*Tapirus bairdii*) in a Costa Rican lowland rainforest. Cuadernos de Investigación UNED 7:259–267

Fragoso JM (1997) Tapir-generated seed shadows: scale-dependent patchiness in the Amazon rain forest. J Ecol 85:519–529

Fragoso JM, Silvius KM, Correa JA (2003) Long-distance seed dispersal by tapirs increases seed survival and aggregates tropical trees. Ecology 84:1998–2006

Fragoso JM, Levi T, Oliveira LFB, Luzar JB, Overman H, Read JM, Silvius KM (2016) Line transect surveys underdetect terrestrial mammals: implications for the sustainability of subsistence hunting. PLoS One 11:e0152659. https://doi.org/10.1371/journal.pone.0152659

Galetti M, Keuroghlian A, Hanada L, Morato MI (2001) Frugivory and seed dispersal by the lowland tapir (*Tapirus terrestris*) in southeast Brazil. Biotropica 33:723–726

García M, Jordan C, O'Farril G, Poot C, Meyer N, Estrada N, Leonardo R, Naranjo EJ, Simons A, Herrera A, Urgilés C, Schank C, Boshoff L, Ruiz-Galeano M (2016) *Tapirus bairdii*. The IUCN red list of threatened species 2016: e.T21471A45173340. https://doi.org/10.2305/IUCN.UK.2016-1.RLTS.T21471A45173340.en

González-Maya JF, Schipper J, Polidoro B, Hoepker A, Zárrate-Charry D, Belant JL (2012) Baird's tapir density in high elevation forest of the Talamanca region of Costa Rica. Integr Zool 7:381–388

Hershkovitz P (1954) Mammals of northern Colombia. Preliminary report 7, Tapirs (genus *Tapirus*), with a systematic review of American species. Proc U S Nat Mus 103:465–496

Hunsaker D, Hahn TC (1969) Vocalization of the South American tapir (*Tapirus terrestris*). Anim Behav 13:69–74

Janzen DH (1981) Digestive seed predation by a Costa Rican Baird's tapir (*Tapirus bairdii*). Biotropica 13:59–63

Keenan RJ, Reams GA, Achard A, de Freitas JV, Grainger A, Lindquist E (2015) Dynamics of global forest area: results from the FAO global forest resources assessment 2015. For Ecol Manag 352:9–20

Kuehn G (1986) Tapiridae. In: Fowler ME (ed) Zoo and wild animal medicine. WB Saunders, London, pp 931–934

Leopold AS (1959) Wildlife of Mexico, the game birds and mammals. University of California Press, Berkeley

Lira I, Briones M, Sánchez G (2014) Abundancia relativa, estructura poblacional, preferencia de hábitat y patrones de actividad del tapir centroamericano Tapirus bairdii (Perissodactyla: Tapiridae) en la Selva de Los Chimalapas, Oaxaca, México. Rev Biol Trop 62:1407–1419

Lizcano DJ, Cavelier J (2000) Densidad poblacional y disponibilidad de hábitat de la danta de montaña (*Tapirus pinchaque*) en los Andes centrales de Colombia. Biotropica 32:165–173

Lizcano DJ, Cavelier J (2004) Using GPS collars to study mountain tapirs (*Tapirus pinchaque*) in the Central Andes of Colombia. Tapir Conserv 13:18–23

Lizcano DJ, Amanzo J, Castellanos A, Tapia A, López-Málaga CM (2016) *Tapirus pinchaque*. The IUCN red list of threatened species 2016: e.T21473A45173922. https://doi.org/10.2305/IUCN.UK.2016-1.RLTS.T21473A45173922.en

March IJ, Naranjo EJ (2005) Tapir (*Tapirus bairdii*). In: Ceballos G, Oliva G (eds) Los mamíferos silvestres de México. CONABIO and Fondo de Cultura Económica, México, pp 496–497

Matola S, Cuarón AD, Rubio-Torgler H (1997) Status and action plan of Baird's tapir (*Tapirus bairdii*). In: Brooks DM, Bodmer RE, Matola S (eds) Tapirs: status survey and conservation action plan. IUCN/SSC Tapir Specialist Group, Gland/Cambridge, pp 29–45

Medici EP (2010) Assessing the viability of lowland tapir populations in a fragmented landscape. PhD dissertation, University of Kent, Canterbury, UK

Ministerio de Ambiente, Vivienda y Desarrollo Territorial (2005) Programa nacional para la conservación del género *Tapirus* en Colombia. Viceministerio de Ambiente, Dirección de Ecosistemas, Bogotá, Colombia

Montenegro OL (1998) The behavior of lowland tapir (Tapirus terrestris) at a natural mineral lick in the Peruvian Amazon. MSc thesis, University of Florida, Gainesville

Montenegro OL (1999) Observaciones sobre la estructura de una población de tapires (Tapirus terrestris) en el sureste de la amazonía peruana. In: Fang TG, Montenegro OL, Bodmer RE (eds) Manejo y conservación de fauna silvestre en América Latina. Instituto de Ecología, La Paz, Bolivia, pp 437–442

Naranjo EJ (1995) Hábitos de alimentación del tapir (Tapirus bairdii) en un bosque lluvioso tropical de Costa Rica. Vida Silv Neotrop 4:32–37

Naranjo EJ (2002) Population ecology and conservation of ungulates in the Lacandon Forest, Mexico. PhD dissertation, University of Florida, Gainesville

Naranjo EJ (2009) Ecology and conservation of Baird's tapir in Mexico. Trop Conserv Sci 4:140–158

Naranjo EJ (2014) Tapir. In: Valdez R, Ortega JA (eds) Ecología y manejo de fauna silvestre en México. Colegio de Postgraduados and New Mexico State University, Texcoco, Mexico, pp 377–387

Naranjo EJ (2018) Baird's tapir ecology and conservation in Mexico revisited. Trop Conserv Sci 11:1–4

Naranjo EJ, Bodmer RE (2002) Population ecology and conservation of Baird's tapir (*Tapirus bairdii*) in the Lacandon Forest, Mexico. Tapir Conserv 11:25–33

Naranjo EJ, Bodmer RE (2007) Source-sink systems of hunted ungulates in the Lacandon Forest, Mexico. Biol Conserv 138:412–420

Naranjo EJ, Cruz E (1998) Ecología del tapir en la Reserva de la Biósfera La Sepultura. Acta Zool Mex 73:111–125

Naranjo EJ, Amador SA, Falconi FA, Reyna-Hurtado RA (2015) Distribución, abundancia y amenazas a las poblaciones de tapir centroamericano (Tapirus bairdii) y pecarí de labios blancos (Tayassu pecari) en México. Therya 6:227–249

Naveda A, de Thoisy B, Richard-Hansen C, Torres DA, Salas L, Wallance R, Chalukian S, de Bustos S (2008) Tapirus terrestris. The IUCN red list of threatened species 2008: e.T21474A9285933. https://doi.org/10.2305/IUCN.UK.2008.RLTS.T21474A9285933.en

Noss AJ, Gardner B, Maffei L, Cuéllar E, Montaño R, Romero-Muñoz A, Sollman R, O'Connell AF (2012) Comparison of density estimation methods for mammal populations with camera traps in the Kaa-Iya del Gran Chaco landscape. Anim Conserv 15:527–535

O'Farrill G, Calme S, González A (2006) *Manilkara zapota*: a new record of a species dispersed by tapirs. Tapir Conserv 15:32–35

O'Farrill G, Galetti M, Campos-Arceiz A (2013) Frugivory and seed dispersal by tapirs: an insight on their ecological role. Integr Zool 8:4–17

Olmos F (1997) Tapirs as seed dispersers and predators. In: Brooks DM, Bodmer RE, Matola S (eds) Tapirs: status survey and conservation action plan. IUCN/SSC Tapir Specialist Group, Gland/Cambridge, pp 3–9

Padilla M, Dowler RC (1994) Tapirus terrestris. Mamm Species 481:1–8

Passos-Cordeiro JL (2004) Estrutura e heterogeneidade da paisagem de uma unidade de conservação no nordeste do Pantanal (RPPN SESC Pantanal), Mato Grosso, Brasil: efeitos sobre a distribuição e densidade de antas (*Tapirus terrestris*) e de cervos-do-pantanal (*Blastocerus dichotomus*). PhD dissertation, Universidade Federal do Rio Grande do Sul, Porto Alegre, Brazil

Poot C, Clevenger AP (2018) Reducing vehicle collisions with the Central American tapir in Central Belize District, Belize. Trop Conserv Sci 11:1–7

Reid FA (1997) A field guide to the mammals of Central America and Southeast Mexico. Oxford University Press, New York

Reyna RA, Sanvicente M, Pérez-Flores J, Carrillo N, Calme S (2016) Insights into the multiannual home range of a Baird's tapir (Tapirus bairdii) in the Maya Forest. Therya 7:271–276

Salas LA, Fuller TK (1996) Diet of the lowland tapir (Tapirus terrestris) in the Tabaro River Valley, southern Venezuela. Can J Zool 74:1444–1451

Serrano-MacGregor I (2017) Daños a los cultivos ocasionados por el tapir centroamericano (Tapirus bairdii) y otra fauna silvestre en el municipio de Calakmul, Campeche, México. MSc thesis, El Colegio de la Frontera Sur, San Cristóbal de Las Casas, Mexico

Tapir Specialist Group (2010) Estrategia nacional para la conservación de los tapires (*Tapirus* spp.) en el Ecuador. IUCN/SSC Tapir Specialist Group, Quito

Terwilliger VJ (1978) Natural history of Baird's tapir on Barro Colorado Island, Panama Canal Zone. Biotropica 10:211–220

Tobler MW (2008) The ecology of the lowland tapir in Madre de Dios, Peru: using new technologies to study large rainforest mammals. PhD dissertation, Texas A&M University, College Station

Tobler MW, Naranjo EJ, Lira-Torres I (2006) Habitat preferences, feeding habits and conservation of Baird's tapir in Neotropical montane oak forests. In: Kappelle M (ed) Ecology and conservation of neotropical montane oak forests. Springer, Berlin, pp 347–361

Tobler MW, Hibert WF, Debeir L, Richard-Hansen C (2014) Estimates of density and sustainable harvest of the lowland tapir *Tapirus terrestris* in the Amazon of French Guiana using a Bayesian spatially explicit capture–recapture model. Oryx 48:410–419

Trolle M, Noss AJ, Passos-Cordeiro JL, Oliveira LFB (2008) Brazilian tapir density in the Pantanal: a comparison of systematic camera-trapping and line-transect surveys. Biotropica 40:211–217

Voss RS, Helgen KM, Jansa SA (2014) Extraordinary claims require extraordinary evidence: a comment on Cozzuol et al. (2013). J Mammal 95:893–898

Williams KD (1984) The central American tapir (*Tapirus bairdii*) in northwestern Costa Rica. PhD dissertation, Michigan State University, East Lansing

Wilson RA, Wilson S (1973) Diet of captive tapirs. Int Zoo Yearbook 13:213–217

Correction to: Distribution and Abundance of White-Tailed Deer at Regional and Landscape Scales at Tehuacán-Cuicatlán Biosphere Reserve, Mexico

Salvador Mandujano, Odalis Morteo-Montiel, Carlos Yáñez-Arenas, Michelle Ramos-Robles, Ariana Barrera-Salazar, Eva López-Tello, Pablo Ramirez-Barajas, Concepción López-Téllez, and Adriana Sandoval-Comte

Correction to: Chapter 4 in: S. Gallina-Tessaro (ed.), *Ecology and Conservation of Tropical Ungulates in Latin America*, https://doi.org/10.1007/978-3-030-28868-6_4

The name of one of the co-authors of Chap. 4 was inadvertently published as Michelle Ramos-García instead of Michelle Ramos-Robles. This has now been amended throughout the book.

The updated online version of this chapter can be found at
https://doi.org/10.1007/978-3-030-28868-6_4

S. Gallina-Tessaro (ed.), *Ecology and Conservation of Tropical Ungulates in Latin America*, https://doi.org/10.1007/978-3-030-28868-6_19

Index

S. Gallina-Tessaro (ed.), *Ecology and Conservation of Tropical Ungulates in Latin America*, https://doi.org/10.1007/978-3-030-28868-6

U

V

Printed in the United States
By Bookmasters